# SEDIMENTATION IN VOLCANIC SETTINGS

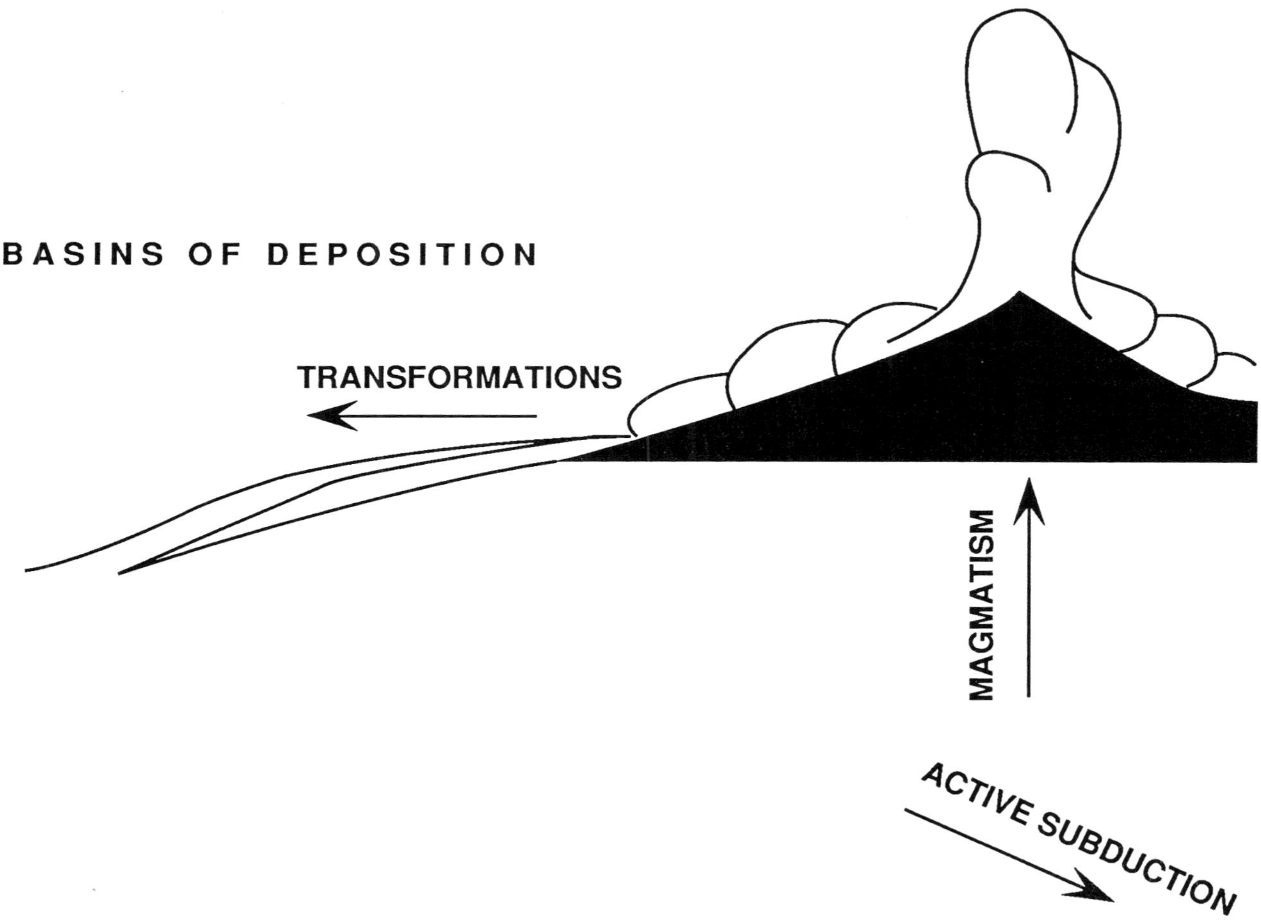


*Edited by*

*Richard V. Fisher, University of California, Santa Barbara, California*
*and*
*Gary A. Smith, Unviersity of New Mexico, Albuquerque, New Mexico*



*Barbara H. Lidz, Editor of Special Publications*
*Special Publication No. 45*



*Tulsa, Oklahoma, U.S.A.* *October, 1991*

A Publication of
SEPM (Society for Sedimentary Geology



ISBN # 0-918985-89-7



Printed in the United States of America

# CONTENTS

## DEDICATION

This volume is dedicated to Harry Glicken, former student and colleague of Richard V. Fisher at the University of California at Santa Barbara (UCSB). Harry was a young scientist of high ideals immersed in the study of volcanoes to the exclusion of nearly all other life pursuits when he was killed on June 3, 1991 at the age of 33 by a high temperature pyroclastic current on Unzen Volcano in Kyushu, Japan. He had been in Japan at Tokyo Metropolitan University for several months doing post-doctoral research. In his own words, Harry wrote, "The fellowship I obtained in Japan will enable me to do just the kind of research that I want and feel most suited for in a land of many volcanoes. It will also enable me to work with and learn from Japanese volcanologists and eminent American volcanologists who may wish to visit Japan. I will try to stay in Japan for a long time."

As a student he was first drawn to a study of volcanoes by his interest in large-scale flow processes observable during the growth of some volcanic forms, but his decision to follow a life career of studying volcanic hazards was the result of the fateful 18 May 1980 eruption of Mount St. Helens, Washington. At that time he was assistant to David Johnston who, at about the same age, also met a tragic end by a pyroclastic current. Harry and David, whose lives were so closely linked, are also the only American volcanologists known to have died in volcanic eruptions. Harry was devastated by David's death and came to the decision to dedicate his life to helping society by increasing the store of knowledge about volcanic processes. He leaves a short body of influencial work that will stand as a monument to this dedication, most notably the extemely detailed study of the debris-avalanche that formed from the partial collapse of Mount St. Helens on 18 May 1980. This study, completed in 1986, has already influenced others' ideas regarding such awesome events and has greatly advanced knowledge about them even though it is still in press as a U.S. Geological Survey Professional Paper. His contribution to the knowledge of volcaniclastic phenomena in this volume also deals with debris-avalanches.

While associated with the U.S. Geological Survey at the David A. Johnston Cascades Volcano Observatory during the 1980's, and though working full time on his Ph.D. dissertation research, Harry unselfishly and tirelessly gave of his time and energy to lead numerous geologists from around the world to observe and discuss origins of the various volcanic deposits formed during the 1980 eruption. He was a unique, dreamy, lovable young man, always with a ready laugh. Harry was following his bliss. He died with his boots on.

## ACKNOWLEDGMENTS

For their considerable help with the papers that appear in this special publication we wish to acknowledge the following reviewers: W. Scott Baldridge, James Boles, David C. Buesch, William Criswell, Richard S. Fiske, William J. Fritz, Harry Glicken, Grant Heiken, Malcolm Howells, Raymond V. Ingersoll, Donald R. Lowe, Neil Lundberg, Gregory Mack, Kathleen Marsaglia, James G. Moore, Alan R. Niem, Geoffrey Orton, Thomas Pierson, Stephen Self, Eugene Smith, Richard Smith, Donald Swanson, Greg A. Valentine, James D. L. White, and John Yarnold.

For indispensable help with the index, the following people are gratefully acknowledged: Kyle Gay, Michael Grubensky, Daniel Larsen, Richard Livacarri, Jean-Luc Schneider, and Judy Lopas Stoopes.

Richard V. Fisher and Gary A. Smith, editors

# VOLCANISM, TECTONICS AND SEDIMENTATION

RICHARD V. FISHER
*Department of Geological Sciences, University of California, Santa Barbara, California 93106*
AND
GARY A. SMITH
*Department of Geology, University of New Mexico, Albuquerque, New Mexico 87131*

## INTRODUCTION

Volcanism provides a unique and locally abundant source of sediment to many of the earth's subaerial and subaqueous basins. In basin and tectonic analyses, volcanic sediments can be treated methodologically like nonvolcanic sediments, but the close association of tectonism and volcanism provides an added dimension to the analytical importance of volcaniclastic sediments. Volcanism occurs at plate margins and in some instances within plates ("hot spots"), and sediments within those environments have a strong, and in some cases, an exclusive component of volcaniclastic particles. The record of tectonism is commonly preserved only within fillings of sedimentary basins, and because volcanism is associated with activity at plate margins, knowledge about volcaniclastic sediments and rocks may be critical for interpreting plate activity and tectonic environments. Moreover, great volumes of clastic materials that rapidly flood sedimentary environments during active volcanism (e.g., 1980 eruption of Mount St. Helens, Lipman and Mullineaux, 1981) can result in disruption of river courses, rapid deposition along deltaic fronts, and therefore increase the number and volumes of turbidity currents in marine basins (Davies and others, 1979; Buesch, this volume; R. Smith, this volume), resulting in sedimentary responses not encountered in nonvolcanic areas where rates of sedimentation do not fluctuate so rapidly (Fig. 1).

In the early 1960s, the term "volcaniclastic" (Fisher, 1961) was quickly accepted into the general literature because the time was ripe for combining the fields of volcanology and sedimentology within a single system to describe sedimentary facies affected or produced by volcanism. The term "volcaniclastic" is defined to include the entire spectrum of clastic materials composed in part or entirely of volcanic fragments, formed by any particle-forming mechanism (e.g., pyroclastic, hydroclastic, epiclastic, autoclastic; see following section), transported by any mechanism, deposited in any physiographic environment or mixed with any other volcaniclastic type or with any nonvolcanic fragment types in any proportion (Fisher, 1966). Thus, any clastic deposit with volcanic particles can be given a nongenetic name in the field without interpreting origins or processes. Genetic interpretations regarding specific particle origins such as pyroclastic or epiclastic, or agents of transport such as wind or running water, can be later applied whenever possible and appropriate. Definition of the term coincided with early discoveries of ocean-floor spreading and plate tectonics, as well as with research and description of deep-sea sediments containing volcaniclastic materials. Recognition of the relationship of tectonism and volcanism and the profound effect that volcanism can have on the sedimentological regime has greatly increased interest in the volcaniclastics. The field is growing dramatically, with the number of papers dealing with volcaniclastics increasing 100-fold in the last two decades and continuing to accelerate. The common belief among many petroleum geologists that regions of volcanic rocks are generally to be avoided as potential hydrocarbon reservoirs (Burns and Ethridge, 1979; Galloway, 1979) has greatly slowed the learning curve about volcaniclastic rocks. But many oil-bearing basins occur within plate margins, and because most of the earth's volcanic action occurs at plate margins, there is hydrocarbon production in volcaniclastic rocks (Iijima and Utada, 1971; Merino, 1975a,b; Crossey and others, 1984; Coffman, 1987; Mathisen and McPherson, this volume).

Volcaniclastic materials are highly varied, ranging from fresh, instantly formed clastic materials rich in metastable glass and minerals, to materials that have experienced long-term weathering, all of which may be later transported by various geomorphic agents. Because of their association with volcanism, which is associated with plate tectonism, hydrocarbon-bearing basins containing volcaniclastic materials can occur in convergent margin settings, both forearc and backarc, and are also found in rift basins (Mathisen and McPherson, this volume). Diagenetic changes in metastable materials of fresh volcanic materials can cause rapid reduction of permeability (Galloway, 1979; Davies and others, 1979; Hunter and Davies, 1979), but also, dissolution of volcaniclastic materials may enhance permeability and lead to increased reservoir quality (Hayes and others, 1976; Loucks and others, 1979, 1984; Mathisen, 1984; Hawlander, 1990). Volcaniclastics not only serve as reservoir rocks in many places, as noted previously, but through facies and basin analysis, they are important in analyzing the dynamics of plates that produce the basins (Dickinson, 1974; Ingersoll, 1978).

In addition to epiclastic volcaniclastic sediment supplied to basins by weathering and erosion of volcanic regions, large volumes of volcaniclastic material are also supplied directly to basins during explosive volcanism. The transport of explosively produced particles is unique, distinct from epiclasts, because they are provided with initial transport energy from explosive expansion rather than exclusively by gravity, and therefore initially are independent of slope or base level of deposition. Airborne clasts from a given explosive eruption can bypass or overtop physiographic barriers and deposit a layer simultaneously in different basins and environments (Fisher and Schmincke, 1984). Thus, in some instances, explosive volcanism offers the only means by which to solve interbasinal correlation and facies prob-

Sedimentation in Volcanic Settings, SEPM Special Publication No. 45

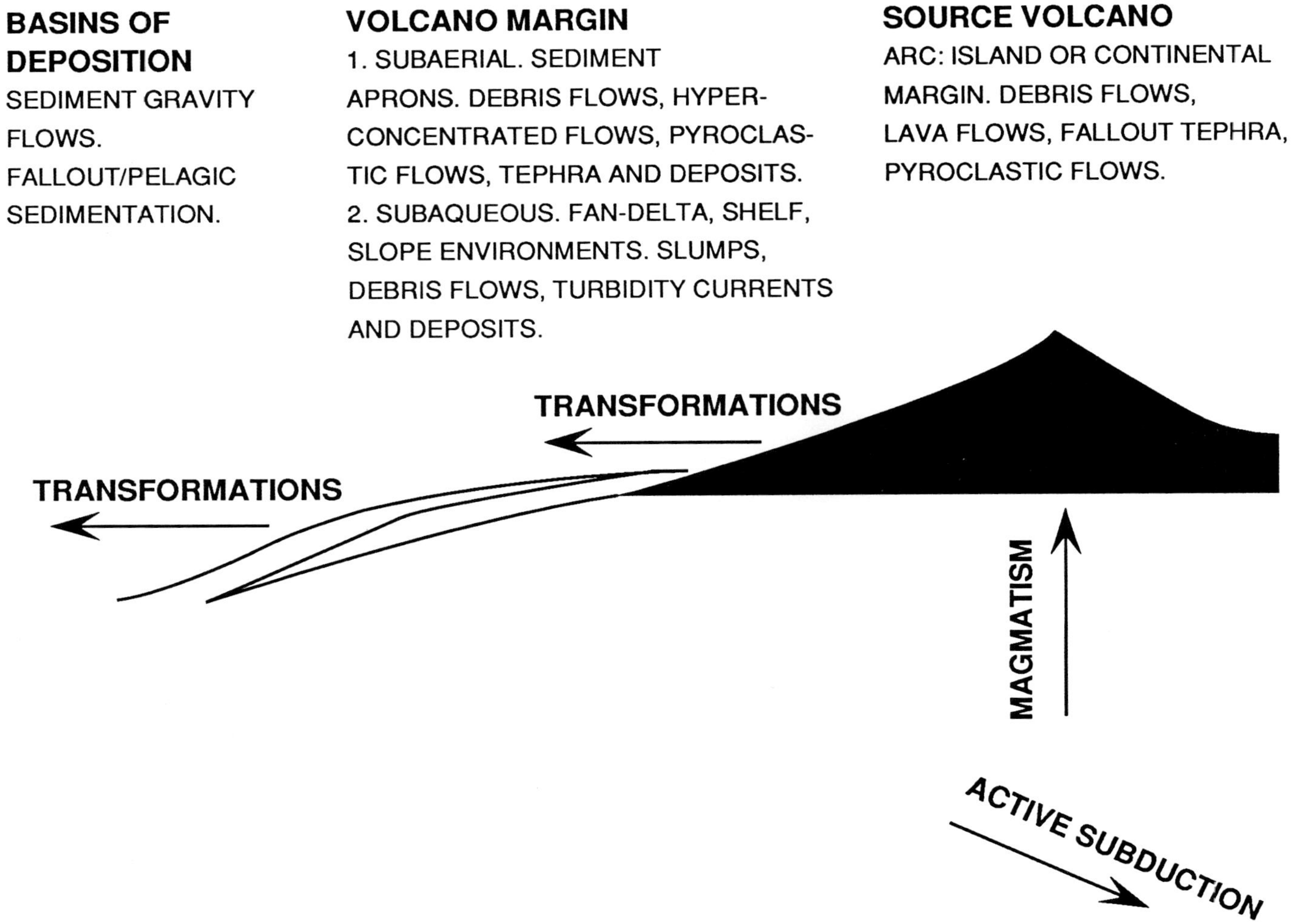


FIG. 1.—Diagram illustrating connection between subduction activity, magmatism, and sedimentation. Subduction leads to volcanic action; large influx of volcaniclastic materials causes increased rates of sedimentation. Stratovolcano composed of lava flows and a large composite volume of lahar, fallout tephra and pyroclastic flow deposits is depicted in diagram. Transfer of volcaniclastics from active volcano to depositional basin occurs directly or in step-wise transfer by flow transformations (see text).

lems by physical rather than paleontological criteria. Volcanic ash layers can be dated and correlated long distances (Wilcox, 1965), commonly known as the field of tephrochronology (Fisher and Schmincke, 1984), and are also useful for stratigraphic correlation in the subsurface (Malm and others, 1979).

## VOLCANICLASTIC PARTICLES

Solving paleotectonic and facies problems in most of the recent or ancient plate tectonic settings requires the establishment of penecontemporaneity of the volcaniclastic components in sedimentary rocks because volcanism is often the signature of tectonism. This may be difficult, however, because newly formed pyroclastic or hydroclastic particles (discussed later), and epiclastic particles derived from weathering of volcanic rocks (lava flows or lithified ash and volcanic breccia) or nonvolcanic rocks, are frequently mixed together by marine or nonmarine transport processes. Penecontemporaneity of volcanism is accomplished by distinguishing between "instant" volcanic particle-forming processes that form pyroclasts and hydroclasts, and long-term particle-forming weathering of rocks that form epiclasts. Careful work requires a knowledge of the different kinds and characteristics of volcaniclastic particles to determine origins unambiguously (Heiken and Wohletz, 1985; and this volume.

Important types of volcaniclastic particles are: pyroclastic, hydroclastic, autoclastic and epiclastic. **Pyroclasts** form from rapidly expanding magma that loses cohesion and breaks into bits when overburden pressures are exceeded. Magmatic expansion creates a glass foam (pumice) that may break into glass shards from broken vesicles (Heiken and Wohletz, this volume). Phenocrysts may be released from the magma, and explosions may cause breakage and incorporation of lithic fragments from vent walls.

**Hydroclasts** form by magma-water interactions that produce chilled glass particles by either explosive or nonexplosive means. Thermal contraction of hot magma or lava in water leads to the formation of shattered, angular, poorly vesiculated shards. Hydroclasts cannot form explosively in deep water because pressures inhibit steam expansion (McBirney, 1963). Explosive disruption by magma-water interactions requires rapid heat exchange, and steam generation requires that the surface area of magma be increased, and this can be accomplished by creating connecting microspaces in the magma large enough for water to enter, primarily by (1) the formation of contraction cracks or (2) vesicles that explosively shatter and spray magma droplets into the water (Wohletz, 1983). Steam explosions within confined areas at the water-magma interface continue to drive water and magma together until the supply of one or the other is gone. The explosive energy comes from the excess heat energy in the magma, which is transferred in fractions of seconds to water and causes vaporization. Features of hydroclasts that reflect the external explosive processes are the abundance of nonvesicular or poorly vesicular glassy pyroclasts that have fracture-bounded, angular shapes. Their physical characteristics can be distinguished by scanning electron microscope (SEM) analyses (Heiken and Wohletz, 1985, and this volume; Marshall, 1987).

**Epiclastic** particles are lithic clasts and crystals derived from any type of pre-existing *rock* by weathering and erosion (Bates and Jackson, 1987, p. 218; Pettijohn and others, 1987, p. 252). If the pre-existing rocks are volcanic, the epiclasts are considered to be one type of volcaniclastic particle. Pyroclastic and hydroclastic particles that are carried to their site of deposition by rivers, wind or other types of transporting agents are not epiclastic because they are not formed by weathering.

A less abundant particle type caused directly from volcanic action includes **autoclastic** fragments formed by mechanical friction of moving lava flows, or gravity crumbling of spines and domes. **Alloclastic** fragments are formed by disruption of pre-existing volcanic rocks by igneous processes beneath the earth's surface, with or without intrusion of fresh magma.

To minimize conceptual confusion about the origin of volcaniclastic particles and contemporaneity of deposition requires that processes that create the particles (e.g., pyroclastic, hydroclastic, epiclastic) need to be clearly separated from processes that transport the particles, e.g., wind, running water, ice, volcanic explosion, and gravity transfer by avalanche. Cas and Wright (1987) contend that pyroclastic or hydroclastic particles reworked by water or other geomorphic agents can be called epiclastic deposits. The terms epiclastic, pyroclastic, and hydroclastic, however, refer to processes by which particles are formed, and they cannot change from one particle type to another merely by changing the agent of transportation. This is a critical, nontrivial conceptual distinction in problems relating penecontemporaneity of volcanism and deposition, and therefore is fundamental to the understanding of volcaniclastic facies. The distinction is also critical to understanding differences in sedimentation in volcanic and nonvolcanic areas. Most nonvolcanic siliciclastic sediments are epiclastic and are, therefore, to varying degrees, supply limited depending upon rates of weathering and erosion in the source area. Many volcaniclastic fragments are generated instantaneously and in large volumes, producing unique facies distributions and geometries not encountered in nonvolcanic epiclastic sediments.

## VOLCANICLASTIC FACIES

The word facies has various meanings and can be used interchangably in the plural or singular (Prothero, 1990). We emphasize here that facies are features of a unit portraying the processes of origin and source, and environments of deposition. Facies studies of rock units may indicate a particular physiographic or tectonic environment. Such a usage fits within the definition of facies as the physical, chemical and biologic variations of three-dimensional rock bodies deposited within a specific interval of geologic time (Pettijohn, 1975).

A useful subdivision for facies discussions deals with position of the rock body relative to source, e.g., **proximal, medial and distal facies.** Such reference to distance is relative, however, and depends upon the scale of the rock body. At Mount St. Helens, for example, the farthest travelled (most distal) blast surge deposit is 24 km from the source, whereas, for the Peach Springs Tuff in California and Arizona, distal exposures are probably greater than 100 km from the source (Valentine and others, 1989; Buesch, this volume). The proximal facies may include the source volcano, but in ancient sedimentary basins the only record of volcanism occurs in the basin, and the proximal facies is defined on the basis of thickness, grain size or rock type (Smith, 1988). These criteria form the basis for defining large-scale facies environments such as volcanoes and their surroundings (Hackett and Houghton, 1989), and various specific facies described by several papers in this book.

The lateral distribution of rock types from the source volcano into adjacent marine or nonmarine basins is determined by the rate at which materials are extruded from a volcano, their total volume and the processes of transportation linked by flow transformations. For example, the growth rate of andesitic stratovolcanoes, with consequent influence upon depositional environments, is on the order of a few hundred to a few thousand years. Stratovolcanoes are composed of great volumes of easily removed fragmental material. Their growth is therefore reflected almost instantly in the sedimentary record of surrounding regions (G. A. Smith, this volume), eruptions that directly supply airborne tephra, ground-hugging pyroclastic flows and surges, and eruption-related debris avalanches, lahars and fluvial materials, or by remobilization of loose material between eruptions. The transfer of these materials from the volcano during active or inactive eruptive events to adjacent basins may be determined by **flow transformations** (Fisher, 1983). A flow transformation is defined as the change from laminar to turbulent behavior (or vice versa) within a sediment gravity flow. These changes in Reynolds number relationships can be caused by (1) density separations within a moving sediment gravity flow caused by gravity segregation of particles (gravity transformations), (2) velocity variations caused by slope changes without much variation in interstitial water or gas content within the body of the current (body transformation), and (3) separations of particles caused by turbulent mixing within the boundary between the ambient fluid and the flow surface (surface transformation). Laminar-turbulent or turbulent-laminar transformations commonly result in changes in transport agents such as from pyroclastic flows or surges to lahars, and lahars to hyperconcentrated flows to normal fluvial transport that occurred during the May 18, 1980, eruption of Mount St. Helens (Scott, 1988). We would add that the sediment from this event that reaches the Astoria fan at the mouth of the Columbia River may eventually be transformed to slumps and to turbidity currents that extend into the deep-sea region of the northern Pacific Ocean. A facies lineage, linked by flow transformations, can be expressed as follows (Fisher and Schmincke, pers. commun.): eruption of *pyroclastic surge or flow* > *debris flow* > *hyperconcentrated flood flow* > *normal fluvial transport* (in the Columbia River) > *stopgap storage of sediment on submarine shelves, slopes, or submarine fans* > *submarine landslides* > *subaqueous debris flows* > *turbidity currents*.

This book is divided into five parts, Part I, **Volcanism as a process of sedimentological significance,** Part II, **Volcaniclastic processes of transport and deposition,** Part III, **Volcaniclastic sedimentation in continental settings,** Part IV, **Volcaniclastic sedimentation in oceanic settings,** and Part V, **Petrology of volcaniclastics.** Thus, the subject matter proceeds from the influence of volcanic processes upon the environment and hydrocarbon potential, followed by the influence of the environment upon depositional and structural patterns in subaerial environments and subaqueous environments and the tectonic regime. The compositional aspects providing criteria for volcaniclastic response to tectonic environment concludes this volume.

The effects of volcanism upon sedimentation can be profound. Whether or not volcanism can cause wholesale changes in climate is an important consideration, for if it can, it can affect many processes that cannot be directly inferred from deposits. For example, if world-wide temperatures drop so as to drastically shorten the growing season and therefore the food supply, mass extinction could take place, but without leaving any substantial evidence of causes within the rock record. Weather changes could increase the severity of rain storms, hence erosion. Loss of vegetation could increase severity of erosion. Drop in temperatures could result in glaciation. Thus, sedimentation regimes could be severely altered by volcanism, even though volcanic deposits could be minimal.

## WHERE FROM HERE?

We have gained considerable experience with volcaniclastic materials over the past 30 years, but the field has undergone considerable growth in the decade following the 1980 eruption of Mount St. Helens. This eruption resulted in accelerated research in explosive volcanic products (Fisher and Schmincke, 1984) and spurred a renewed interest in volcaniclastic materials as they relate to plate tectonic boundaries and explosive volcanism in general (Cas and Wright, 1987). Since the early 1970s a loosely defined field called "sedimentary tectonics" has emerged. We prefer the name "sedimentation and tectonics" to exclude deformation by shale diapirs, salt domes, and other local deformational processes.

A large part of the field of sedimentation and tectonics includes studies of volcaniclastic sedimentation, largely because of the direct association of tectonism, volcanism and sedimentation. This book attempts to illuminate the field and to present its salient features to sedimentologists not generally versed in volcaniclastic particles, deposits or facies.

Where the field goes from here is anyone's guess. As in all fields of science, directions of knowledge are unpredictably determined by the excellence of published research and innovative ideas of individual researchers, and, in the geological sciences, by unforeseen natural events that give researchers new insights, as did the 1980 eruption of Mount St. Helens. It is our contention that volcaniclastic sedimentology is emerging as one of the more powerful stratigraphic tools in solving many of the problems related to plate genesis, development and movements.

## REFERENCES


BATES, R. L., AND JACKSON, J. A., 1987, Glossary of Geology: American Geologic Institute, Alexandria, Virginia, 788 p.

BURNS, L. K., AND ETHRIDGE, F. G., 1979, Petrology and diagenetic effects of lithic sandstones: Paleocene and Eocene Umpqua Formation, southwest Oregon, *in* Scholle, P. A., and Schluger, P. R., eds., Aspects of Diagenesis: Society of Economic Paleontologists and Mineralogists Special Publication 26, p. 307–317.

CAS, R. A. F., AND WRIGHT, J. V., 1987, Volcanic successions, modern and ancient: Allen and Unwin, London, 528 p.

COFFMAN, R. L., 1987, Laumontization and its relationship to carbonate cementation and dissolution within Santa Fe Springs oil field, Los Angeles Basin, California: American Association of Petroleum Geologists Bulletin, v. 71, p. 540.

CROSSEY, L. J., FROST, R. B., AND SURDAM, R. C., 1984, Secondary porosity in laumontite-bearing sandstones, *in* McDonald, D. A., and Surdam, R. C., eds., Clastic Diagenesis: American Association of Petroleum Geologists Memoir 37, p. 225–237.

DAVIES, D. K., ALMON, W. R., BONIS, S. B., AND HUNTER, B. E., 1979, Deposition and diagenesis of Tertiary-Holocene volcaniclastics, Guatemala, *in* Scholle, P. A., and Schluger, P. R., eds., Aspects of Diagenesis: Society of Economic Paleontologists and Mineralogists Special Publication 26, p. 281–306.

DICKINSON, W. R., 1974, Sedimentation within and beside ancient and modern magmatic arcs, *in* Dott, R. H., Jr., and Shaver, R. H., eds., Modern and Ancient Geosynclinal Sedimentation: Society of Economic Paleontologists and Mineralogists Special Publication 19, p. 230–239.

FISHER, R. V., 1961, Proposed classification of volcaniclastic sediments and rocks: Geological Society of America Bulletin, v. 72, p. 1409–1414.

FISHER, R. V., 1966, Rocks composed of volcanic fragments and their classification: Earth-Science Reviews, v. 1, p. 287–298.

FISHER, R. V., 1983, Flow transformations in sediment gravity flows: Geology, v. 11, p. 273–274.

FISHER, R. V., AND SCHMINCKE, H.-U., 1984, Pyroclastic rocks: Berlin, Springer-Verlag, 472 p.

GALLOWAY, W. E., 1979, Diagenetic control of reservoir quality in arc-related sandstones: Implication for petroleum exploration: Society of Economic Paleontologists and Mineralogists Special Publication 26, p. 251–262.

HACKETT, W. R., AND HOUGHTON, B. F., 1989, A facies model for a Quaternary andesitic composite volcano: Ruapehu, New Zealand: Bulletin of Volcanology, v. 51, p. 51–68.

HAWLANDER, H. M., 1990, Diagenesis and reservoir potential of volcanogenic sandstones—Cretaceous of the Surat Basin, Australia: Sedimentary Geology, v. 66, p. 181–195.

HAYES, J. B., HARMS, J. C., AND WILSON, T., 1976, Contrasts between braided and meandering stream deposits, Beluga and Stering Formations (Tertiary), Cook Inlet, Alaska, *in* Miller, T. P., ed., Recent and Ancient Sedimentary Environments in Alaska: Anchorage: Alaska Geological Society, p. J1–J27.

HEIKEN, G., AND WOHLETZ, K. H., 1985, Volcanic Ash: University of California Press, Berkeley, 246 p.

HUNTER, B. E., AND DAVIES, D. K., 1979, Distribution of volcanic sediments in the Gulf coastal province—significance to petroleum geology: Transactions, Gulf Coast Association of Geological Societies, v. 29, p. 147–155.

IIJIMA, A., AND UTADA, M., 1971, Present-day zeolitic diagenesis in the Neogene geosynclinal deposits in Niigata oil field, Japan: American Chemical Society, Advances in Chemistry Series, v. 101, p. 342–349.

INGERSOLL, R. V., 1978, Petrofacies and petrologic evolution of the late Cretaceous fore-arc basin, northern and central California: Journal of Geology, v. 86, p. 335–352.

LIPMAN, P. W., AND MULLINEAUX, eds., 1981, The 1980 eruptions of Mount St. Helens: U.S. Geological Survey Professional Paper 1250, 844 p.

LOUCKS, R. G., DODGE, M. M., AND GALLOWAY, W. E., 1979, Importance of secondary leached porosity in lower Tertiary sandstones along the Texas Gulf Coast: Transactions, Gulf Coast Association of Geological Societies, v. 29, p. 164–171.

LOUCKS, R. G., DODGE, M. M., AND GALLOWAY, W. E., 1984, Regional controls on diagenesis and reservoir quality in Lower Tertiary sandstones along the Texas Gulf Coast, *in* McDonald, D. A., and Surdam, R. C., eds., Clastic Diagenesis: American Association of Petroleum Geologists Memoir 37, p. 15–45.

MALM, O. A., FURNES, H., AND BJORLYKKE, K., 1979, Volcaniclastics of Middle Jurassic age in the Statford oil field of the North Sea: Neues Jahrbuch Geologie und Palaeontologie Monatatshefte, v. 10, p. 607–618.

MARSHALL, J. R., ed., 1987, Clastic Particles: Scanning Electron Microscopy and Shape Analysis of Sedimentary and Volcanic Clasts: Van Nostrand Reinhold, New York, 346 p.

MATHISEN, M. E., 1984, Diagenesis of Plio-Pleistocene nonmarine sandstones, Cagayan Basin, Philippines: Early development of secondary porosity in volcanic sandstones, *in* McDonald, D. A., and Surdam, R. C., eds., Clastic Diagenesis: American Association of Petroleum Geologists Memoir 37, p. 177–193.

MCBIRNEY, A. R., 1963, Factors governing the nature of submarine volcanism: Bulletin Volcanologique, v. 26, p. 455–469.

MERINO, E., 1975a, Diagenesis in Tertiary sandstones from Kettleman North Dome, California, I. Diagenetic mineralogy: Journal of Sedimentary Petrolology, v. 45, p. 320–336.

MERINO, E., 1975b, Diagenesis in Tertiary sandstones from Kettleman North Dome, California, II. Interstitial solutions: distributions of aqueous species at 100°C and chemical relation to the diagenetic mineralogy: Geochimica Cosmochimica Acta, v. 39, p. 1629–1645.

PETTIJOHN, F. J., 1975, Sedimentary Rocks: 3rd edition, Harper and Row, Publishers, New York, 628 p.

PETTIJOHN, F. J., POTTER, P., AND SIEVER, R., 1987, Sand and Sandstone: 2nd edition, Springer-Verlag, New York, 553 p.

PROTHERO, D. R., 1990, Interpreting the Stratigraphic Record: W. H. Freeman and Company, New York, 410 p.

SCOTT, K. M., 1988, Origins, behavior, and sedimentology of lahars and lahar-runout flows in the Toutle-Cowlitz River system: U.S. Geological Survey Professional Paper 1447-A, 74 p.

SMITH, G. G., 1988, Sedimentology of proximal to distal volcaniclastics dispersed across an active foldbelt: Ellensburg Formation (late Miocene, central Washington): Sedimentology, v. 35, p. 953–977.

VALENTINE, G. S., BUESCH, D. C., AND FISHER, R. V., 1989, A volcanological interpretation of basal layered deposits of the Peach Springs Tuff, Northwestern Arizona, U.S.A.: Bulletin of Volcanology, v. 51, p. 395–414.

WILCOX, R. E., 1965, Volcanic ash chronology, *in* Wright, H. E., Jr., and Frey, D. G., eds., Quaternary of the United States: Princeton University Press, Princeton, New Jersey, p. 807–816.

WOHLETZ, K. H., 1983, Mechanisms of hydrovolcanic pyroclastic formation: grain-size, scanning electron microscopy, and experimental studies: Journal of Volcanology and Geothermal Research, v. 17, p. 31–63.

# PART I
# VOLCANISM AS A PROCESS OF SEDIMENTOLOGICAL SIGNIFICANCE

Volcanism is a fundamental process in the development of earth and its atmosphere. It affects the atmosphere, hydrosphere, biosphere, lithosphere and surface environments by introducing abundant clasts into the sedimentological systems. Traditionally, volcanic regions have been spurned in hydrocarbon exploration, but modern work indicates this approach is too narrow. Part I discusses atmospheric aspects of volcanism, particle types and identification, and hydrocarbon exploration in volcaniclastic regions.

# VOLCANISM, CLIMATIC CHANGE, AND THE GEOLOGIC RECORD

MICHAEL R. RAMPINO

*Earth Systems Group, New York University, New York, New York 10003, and NASA, Goddard Space Flight Center, Institute for Space Studies, New York, New York 10025*

ABSTRACT: The geologic record shows possible connections between volcanism and climate change on various spatial and temporal scales, from documented seasonal and regional climatic effects of historic eruptions to the long-term impact of episodes of increased global volcanism/tectonism on atmospheric composition, climate, and life. The geologic record of volcanism, in the form of volcanic rocks, widespread ash layers, disseminated ash in sediments, and alteration products, provides information on the size, frequency, and location of past eruptions, the composition of the erupted magma, and amounts of climatically important volatiles released ($CO_2$, $SO_2$, HCl). These may be compared with climatic records from oxygen-isotope measurements, faunal/floral analyses, sea-level changes, and so forth. Recent eruptions and their climatic effects can be traced in proxy climate records such as polar ice cores, tree rings, lakes and bogs.

Historic eruptions (up to about 50 $km^3$ of erupted magma, e.g., Tambora AD 1815) that produced widespread clouds of sulfuric acid aerosols are associated with small (<1°C) coolings on a hemispheric to global scale that last from 1 to 3 years after the eruptions. Coolings are accentuated at high latitudes. The climatic effects are generally proportional to the amount of sulfur-rich gases injected into the stratosphere, where they are converted into aerosols. Seasonal and regional effects, resulting from perturbations of atmospheric circulation patterns, may be considerably greater, for example, climate excursions during "the year without a summer" in AD 1816.

Many prehistoric eruptions were much larger (>1,000 $km^3$ of magma, e.g., Toba 75 ka), released more dust and sulfur-rich aerosols (more than $10^3$ megatons), and may have triggered significant global cooling. There is some evidence for correlation between climate cooling and enhanced volcanism on time scales from decades to $10^6$ years, but problems in dating the episodes make it difficult to establish cause-and-effect relationships. Climate cooling for longer than the stratospheric residence time of the aerosols (a few years) requires that positive feedback mechanisms, such as snow/ice-albedo feedback, come into play, or that climatically significant eruptions in the past were more closely spaced in time than in recent centuries.

On long time scales ($10^7$–$10^8$ years), release of $CO_2$ through subduction-zone volcanism is a major control of climate. Intervals of rapid ocean-floor accretion (and subduction) correlate with times of increased volcanism, high sea levels and warm global climates. Accretion-rate data can be used in biogeochemical models to estimate paleoatmospheric $CO_2$ levels. Increased levels of atmospheric $CO_2$ and warm climate, so-called "Greenhouse Episodes", are predicted for those periods in the Phanerozoic showing evidence of enhanced volcanic/tectonic activity.

## INTRODUCTION: THE VOLCANO/CLIMATE PROBLEM

Volcanic activity manifests itself in a variety of styles, from relatively quiet lava effusion to devastatingly explosive eruptions. Explosive volcanic eruptions, and perhaps very large effusive eruptions (basalt fissure eruptions to flood basalt eruptions) are capable of lofting fine volcanic ash particles (<2-μm-diameter dust) and gases into the stratosphere (Rampino and others, 1988). The sulfurous gases (mainly sulfur dioxide) are converted into small droplets (aerosols) of sulfuric acid through photochemical reactions with stratospheric water vapor. The sulfuric acid aerosols and volcanic dust absorb and scatter some of the incoming solar radiation, so that less sunlight gets through to heat the Earth's surface (Toon and Pollack, 1980). Aerosols from a typical eruption remain in the stratosphere, in diminishing amounts, for a few years (an *e*-folding time of about 1 year for most eruptions studied to date), whereas volcanic dust particles, being larger, are mostly removed through aggregation and fallout in a few months. As the aerosols form, stratospheric winds carry them around the globe latitudinally, and the aerosol cloud spreads toward the poles (Toon and Pollack, 1980). The distribution of the aerosol cloud is thus determined by stratospheric circulation patterns, which change with time.

High-latitude eruptions in either hemisphere produce aerosol clouds that remain largely at high latitudes, whereas equatorial eruptions can be effective in dispersing aerosols into both hemispheres, thus having a larger global effect (Lamb, 1970). Because of the weak vertical temperature gradient in the stratosphere, there is little vertical mixing, and the volcanic aerosols tend to stay at the altitude to which they were carried by the volcanic eruption, forming a definite aerosol cloud layer. In the 1982 El Chichón eruption, for example, the aerosol layer remained centered at 26 km, the maximum height of the volcanic eruption column (Rampino and Self, 1984a).

A survey of historical eruptions shows many cases where the stratospheric aerosol clouds could be detected through the reduction in the amount of solar radiation reaching the ground (Toon and Pollack, 1980; Fig. 1), and by unusual optical phenomena and brilliant sunrise/sunsets (see Stothers and Rampino, 1983; Rampino and others, 1988). In the case of the El Chichón event, the best observed stratospheric cloud to date, the observatory at Mauna Loa on Hawaii recorded a 20 percent decrease in direct solar radiation, and satellite measurements of infrared emissions to space were used to track the aerosol cloud as it moved around the world. El Chichón may have also released significant amounts of HCl, which might affect stratospheric ozone (Devine and others, 1984).

The amount of sulfur-rich gases released from the magma is a critical factor in the short-term climatic impact of any eruption. Devine and others (1984) have developed petrographic methods to determine the original volatile content of erupted magmas, and the amounts released during the eruption (Table 1). In order to do so, they must estimate the volume of the eruption, and measure the volatile content of glass inclusions within late-formed phenocrysts, in which the original magma volatile content may be preserved. Analysis of glassy volcanic ash formed during the eruption provides an estimate of the volatiles remaining in the degassed magma. The difference gives an estimate of the amount of volatiles released during the eruptions (Palais

Sedimentation in Volcanic Settings, SEPM Special Publication No. 45

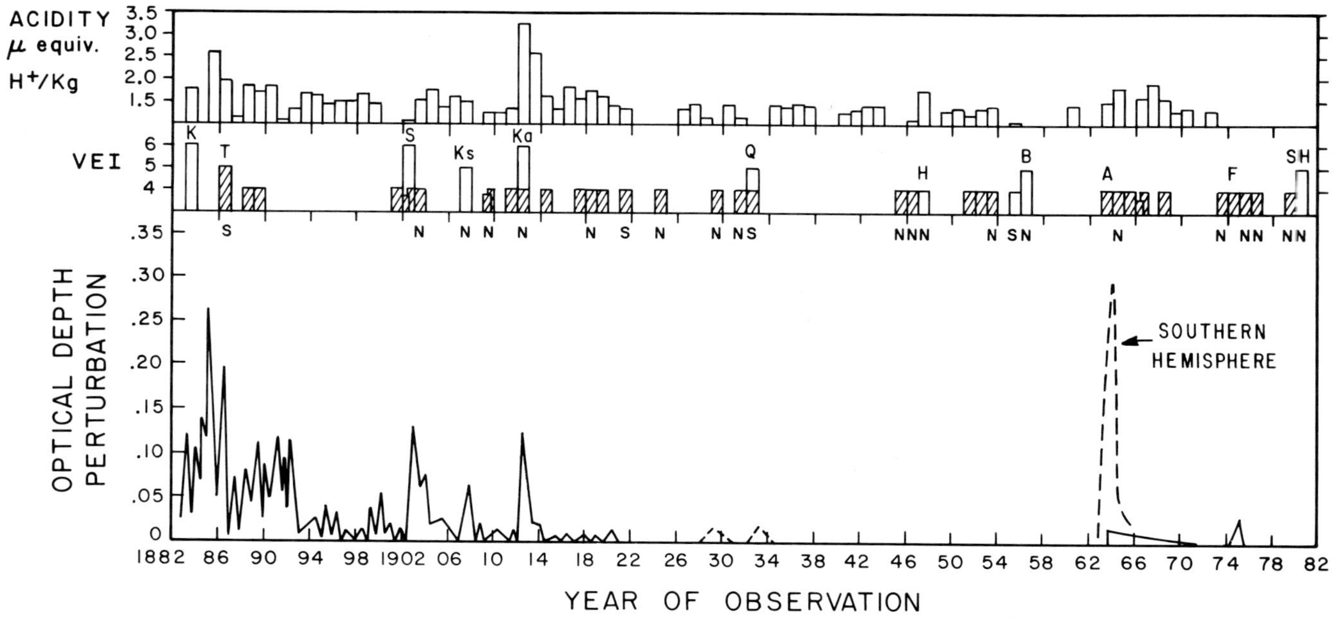


FIG. 1.—Northern hemisphere stratospheric optical depth (1882–1980) (a measure of aerosol loading, see text) (after Toon and Pollack, 1980, with additional data), the record of large explosive eruptions, and acidity record in Greenland ice cores (Hammer and others, 1981). The volcanic record is for eruptions with a VEI > 4 (Newhall and Self, 1982). Eruptions are also divided into two categories: low $SiO_2$ and possibly high sulfur (cross-hatched columns) and high $SiO_2$ and low sulfur (open columns). S under eruption indicates southern hemisphere eruption (>10°S), N under eruption indicates relatively high northern latitude eruption (>40°N). Some of the largest and best known eruptions are labeled K, Krakatau (1883); T, Tarawera (1886); S, Santa Maria (1902); Ks, Ksudach (1907), Ka, Katmai (1912); Q, Quizapu (1932); H, Hekla (1947); B, Bezymianny (1956), A, Agung (1963); F, Fuego (1974); SH, St. Helens (1980) (after Rampino and Self, 1982). The sulfur-rich 1982 eruption of El Chichón (not shown) produced peak optical depths of about 0.15 in the northern hemisphere (Rampino and Self, 1984a).

and Sigurdsson, 1989). For recent eruptions, the amount of sulfur volatiles released can be estimated from observations of the opacity of the aerosol clouds, and from measurements of the acidity in yearly layers in ice cores that record the fallout of the acid from the atmosphere (Table 1; Rampino and Self, 1984b).

Whereas it has been reasonably well established that volcanic eruptions can lead to cooling of the climate of a few years duration (Self and Rampino, 1988), a major question remains as to the ultimate effects of large eruptions or groups of eruptions on short-term climate and longer term climatic change. Long-term effects on climate require some sort of positive feedback mechanism to extend and amplify the cooling caused by the short-lived volcanic aerosol cloud, e.g., the enhanced albedo of increased snow and ice cover following eruptions. If such feedback mechanisms are important, then volcanic eruptions may be an important forcing for climate cooling in the geologic record.

## VOLCANOES AND SHORT-TERM CLIMATE

Connections between volcanic eruptions and climatic change have been suggested on time scales ranging from days to $10^8$ years. To investigate the connection on short time scales, one approach has been to compare the times of historic volcanic eruptions with changes in yearly, monthly

TABLE 1.—COMPARISON OF ERUPTION CHARACTERISTICS

| Eruption | VEI | Volume of magma erupted ($km^3$) | Eruption column height (km) (c = calculated o = observed) | Stratospheric $H_2SO_4$ aerosols (g) (from optical depth) | Stratospheric $H_2SO_4$ aerosols (g) (from ice-core acidity) | Northern Hemisphere $\Delta T$ (°C) |
|---|---|---|---|---|---|---|
| Tambora (8°S) 1815 | 7 | ≥50 | >40 (c) | $2 \times 10^{14}$ | $1.5 \times 10^{14}$ | −0.4 to −0.7 |
| Krakatau (6°S) 1883 | 6 | ≥10 | >40 (c) | $5 \times 10^{13}$ | $5.5 \times 10^{13}$ | −0.3 |
| Santa Maria (15°N) 1902 | 6 | ~9 | >30 (c, o) | $\leq 2 \times 10^{13}$ | $\leq 2 \times 10^{13}$ | −0.4 |
| Katmai (58°N) 1912 | 6 | 15 | >27 (c) | $\leq 2 \times 10^{13}$ | $\leq 3 \times 10^{13}$ | −0.2 |
| Mt St Helens (46°N) 1980 | 5 | 0.35 | 22 (o) | $\sim 3 \times 10^{11}$ | — | 0 to −0.1 |
| Agung (8°S) 1963 | 4 | 0.3 to 0.6 | >18 (c)<br>13 (o) | $1 - 2 \times 10^{13}$ | | −0.3 |
| El Chichón (17°N) 1982 | 4 | 0.3 to 0.35<br>0.3 (tephra) | 26 (o) | $1 - 2 \times 10^{13}$ | — | −0.4 to −0.6 |
| Laki (64°N) 1783 | 4 | 12.3 (lava) | ? | — | $\leq 1 \times 10^{14}$ | ~−1.0 |

or seasonal surface temperatures on regional, hemispheric and global scales. These studies involve either a direct comparison of the times of significant eruption years with temperature records (e.g., Rampino and Self, 1982, 1984b; Stothers, 1984a; Angell and Korshover, 1985; Mass and Portman, 1989; Table 1) or a study of composited temperature records for a number of years (months, seasons) before and after a chosen set of eruptions (e.g., Mass and Schneider, 1977; Self and others, 1981; Taylor and others, 1980). Eruptions are usually selected using various measures of volcanic intensity including the VEI (Volcanic Explosivity Index) of Newhall and Self (1982), or the DVI (Dust Veil Index) after Lamb (1970), or on the basis of the stratospheric aerosol loading determined directly by observations or indirectly from the acidity record of ice cores (Rampino and Self, 1982, 1984b). The various studies give similar results—the composites show a Northern Hemisphere cooling of 0.2 to 0.3°C for 1 to 3 years after eruptions for a number of eruptions grouped together (Fig. 2), and individual volcanic events that produced significant aerosol clouds such as Krakatau, 1883, or Tambora, 1816, are followed by Northern Hemisphere coolings of 0.3 to 0.7°C for 1 to 3 years after the eruption (Table 1; Baldwin and others, 1976; Rampino and Self, 1984b; Stothers, 1984a; Angell and Korshover, 1985; Ellsaesser, 1986). Zonally, the cooling is amplified at high latitudes (Self and others, 1981), and regional temperatures show more variability.

Monthly and seasonal temperature data also give evidence of a volcanic signal. Bradley (1988) found that several of the larger eruptions of the past 100 years are followed by significant negative anomalies in summer and fall temperatures. Temperature decreases after major eruptions seem to be abrupt and short lived (1 to 3 months), with a recurrence of the cooling about 12 and 24 months after the eruptions (c.f., Sear and others, 1987). The maximum effect of about 0.4°C occurs in the summer and fall months immediately following the eruptions, and falls off in the same seasons over the next 2 years.

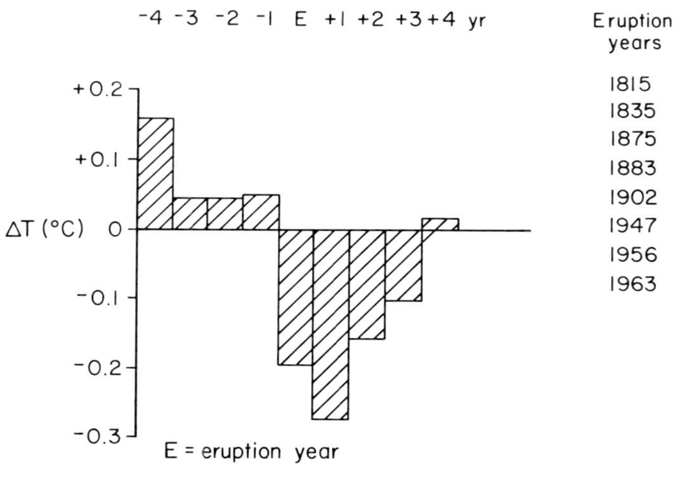


FIG. 2.—Composite plot of temperature anomaly ΔT(°C) for the northern hemisphere for 4 years before and after eruption years of some volcanic events since 1800 (Self and others, 1981).

Mass and Portman (1989) recently suggested that the volcanic signal in surface temperatures is smaller than that found in previous studies. In the last century, it appears to be limited to those eruptions that created the densest aerosol clouds, and could be enhanced by subtracting out other sources of interannual variability, e.g., the El Niño/Southern Oscillation (Angell, 1988). Mass and Portman (1989) stressed that the volcanic "signal" of 0.1 to 0.2°C that they detected is of the same order as "background" temperature variations in nonvolcanic years, and found no evidence of large coolings in the first few months after the eruptions. However, some of the eruptions they studied were quite small, and one would expect little climatic cooling in their aftermath. Stratospheric aerosol clouds may have some effect on the ENSO phenomena, either triggering them, or intensifying already existing ENSO patterns (Handler, 1984; Handler and Andsager, 1990). Handler (1986) has also suggested a connection between stratospheric aerosols and the strength of the yearly Indian monsoonal precipitation, but these proposals remain controversial (Nicholls, 1990). Mass and Portman (1989), for example, report no significant precipitation anomalies after the historic volcanic eruptions in their study.

Climate model experiments have also been used to estimate the effects of volcanic aerosols on short term. In one study, Hansen and others (1978) chose one of the best-monitored eruptions at the time, the 1963 eruption of Agung volcano on the island of Bali. Their observational temperature data showed a 0.5°C cooling at the surface in the tropics and a stratospheric warming (from the presence of the aerosols) of more than 6°C in the 2 years after the eruption. Using a simple radiation-balance model, in which they simulated an aerosol cloud in the tropics, they were able to reproduce this basic pattern of temperature change in the tropics and subtropics (Hansen and others, 1978). More sophisticated 2-D and 3-D General Circulation Models (e.g., Batten, 1974; Chou and others, 1984; McCracken and Luther, 1984) and coupled ocean/atmosphere models (Walsh and Pittock, 1990) have also been applied to the volcano-climate problem with generally similar results.

Natural limits to the atmospheric effects of any volcanic eruption may exist (Turco and others, 1983; Rampino and Self, 1982). Pinto and others (1989) have recently proposed self-limiting physical and chemical effects in eruption clouds. Model results (using up to 200 Mt, or 2 × $10^{14}$ g of aerosols) suggest that aerosol microphysical processes of condensation and coagulation produce larger aerosols as the $SO_2$ injection rate is increased (rather than a larger number of particles of the same small size). Larger particles have a smaller optical depth per unit mass than do smaller ones; they also settle out of the stratosphere at a faster rate, restricting the total number of particles in the stratospheric cloud. These processes may act to moderate the impact of volcanic aerosol clouds on the Earth's radiation budget and climate, and suggest that volcanic aerosol effects might be self limiting, so that extrapolation of volcanic effects on the atmosphere from known, small eruptions to larger prehistoric eruptions may not be straightforward.

The above studies have attempted to "isolate" the volcanic signal in noisy temperature data. This assumes that it *is* possible to isolate a distinct volcanic signal in a record that may have a combination of forcings (ENSO, solar variability, random fluctuations, volcanism) that all interact. The key to discovering the greatest effects of volcanoes on short-term climate, however, may be to concentrate on temperatures in regions where the effects of volcanic aerosol clouds may be amplified by perturbed atmospheric circulation patterns. This is especially true in subpolar and mid-latitude areas affected by changes in the position of the polar front. Such climatic perturbations can be detected in surface temperatures, and in proxy evidence such as frost-damage rings, decreases in tree-ring widths, changes in the treeline, weather anomalies such as unusually cold springs and summers, severity of sea-ice in polar and subpolar regions, and poor grain yields and crop failures (for a review, see Rampino and others, 1988). For some eruptions, such as Tambora, 1815, these kinds of proxy and anectdotal information have been summarized in great detail in a number of papers and books (e.g., Stothers, 1984a; Stommel and Stommel, 1983; Harrington, 1991). These studies lead to the general conclusion that regional and seasonal effects on climate, sometimes quite severe, may be the major impact of large historical volcanic aerosol clouds.

*Tambora and 1816: A Test Case?—*

In addition to searching for the small climatic effects of small historic eruptions in climate data sets, it may be instructive to look in detail at the strongest volcanic perturbation in recent history, the Tambora eruption of April, 1815, and the events that followed it. Stothers (1984a) estimates a peak optical depth $\tau$ of about 1.0 (optical depth is a measure of the attenuation of incident light, $\tau = -\ln (I/I_0)$ where $I_0$ and $I$ are the initial and final light intensity, respectively; optical depth can be related to aerosol loading by the relationship $M = 1.5 \times 10^{14}\, \tau$ in grams, where $M$ is the mass of stratospheric aerosol; Stothers, 1984b) following the Tambora eruption, and the presence of large acidity peaks in ice cores from Greenland and Antarctica argues for a global dispersal of the aerosol cloud of some $10^{14}$ g (100 megatons). (This may be compared with peak optical depths of about 0.15 for El Chichón in 1982, and 0.3 for Krakatau in 1883.)

In most studies of weather records or proxy data, the year 1816, following the Tambora eruption, displays the strongest signal of possible "volcano weather" in historical times. For example, in a key area in the Eastern Hudson Bay region in midsummer 1816, the reduction in mean daily temperature from the long-term average was about 5 to 6°. It has been suggested that a reduction in mean daily temperature in this range could lead to the formation of perennial snow cover in northern Canada (Wilson, 1985). The negative anomaly in July 1816 brought the absolute mean temperature to below 5°C, more than 2° lower than the modern record in 1965. The median height of the freezing level above eastern Hudson Bay in July 1965 (2 years after the Agung eruption) was about 600 to 700 m lower than the modern 10-year value. This suggests that the −6°C anomaly in 1816 might have produced a 1,000-m drop, to bring the freezing level to within 1,500 m of the surface (Wilson, 1985). Such conditions were maintained for only 2 years, 1816 and 1817, but one may wonder how many years of consecutive extreme seasonal weather would be required to establish an effective snow cover, and bring snow and ice-albedo feedback into play to produce cooling over a longer time period.

Frequent and extended snowfall also took place in the region in 1816–1817—in 1816, there were only five weeks without measurable snowfall (mid-July to mid-August). Furthermore, the atmospheric circulation pattern was dominated by a high-pressure area over the surface of Hudson Bay, which remained ice covered through the summer of 1816 (Wilson, 1985). Catchpole and Faurer (1983) show sea-ice patterns in 1816 that are also consistent with a highly meridional atmospheric circulation pattern over eastern North America, showing strong north to northwest winds in July and August of 1816. Northwesterly winds brought cold air southward into the northeastern United States, bringing the "Year without a Summer". The meridional circulation pattern also brought cold weather to western Europe, and there is some evidence that, between these two atmospheric waves, warmer weather dominated, with an opening of the usually ice-covered Greenland Sea between 74° and 80°N (Wilson, 1985).

It should be pointed out, however, that the cold weather in the Hudson Bay area began before April 1815, with unusually cold years beginning in 1811/1812, so that the Tambora eruption cannot be the only cause of the climate shift. Perhaps the coincidence of the Tambora eruption with a time of low sunspot numbers (indicating lower solar activity) led to the unusually cold weather of the 1811–1817 period (Wilson, 1985).

Frost rings in trees in the Western United States also show a correspondence with years of volcanic eruptions (LaMarche and Hirschboeck, 1984). These may indicate outbreaks of exceptionally cold weather during the summer months, again possibly the result of increased meridional circulation. Light rings in subarctic trees from northern Quebec also show a correspondence with volcanic eruptions; for example, damage is most widespread in 1816–1817 (Filion and others, 1986). Tree rings have also been used to create synoptic summer temperature patterns for Europe, which show significantly cooler summers from 1812 and 1816 (Briffa and others, 1988). The same kinds of data can be assembled for the other major volcanic aerosol clouds of the last 200 years. If one can establish patterns that appear to be characteristic of "volcanic" perturbation, then it may be possible to detect such patterns even for the smaller eruptions, and to see how well climate models simulate such perturbations. It may be necessary not only to model the volcanic perturbations, but also to include the effects of ENSO events and possible solar variations in the model runs.

## TOBA, 75 ka: VOLCANIC WINTER?

One problem with studying historical eruptions is that they are quite small compared to those in the geologic record. The largest explosive eruption in historic times was probably the Tambora, 1815, explosion, with about 50 km$^3$ of erupted magma. This can be compared with large ignimbrite-forming eruptions such as the Toba event of 75 ka, which erupted more than 2,800 $km^3$ of magma (Rose and Chesner, 1987). (For effusive eruptions, the largest historic event was Laki, 1783, which produced 12 $km^3$ of basalt, compared with great flood-basalt eruptions like the Roza flow in the Columbia River Basalt Group (14 Ma), composed of 700 $km^3$ of basaltic lava in a single flow [Devine and others, 1984; Rampino and others, 1988].)

By simply scaling up from smaller eruptions, it has been estimated that the Toba eruption could have produced up to 20,000 Mt of fine (<2 μm) dust, and 1,000 to 5,000 Mt of sulfuric acid aerosols, as compared with 1,500 Mt of dust and 200 Mt of aerosols for Tambora (Rampino and others, 1988). The Toba eruption is known to correlate with a dramatic cooling and increase in ice volume that defines the oxygen-isotope Stage 4/5 boundary in deep-sea cores (Fig. 3), dated at 75 ka (the Toba ash occurs in deep-sea cores in the Indian Ocean coincident with the Stage 4/5 boundary; Ninkovich and others, 1978), and this has been described as the "major transition into the Würm glaciation" for temperate latitudes of the northern hemisphere (south of 65°N); Ruddiman, 1977a; Ruddiman and others, 1980). Summer sea-surface temperatures in the North Atlantic may have decreased about 10°C, the greatest single decrease of the last glacial cycle, with significant decreases in winter sea-surface temperature and salinity showing a strong polar water advance at the time (Sancetta and others, 1972; Ruddiman, 1977b). The first significant peak in ice-rafted detritus in the North Atlantic during the last glacial cycle occurs at the same time (Ruddiman, 1977b; Heinrich, 1988). Kennett and Huddlestun (1972) also note a severe cooling at about 75 to 73 ka in the Gulf of Mexico.

This cooling is also evident in the $\partial^{18}O$ record in polar ice cores (Paterson and others, 1977; Dansgaard and others, 1971; Broecker and others, 1988; Jouzel and others, 1987). The Vostok ice core shows marked decreases in Antarctic temperatures (about 5°C), atmospheric $CO_2$, and a sharp increase in windblown dust at the time, indicating sparse vegetation and increased wind strength (Barnola and others, 1987; Jouzel and others, 1987). Ice sheets grew rapidly, producing a drop in sea level of about 60 m in a few thousand years or less (Ruddiman, 1977a,b), more than half of the total drop in sea level during the last glacial stage. A rapid change from temperate to boreal forests is seen in detailed pollen records in western Europe at the same time (Woillard, 1979).

Could the Toba eruption have played some role in the dramatic climatic changes seen at 75 ka, which initiated the last glacial cycle? Rapid growth of snow and ice at high latitudes might create the kinds of positive feedback that could trigger a glaciation (Barry and others, 1975; Bray, 1976; Flohn, 1979; Harvey, 1989). As noted above, several researchers have suggested that a reduction in summer daily mean temperatures over the central plateau area of Quebec and Labrador, perhaps accompanied by increased snowfall, would favor the formation of perennial snow cover (Loewe, 1971; Williams, 1979; Wilson, 1985) and could lead to instantaneous glacierization of the high plateaus of the Canadian Arctic within a few hundred years (Ives and others, 1975; Koerner, 1980).

A stratospheric aerosol burden of $10^3$ Mt ($10^{15}$ g) would produce an aerosol optical depth of about 10, roughly 10 times greater than the Tambora eruption (we are ignoring here the influence of the shorter residence time dust on the climate, which may be significant). If northern hemispheric temperatures vary in proportion to the 1/3 power of estimates of sulfuric acid aerosol yield, as suggested by Devine and others (1984) and Palais and Sigurdsson (1989), then extending their relationship to eruptions producing $10^{15}$ g of aerosols, the hemispheric temperature decrease after Toba could have been about 3 to 5°C. The northern hemisphere $-\Delta T$ should have been amplified at high latitudes (Self and others, 1981), perhaps sufficient to bring the freezing level over the Canadian plateaus down to the surface in summer, and favoring the growth of permanent snowfields. The information available to date, although incomplete, suggests that a super-eruption that produces on the order of $10^{15}$ g

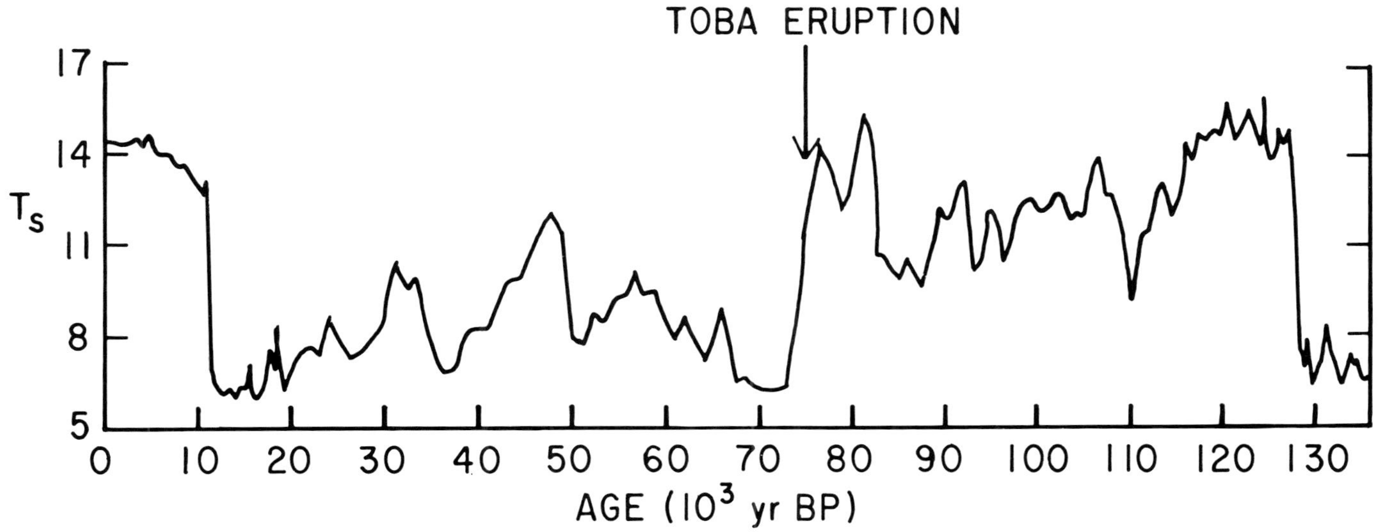


FIG. 3.—Faunal index for estimated summer sea-surface temperature in the North Atlantic for the past 130 ky (after Sancetta and others, 1972), and the date of the Toba eruption (Ninkovich and others, 1978).

of stratospheric aerosols might be capable of creating conditions favorable for the initiation of a glaciation.

## VOLCANISM AND LONG-TERM CLIMATE CHANGE

A number of authors have claimed correlation between volcanic eruptions and glacial fluctuations on $10^1$- to $10^4$-year time scales. For example, Bradley and England (1978) and Porter (1981, 1986) proposed that second-order "Little Ice Age" glacial advances during the Holocene could have been driven by eruptions, since the glacial advances correlate with peaks of acidity in polar ice cores. Greenland ice cores show an overall increase in acidity levels during the Little Ice Age of the 17th and 18th Centuries (Hammer and others, 1980). Porter (1981) estimates that a global cooling of 1°C could lead to a snowline depression sufficient to cause glacier advances equivalent to those of the last several centuries, but this seems to be greater than the measured volcanic perturbations of the past few hundred years. Bray (1974, 1976, 1977) proposed that large eruptions preceded glacial periods, with the glaciations triggered by atmospheric aerosol clouds. However, in this case, because of the inaccuracies in dating, it is often difficult to determine which came first, the eruptions or the glaciation (Rampino and others, 1979). Paterne and others (1990) recently reported that explosive volcanism in the Mediterranean over the last 200 ka shows a periodicity of 23 ka, matching the Earth's precessional cycle. This suggests that the same orbital parameters that force the glacial-interglacial cycle may be forcing volcanism, perhaps through changes in crustal loading as a result of sea-level variations.

On longer time scales, studies by Kennett and others (1977) have delineated pulses of explosive volcanic activity in the Circum-Pacific region over the last 30 Ma from studies of age determinations (mostly K/Ar dates) on igneous rocks, and counts of ash layers in deep-sea sediment cores. They find significant pulses of volcanism in the Plio-Pleistocene (about 2 Ma), latest Miocene to Early Pliocene (6 to 3 Ma), Late Miocene (11 to 8 Ma) and Middle Miocene (16–14 Ma). Combined with other data (e.g., Hein and others, 1978), it seems that pulses of widespread (perhaps global) explosive volcanism took place near 0.5, 2.5, 5, 10, 15, 20 and 40 Ma.

Some of these spurts in volcanism correlate with times of apparent pulses of plate motion and sea-level changes (Masuda, 1986; Rampino and Stothers, 1987), so the three may be linked. These volcanic pulses can be also be correlated with times of global cooling and ice formation as seen in oxygen isotope data for the Cenozoic (e.g., Savin and others, 1975; Table 2), but here again the lack of sufficient accuracy in dating makes it difficult to establish any cause-and-effect relationship.

TABLE 2.—DATES (MA) OF MARINE EPISODES IN THE CENOZOIC RECORD VERSUS VOLCANIC EPISODES (SEE TEXT). DATES AFTER RAMPINO AND STOTHERS (1987)

| Volcanic Episodes | Geologic boundaries | $\delta^{18}O$ increases | Hiatus increases | Major unconformities | Minor unconformities |
|---|---|---|---|---|---|
| 0.5 | 1.6 | 0.9 | — | — | 3.8 |
| 2.5 | 3.4 | 2.4 | 3.5 | 2.8 | 4.2 |
| 5 | 5.3 | 5.5 | — | 6.6 | 5.2 |
| 10 | 11.2 | 10 | 11 | 9.8 | 6.0 |
| 15 | 16.6 | 15 | 17.5 | — | 13 |
| 20 | 23.7 | 24.5 | — | 22.5 | 15.5 |
| — | 30.0 | 29.5 | — | 29 | 16.5 |
| — | 36.6 | 35.5 | 38 | — | 19 |
| 40 | 40.0 | 41 | 42.5 | 40 | 22 |
| — | 52.0 | 50 | 54 | 49.5 | 24 |
| — | 57.8 | — | — | 60 | 26 |
| 62 | 63.6 | — | — | — | 37 |
| 66 | 66.4 | — | 65 | — | 45.5 |
| | | | | | 52 |
| | | | | | 53.5 |
| | | | | | 56 |
| | | | | | 59 |

## LONG TIMESCALES OF VOLCANISM, ATMOSPHERIC COMPOSITION AND CLIMATE

A number of researchers have suggested that the Earth experiences alternating periods of enhanced and lessened volcanism over millions of years and longer (e.g., Vogt 1972; Ronov, 1976; 1983; Rea and Vallier, 1982; Budyko and others, 1987; Sheridan, 1987; Rampino and Stothers, 1988). Several studies in the past few years have suggested a direct connection between global climate on long time scales and volcanism/tectonism through variation in the amount of carbon dioxide in the atmosphere and the resulting greenhouse warming (Berner and others, 1983; Lasaga and others, 1985; Walker and others, 1981; Berner, 1990). Atmospheric carbon dioxide content over $10^7$ and $10^9$ years is now seen as the result of two competing processes, removal of $CO_2$ from the atmospheric through weathering of silicate minerals, and influx of atmospheric $CO_2$ from subduction-zone volcanism (Walker and others, 1981; Volk, 1987).

A number of studies have attempted to determine quantitatively the rates of tectonism and volcanism over the last 600 Ma (the Phanerozoic). For example, variations of the rates of seafloor spreading (crustal accretion rates) for the past 180 Ma have been determined directly from studies of linear magnetic anomalies (Rich and others, 1986). Ocean-ridge volume can be estimated from seafloor accretion rates since the depth of the cooling and subsiding ocean plate is a function of its age (Parsons and Sclater, 1977). Changes in ocean-ridge volume should cause changes in global sea level. The first-order Phanerozoic sea-level curves of Vail and co-workers (Vail and others, 1977; Vail and others, 1985; Haq and others, 1987) have thus been used as a proxy for ocean-ridge volume and hence accretion rate over the Phanerozoic (Abbott, 1984; Heller and Angevine, 1985). Gaffin (1987a) performed a formal mathematical inversion of the first-order Vail curve to derive a record of global accretion rate for the past 600 Ma (Fig. 4). He found that although the ridge-volume and seafloor accretion rate are temporally related, they are not simply proportional, and that the highest stand of sea level lags the highest rates of accretion by about 10 Ma (Gaffin, 1987b).

A comparison of the first-order sea-level curve of Vail and others (1985) and the global seafloor accretion rate from Gaffin (1987a) shows high rates of accretion and high sea level in the Cretaceous and in the middle Paleozoic back to the Ordovician, with lower spreading rates and sea level in the late Paleozoic and early Mesozoic. Figure 4 shows

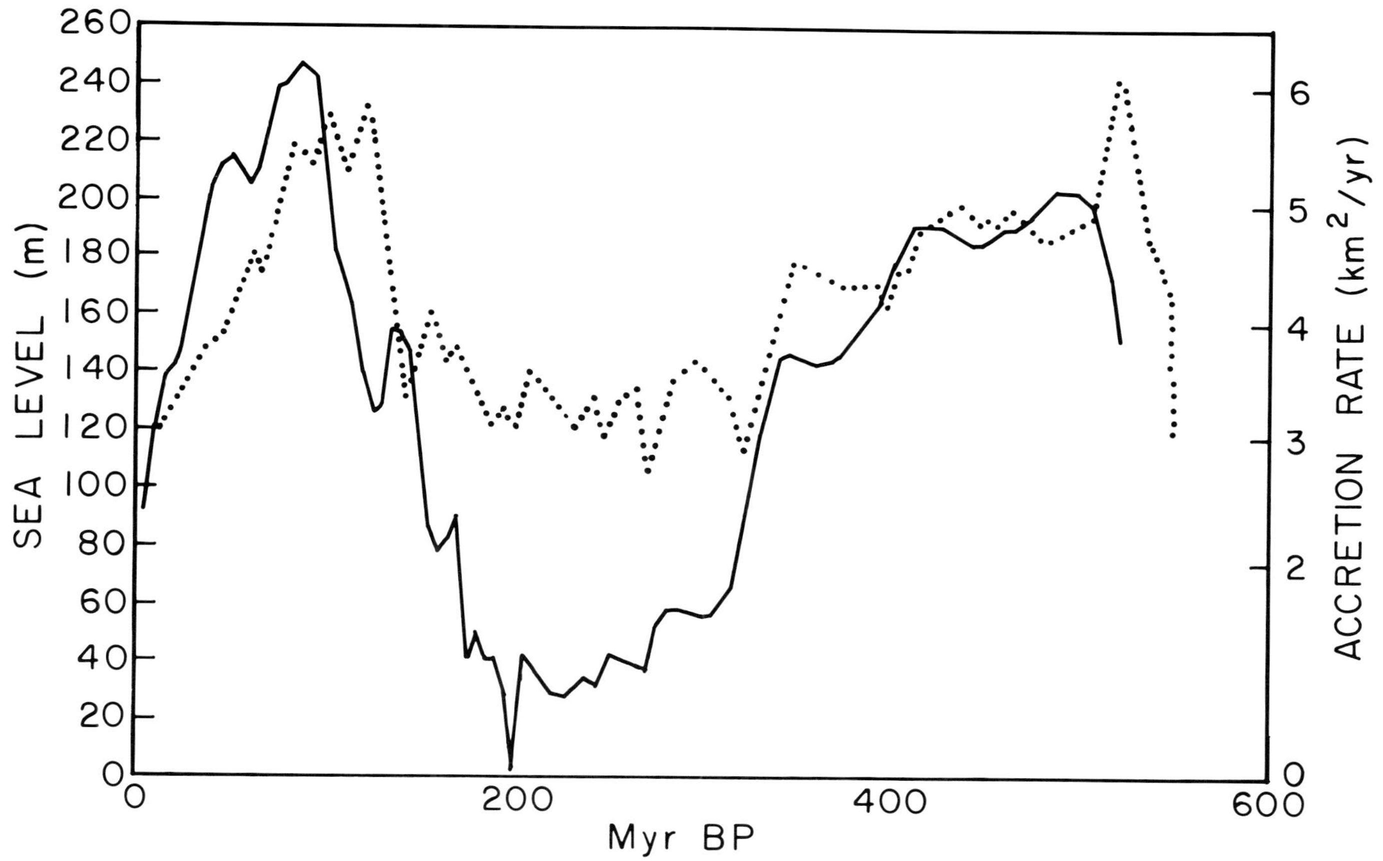


FIG. 4.—A comparison of first-order Phanerozoic sea level (solid line, after Vail and others, 1977), and ocean-floor accretion rate (dotted line, after Gaffin 1987a).

that according to Gaffin's results, the rate of seafloor accretion seems to have varied by a factor of two over the Phanerozoic. If one scales subduction-zone volcanism with accretion rates, then the calculated values of seafloor accretion rate can be used in biogeochemical cycle models to estimate the carbon dioxide content of the atmosphere. Such estimates of atmospheric $CO_2$ can be compared with values of atmospheric $CO_2$ suggested by studies of the composition of abiotic carbonate sediments over the Phanerozoic (Sandberg, 1983; Wilkinson and Given, 1986).

Berner (1990) gives early Phanerozoic estimates of up to 18 times present atmospheric $pCO_2$ (with values possibly much higher), and a low in atmospheric $pCO_2$ approaching modern values during the Permo-Carboniferous. The predicted periods of high atmospheric $CO_2$ agree with proposed "Greenhouse Episodes" of warm global climate (Berner, 1990). Atmospheric $CO_2$ during the mid-Cretaceous was estimated at most likely about six times present $CO_2$, but with possible values up to 12 times present $CO_2$.

Ronov (1976, 1983; Budyko and others, 1987) compiled data on the volumes of volcanic rocks by Period and Epoch during the Phanerozoic (Fig. 5). This curve shows well-developed peaks of volcanism in the Ordovician, Devonian, early Permian, Triassic and Cretaceous Periods. The volcanic peaks may be compared with the data in Figure 4. Peaks in volcanism show a general agreement with times of rapid accretion rates and high sea level. The main exception is the volcanic peak in the Triassic, a time of apparent low sea level and perhaps slow spreading rates. According to Ronov (1976), however, this peak is made up largely of "terrestrial volcanogenic associations of the platforms" and hence is probably rift-related volcanism and not a peak in subduction-zone volcanism, and therefore might not represent a period of enhanced ocean-floor accretion rate.

Studies of clay minerals in shales have also revealed the fluctuations in the importance of volcanism as a source of fine-grained sediment in depositional basins. Nadeau and Reynolds (1981) have shown that the majority of Cretaceous marine shales from the Western Interior of North America had a volcanic origin. These Cretaceous marine sediments are primarily composed of interstratified illite/smectite (I/S), probably originating as volcanic ash. At times when sedimentation rates of volcanic debris are very high, the volcanic material has much shorter marine residence times and bentonite beds of pure smectite are formed. It is thus possible to examine shales from the geologic record in order to determine the importance of explosive volcanism at various times. It seems that, according to clay mineralogy, explosive volcanism was strongest in the early to mid-Pa-

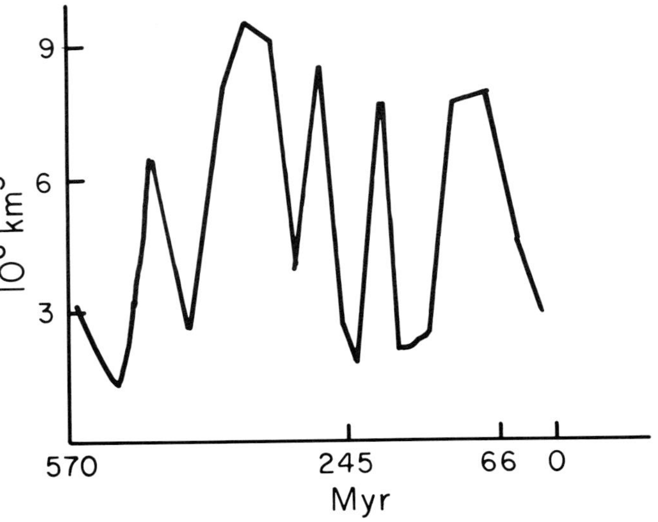


FIG. 5.—Estimate of the volume of volcanic rocks during the Phanerozoic (after Ronov, 1983).

leozoic, and during the Cretaceous. This correlates well with the times of increased spreading rate, subduction-zone volcanism, and high sea level, as shown in Figures 4 and 5.

What might cause these long-term fluctuations in volcanism/tectonism? Perhaps they are self driven. Anderson (1982, 1984) developed a model in which the thick continental "tectosphere" beneath a supercontinent acts as an effective insulator with respect to heat flow from the mantle. At times when the continents were clustered in Pangaea configurations (in the latest Paleozoic/earliest Mesozoic, for example), the heat flow would be bottled up, leading to uplift, rifting and eventual break-up of the supercontinent. This would be followed by a period of rapid accretion rates, high sea levels, and increased subduction-zone volcanism (Vogt, 1972, Sheridan, 1987). Worsley and coworkers, in a series of papers, have expanded on this idea to create cyclical episodes of continental dispersal and reorganization based on the difference in heat flow through supercontinent and ocean basin (Worsley and others, 1984, 1985; Nance and others, 1986). In this model, the continental configuration becomes involved in the driving forces for plate movement—the boundary conditions of continental arrangement is the control over the opening and closing of an ocean roughly in the position of today's Atlantic. This mechanism drives the global volcanic/eustatic super-cycle, which controls atmospheric and oceanic composition through volcanic input of carbon dioxide, sea level, weathering rates, carbon burial, and climate change (including major ice ages) involving numerous biogeochemical feedback mechanisms (e.g., burial of carbon controls atmospheric oxygen levels; climate affects weathering rates, which control removal of $CO_2$ from the atmosphere, etc.). Rampino and Stothers (1984) and others have suggested an external forcing, showers of large-body impacts, as a driving force for plate tectonic cyclicity. Whatever the ultimate driving force, it seems that volcanism has varied by a factor of four over the Phanerozoic (Fig. 5), in step with global seafloor accretion rates (which have varied by about a factor of two; Gaffin, 1987a) and eustatic sea level. Thus, long-term changes in the Earth's climate, with megacycles of about 30 and 300 Ma (Fisher, 1984; Imbrie and Shackleton, 1990), may be the result of variations in $CO_2$ resulting from cyclic global tectonism.

ACKNOWLEDGMENTS

Partial support was provided by the Center for Global Habitability at Columbia University. I thank K. Caldeira, S. Gaffin, J. Palais, C. Sagan, H. Sigurdsson, R. B. Stothers and T. Volk for helpful discussions, and S. Self for a review.

REFERENCES

ABBOTT, D. H., 1984, Archaean plate tectonics revisited 2. Paleo-sea level changes, continental area, oceanic heat loss and the area-age distribution of the ocean basins: Tectonics, v. 3, p. 709–722.

ANDERSON, D. L., 1982, Hotspots, polar wander, Mesozoic convection and the geoid: Nature, v. 297, p. 391–393.

ANDERSON, D. L., 1984, The earth as a planet: Paradigms and paradoxes: Science, v. 223, p. 347–355.

ANGELL, J. K., 1988, Impact of El Niño on the delineation of tropospheric cooling due to volcanic eruptions: Journal of Geophysical Research, v. 93, p. 3697–3704.

ANGELL, J. K., AND KORSHOVER, J., 1985, Surface temperature changes following the six major volcanic episodes between 1780–1980: Journal of Climate and Applied Meteorology, v. 24, p. 937–951.

BALDWIN, B., POLLACK, J. B., SUMMERS, A., TOON, O. B., SAGAN, C., AND VAN CAMP, W., 1976, Stratospheric aerosols and climatic change: Nature, v. 263, p. 551–555.

BARNOLA, J. M., RAYNAUD, D., KOROTKEVICH, Y. S., AND LORIUS, C., 1987, Vostok ice core provides 160,000 year record of atmospheric $CO_2$: Nature, v. 329, p. 408–414.

BARRY, R. G., ANDREWS, J. T., AND MAHAFFY, M. A., 1975, Continental ice sheets: Conditions for growth: Science, v. 190, p. 979–981.

BATTEN, E. S., 1974, The atmospheric response to a stratospheric dust cloud as simulated by a general circulation model: DARPA Report R-1324-ARPA, Rand Corp., Santa Monica, 17 p.

BERNER, R. A., 1990, Atmospheric carbon dioxide levels over Phanerozoic time: Science, v. 249, p. 1382–1386.

BERNER, R. A., LASAGA, A. C., AND GARRELS, R. M., 1983, The carbonate-silicate geochemical cycle and its effect on atmospheric carbon dioxide over the past 100 million years: American Journal of Science, v. 283, p. 641–683.

BRADLEY, R. S., 1988, The explosive volcanic eruption signal in northern hemisphere continental temperature records: Climatic Change, v. 12, p. 221–243.

BRADLEY, R. S., AND ENGLAND, J., 1978, Influence of volcanic dust on glacier mass balance at high latitudes: Nature, v. 271, p. 736–738.

BRAY, J. R., 1974, Volcanism and glaciation during the past 40 millenia: Nature, v. 252, p. 679–680.

BRAY, J. R., 1976, Volcanic triggering of glaciation: Nature, v. 260, p. 414–415.

BRAY, J. R., 1977, Pleistocene volcanism and glacial initiation: Science, v. 197, p. 251–254.

BRIFFA, K. R., JONES, P. D., AND SCHWEINGRUBER, F. H., 1988, Summer temperature patterns over Europe: A reconstruction from 1750 AD based on maximum latewood density indices of conifers: Quaternary Research, v. 30, p. 36–52.

BROECKER, W. S., AND FIVE OTHERS, 1988, Can the Greenland climatic jumps be identified in records of ocean and land?: Quaternary Research, v. 30, p. 1–6.

BUDYKO, M. I., RONOV, A. B., AND YANSHIN, A. L., 1987, History of the Earth's atmosphere: Springer-Verlag, Berlin, 139 p.

CATCHPOLE, A. J. W., AND FAURER, M.-A., 1983, Summer sea ice severity in Hudson Strait, 1751–1870: Climatic Change, v. 5, p. 115–139.

CHOU, M.-D., PENG, L., AND ARKING, A., 1984, Climate studies with a multi-layer energy balance model. Part III: Climatic impact of stratospheric volcanic aerosols: Journal of the Atmospheric Sciences, v. 41, p. 759–767.

DANSGAARD, W., JOHNSON, S. J., CLAUSEN, H. B., AND LANGWAY, C. C., Jr., 1971, Climatic record revealed by Camp Century ice core, *in* Turekian, K. K., ed., The Late Cenozoic Glacial Ages: Yale University Press, New Haven, p. 37–56.

DEVINE, J. D., SIGURDSSON, H., DAVIS, A. N., AND SELF, S., 1984, Estimates of sulfur and chlorine yield to the atmosphere from volcanic eruptions and potential climatic effects: Journal of Geophysical Research, v. 89, p. 6309–6325.

ELLSAESSER, H. W., 1986, Comments on "Surface temperature changes following the six major volcanic episodes between 1780 and 1980": Journal of Climate and Applied Meteorology, v. 25, p. 1184–1185.

FILION, L., PAYETTE, S., GAUTHIER, L., AND BOUTIN, Y., 1986, Light rings in subarctic conifers as a dendrochronological tool: Quaternary Research, v. 26, p. 272–279.

FISCHER, A. G., 1984, The two Phanerozoic supercycles, *in* Berggren, W., and Van Couvering, J., eds., Catastrophes in Earth History: Princeton, Princeton University Press, p. 129–150.

FLOHN, H., 1979, On time scales and causes of abrupt climate change: Quaternary Research, v. 12, p. 135–149.

GAFFIN, S., 1987a, Ridge volume dependence on seafloor generation rate and inversion using long term sealevel change: American Journal of Science, v. 287, p. 596–611.

GAFFIN, S., 1987b, Phase difference between sea level and magnetic reversal rate: Nature, v. 329, p. 316–319.

HAMMER, C. U., CLAUSEN, H. B., AND DANSGAARD, W., 1980, Greenland ice sheet evidence of post-glacial volcanism and its climatic impact: Nature, v. 288, p. 230–235.

HANDLER, P., 1984, Possible associations of stratospheric aerosols and El Niño type events: Geophysical Research Letters, v. 11, p. 1121–1124.

HANDLER, P., 1986, Stratospheric aerosols and the Indian monsoon: Journal of Geophysical Research, v. 91, p. 14,475–14,490.

HANDLER, P., AND ANDSAGER, K., 1990, Volcanic aerosols, El Niño and the Southern Oscillation: International Journal of Climatology, v. 10, p. 413–424.

HANSEN, J. E., WANG, W. C., AND LACIS, A. A., 1978, Mount Agung eruption provides test of a global climatic perturbation: Science, v. 199, p. 1065–1068.

HAQ, B. U., HARDENBOL, J., AND VAIL, P. R., 1987, Chronology of fluctuating sea levels since the Triassic: Science, v. 235, p. 1156–1167.

HARRINGTON, C. R., ed., 1991, The Year Without a Summer? World Climate in 1816: Ottawa, Canadian National Museum of Natural Sciences.

HARVEY, L. D., D., 1989, Milankovitch forcing, vegetation feedback, and North Atlantic deep-water formation: Journal of Climate, v. 2, p. 800–815.

HEIN, J. R., SCHOLL, D. W., AND MILLER, J., 1978, Episodes of Aleutian ridge explosive volcanism: Science, v. 199, p. 137–141.

HEINRICH, H., 1988, Origin and consequences of cyclic ice rafting in the Northeast Atlantic Ocean during the past 130,000 years: Quaternary Research, v. 29, p. 142–152.

HELLER, P. L., AND ANGEVINE, C. L., 1985, Sea level cycles during the growth of Atlantic type oceans: Earth and Planetary Science Letters, v. 75, p. 417–426.

IMBRIE, J., AND SHACKLETON, N. J., 1990, The $\partial^{18}O$ spectrum of oceanic deep water over a five-decade band: Climatic Change, v. 16, p. 217–230.

IVES, J. D., ANDREWS, J. T., AND BARRY, R. G., 1975, Growth and decay of the Laurentide ice sheet and comparison with Fennoscandia: Naturwissenschaften, v. 62, p. 118–125.

JOUZEL, J., AND SIX OTHERS, 1987, Vostok ice core: A continuous isotope temperature record over the last climatic cycle (160,000 years): Nature, v. 329, p. 403–408.

KENNETT, J. P., AND HUDDLESTUN, P., 1972, Abrupt climatic change at 90,000 Yr BP: Faunal evidence from the Gulf of Mexico: Quaternary Research, v. 2, p. 384–395.

KENNETT, J. P., MCBIRNEY, A. R., AND THUNELL, R. C., 1977, Episodes of Cenozoic volcanism in the circum-Pacific region: Journal of Volcanology and Geothermal Research, v. 2, p. 145–163.

KOERNER, R. M., 1980, Instantaneous glacierization, the rate of albedo change, and feedback effects at the beginning of an ice age: Quaternary Research, v. 13, p. 153–159.

LAMARCHE, V. C., AND HIRSCHBOECK, K. K., 1984, Frost rings in trees as records of major volcanic eruptions: Nature, v. 307, p. 121–126.

LAMB, H. H., 1970, Volcanic dust in the atmosphere; with its chronology and assessment of its meteorological significance: Philosophical Transactions of the Royal Society of London A, v. 266, p. 425–533.

LASAGA, A. C., BERNER, R. A., AND GARRELS, R. M., 1985, An improved geochemical model of the atmospheric $CO_2$ fluctuations over the past 100 million years: American Geophysical Union Monograph 32, p. 397–411.

LOEWE, F., 1971, Considerations of the origin of the Quaternary ice sheet of North America: Arctic and Alpine Research, v. 3, pp. 331–344.

MASS, C. F., AND PORTMAN, D. A., 1989, Major volcanic eruptions and climate: A critical evaluation: Journal of Climate, v. 2, p. 566–593.

MASS, C. F., AND SCHNEIDER, S. H., 1977, Statistical evidence on the influence of sunspots and volcanic dust on long-term temperature records: Journal of the Atmospheric Sciences, v. 34, p. 1995–2008.

MASUDA, F., 1986, "Vail sea-level curve" records of the ages of plate motion changes: Annual Report of the Institute of Geoscience, the University of Tsukuba, no. 12, p. 64–67.

MCCRACKEN, M. C., AND LUTHER, F. M., 1984, Preliminary estimate of the radiative and climatic effects of the El Chichón eruption: Geofisica Internationale Volc., v. 23, p. 385–401.

NADEAU, P. H., AND REYNOLDS, R. C., Jr., 1981, Volcanic components in pelitic sediments: Nature, v. 294, p. 72–74.

NANCE, R. D., WORSLEY, T. R., AND MOODY, J. B., 1986, Post-Archean biogeochemical cycles and long-term episodicity in tectonic processes: Geology, v. 14, p. 514–518.

NEWHALL, C. G., AND SELF, S., 1982, The Volcanic Explosivity Index (VEI): An estimate of explosive magnitude of historic eruptions: Journal of Geophysical Research, v. 87, p. 1231–1238.

NICHOLLS, N., 1990, Low-latitude volcanic eruptions and the El Niño/ Southern Oscillation: A reply: International Journal of Climatology, v. 10, p. 425–429.

NINKOVICH, D., SHACKLETON, N. J., ABDEL-MONEN, A. A., OBRADOVICH, J. D., AND IZETT, G., 1978, K-Ar age of the late Pleistocene eruption of Toba, north Sumatra: Nature, v. 276, p. 574–577.

PALAIS, J. M., AND SIGURDSSON, H., 1989, Petrologic evidence of volatile emissions from major historic and pre-historic volcanic eruptions: American Geophysical Union, Geophysical Monograph 52, p. 31–53.

PARSONS, B., AND SCLATER, J. G., 1977, An analysis of the variation of ocean floor bathymetry and heat flow with age: Journal of Geophysical Research, v. 87, p. 289–302.

PATERNE, M., LABEYRIE, J., GUICHARD, F., MAZAUD, A., AND MAITRE, F., 1990, Fluctuations of the Campanian explosive volcanic activity (South Italy) during the past 190,000 years, as determined by marine tephrochronology: Earth and Planetary Science Letters, v. 98, p. 166–174.

PATERSON, W. S. B., AND SEVEN OTHERS, 1977, An oxygen-isotope climatic record from the Devon Island ice cap, Arctic Canada: Nature, v. 266, p. 508–511.

PINTO, J. P., TURCO, R. P., AND TOON, O. B., 1989, Self-limiting physical and chemical effects in volcanic eruption clouds: Journal of Geophysical Research, v. 94, p. 11,165–11,174.

PORTER, S. C., 1981, Recent glacier variations and volcanic eruptions: Nature, v. 291, p. 139–142.

PORTER, S. C., 1986, Pattern and forcing of northern hemisphere glacier variations during the last millennium: Quaternary Research, v. 26, p. 27–48.

RAMPINO, M. R., AND SELF, S., 1982, Historic eruptions of Tambora (1815), Krakatau (1883), and Agung (1963), their stratospheric aerosols and climatic change: Quaternary Research, v. 18, p. 127–143.

RAMPINO, M. R., AND SELF, S., 1984a, The atmospheric effects of El Chichón: Scientific American, v. 250, p. 48–57.

RAMPINO, M. R., AND SELF, S., 1984b, Sulphur-rich volcanic eruptions and stratospheric aerosols: Nature, v. 310, p. 677–679.

RAMPINO, M. R., AND STOTHERS, R. B., 1984, Geological rhythms and cometary impacts: Science, v. 226, 1427–1431.

RAMPINO, M. R., AND STOTHERS, R. B., 1987, Episodic nature of the Cenozoic marine record: Paleoceanography, v. 2, p. 255–258.

RAMPINO, M. R., AND STOTHERS, R. B., 1988, Flood basalt volcanism during the past 250 million years: Science, v. 241, p. 663–668.

RAMPINO, M. R., AND SELF, S., AND FAIRBRIDGE, R. W., 1979, Can rapid climatic change cause volcanic eruptions?: Science, v. 206, p. 826–829.

RAMPINO, M. R., SELF, S., AND STOTHERS, R. B., 1988, Volcanic winters: Annual Reviews of Earth and Planetary Sciences, v. 16, p. 73–99.

REA, D., AND VALLIER, T., 1982, Two Cretaceous volcanic episodes in the western Pacific Ocean: Geological Society of America Bulletin, v. 94, p. 1430–1437.

RICH, J. E., JOHNSON, G. L., JONES, J. E., AND CAMPSIE, J., 1986, A significant correlation between fluctuations in seafloor spreading rates and evolutionary pulsations: Paleoceanography, v. 1, p. 85–95.

RONOV, A. B., 1976, Global carbon geochemistry, volcanism, carbonate accumulation and life: Geochemistry International 1976, p. 172–195.

RONOV, A. B., 1983, The Earth's Sedimentary Shell: Washington, D.C., The American Geological Institute, 80 p.

ROSE, W. I., Jr., AND CHESNER, C. A., 1987, Dispersal of ash in the great Toba eruption, 75 ka: Geology, v. 15, p. 913–917.

RUDDIMAN, W. F., 1977a, Late Quaternary deposition of ice-rafted sand in the subpolar North Atlantic: Geological Society of America Bulletin, v. 88, p. 1813–1827.

RUDDIMAN, W. F., 1977b, North Atlantic ice rafting: A major change at 75,000 years before present: Science, v. 196, p. 1208–1211.

RUDDIMAN, W. F., MCINTYRE, A., NIEBLER-HUNT, V., AND DURAZZI, J. T., 1980, Oceanic evidence for the mechanism of rapid northern hemisphere glaciation: Quaternary Research, v. 13, p. 33–64.

SANCETTA, C., IMBRIE, J., KIPP, N. G., MCINTYRE, A., AND RUDDIMAN, W. F., 1972, Climatic record in North Atlantic deep-sea core V23-82: Comparison of the last and present interglacials based on quantitative time series: Quaternary Research, v. 2, p. 363–367.

SANDBERG, P., 1983, An oscillating trend in Phanerozoic non-skeletal carbonate mineralogy: Nature, v. 305, p. 19–22.

SAVIN, S. M., DOUGLAS, R. G., AND STEHLI, F. G., 1975, Tertiary marine paleotemperatures: Geological Society of America Bulletin, v. 86, p. 1499–1510.

SEAR, C. B., KELLY, P. M., JONES, P. D., AND GOODNESS, C. M., 1987, Global surface temperature responses to major volcanic eruptions; Nature, v. 320, p. 365–367.

SELF, S., AND RAMPINO, M. R., 1988, The relationship between volcanic eruptions and climate change: Still a conundrum? EOS, Transactions of the American Geophysical Union, v. 69, p. 85–86.

SELF, S., AND RAMPINO, M. R., AND BARBERA, J. J., 1981, The possible effects of large 19th and 20th century volcanic eruptions on zonal and hemispheric surface temperatures: Journal of Volcanology and Geothermal Research, v. 11, p. 41–60.

SHERIDAN, R. E., 1987, Pulsation tectonics as the control of long-term stratigraphic cycles: Paleoceanography, v. 2, p. 97–118.

STOMMEL, H., AND STOMMEL, E., 1983, Volcano Weather: Newport, Rhode Island: Seven Seas Press, 177 p.

STOTHERS, R. B., 1984a, The great eruption of Tambora and its aftermath: Science, v. 224, p. 1191–1198.

STOTHERS, R. B., 1984b, Mystery cloud of AD 536: Nature, v. 307, p. 344–345.

STOTHERS, R. B., AND RAMPINO, M. R., 1983, Volcanic eruptions in the Mediterranean before A.D. 630 from written and archaeological sources: Journal of Geophysical Research, v. 88, p. 6357–6371.

TAYLOR, B. L., GAL-CHEN, T., AND SCHNEIDER, S. H., 1980, Volcanic eruptions and long-term temperature records: An empirical search for cause and effect: Quarterly Journal of the Royal Meteorological Society, v. 196, p. 175–199.

TOON, O. B., AND POLLACK, J. B., 1980, Atmospheric aerosols and climate: American Scientist, v. 68, p. 268–278.

TURCO, R. P., TOON, O. B., WHITTEN, R. C., HAMILL, P., AND KEESEE, R. G., 1983, The 1980 eruptions of Mount St. Helens: Physical and chemical processes in the stratospheric clouds: Journal of Geophysical Research, v. 88, p. 5299–5319.

VAIL, P. R., HARDENBOL, J., AND TODD, R. G., 1985, Jurassic unconformities, chronostratigraphy, and sea level changes from seismic stratigraphy and biostratigraphy: American Association of Petroleum Geologists Memoir 36, p. 129–144.

VAIL, P. R., MITCHUM, R. M., AND THOMPSON, S., 1977, Seismic stratigraphy and global changes in sea level: American Association of Petroleum Geologists Memoir 26, p. 83–97.

VOGT, P. R., 1972, Evidence for global synchronism in mantle plume convection, and possible significance to geology: Nature, v. 240, p. 338–342.

VOLK, T., 1987, Feedbacks between weathering and atmospheric $CO_2$ over the last 100 million years: American Journal of Science, v. 287, p. 763–779.

WALKER, J. C. G., HAYS, P. B., AND KASTING, J. F., 1981, A negative feedback mechanism for the long-term stabilization of the Earth's surface temperature: Journal of Geophysical Research, v. 86, p. 9776–9782.

WALSH, K., AND PITTOCK, A. B., 1990, The sensitivity of a coupled atmospheric-oceanic model to variations in the albedo and absorptivity of a stratospheric aerosol layer: Journal of Geophysical Research, v. 95, p. 9941–9950.

WILKINSON, B. H., AND GIVEN, R. K., 1986, Secular variation in abiotic marine carbonates: Constraints on Phanerozoic atmospheric carbon dioxide contents and oceanic Mg/Ca ratios: Journal of Geology, v. 94, p. 321–333.

WILLIAMS, L. D., 1979, An energy balance model of potential glacierization of northern Canada: Arctic and Alpine Research, v. 11, p. 443–456.

WILSON, C., 1985, The Little Ice Age on Eastern Hudson/James Bay: the Summer weather and climate at Great Whale, Fort George and Eastmain, 1814–1821, as derived from Hudson's Bay Company records: Syllogeus, no. 55, p. 147–190.

WOILLARD, G., 1979, Abrupt end of the last interglacial 5.5 in north-east France: Nature, v. 281, p. 558–562.

WORSLEY, T. R., MOODY, J. B., AND NANCE, R. D., 1985, Proterozoic to Recent tectonic tuning of biogeochemical cycles: American Geophysical Union Monograph 32, p. 561–572.

WORSLEY, T. R., NANCE, R. D., AND MOODY, J. B., 1984, Global tectonics and eustasy for the past 2 billion years: Marine Geology, v. 58, 373–400.

# FRAGMENTATION PROCESSES IN EXPLOSIVE VOLCANIC ERUPTIONS

GRANT HEIKEN AND KENNETH WOHLETZ
*Los Alamos National Laboratory, Los Alamos, New Mexico 87545*

ABSTRACT: A remarkable variety of pyroclast types is produced in explosive volcanic eruptions, each reflecting the many factors controlling the eruption energy and eruption sequence, including viscosity, gas content, and phenocryst content of the magma. Fragmentation during a volcanic eruption and the resulting pyroclast characteristics can also be linked to external water affecting the eruption process. Volcanic ash characteristics can be used to infer the eruption type and emplacement processes.

## INTRODUCTION

The products of explosive volcanic eruptions vary in volume from 1 $m^3$ at Krafla, Iceland, where a geothermal well briefly erupted in 1977, to over 1,000 $km^3$ during formation of the Earth's largest calderas (e.g., several of the youngest calderas of Yellowstone National Park, U.S.A.). Volcanic ashes produced during these eruptions exhibit as much diversity as sandstones. As for sandstones, where the provenance and depositional environment can be interpreted by a study of thin sections and grain size, volcanic ashes can provide information about the composition, rheological properties, and volatile content of source magmas. Information can also be derived from volcanic ashes concerning fragmentation processes that occur during eruptions and transport processes before deposition of the ashes.

The diversity of conditions controlling an explosive eruption produces a large variety of volcanic ash types. Examination of stratigraphic position, particle compositions, and particle shapes can be used by the sedimentologist to establish criteria for correlation of ash deposits; the field of *tephrochronology* is devoted to ways of using ash beds as stratigraphic markers. The type of explosive eruption also controls, in part, the grain size and susceptibility of glassy components to alteration, parameters that are important to the sedimentologist.

Observations of fragmentation processes within or near a vent during a volcanic eruption are limited. We can get reasonably close to only the smallest (and generally the most mafic) eruptions of juvenile material to observe the processes that form volcanic ash. These eruptions are, however, of minimal interest to the sedimentologist, for they produce only small, local deposits. Only the least energetic explosive eruptions of andesitic to rhyolitic magmas have been observed; from afar the view of even these eruption columns is distorted by turbulent clouds of ash, gases, and condensed water vapor. Numerical models proposed for magma fragmentation and formation of pyroclastic debris are at best first approximations—at least until we find some way to emplace instruments that will survive within the throat of a volcano in full eruption. To understannd the processes that produce volcanic ash, we must rely on linking eruption observations to interpretations of the volcanic ashes themselves, on numerical eruption models based on the physics of magma evolution, migration, and eruption, and on laboratory or field experiments.

The purpose of this paper is to provide the reader with an introduction to the explosive eruption processes that produce *tephra*—i.e., lithic and vitric volcanic ash, pumice pyroclasts, bombs, and lithic clasts (xenoliths). For a more complete view of the spectrum of volcanic ashes from smaller eruptions and fragmentation processes in explosive volcanic eruptions, see *Pyroclastic Rocks* (Fisher and Schmincke, 1984) or *Volcanic Ash* (Heiken and Wohletz, 1985).

## EXPLOSIVE FRAGMENTATION OF SILICIC MAGMAS

The formation and growth of vesicles must be understood before evaluating eruption processes that produce pumice and volcanic ash. Verhoogen (1951), in an elegant analysis of bubble nucleation, growth and rise in the magma, concluded that volcanic ash formed when expanding bubbles coalesced. That is most likely correct for lava fountains of low-viscosity magmas, but not for the formation of pumice and volcanic ash of silicic compositions; pumice cannot exist if all of the bubbles have coalesced and disintegrated. McBirney and Murase (1970) showed that the rate of bubble rise within a magma is far too slow to be responsible for pyroclast formation by bubble coalescence. Rittmann (1936) and Bennett (1974) proposed that vesicular magma is disrupted by shock decompression before bubble coalescence can occur.

The distinctive shapes of pumice pyroclasts are controlled by the history of vesicle growth, presence or absence of phenocrysts as nucleation points, and degree (nature, type) of fragmentation of the vesicular magma at the time of eruption. There may be as many histories and size and shape distributions of vesicles within pumices as there are volcanic eruptions, but most vesicles are interpreted as having grown either continuously during shallow magma emplacement and eruption (one generation) or discontinuously, before and during eruption (two generations) (Heiken and Eichelberger, 1980; Toramaru, 1989). The following hypothetical example of pumice pyroclast formation demonstrates what we can infer about the eruption process, based on the vesicle characteristics, and grain shapes and sizes of the eruption products, pumice and volcanic ash:

> A rhyolitic magma containing 2 to 3% $H_2O$ may begin to vesiculate at depths of 1 to 3 km, where gas pressure exceeds lithostatic pressure (Murase and McBirney, 1973). If $CO_2$ is present, it may exsolve and form vesicles at greater depths (Holloway, 1976). Uniform nucleation of bubbles may begin if the magma contains few phenocrysts (Fig. 1A). If the magma contains abundant phenocrysts, nucleation and bubble growth on phenocryst surfaces may produce pumice

Sedimentation in Volcanic Settings, SEPM Special Publication No. 45

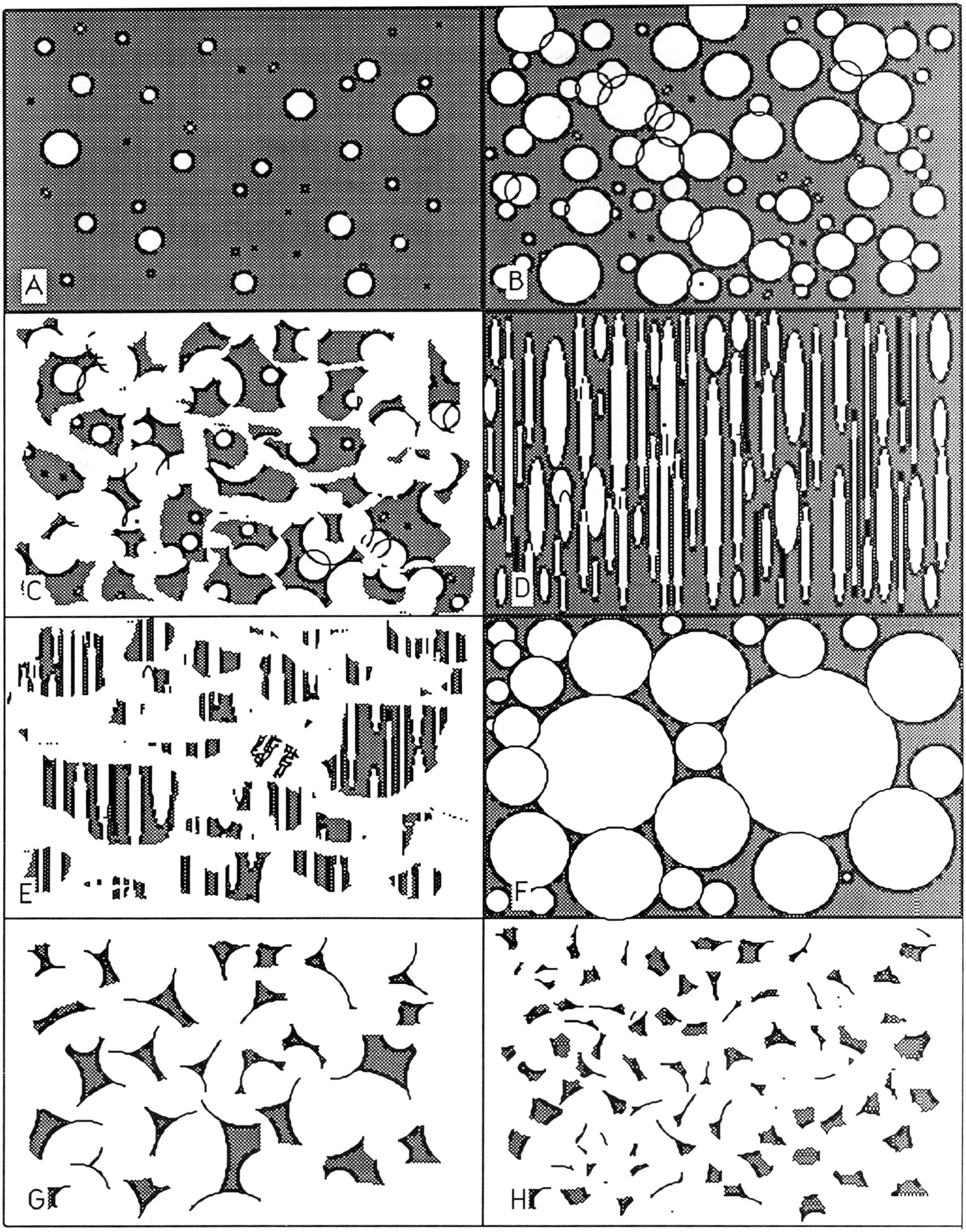
A
B
C
D
E
F
G
H

fragments containing vesicle clusters and elongate, thin vesicles oriented normal to phenocryst surfaces (Fig. 2). Elongate vesicles clustered around phenocrysts or glomeroporphyritic clots may coalesce and form large cavities in a pumice pyroclast (Fig. 3).

As magma approaches the Earth's surface, uniform vesicle growth may produce a foam consisting of evenly spaced bubbles (with the exception of larger clusters of vesicles around phenocrysts; Fig. 1B). Isothermal expansion of bubbles and surface tension of thinning walls between bubbles may increase the bulk viscosity of the magma considerably and give it a finite yield strength, making it more brittle. High viscosity prevents significant buoyant bubble rise.

If vesiculation occurs at shallow depths, near the mouth of a vent, the pumice pyroclasts may contain nearly spherical vesicles. There is no time before fragmentation for vesicles to be elongated by flow into tube-like shapes. The ashes produced in this type of eruption consist of arcuate, blocky, glassy pyroclasts (Fig. 1C).

After fractures in country rock overlying a rising magma body are opened to the surface, the magma flow rate increases. If this flow rate exceeds that of rising bubbles, the vesicles are sheared by flow into long tubes or pita-bread-like shapes (Fig. 1D). Shapes and aspect ratios of tube-like vesicles are dependent on temperature and viscosity, flow rate of rising magma, and magmatic gas content. If indeed there is an early generation of vesicle growth that forms large cavities consisting of coalesced vesicles, these cavities can retain their shape and act like lithic clasts or

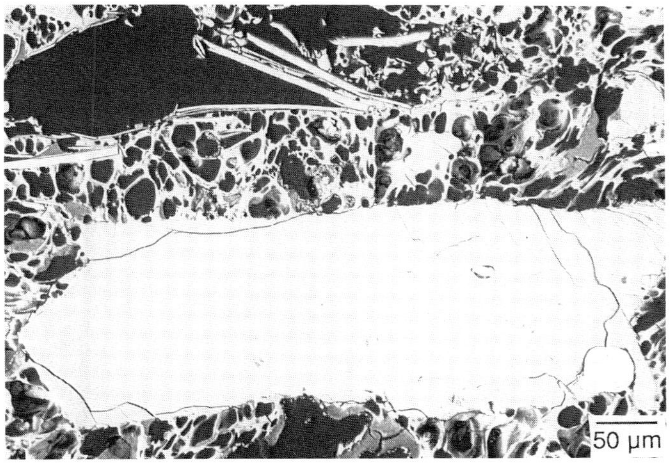


FIG. 2.—Scanning electron micrograph (SEM) of a polished thin section of a pumice pyroclast from the Holocene dacitic dome cluster of El Soche, Ecuador (sample from Bernardo Beate, INECEL). Vesicles around a sanidine phenocryst have grown radially from the phenocryst surface. This radial texture was deformed slightly by flow within the pumiceous magma before eruption. Overall vescularity of this pumice is 43%.

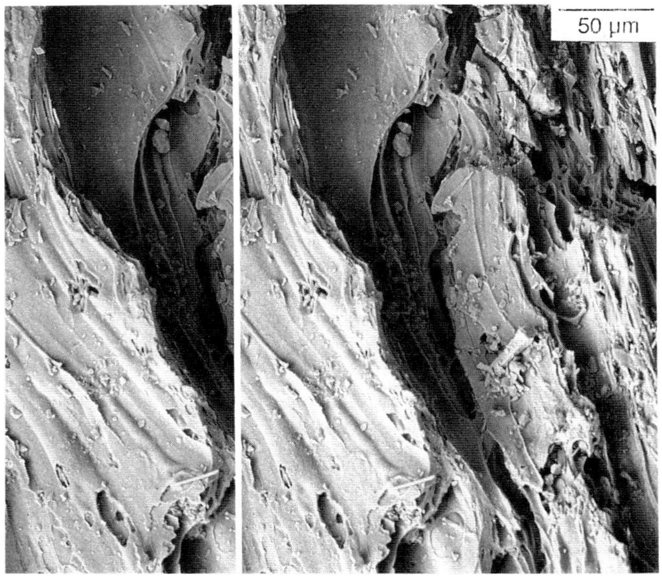


FIG. 3.—Scanning electron stereomicrograph of a pumice pyroclast erupted from an intracaldera dome complex, Toledo cadera, Jemez Mountains, New Mexico. Vesicles were stretched and flattened by flow of the pumiceous magma before it was fragmented into pumice and ash particles by expansion waves moving downward into the vent. Coalesced vesicles form elongate cavities in this image. On the walls of the elongate cavites are rib-like septae, which are remnants of collapsed vesicle walls common to the cluster.

FIG. 1.—Diagrams to illustrate possible sequences of events leading to vesicle (bubble) growth and fragmentation during eruptions of phenocryst-poor silicic magmas. The hypothetical scale is 0.5 mm for the width of each illustration. (A) At a depth of somewhere between 1 and 3 km, in a magma containing 2 to 3% $H_2O$, volatiles begin to come out of solution to form vesicles. (B). Vesicle growth continues as the magma rises slowly toward the surface. The earliest generation of vesicles may now be touching and will coalesce to form cavities as much as several centimeters in diameter (see also Fig. 3). (C) Should the pumiceous melt be fragmented during eruption by an expansion wave passing down the vent before having been sheared by flow, the resulting vitric ash will consist of arcuate, blocky glassy pyroclasts. (D) Most pumiceous magmas are sheared by flow, perhaps during accelerated movement of the magma as eruption begins. Spherical vesicles are sheared into highly elongate pipe-like forms or flattened tubes, depending on the nature of the flow. Pumiceous magmas containing vesicles with the highest aspect ratios produce pumice clasts with a silky appearance, reflecting lower overall lower viscosity and higher flow rate within the rising magma. (E) Fragmentation of a pumiceous magma by expansion wave produces blocky, angular pumice clasts and hollow needle-like shards. The vesiculated magma must act as a brittle mass, with very high viscosity, before common pumice pyroclasts can be formed and erupted. (F) Should gas contents be high enough and viscosities low enough, large spherical vesicles can develop, with bulk vesicularities of over 70%. Nearly all of the bubbles are touching and vesicle walls are very thin. This scenario may be the case for peralkaline magmas, whose viscosities are low. (G) Fragmentation during eruption of a highly inflated pumiceous magma forms large, curved, plate-like, glassy pyroclasts. Such pyroclasts are as delicate as the shards produced by breaking a glass Christmas-tree ornament. (H) If a pumiceous, volatile-rich silicic magma encounters ground water or shallow surface water during the eruption, magma/water interactions generate steam and enormous energy release, which further fragments the vesiculated melt. The very fine ashes produced in these eruptions are referred to as *phreatomagmatic* or *phreatoplinian*.

phenocrysts, with the second generation of elongate vesicles flowing around them (Fig. 4).

The more or less brittle mass of vesiculated magma is disrupted by a large pressure differential at the magma/atmosphere interface. A shock wave moves ahead of the pyroclasts out of the vent, while an expansion wave propagates down into the inflated magma and allows its rapid decompression. An expansion wave disrupts the nearly solid foam, fragments it into particles, and accelerates the particles out of the vent (Ritmann, 1936; Bennett, 1974; Wohletz and others, 1984). Most elongate pumice pyroclasts are broken from a foam parallel to and normal to elongate vesicles and, by inference, to the flow direction. This observation supports the interpretation that an expansion wave passed down the vent, normal to vent walls and direction of flow (Fig. 1E). The explosive eruption process for silicic magmas is likely driven by this pressure differential and *not* by bubble coalescence.

Blocky, angular pumice pyroclasts, less than 10 cm in diameter, are usually cool enough when deposited to retain their angular shapes. Larger pumice bombs deform on landing and may continue to vesiculate, with bomb surfaces subsequently broken by "breadcrust" cracks. Some bombs are hot enough to deform plastically after deposition and others may break into angular pyroclasts during cooling (Walker, 1969).

Peralkaline (and some calc-alkaline) silicic magmas are characterized by high eruption temperatures, low viscosities and high gas contents (Schmincke, 1974). Before fragmentation and eruption, vesicles are large and can be mostly spherical at the time of eruption (Fig. 1F). Vesicle walls may break and the wall remnants form shards that are broadly curved, plate-like, or Y-shaped (Fig. 1G). Some of the pyroclasts in these eruptions are small hollow spheres.

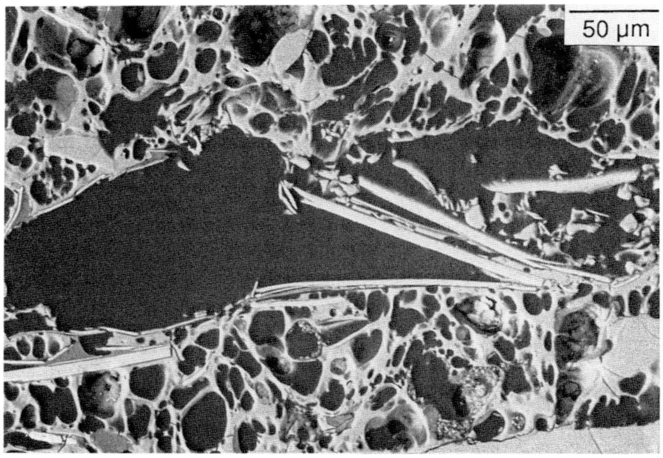


FIG. 4.—SEM of a thin section of pumice from El Soche, Ecuador (also shown in Fig. 2). A cavity, consisting of coalesced vesicles has also expanded and pried apart a biotite phenocryst. The bimodality of vesicle volumes in pumice pyroclasts is usually caused by a normal distribution of vesicle volumes plus cavities consisting of coalesced vesicles in which vesicle walls have collapsed.

### LARGE-VOLUME ERUPTIONS

Large-volume eruptions (>$10^9$ $m^3$ of ejecta), nearly always associated with the formation of calderas, are important to the sedimentologist, since the ashfall deposits mantle tens of thousands of square kilometers and greatly alter the subsequent regional sedimentation processes. These widespread ash beds are also important to the stratigrapher as marker horizons that can simultaneously blanket multiple sedimentary facies in a basin.

#### *Silicic Tephra*

Coarse pumice pyroclasts from *Plinian* eruptions usually fall within tens of kilometers of the vent area, although the ash particles produced by very energetic eruptions or those carried along with pyroclastic flows can be found hundreds of kilometers from their source (a *Plinian* eruption is the paroxysmal ejection of large volumes of ash and pumice in a well-defined eruption column; the resulting dispersal of pumice covers an area of over 500 $km^2$). *"Co-ignimbrite"* ashes are fine-grained ashes that form the upper part of pyroclastic flows; they are made up mostly of the lighter vitric pyroclasts, fractionated during rise, collapse, and outflow of the eruption column. Co-ignimbrite ashes and finer grained portions of Plinian eruption clouds comprise the most widespread ashfall beds that are used as stratigraphic markers.

Glass shards are pieces of vesicle walls; their shapes depend upon vesicle sizes and shapes within the magma being erupted and the eruption processes responsible for the fragmentation of that magma. Shards exhibit a wide range of physical characteristics, ranging from slightly curved, thin glass plates broken from large, thin-walled spherical vesicles, to hollow needles broken from pumiceous melts containing vesicles stretches by flow (Figs. 1C, 1G, 5).

#### *Phreatomagmatic Silicic Tephra*

*Phreatomagmatic (hydrovolcanic)* eruptions occur when magma comes into contact with ground water or the shallow surface water in lakes, marshes, and littoral areas. Formation and collapse of steam films on melt surfaces causes deformation and fragmentation of the melt and propagation of strong stress waves that further fragment the melt (Wohletz, 1983). The rapid superheating of water results in far more explosive conditions than normal boiling, and very energetic, steam-rich eruptions occur. When an eruption is driven by both vesiculation and magmatic overpressures and magma/water interactions, the Earth's most violent eruptions occur. These eruptions, sometimes called *phreatoplinian* (Self and Sparks, 1978), produce very fine-grained volcanic ash, composed mostly of bits of individual vesicle walls, finely fragmented during magma/water interactions (Figs. 1H, 6). These deposits may have median grain sizes as small as 40 μm (Self and Sparks, 1978). We now recognize that the very fine-grained components (<10 μm) of such eruptions stay in stratospheric suspensions long enough to be carried thousands of kilometers. Such components may

FIG. 5.—SEM stereopair of rhyolitic glass shards, Mazama ash. The Mazama ash is the fallout deposit formed during the eruption at 7 ka that formed Crater Lake caldera in Oregon. This sample was collected in Steptoe Canyon, Washington. The shards were the common junctions of walls between adjacent vesicles. In thin sections, many of these shards would be Y-shaped or curved plates.

FIG. 6.—Rhyolitic phreatomagmatic ash from intracaldera eruptions, Toledo caldera, Jemez Mountains, New Mexico. These very fine-grained, blocky pyroclasts are believed to have formed during the explosive interaction between a vesiculated, erupting magma and water in a caldera lake.

even be spherical and form anomalous sedimentary layers far from any known volcanism.

### *Peralkaline Silicic Tephras*

Silicic magmas containing abundant $Na_2O$ and $K_2O$ and depleted in $Al_2O_3$ are generally much more fluid and hotter than other silicic magmas. Volcanic ash formed during eruptions of these magmas shows distinctive particle shapes that include hollow glass spheres and tricuspate shards formed by rupture of very large, thin-walled vesicles (Fig. 7). Post-

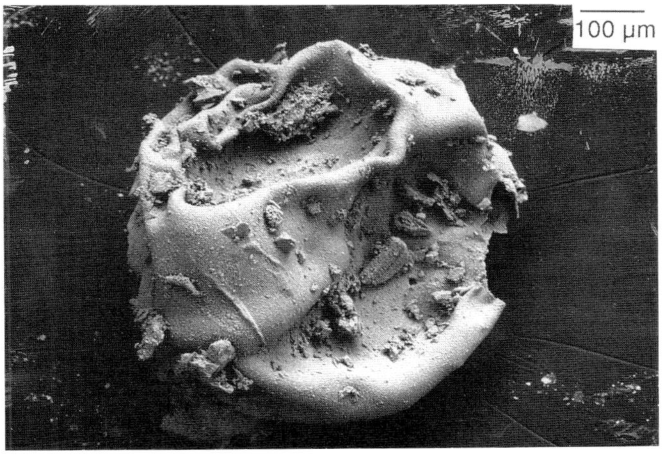


FIG. 7.—SEM of a partly collapsed globular shard from the peralkaline Green Tuff of Pantelleria, Italy. The surface embayments are deformed vesicles, which along with other smooth textural features, are a result of the relatively high fluidity of the magma.

depositional deformation and welding of the ash deposits is common (Schmincke, 1974).

## SMALL-VOLUME EXPLOSIVE ERUPTIONS

Most of the Earth's subaerial volcanoes (on land and in shallow littoral areas) contain less than 0.1 $km^3$ of pyroclastic debris (some eruptions have only a very small explosive component; e.g., basaltic fissure eruptions produce mostly lava flows, but have small ramparts composed of pyroclastic spatter). These volcanoes range in composition from basalt to high-silica rhyolite and in form from small tuff rings and cinder cones to impressive composite cones. These eruptions have produced volcanic ashes by a variety of fragmentation mechanisms.

### *Vulcanian Eruptions and Composite Cones (Stratovolcanoes)*

*Vulcanian* eruptions are those characterized by intermittent explosions that deposit lithic ashes composed of older solidified lava clasts. These eruptions are most likely caused by overpressures developed within blocked vents. The eruption mechanism is not clear, but consists of alternating magmatic and phreatomagmatic processes (Nairn and Self, 1978; Self and others 1979; Frazzetta and others 1983). Sometimes the eruption will shift back and forth between Vulcanian and Strombolian (see later discussion), controlled by either a blocked or open vent. Vulcanian eruptions are associated with some cinder cones, craters in composite cones, and lava domes. Coarser materials erupted include large blocks and bombs. Vulcanian eruptions can produce eruption clouds that rise several kilometers above the vent and small pyroclastic flows and surges. Because the original material is rich in water vapor, the pyroclastic flow deposits may grade into lahar (volcanic mudflow) deposits along the lower flanks of a volcano (Frazzetta, and others 1983).

The ashes from Vulcanian eruptions consist of mixtures of juvenile and nonjuvenile fragments (Fig. 8). The juvenile components may be vesicular (mostly magmatic) or blocky and nonvesicular (mostly phreatomagmatic; Heiken and Wohletz, 1985). Nonjuvenile components are interpreted as fragments of a lava plug from the conduit, bits of dome lava, or fragments of the conduit and crater wall.

At Ngauruhoe volcano, New Zealand, Nairn and Self (1978) interpret the cannon-like explosions and subsequent pyroclastic falls and flows as having been caused by pulverization of a lava plug in the conduit by a combination of magma degassing and vaporization of ground water.

### *Strombolian Eruptions and Cinder Cones*

Explosive bursts of solidified and partly solidified bombs, blocks, and ash are known as *Strombolian* activity, named after Stromboli, a volcano in the Aeolian Islands, Italy. Well-documented Strombolian eruptions consist of "weak to violent ejection of partly-fluid blobs" (Macdonald, 1972; Self and others, 1974; McGetchin and others, 1974). All pyroclasts fall ballistically around the vent, building a cinder (scoria) cone; tephra fallout beds accumulate until their angle of repose is exceeded, after which avalanches cascade down the flanks and back into the crater (McGetchin and others, 1974). Of the tephra erupted, about 50% is deposited in the cinder cone and 50% as fallout layers around the cone (Heiken, 1978). Activity at cinder cones can rapidly switch between lava fountaining, Strombolian bursts, and Vulcanian eruptions (discussed earlier). Interbedded with the loose scoria fall and avalanche beds of cinder cones are layers of welded scoria and finer grained phreatomagmatic tephra; even Stromboli Volcano has experienced periods of phreatomagmatic activity.

Pyroclasts in Strombolian deposits range from irregular, smooth-skinned, vesicular sideromelane droplets (basaltic glass) to blocky, crystalline, poorly vesicular tachylite pyroclasts (Fig. 9). This spectrum of textural types is represented in all size categories, from large bombs to scoria to fine ash.

### *Lava Fountains*

Low-viscosity basaltic magmas erupt as lava fountains and as basaltic lava flows. The fountains range from a few meters to over 1 km in height and deposit welded spatter (bombs and ash) as a circular or oval apron around a central vent or as ridges parallel to a fissure vent. A spray of low-viscosity ($<10^3$ poise) basaltic liquid is driven by expansion of magmatic gases, which form large bubbles, which collapse. Basaltic glass pyroclasts from lava fountains range in size from vesicular bombs a meter or more across to spheres a few μm in diameter (Fig. 10). Coarser pyroclasts are deposited within a few hundred meters of the vent and finer ash, including filamentous Pele's hair (natural fiberglass), is swept downwind and deposited as ash fall. Nearly all fountaining is accompanied by lava flows; a high rate of lava fountaining can also form spatter deposits that begin to flow as rootless lava flows.

Structures associated with lava fountains are spatter ramparts, from a few meters to more than 100 m high, and cinder cones composed of mixed welded spatter and basaltic ash.

### *Phreatomagmatic Eruptions, Tuff Rings and Tuff Cones*

Where explosive eruptions of small volumes of magma occur in wet environments, such as those for phreatoplinian eruptions, most of the magma is erupted as very fine ash. Abundant ground or surface water can mix with the magma just prior to the eruption. The mixing of water and magma

FIG. 8.—Volcanic ash from a Vulcanian eruption of the andesitic cone Volcán Pacaya, Guatemala, March 7, 1989. During several months prior to this eruption, intermittent eruptions of Pacaya appeared to be Strombolian (bursts of lava clots). This eruption followed an intense tropical storm with heavy rainfall. The ash is a mixture of blocky, phreatomagmatic pyroclasts, formed by magma/water interaction, and crystalline lava lithic clasts broken from cooling magma in the conduit by steam explosions. (Collected by Andrew Adams.)

FIG. 9.—SEM of an ash grain erupted from Stromboli Volcano, Italy. Note its vesicularity and angular outline.

FIG. 10.—SEM of ash deposited during lava fountaining at Mauna Ulu, Kilauea Volcano, Hawaii. Fountains of low-viscosity basaltic magma produce a spray of droplets in which vesiculation sometimes continues after droplets are formed. If droplets are intact, they typically have surfaces rounded by surface tension. The pyroclast in the center was a droplet with a tapered tail folded back onto itself. Also present are shards broken from highly inflated vesicular droplets that either burst or broke on hitting the ground surface; they are made up of arcuate vesicle walls. Also present, but not clearly shown in this SEM, are glass spheres.

FIG. 11.—SEM of fine-grained basaltic phreatomagmatic ash that, in some instances, may be deposited over a wide area. The pyroclasts are typically blocky and have low vesicularity. Very fine particles adhere to pyroclast surfaces.

is driven by fluid instabilities along the margins of the magma body, where water or saturated sediment slurries are driven by high vapor pressures into the conduit. These instabilities efficiently break up the melt while magma and water churn and mix. Such mixing may also be enhanced if growing vesicles are present in the magma. The resulting steam explosions form typical volcanic landforms characterized by wide craters surrounded by low-profile rings or cones of ash that are rapidly consolidated to form tuff. Because so much fine ash and steam are generated during phreatomagmatic eruptions, large portions of the erupted material may be carried by winds over relatively wide areas. The ashes are distinctive because of their typically blocky and poorly vesicular shapes (Fig. 11), the presence of admixed sedimentary rock fragments, and accretionary lapilli (pisolites) of fine-grained ash (Fig. 12).

FIG. 12.—A polished thin section through an accretionary lapillus from the eruption of Vesuvius in AD 79. Layers of very fine ash formed accretionary coats around a core of a larger blocky ash grain. The "vesicles" between ash grains were most likely formed by steam trapped between layers of rapidly accreting fine ash.

### *Phreatic Eruptions*

Perturbations of a hydrothermal system by injection of new magma at depth, changes in ground-water level, or tectonic activity may cause explosive steam eruptions without the eruption of juvenile tephra. This type of activity is possible at many types of volcanoes and even in geothermal areas with no volcanic activity; however, it is common in mature composite cones. Phreatic eruptions may occur with no subsequent activity (Soufrière de Guadeloupe, 1976–1977) or be precursors to a significant magmatic eruption (Mount St. Helens, 1980). The eruptions consist of intermittent or continuous explosive steam bursts and can form large craters and may be accompanied by large-scale volcanic collapse (Bandai-type eruptions; Siebert and others, 1987).

Phreatic deposits observed at Soufrière de Guadeloupe are thin, fine-grained ash beds, large blocks near the vent and some small volcanic mudflows (Wohletz and Crowe, 1978). These deposits are composed of hydrothermally altered or weathered lithic clasts and mud (Heiken and Wohletz, 1985).

## CONCLUSIONS

There are four basic mechanisms of fragmentation in volcanic eruptions: (1) the decompression of rising magma, vesicle (bubble) growth, and subsequent comminution by an expansion wave *(magmatic)*, (2) the explosive mixing of magma with ground or surface water *(hydrovolcanic or phreatomagmatic)*, (3) fragmentation of country rock by rapid expansion of steam and/or hot water *(phreatic)*, and (4) tephra abrasion during transport by collision of grains with each other and the substrate. The fourth mechanism is especially important during gravitational collapse of the volcanic edifice and formation of avalanches, such as that which occurred during the 1980 eruption of Mount St. Helens. The variations of eruption style and the characteristics of volcanic ashes produced in explosive eruptions depend on factors such as magmatic temperature, volatile content, viscosity and phenocryst content, the ratio of magma and meteoric water, and physical properties of rock enclosing the vent. The type of fragmentation mechanism can also ultimately affect the means of transportation and deposition (e.g., pumice falls, surges, and pyroclastic flows), post-depositional changes such as welding, and the ease with which an ash deposit is chemically altered.

Although the variety of pyroclasts is great, a general classification can be made on the basis of grain morphology, size, vesicle characteristics, and type of weathering (see, e.g., Heiken and Wohletz, 1985). Ideally, this characterization should be made with abundant field observations, but that is not always possible for isolated samples from drill cores or thin ash falls far from their source.

ACKNOWLEDGMENTS

This review was supported by the U.S. Department of Energy, Office of Basic Energy Sciences, and done under their auspices. Comments on the manuscript by Harry Glicken, R. V. Fisher, and Robert Raymond are greatly appreciated.

REFERENCES

BENNETT, F. D., 1974, On volcanic ash formation: American Journal of Science, v. 274, p. 648–661.

FISHER, R. V., AND SCHMINCKE, H.-U., 1984, Pyroclastic Rocks: Springer-Verlag, Heidelberg, 472 p.

FRAZZETTA, G., LAVOLPE, L., AND SHERIDAN, M. F., 1983, Evolution of the Fossa Vent, Vulcano: Journal of Volcanology and Geothermal Research, v. 17, p. 329–360.

HEIKEN, G. 1978, Characteristics of tephra from Cinder Cone, Lassen Volcanic National Park, California: Bulletin Volcanologique, v. 41, p. 119–130.

HEIKEN, G., AND EICHELBERGER, J. C., 1980, Eruptions at Chaos Crages, Lassen Volcanic National Park, California: Journal of Volcanology and Geothermal Research, v. 7, p. 443–481.

HEIKEN, G., AND WOHLETZ, K., 1985, Volcanic Ash: University of California Press, Berkeley, 246 p.

HOLLOWAY, J. R., 1976, Fluids in the evolution of granitic magmas: Consequencies of finite $CO_2$ solubility: Geological Society of America Bulletin, v. 87, p. 1513–1518.

MACDONALD, G. A., 1972, Volcanoes: Prentice-Hall, Englewood Cliffs, 510 p.

MCBIRNEY, A. R., AND MURASE, T., 1970, Factors governing the formation of pyroclastic rocks: Bulletin Volcanologique, v. 34, p. 372–384.

MCGETCHIN, T. R., SETTLE, M., AND CHOUET, B. A., 1974, Cinder cone growth modeled after Northeast Crater, Mount Etna, Sicily: Journal of Geophysical Research, v. 79, p. 3257–3272.

MURASE, T. AND MCBIRNEY, A. R., 1973, Properties of some common igneous rocks and their melts at high temperatures: Geological Society of America Bulletin, v. 84, p. 3563–3592.

NAIRN, I. A., AND SELF, S., 1978, Explosive eruptions and pyroclastic avalanches from Ngauruhoe in February, 1975: Journal of Volcanology and Geothermal Research, v. 3, p. 39–60.

RITTMANN, A., 1936, Vulkane und ihre Tätigkeit: Ferdinand Enke Verlag, Stuttgart, 188 p.

SCHMINCKE, H.-U., 1974, Volcanological aspects of paralkaline silicic welded ash-flow tuffs: Bulletin Volcanologique, v. 38, p. 594–636.

SELF, S., AND SPARKS, R. S. J., 1978, Characteristics of widespread pyroclastic deposits formed by the interaction of silicic magma and water: Bulletin Volcanologique, v. 41, p. 196–212.

SELF, S., SPARKS, R. S. J., BOOTH, B., AND WALKER, G. P. L., 1974, The 1973 Heimaey Strombolian scoria deposit, Iceland: Geological Magazine, v. 111, p. 534–548.

SELF, S., WILSON, L., AND NAIRN, I. A., 1979, Vulcanian eruption mechanisms: Nature, v. 277, p. 440–443.

SIEBERT, L., GLICKEN, H., AND UI, T., 1987, Volcanic hazards from Bezymianny- and Bandai-type eruptions: Bulletin of Volcanology, v. 49, p. 435–459.

TORAMARU, A., 1989, Vesiculation process and bubble size distributions in ascending magmas with constant velocities: Journal of Geophysical Research, v. 94, p. 17,523–17,542.

VERHOOGEN, J., 1951, Mechanisms of ash formation: American Journal of Science, v. 249, p. 729–739.

WALKER, G. P. L., 1969, The breaking of magma: Geological Magazine, v. 106, p. 166–173.

WOHLETZ, K. H., 1983, Mechanisms of hydrovolcanic pyroclast formation: Size, scanning electron microscopy, and experimental results: Journal of Volcanology and Geothermal Research, v. 17, p. 31–63.

WOHLETZ, K. H., AND CROWE, B. M., 1978, Development of lahars during the 1976 activity of La Soufrière de Guadeloupe: Geological Society of America Abstracts with Programs, v. 10, p. 154.

WOHLETZ, K. H., MCGETCHIN, T. R., SANDFORD, M. T., III, AND JONES, E. M., 1984, Hydrodynamic aspects of caldera-forming eruptions: Numerical models: Journal of Geophysical Research, v. 89, p. 8269–8286.

# VOLCANICLASTIC DEPOSITS: IMPLICATIONS FOR HYDROCARBON EXPLORATION

MARK E. MATHISEN AND JOHN G. McPHERSON

*Mobil Research and Development Corporation, Dallas Research Laboratory, P.O. Box 819047, Dallas, Texas 75381-9047*

ABSTRACT: Recent developments in sedimentology, diagenesis, and hydrocarbon exploration suggest that the recognition and interpretation of volcaniclastic sediments can significantly influence exploration methods, and the prediction of reservoir geometry and quality in volcaniclastic sequences.

Volcaniclastic sediments are characterized by predictable changes in composition, texture, geometry and distribution, which can be used during both geologic and seismic interpretation. Interpretations based on volcaniclastic sediments help to better define the volcano-tectonic and paleogeographic setting that controls the deposition of associated siliciclastic and/or carbonate reservoirs. Volcaniclastics are also important to an understanding of the thermal history of a sedimentary basin and its deposits, and to evaluations of source-rock maturity and reservoir diagenesis. Seismic lithostratigraphic modeling and facies interpretations can be used to differentiate high-impedance volcaniclastic facies from associated siliciclastic deposits. In additon, laterally continuous "marker bed" pyroclastic-fall and large-volume ignimbrite deposits, characterized by continuous reflections, can be differentiated from discontinuous pyroclastic-flow deposits and lahars characterized by more discontinuous reflections.

The characteristics of reservoirs in volcaniclastic sequences are controlled by the volcano-tectonic setting, eruptive mechanism, depositional system, composition, age, diagenesis, and thermal and burial history. Volcaniclastic lithofacies typically have predictable distribution patterns that control reservoir geometry. The reservoir quality of volcaniclastic sediments is controlled primarily by their diagenetic history, because volcaniclastics are composed largely of reactive and unstable minerals, with the potential for rapid and extensive changes during burial diagenesis. The abundance of unstable minerals commonly leads to the destruction of porosity by cemetation and compaction processes, but may also enhance porosity by grain dissolution. Successful efforts to find hydrocarbons in volcaniclastic deposits will depend on the coincidence of porosity preservation and generation processes, with the timing of hydrocarbon migration and entrapment.

## INTRODUCTION

Exploration in the rift basins of Northern Europe and Africa (Schull, 1988; Brooks and Glennie, 1987; Schott, 1982) and along convergent margins of Asia and South America (Fitzgerald and others, 1990; Bee, 1982) has led to major hydrocarbon discoveries in sedimentary sequencies that also contain volcaniclastic deposits. Volcaniclastics are commonly thought to have poor reservoir potential (Surdam and Boles, 1979; Galloway, 1979) and are often avoided during the development of exploration plays (Gage and Wing, 1980; Kingston and others, 1985). Recent developments in volcaniclastic sedimentology, volcaniclastic diagenesis, and exploration technology suggest, however, that the recognition and interpretation of volcaniclastic sediments can actually benefit hydrocarbon exploration efforts.

The major characteristics of volcaniclastic deposits that affect hydrocarbon exploration methods and the interpretation of reservoir quality are discussed herein. These characteristics need to be considered during the development of convergent-margin and rift-basin exploration plays. Since a relatively small amount of volcanic detritus can have a signficiant impact on sandstone diagenesis and reservoir quality, the term "volcaniclastic" will be used to refer to a clastic rock containing volcanic material in whatever proportion, and without regard to its origin.

## EXPLORATION METHODS

Volcaniclastic sediments typically display systematic changes in composition, texture, geometry and distribution which can be used during both geologic and seismic interpretation. Regional geology data, outcrops and cores provide the primary data for analyzing volcaniclastic deposits (Campbell, 1989). Wireline logs, when available and calibrated with core data, can be used to quantify variations in volcaniclastic facies. With both geologic and well-log data, useful and sophisticated models for volcaniclastic facies distribution, the volcano-tectonic and paleogeographic setting, thermal history, source-rock maturity, and reservoir quality can often be developed. Given seismic data of good quality and resolution, lithostratigraphic modeling and interpretation can also be completed to document the seismic response and infer the distribution of volcaniclastic facies.

### *Volcaniclastic Facies Models*

Research in volcaniclastic sedimentology has led to the definition of autoclastic, hydroclastic, pyroclastic and epiclastic deposits (Suthren, 1985). Autoclastic and hydroclastic deposits are closely associated with volcanic flows and may contribute to some of the known reservoirs in fractured volcanic sequences (Arpandi and Patmosukismo, 1975). Volumetrically, however, pyroclastic and epiclastic deposits are more abundant and more often associated with hydrocarbon reservoirs.

Pyroclastic deposits have been divided into three major facies; fall, flow and surge (Fig. 1; Wright and others, 1980; Fisher and Schmincke, 1984; Cas and Wright, 1987). Two types of pyroclastic deposits can be recognized from outcrops and cores, and used as a reference during hydrocarbon exploration: pyroclastic-fall and pyroclastic-flow deposits. Both can be thick enough over large distances (>100 km) from the source vent to form reservoirs or be associated with reservoirs in epiclastic sediments. Pyroclastic-surge deposits, in contrast, are of more limited use during exploration. Base-surge deposits, although thick (>100 m), are typically confined to the immediate vicinity of phreatomagmatic and phreatic eruptions, limited areas which, because of high temperatures, probably have negligible hydrocarbon potential. Ground-surge deposits, in contrast, may be spread over large areas but are generally thin (<1 m). For the purposes of this paper, surge deposits will be undifferentiated from flow deposits.

Sedimentation in Volcanic Settings, SEPM Special Publication No. 45
 ISBN 0-918985-89-7

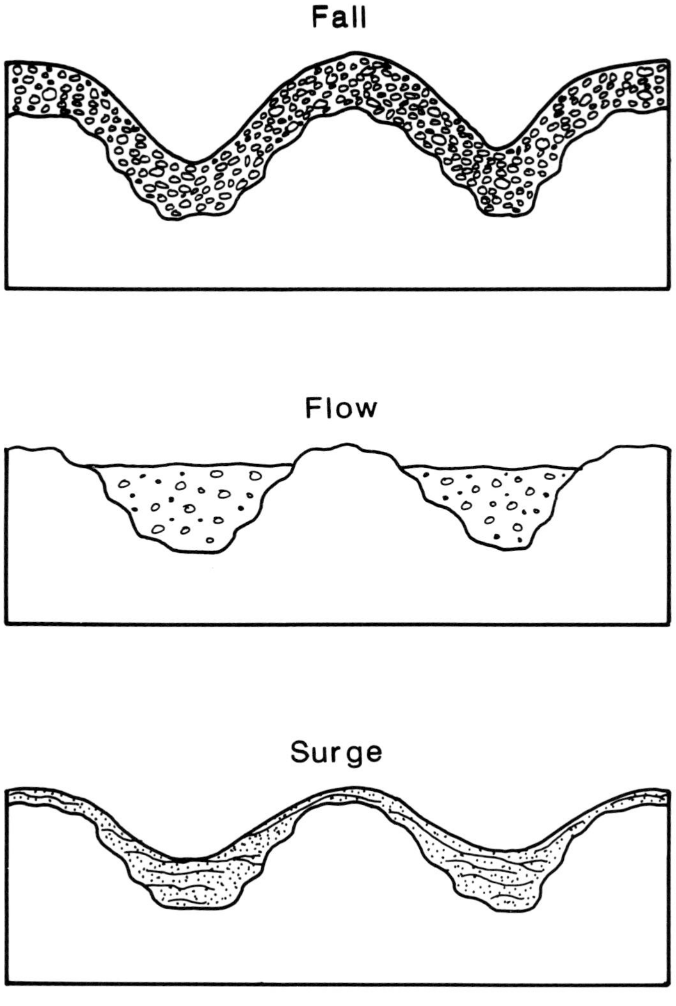


FIG. 1.—Three main types of pyroclastic deposits and typical variations in geometry (after Wright and others, 1980).

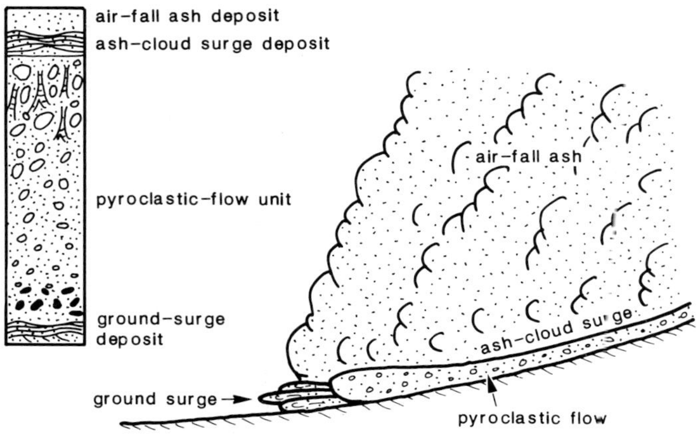


FIG. 2.—Idealized section of pyroclastic-flow deposit (pumice-flow or ignimbrite). Based on Sparks and others (1973) and Fisher (1979).

Pyroclastic-flow deposits, the result of gravity-driven mass flows from explosive eruptions, consist of a distinctive sequence of deposits dominated by a massive flow unit (Fig. 2) (Sparks and others, 1973; Fisher, 1979). Large-volume pumiceous pyroclastic-flow deposits, "ignimbrites," are especially important to recognize during exploration. Since these deposits are widespread and often mantle topography (Fisher and Schmincke, 1984, p. 194), they can form valuable stratigraphic marker beds. Care must be used, however, since they may be difficult to differentiate from reworked deposits such as lahars, which are confined to valleys and therefore are not as useful as marker beds.

Pyroclastic-fall deposits are characterized by uniform composition, good sorting, mantle bedding, widespread distribution, and are often suitable as stratigraphic datums during exploration. For example, thin tuffs or tuffaceous units have been used for detailed correlation across the Statfjord Field (Malm and others, 1979). Ash beds (bentonite marker beds) are a key stratigraphic tool in the exploration of the frontier sands of the Upper Cretaceous of the Western Interior basin of the United States (Weimer, 1966; Asquith, 1970; Kauffman, 1985; Pratt and others, 1985). Volcanic tuffs up to 300 m thick have been used as marker beds across the Tupungato oil field of Argentina (Baldwin, 1944). When reservoirs occur in fall deposits or ignimbrites, the lateral continuity of the deposits can also be used to infer reservoir continuity.

Lahars, volcanic debris-flows, are important epiclastic deposits that originate in many ways (Smith, this volume). They are typically confined to valleys (Fig. 1) and are therefore of limited use as stratigraphic datums. More extensive volcanic debris-flow deposits occur, however, ranging from coalescing aprons (Palmer and Walton, 1990), to thick sequences, such as the Absaroka laharic breccias, which cover thousands of square kilometers (Parsons, 1969).

### *Volcano-Tectonic and Paleogeographic Setting*

Interpretations based on volcaniclastics help to define better the volcano-tectonic and paleogeographic setting that may have controlled the deposition of associated siliciclastic and/or carbonate sequences. In many areas, the erosion of volcanic highs has been so extensive that volcaniclastics within the basin fill provide the most complete or only data set for reconstructing the volcano-tectonic, sedimentologic and paleogeographic history (Smith and others, 1988). Interpretations of volcaniclastics are based on documentation of systematic changes in grain size, composition, bed thickness, sedimentary structures and vertical sequences with distance from the volcanic source. Armentrout and Suek (1985), for example, documented how studies of volcaniclastics contributed to an interpretation of the Cenozoic volcano-tectonic and paleogeographic history of the Oregon-Washington area. This interpretation was then used to locate excellent reservoir-quality sandstones in an area previously thought to have low reservoir potential.

Studies of volcaniclastics along convergent margins serve as examples of the volcano-tectonic and paleogeographic interpretations possible in both forearc and backarc basins. Based on studies of an active forearc basin, Vessel and

Davies (1981) described the characteristics and distribution of proximal, medial and distal nonmarine volcaniclastic facies that can be used as a model for interpreting ancient deposits (Fig. 3). Similar proximal-to-distal facies relationships occur in ignimbrites (Fig. 4; Wright and others, 1981). Documentation of proximal-to-distal facies changes has been used to interpret the volcano-tectonic and paleogeographic setting of interbedded pyroclastic and fluvial sediments in backarc settings (Mathisen and Vondra, 1983; Smith, 1988). In addition to regional proximal-to-distal variations, the complex juxtaposition of vent, proximal, medial and distal facies was recently documented by Riggs and Busby-Spera (1990) and used to infer that the Early Jurassic magmatic arc in southern Arizona was low standing and extensional in character. In marine environments of backarc/marginal basins, studies of volcaniclastics have documented the characteristics of asymmetric volcaniclastic aprons that can be used to infer the history of arc systems (Carey and Sigurdson, 1984; Farquharson and others, 1984; Busby-Spera, 1988). The complete evolution of an arc system from an incipient to mature arc, followed by backarc- or marginal-basin extension, was interpreted from volcano-sedimentary facies described by Mueller and others, (1989).

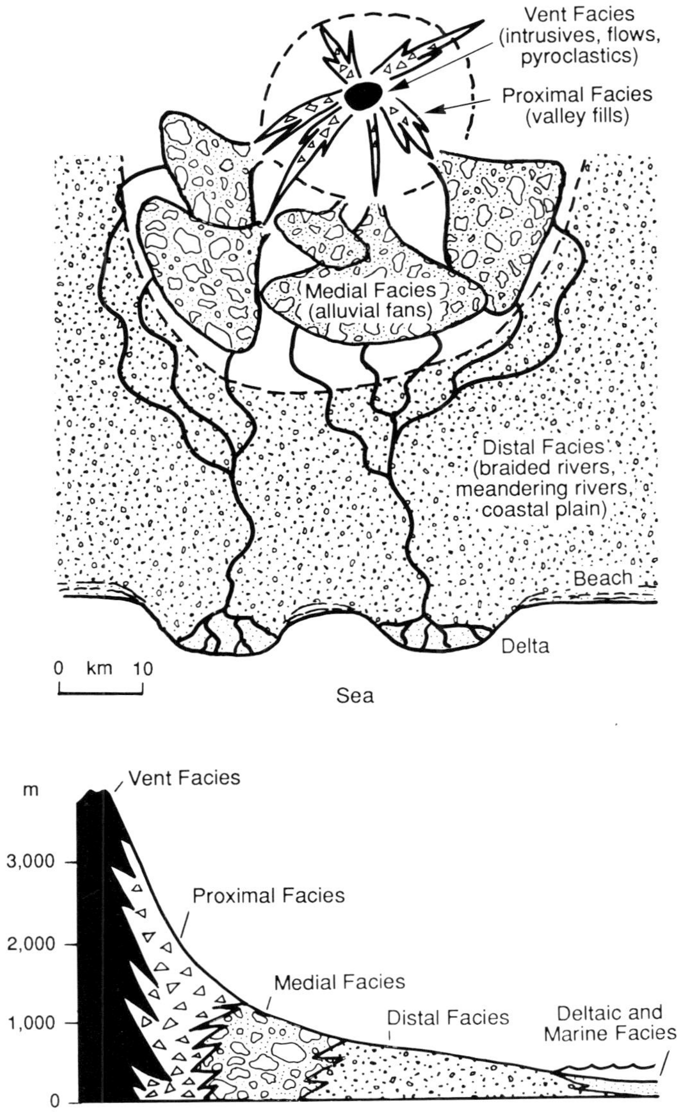


FIG. 3.—Volcaniclastic facies trends illustrating proximal-medial-distal facies variations in a high-relief andesitic system (after Vessel and Davies, 1981).

Rift-basin and continental volcaniclastics display some important similarities to convergent margin volcaniclastics, but are characterized by important differences. They exhibit considerable complexity as a product of sub-basin development through extensive intra-rift faulting (Fig. 5). Faulting causes rapid lateral variations in facies and thickness, whereas volcanism tends to subdivide and multiply the number of basins by diverting or damming drainage (Baker, 1986). Despite the complexity, proximal-to-distal trends in rift-basin volcaniclastic sediments have been demonstrated (McPherson and others, 1986; Pickford, 1986). In addition, cyclic variations in volcaniclastic facies have been observed and used to interpret the growth and denudation of source volcanoes (Waresback and Turbeville, 1990). Continental volcaniclastic deposits commonly have facies variations similar to those found in high-relief convergent-margin andesitic terranes (Fritz and Harrison, 1985). Documentation of these volcaniclastic facies characteristics by numerous workers has been important to the development of Rocky Mountain tectonic and paleogeographic interpretations (Dickinson and others, 1988).

### *Thermal History and Source Rock Potential*

Volcaniclastic sediments can be of significant value to the thermal-history interpretation of a basin, and to studies of temperature-dependent source-rock maturity and reservoir diagenesis. Areas with volcaniclastics have commonly experienced high heat flow (Klemme, 1980; Asakawa and others, 1980; Ritter, 1986). Since interpretation of temperature is critical to source-rock maturity interpretations, documentation of volcaniclastics in an exploration area is important to the evaluation of source rocks. Temperature interpretations are also important because the reservoir quality of feldspathic sandstone is known to decrease with depth at a greater rate in areas with high temperature gradients (Galloway, 1974; Dixon and Kirkland, 1985). A direct relationship between volcaniclastic deposits and source rocks has been proposed by Zimmerle (1985), who suggests that episodic volcanism has contributed to the formation of organic-rich source rocks by the generation of anoxia, and preservation of organic matter. Studies of volcaniclastics have also contributed to documentation of floral and faunal assemblages and the corresponding interpretations of paleoclimate and paleoecology (Fritz and Harrison, 1985; Frostick and others, 1986; Lockley and Rice, 1990), topics that may be important for the evaluation of hydrocarbon source potential.

### *Seismic Facies Characteristics*

Seismic data can be used to define more accurately the distribution and rock properties of volcaniclastic deposits.

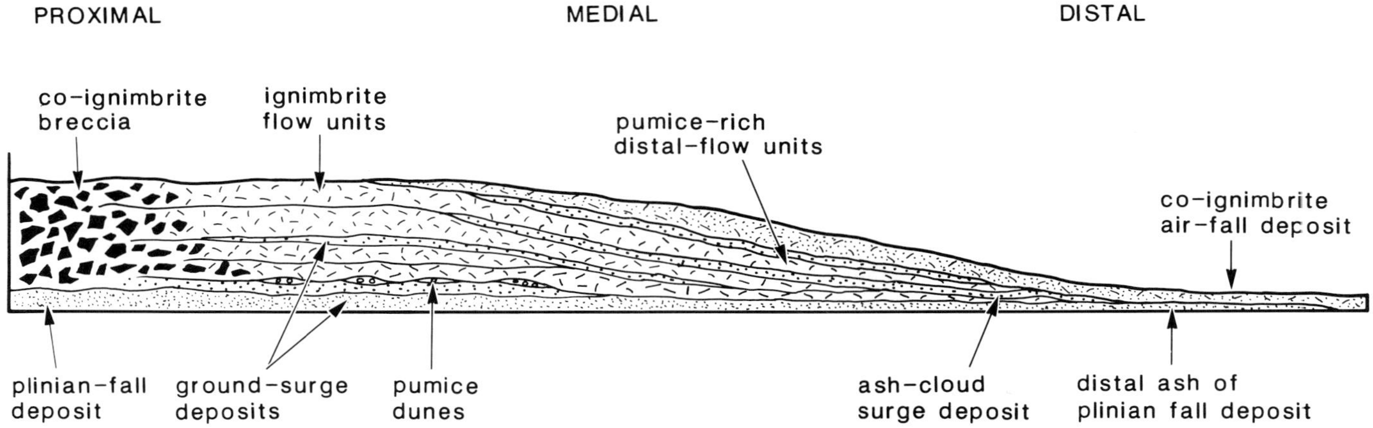


FIG. 4.—Proximal-to-distal ignimbrite facies variations (after Wright and others, 1981).

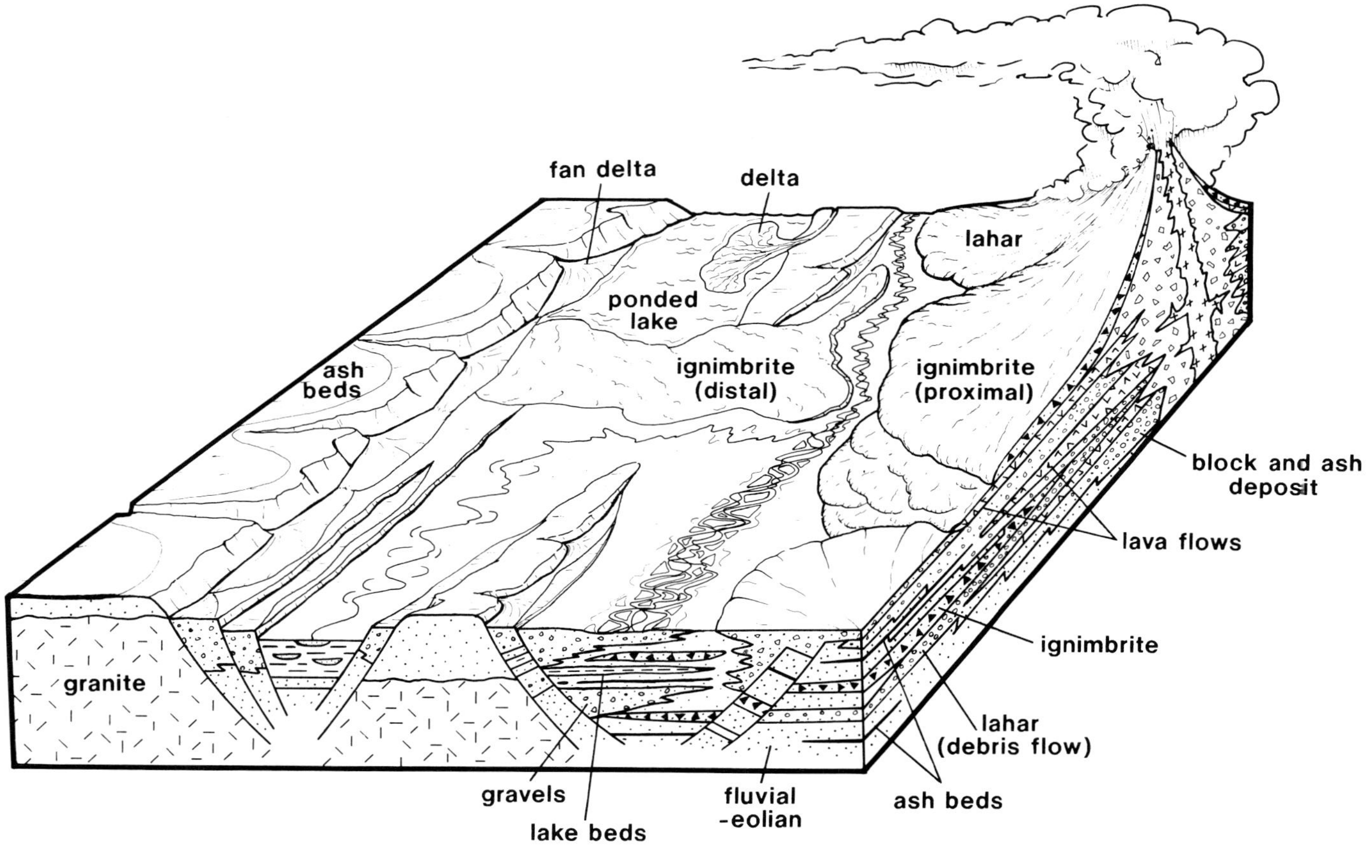


FIG. 5.—Schematic figure of volcaniclastic facies in an active rift setting. Intra-rift faulting and sub-basin development cause complex variations in lithofacies distribution.

Volcanic and pyroclastic rocks are usually characterized by higher velocities and densities than associated siliciclastic sediments. Seismic lithostratigraphic modeling and facies interpretations can therefore be used to differentiate "high-impedance" pyroclastic facies from associated "low-impedance" epiclastic deposits. The high impedance of pyroclastic deposits is most commonly caused by the more extensive cementation, compared with associated siliciclastic or epiclastic sediments, or welding following emplacement. In addition to lithology, variations in volcaniclastic facies distribution and geometry affect seismic-facies characteristics.

Successful seismic facies interpretation is dependent on the availability of "good-quality" seismic data. In many basins, the data are of sufficient quality to permit lithostratigraphic interpretation. Problems with data quality occur, however, in regions with extensive shallow volcanics (Young and Lucas, 1988). Interpretation in these areas may be limited.

Several examples of proximal seismic facies are discussed in the literature. Jansa and Pepiper (1988) have documented the distinctive "cone-like" seismic response of a remnant volcanic cone along the eastern American continental margin. In Japan, volcanic mounds of similar size and composition form oil reservoirs (Kujiraoka, 1980). The hydrocarbon-bearing Jatibarang Volcanic Formation of Indonesia, a sequence of graben-filling volcaniclastics (tuffs) and volcanic flows up to 1,200 m thick, is characterized by well-defined seismic horizons at the top and base (Arpandi and Patmosukismo, 1975). These seismic horizons should be useful for interpreting formation thickness variations, which are known to occur. Seismic data may also be useful for interpreting proximal-to-distal volcaniclastic-facies trends and reservoir properties.

The characteristics of distal volcaniclastic facies suggest that some distal deposits can be interpretated from seismic data. Laterally continuous pyroclastic-fall deposits are often thick enough to be resolved, and characterized by more extensive cementation, and correspondingly higher impedance, than associated epiclastic deposits. Assuming that properties of the enclosing sediments do not change, continuous reflections can be expected. Discontinuous reflections, in contrast, can be expected from "valley-filling" pyroclastic-flow deposits and lahars.

Ignimbrites are a major exception to the above generalization on the seismic response of pyroclastic-flow deposits. Ignimbrites are large-volume deposits (>1,000 km$^3$) that can travel great distances from the source (>100 km) and also mantle moderate amounts of topography (Cas and Wright, 1987). Since they can extend great distances, ignimbrites are commonly associated with nonvolcanic sequences, some of which have reservoir potential. Recognition of ignimbrites in these areas is important because the seismic impedance can be high enough, due to poor sorting and cementation, to produce "seismic marker beds".

In Germany, graben-filling Rotliegendes gas sands (Drong and others, 1982) overlie and interfinger with volcanic breccias, including widespread ignimbrites (Gast, 1988), which are seismically resolvable. Synthetic seismograms indicate that ignimbrites can have a marked affect on the seismic response of a reservoir sandstone (Fig. 6). The actual seismic response, however, will vary with ignimbrite velocity and thickness (Fig. 7). With good-quality seismic data, geologically inferred interpretations of ignimbrite continuity can be confirmed. If ignimbrites can be used as marker beds, the continuity of under and overlying reservoirs may also be interpreted. Mathisen and Budny (1990) have documented that similar lithostratigraphic interpretations are possible with seismic data from the Rotliegendes of Germany.

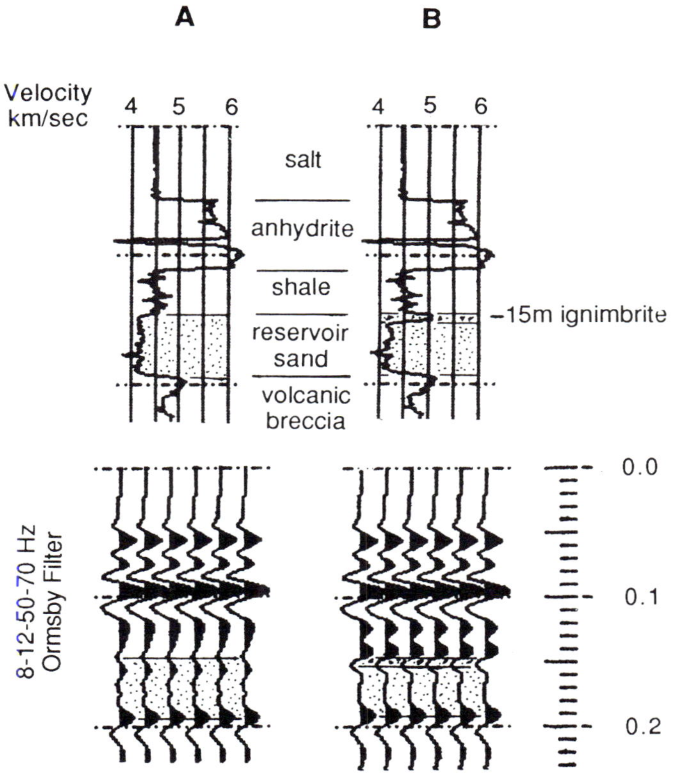


FIG. 6.—The modeled seismic response of a reservoir sandstone overlain by shale (A), versus a reservoir sandstone overlain by ignimbrite (B). The ignimbrite has higher velocity than the associated shale and causes higher amplitude reflections at the top of the sandstone.

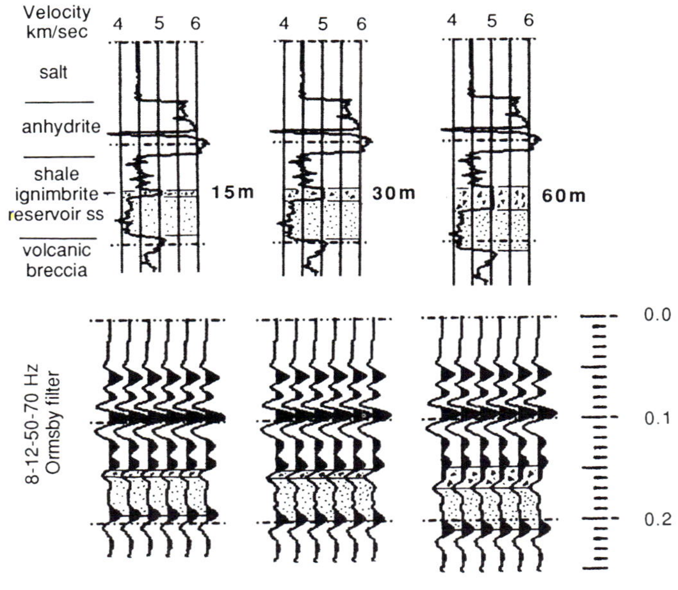


FIG. 7.—The effect of ignimbrite thickness variations on ignimbrite/reservoir sand reflection character. The pronounced ignimbrite peak and trough at 15 m increases slightly at 30 m. At 60 m, amplitude decreases and the peak grades to a doublet.

## RESERVOIR QUALITY

The reservoir quality of volcaniclastic deposits, as with other siliciclastic sediments (Hayes, 1979), is dependent on factors such as the depositional system, composition, age, and thermal and burial history. The depositional system is the initial control of reservoir quality. Depositional processes control the sediment distribution and texture, both of which determine or greatly influence reservoir quality. Since volcaniclastic sediments are largely composed of reactive and unstable minerals or glass, rapid and extensive changes tend to occur during burial diagenesis. Because these diagenetic changes often occur early in the burial history of volcaniclastic deposits, and at shallow depths, reservoir quality is commonly dominated by diagenetic processes.

### *Volcaniclastic Facies*

Before the overprint of diagenesis, the reservoir quality of volcaniclastics, as with other sandstones, is largely a measure of grain size and sorting characteristics (Beard and Weyl, 1973). Porosity is highly dependent upon sorting, whereas permeability is most influenced by grain size. The highest reservoir-quality facies will generally be the coarsest and the best sorted.

Proximal-to-distal facies variations in volcaniclastics cause important trends in textural properties that directly influence reservoir quality. Well-sorted distal facies, for example, generally have better reservoir quality than more poorly sorted proximal facies. The depositional mechanism of volcaniclastic facies also directly influences reservoir quality. Well-sorted pyroclastic-fall deposits, for example, have better reservoir quality than poorly sorted pyroclastic-flow deposits. Tuffs are therefore often cited as the dominant reservoir lithology in volcaniclastic sequences (Baldwin, 1944; Asakawa and others, 1980). Because of the sorting—reservoir potential relationship, the pronounced variations in sorting used to differentiate fall and flow deposits (Walker, 1971) can also be used to infer facies-dependent reservoir-quality variations in pyroclastic rocks.

Proximal-to-distal facies variations also influence sediment composition, which partly controls diagenetic processes and reservoir quality in pyroclastic deposits. A decrease in lithics and increase in crystal and unstable vitric detritus with distance can increase either the cementation or grain dissolution potential. The reduction of volcanic components in distal volcaniclastics, principally by reworking, will result in an increase in the more stable mineral assemblages. This reduces the likelihood of early pore-filling cementation, thereby maintaining good reservoir quality. Conversely, the reduction in the unstable minerals may limit reservoir quality by decreasing the potential for secondary-porosity development.

The high temperature of emplacement that is involved in the formation of welded tuffs may actually create reservoir rock. French and Freeman (1979) have documented that substantial fractures, which developed during cooling as well as during later weathering and faulting, permit the production of 600 to 1,200 barrels of oil per day from a densely welded Tertiary ash-flow tuff in Nevada.

### *Volcaniclastic Diagenesis*

The current literature presents a conflicting account of volcaniclastic diagenesis and its effect on reservoir quality. All studies document early diagenetic changes in rock properties due to the unstable mineralogy. Whereas many studies emphasize that these changes destroy reservoir quality and imply that areas with volcaniclastics have no significant reservoir potential, other studies have concluded that volcaniclastic diagenesis may contribute to the enhancement or preservation of reservoir quality.

Porosity occlusion by compaction and cementation commonly causes a relatively rapid reduction of reservoir quality in volcaniclastic sandstones. Galloway (1974, 1979) noted three major stages in the diagenesis of marine sandstones from arc-related basins: early calcite pore fill, clay rims and coats, and phyllosilicate or zeolite pore fill. Davies and others (1979) and Surdam and Boles (1979) have documented similar progressive increases in the cementation of volcaniclastics, and the factors involved. They concluded that fluid flow and fluid composition are some of the dominant controls of diagenesis, along with depth, temperature, and pressure. Another factor relevant to reservoir considerations is that the clay minerals derived from volcaniclastics not only reduce reservoir quality, but can also cause drilling and stimulation problems (Hunter and Davies, 1979).

In addition to porosity occlusion, the occurrence of volcanic detritus may contribute to the development of dissolution porosity, which can enhance reservoir quality. Extensive dissolution of volcaniclastic detritus has been documented to occur in fluvial sequences (Hayes and others, 1976; Mathisen, 1984), desert alluvium (Walker and others, 1978), and paralic/shallow-marine deposits (Hawlander, 1990). Seeman and Scherer (1984) surveyed the reservoir quality of numerous volcaniclastic reservoirs from different basins and concluded that volcaniclastic sandstones may act as good reservoir rocks given favorable circumstances such as devitrification at shallow depth, framework grain dissolution, fracturing, and reworking. At greater depth, Hawlander (1990) has suggested that the expected destruction of early dissolution porosity by compaction processes may be limited by the corresponding early cement formation, which creates a stronger rock frame. Loucks and others (1979, 1984) and other workers of Gulf Coast sandstones have documented that volcanic framework grain dissolution can actually contribute to porosities as high as 30%, and correspondingly high permeabilities, at 15,000 ft. The dissolution process is also known to affect volcaniclastics metamorphosed to the zeolite facies. Vernik (1990) recently described the formation of secondary porosity and permeability due to hydrothermal metasomatic processes in laumontite tuffs. In addition to the influence on reservoir quality, Mathisen (1984) noted that grain dissolution can be so extensive that interpretations of provenance and tectonic setting are distorted.

Integration of available data concerning the diagenesis of volcaniclastics makes it possible to generate a schematic model of volcaniclastic diagenesis that includes the dissolution process and the corresponding increase or preservation of porosity (Fig. 8). The depositional system is an im-

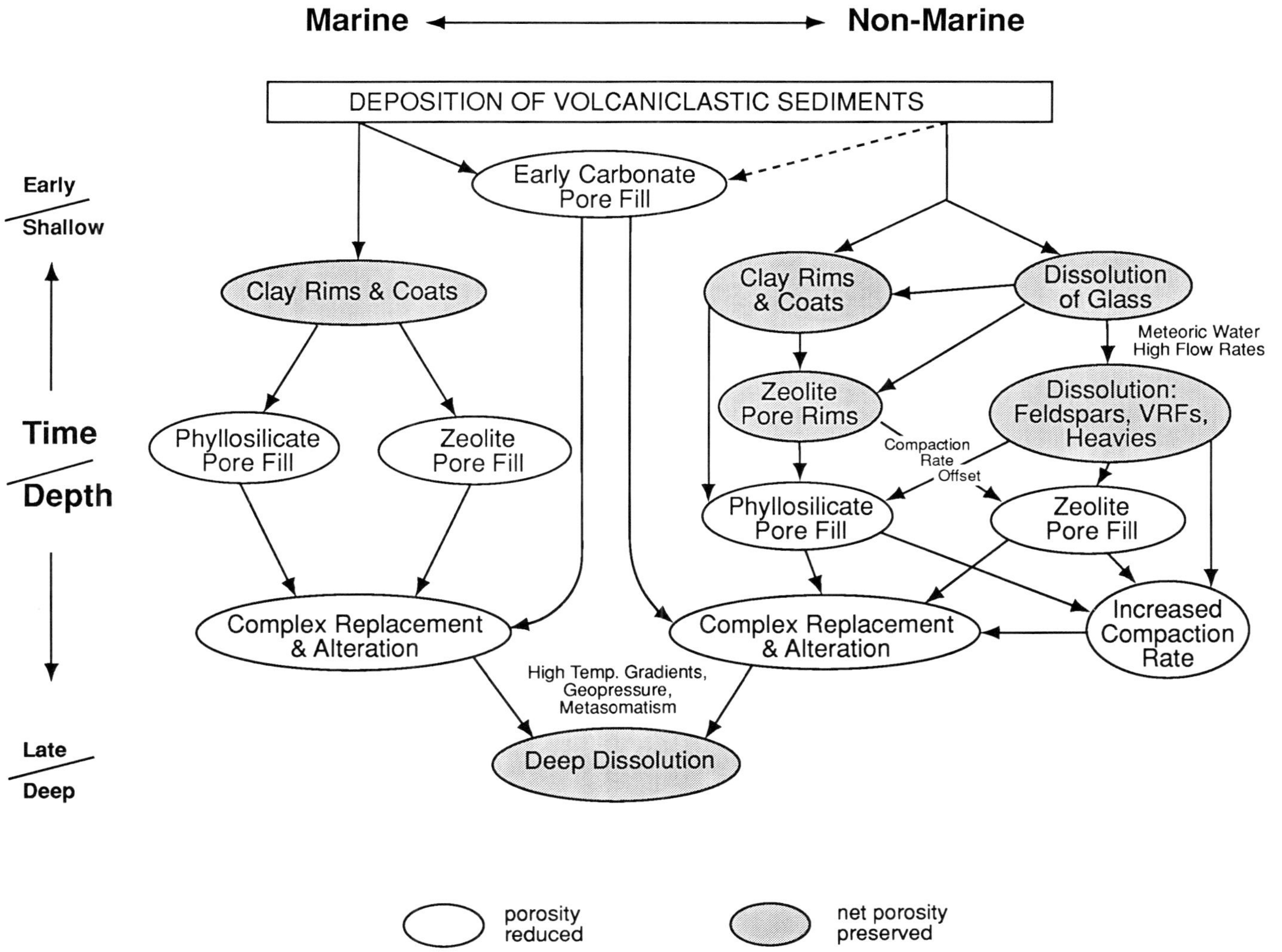


FIG. 8.—Schematic summary of factors controlling the diagenesis of volcaniclastic sediments. The shaded factors contribute to primary porosity preservation or secondary porosity generation, factors that preserve net porosity with depth.

portant initial control of diagenesis. Nonmarine sediments, for example, lack the pervasive early carbonate cement that is common in marine sediments (Blatt, 1979). Nonmarine sediments are also affected by high rates of porewater flow, which contribute to the dissolution of unstable grains and removal of dissolved ions (Hayes and others, 1976; Mathisen, 1984). Arid eolian/alluvial sediments, however, may be characterized by limited porewater flow, which can reduce reaction rates and inhibit the early diagenetic alteration of volcaniclastic sandstones (Stanley and Benson, 1979). At greater depths, where hydrocarbon generation and trapping may occur, the early pore-lining zeolite or carbonate cements that form in volcaniclastic sandstones may actually retard compaction and preserve dissolution porosity (Hawlander, 1990). The generation of dissolution porosity also occurs at depth in volcaniclastic sequences and has been attributed to elevated geothermal gradients and geopressure (Klass and others, 1981). Successful efforts to find hydrocarbons in volcaniclastic deposits will depend on the coincidence of porosity preservation and generation processes, with the timing of hydrocarbon migration and entrapment.

The reservoir quality of any lithology, including volcaniclastics, is dependent on permeability as well as porosity. Several factors may limit permeability. Authigenic clay, a common product of volcaniclastic sandstone diagenesis, can, even in limited amounts, significantly degrade reservoir quality by reducing permeability despite the presence of substantial porosity. As a result of grain dissolution processes, some volcaniclastic sandstones have high moldic porosity, ineffective porosity characterized by low permeability. The production of authigenic clays and the tendency for moldic porosity to develop increases with age and depth of the deposit. The available data suggest, however, that at shallow to moderate depths, nonmarine to paralic volcaniclastic deposits can have significant reservoir potential.

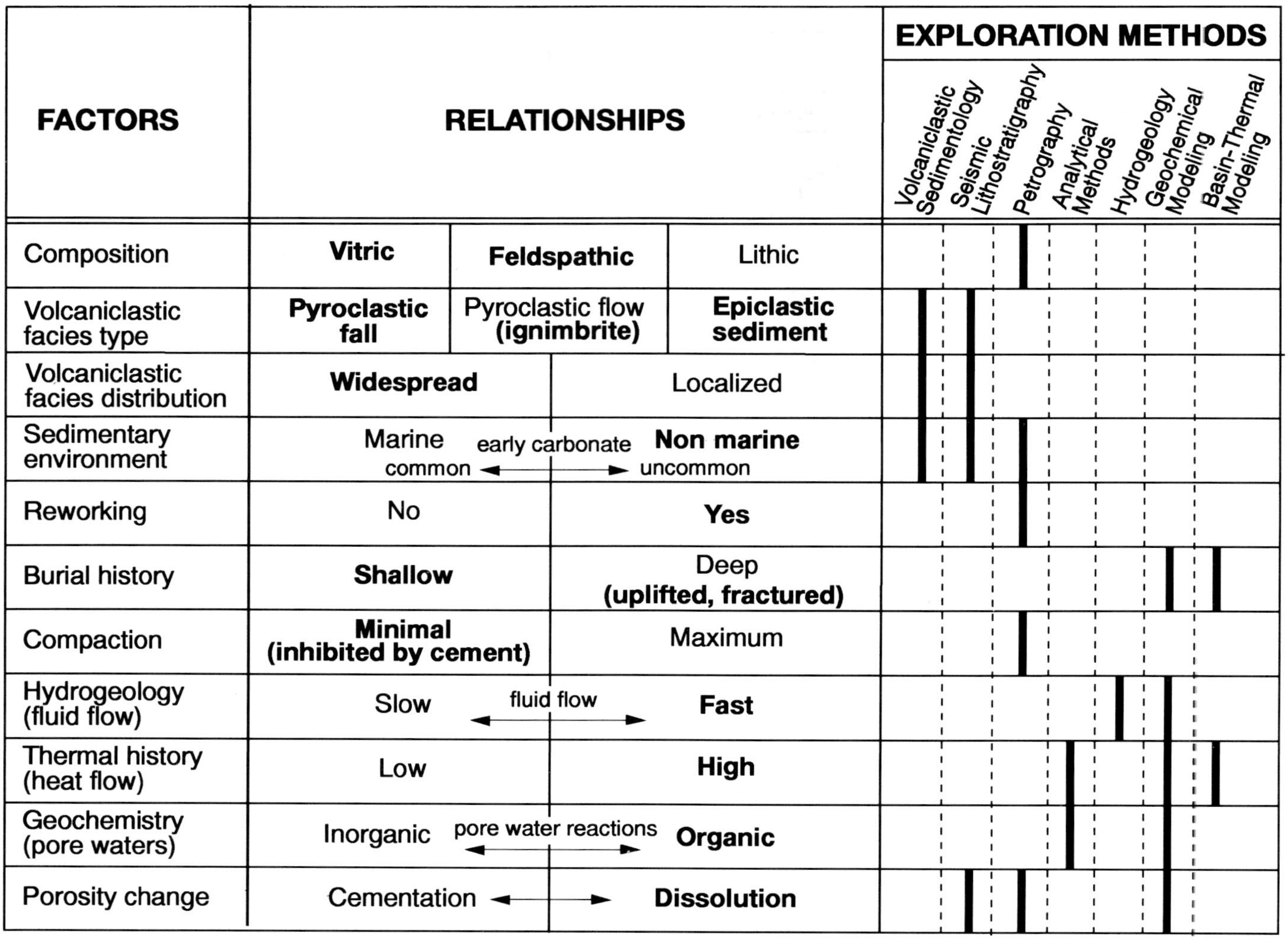


FIG. 9.—Factors controlling the properties of volcaniclastic sandstone reservoirs, the principal relationships, and the most appropriate exploration methods. The factors most favorable for reservoir development are in bold-face type.

## CONCLUSIONS

Volcaniclastic deposits are widespread in many convergent-margin and rift basins that have hydrocarbon production or potential (Asakawa and others, 1980; Stewart and Durkee, 1985; Sheehan and others, 1990). The characteristics of these sediments should be considered during the selection of exploration methods and the interpretation of reservoir potential. As summarized in Figure 9, volcaniclastic sedimentology and seismic lithostratigraphic analysis form a basis for interpreting the distribution and porosity of volcaniclastic facies. Petrographic studies of such rocks are essential to document the composition, amount of reworking, compaction, and the dissolution and cementation relationships. Analytical methods can then be used to document porewater reactions and evaluate thermal history. Hydrogeology, geochemical, and basin-modeling studies are also enhanced using data from volcaniclastic deposits.

The available models of volcaniclastic lithofacies distribution and porosity/permeability development provide a broad foundation for reservoir geometry and quality prediction. These models need to be refined and incorporated into exploration plays in areas with volcaniclastics.

## ACKNOWLEDGMENTS

We are grateful for the reviews provided by J. M. Armentrout, T. W. Cooley, T. H. McCulloh, and S. E. Sommer. We thank Mobil Research and Development Corporation for permission to publish this chapter.

## REFERENCES

ARMENTROUT, J. M., AND SUEK, D. H., 1985, Hydrocarbon exploration in western Oregon and Washington: American Association of Petroleum Geologists Bulletin, v. 69, p. 627–643.

ARPANDI, D., AND PATMOSUKISMO, S., 1975. The Cibulakan Formation as one of the most prospective stratigraphic units in the north-west Java basinal area. Proceedings, Indonesian Petroleum Association, p. 181–207.

ASAKAWA, T., S. SATO, AND ISHIWADA, Y., 1980, Contrast in hydrocarbon potential between fore-arc and back-arc basins of Tohoku Island Arc, Japan, *in* Halbouty, M. T., ed., Giant Oil and Gas Fields of the Decade, 1968–78: American Association of Petroleum Geologists Memoir 30, p. 223–228.

ASQUITH, D. O., 1970, Depositional topography and major marine environments, Late Cretaceous, Wyoming: American Association of Petroleum Geologists Bulletin, v. 54, p. 1184–1224.

BAKER, B. H., 1986, Tectonics and volcanism of the southern Kenya rift valley and its influence on rift sedimentation, *in* Frostick, L. E., Renault, R. W., Reid, I., and Tiercelin, J. J., eds. Sedimentation in the African Rifts: Geological Society of London Special Publication 25, p. 45–57.

BALDWIN, H. L., 1944, Tupungata oil field, Mendoza, Argentina: American Association of Petroleum Geologists Bulletin, v. 28 p. 1455–1484.

BEARD, D. C., AND WEYL, P. K., 1973, Influence of texture on porosity and permeability of unconsolidated sand: American Association of Petroleum Geologists Bulletin, v. 57, p. 349–369.

BEE, O. J., 1982, The Petroleum Resources of Indonesia: Oxford University Press, Oxford, 256 p.

BLATT, H., 1979, Diagenetic processes in sandstones, *in* Scholle, P. A., and Schluger, P. R., eds., Aspects of Diagenesis: Society of Economic Paleontologists and Mineralogists Special Publication 26, p. 141–157.

BROOKES, J., AND GLENNIE, K., 1987, Petroleum Geology of North West Europe: Graham and Trotman, London, 1219 p.

BUSBY-SPERA, C. J., 1988, Evolution of a Middle Jurassic back-arc basin, Cedros Island Baja, California: Evidence from a marine volcaniclastic apron: Geological Society of America Bulletin, v. 100, p. 218–233.

CAMPBELL, N. P., 1989, Structural and stratigraphic interpretation of rocks under the Yakima fold belt, Columbia Basin, based on recent surface mapping and well data, *in* Reidel, S. P., and Hooper, P. R. eds., Volcanism and Tectonism in the Columbia River Flood-Basalt Province: Geological Society of America Special Paper 239, p. 209–222.

CAREY, S., AND SIGURDSSON, H., 1984, A model of volcanogenic sedimentation in marginal basins, *in* Kokelaar, B. P., and Howells, M. F., eds., Marginal Basins Geology: Geological Society, Blackwell, London, p. 37–58.

CAS, R. A. F., AND WRIGHT, J. V., 1987, Volcanic Successions, Modern and Ancient: Allen and Unwin, London, 528 p.

DAVIES, D. K., ALMON, W. R., BONIS, S. B., AND HUNTER, B. E., 1979, Deposition and diagenesis of Tertiary-Holocene volcaniclastics, Guatemala, *in* Scholle, P. A., and Schluger, P. R., eds., Aspects of Diageneses: Society of Economic Paleontologists and Mineralogists Special Publication 26, p. 281–306.

DICKINSON, W. R.. KLUTE, M. A., HAYES, M. J., JANECKE, S. U., LUNDIN, E. R., MCKITTRICK, M. A., AND OLIVARES, M. D., 1988, Paleogeographic and paleotectonic setting of Laramide sedimentary basins in the central Rocky Mountain region: Geological Society of America Bulletin, v. 100, p. 1023–1039.

DIXON, S. A., AND KIRKLAND, D. W., 1985, Relationship of temperature to reservoir quality for feldspathic sandstones of southern California, *in* McNulty, C. L., and McPherson, J. G., eds., Transactions, American Association of Petroleum Geologists Southwest Section, p. 82–99.

DRONG, H. J., PLEIN, E., SANNEMANN, D., SCHUEPBACH, M. A. AND ZIMDARS, J., 1982, Der Schneverdingen-Sandstein des Rotliegenden-eine aolishe Sedimentfullung alter Grabenstrukturen: Deutsche geologische gesellschaft Zeitschrift, v. 133, p. 699–725.

FARQUHARSON, G. W., HAMER, R. D., AND INESON, J. R., 1984, Proximal volcaniclastic sedimentation in a Cretaceous back-arc basin, northern Antarctic Peninsula, *in* Kokelaar, B. P., and Howells, M. F., eds., Marginal Basins Geology: Geological Society, Blackwell, London, p. 219–229.

FISHER, R. V., 1979, Models for pyroclastic surges and pyroclastic flows: Journal of Volcanology and Geothermal Research, v. 6, p. 305–318.

FISHER, R. V., AND SCHMINCKE, H.-U., 1984, Pyroclastic Rocks: Springer-Verlag, Berlin, 472 p.

FITZGERALD, M. G., MITCHUM, R. M. Jr., ULIANA, M. A., AND BIDDLE, K. T., 1990, Evolution of the San Jorge basin, Argentina: American Association of Petroleum Geologists Bulletin, v. 74, p. 879–920.

FRENCH, D. E., AND FREEMAN, K. J., 1979, Volcanics yield another oil find: World Oil, May, p. 58–63.

FRITZ, W. J., AND HARRISON, S., 1985, Early Tertiary volcaniclastic deposits of the Northern Rocky Mountains, *in* Flores, R. M., and Kaplan, S. S., eds., Cenozoic Paleogeography of West-Central United States, Symposium 3: Rocky Mountain Section, Society of Economic Paleontologists and Mineralogists, p. 383–402.

FROSTICK, L. E., RENAUT, R. W., REID, I., AND TIERCELIN, J. J., 1986, Sedimentation in the African Rifts: Geological Society Special Publication 25, Blackwell, Oxford, 382 p.

GAGE, M. S., AND WING, R. S., 1980, Southeast Asian basin-types versus oil opportunities: Proceedings, Indonesian Petroleum Association, p. 124–147.

GALLOWAY, W. E., 1974, Deposition and diagenetic alteration of sandstone in northeast Pacific arc-related basins: Implications for graywacke genesis: Geological Society of America Bulletin, v. 85, p. 379–390.

GALLOWAY, W. E., 1979, Diagenetic control of reservoir quality in arc-derived sandstones: Implications for petroleum exploration, *in* Scholle P. A., and Schluger, P. R., eds., Aspects of Diagenesis: Society of Economic Paleontologists and Mineralogists Special Publication 26, p. 251–262.

GAST, R. E., 1988, Rifting im Rotliegenden Niedersachsens: Die Geowissenschaften v. 6, p. 115–122.

HAWLANDER, H. M., 1990, Diagenesis and reservoir potential of volcanogenic sandstones-Cretaceous of the Surat Basin, Australia: Sedimentary Geology, v. 66, p. 181–195.

HAYES, J. B., 1979, Sandstone diagenesis—The hole truth, *in* Scholle, P. A., and Schluger, P. R., eds., Aspects of Diagenesis: Society of Economic Paleontologists and Mineralogists Special Publication 26, p. 127–139.

HAYES, J. B., HARMS, J. C., AND WILSON, T., 1976, Contrasts between braided and meandering stream deposits, Beluga and Sterling Formations (Tertiary), Cook Inlet, Alaska, *in* Miller, T. P., ed., Recent and Ancient Sedimentary Environments in Alaska: Anchorage, Alaska Geological Society, p. J1–J27.

HUNTER, B. E., AND DAVIES, D. K. 1979, Distribution of volcanic sediments in the Gulf coastal province—significance to petroleum geology: Transactions, Gulf Coast Association of Geological Societies, v. 29, p. 147–155.

JANSA, L. F., AND PE-PIPER, G., 1988, Middle Jurassic to Early Cretaceous igneous rocks along eastern North American continental margin: American Association of Petroleum Geologists Bulletin, v. 72, p. 347–366.

KAUFFMAN, E. G., 1985, Cretaceous evolution of the Western Interior basin of the United States, *in* Pratt, L. M., Kauffman, E. G., and Zelt, F. B., eds., Fine-grained Deposits and Biofacies of the Cretaceous Western Interior Seaway: Evidence of Cyclic Sedimentary Processes: Society of Economic Paleontologists and Mineralogists Guidebook 9, p. iv–xiii.

KINGSTON, D. R., DISHROOM, C. P., AND WILLIAMS, P. A., 1985, Hydrocarbon plays, global basin classification: Oil and Gas Journal, May 6, p. 265–270.

KLASS, M. J., KERSEY, D. G., BERG, R. R., AND TIEH, T. T., 1981, Diagenesis and secondary porosity in Vicksburg sandstones, McAllen ranch field, Hidalgo County, Texas: Transactions, Gulf Coast Association of Geological Societies, v. 31, p. 115–123.

KLEMME, H. D., 1980, Petroleum basins-classifications and characteristics: Journal of Petroleum Geology, v. 3, p. 187–207.

KUJIRAOKA, A., 1980, Volcanic activity and the mode of occurrence of oil and gas in Japan: United Nations International Meeting on Petroleum Geology, Beijing, China (no page numbers).

LOCKLEY, M. G., AND RICE, A., 1990, Volcanism and fossil biotas: Geological Society of America Special Paper 244, 125 p.

LOUCKS, R. G., DODGE, M. M., AND GALLOWAY, W. E., 1979, Importance of secondary leached porosity in lower Tertiary sandstones along the Texas Gulf coast: Transactions, Gulf Coast Association of Geological Societies, v. 29, p. 164–171.

LOUCKS, R. G., DODGE, M. M., AND GALLOWAY, W. E., 1984, Regional controls on diagenesis and reservoir quality in Lower Tertiary sandstones along the Texas Gulf Coast, *in* McDonald, D. A., and Surdam, R. C., eds., Clastic Diagenesis: American Association of Petroleum Geologists Memoir 37, p. 15–45.

MALM, O. A., FURNES, H., AND BJORLYKKE, K., 1979, Volcaniclastics of Middle Jurassic age in the Statfjord oil field of the North Sea: Neues Jahrbuch Geologische Palaeontologie Monatatshefte, v. 10, p. 607–618.

MATHISEN, M. E., 1984, Diagenesis of Plio-Pleistocene nonmarine sandstones, Cagayan Basin, Philippines: Early development of secondary porosity in volcanic sandstones, *in* McDonald, D. A., and Surdam, R. C., eds., Clastic Diagenesis: American Association of Petroleum Geologists Memoir 37, p. 177–193.

MATHISEN, M. E., AND BUDNY, M., 1990, Seismic litho-stratigraphy of deep subsalt Permocarboniferous gas reservoirs, Northwest German basin: Geophysics, v. 55, p. 1357–1365.

MATHISEN, M. E., AND VONDRA, C. F., 1983, The fluvial and pyroclastic deposits of the Cagayan basin, Northern Luzon, Philippines—an example of nonmarine volcaniclastic sedimentation in an interarc basin: Sedimentology, v. 30, p. 369–392.

MCPHERSON, J. G., WARESBACK, D. B., AND SELF, S., 1986, Volcaniclastic alluvial fan sedimentation: American Association of Petroleum Geologists Bulletin, v. 70, p. 619.

MUELLER, W., CHOWN, E. H., SHARMA, K. N. M., TAIT, L., AND ROCHELEAU, M., 1989, Paleogeographic and paleotectonic evolution of a basement-controlled Archean supracrustal sequence, Chibougamau-Caopatina, Quebec: Journal of Geology, v. 97, p. 399–420.

PALMER, B. A., AND WALTON, A. W., 1990, Accumulation of volcaniclastic aprons in the Mount Dutton Formation (Oligocene-Miocene), Marysvale volcanic field, Utah: Geological Society of America Bulletin, v. 102, p. 734–748.

PARSONS, W. H., 1969, Criteria for the recognition of volcanic breccias: Review: Geological Society of America Memoir 115, p. 263–304.

PICKFORD, M., 1986, Sedimentation and fossil preservation in the Nyanza Rift System, *in* Frostick, L. E., Renaut, R. W., Reid, I., and Tiercelin, J. J., eds., Sedimentation in the African Rifts: Geological Society of London Special Publication 25, p. 345–362.

PRATT, L. M., KAUFFMAN, E. G., AND ZELT, F. B., 1985, Fine-grained deposits and biofacies of the Cretaceous Western Interior Seaway; Evidence of cyclic sedimentary processes: Society of Economic Paleontologists and Mineralogists Guidebook 9, 375 p.

RIGGS, N. R., AND BUSBY-SPERA, C. J., 1990, Evolution of a multi-vent volcanic complex within a subsiding arc graben depression: Mount Wrightson Formation, Arizona: Geological Society of America Bulletin, v. 102, p. 1114–1135.

RITTER, U., 1986, Heat flow during the Carboniferous and Mesozoic of the Northwest German Basin: Geologische Rundschau v. 75, p. 293–300.

SCHOTT, W., 1982, Crude oil and natural gas in the Federal Republic of Germany: Development and prospects: Journal of Petroleum Geology, v. 3, p. 235–266.

SCHULL, T. J., 1988, Rift basins of interior Sudan: Petroleum exploration and discovery: American Association of Petroleum Geologists Bulletin, v. 72, p. 1128–1142.

SEEMANN, U., AND SCHERER, M., 1984, Volcaniclastics as potential hydrocarbon reservoirs: Clay Minerals, v. 9, p. 457–470.

SHEEHAN, C. A., PENFIELD, G. T., AND MORALES, E., 1990, Costa Rica geologic basins lure wildcatters: Oil and Gas Journal, April 30, p. 73–79.

SMITH, G. A., 1988, Sedimentology of proximal to distal volcaniclastics dispersed across an active foldbelt: Ellensburg Formation (Late Miocene), central Washington: Sedimentology, v. 35, p. 953–977.

SMITH, G. A., CAMPBELL, N. P., DEACON, M. W., AND SHAFIQULLAH, M., 1988, Eruptive style and location of volcanic centers in the Miocene Washington Cascade range: Reconstruction from the sedimentary record: Geology, v. 16, p. 337–340.

SPARKS, R. S. J., SELF, S., AND WALKER, G. P. L., 1973, Products of ignimbrite eruption: Geology, v. 1, p. 115–118.

STANLEY, K. O., AND BENSON, L. V., 1979, Early diagenesis of high plains Tertiary vitric and arkosic sandstone, Wyoming and Nebraska, *in* Scholle, P. A., and Schluger, P. R., eds., Aspects of Diagenesis: Society of Economic Paleontologists and Mineralogists Special Publication 26, p. 401–423.

STEWART, W. D., AND DURKEE, E. F., 1985, Petroleum potential of the Papuan basin: Oil and Gas Journal, April 1, p. 151–160.

SURDAM, R. C., AND BOLES, J. R., 1979, Diagenesis of volcanic sandstones, *in* Scholle, P. A., and Schluger, P. R., eds., Aspects of Diagenesis: Society of Economic Paleontologists and Mineralogists Special Publication 26, p. 227–242.

SUTHREN, R. J., 1985, Facies analysis of volcaniclastic sediments: A review, *in* Brenchley, P. J., and Williams, B. P. J., eds., Sedimentology: Geological Society of London, Blackwell, Oxford, p. 123–146.

VERNIK, L., 1990, A new type of reservoir rock in volcaniclastic sequences: American Association of Petroleum Geologists Bulletin, v. 74, p. 830–836.

VESSELL, R. K., AND DAVIES, D. K., 1981, Nonmarine sedimentation in an active fore arc basin, *in* Ethridge, F. G., and Flores, R. M., eds., Recent and Ancient Nonmarine Depositional Environments: Models for Exploration: Society of Economic Paleontologists and Mineralogists Special Publication 31, p. 31–45.

WALKER, G. P. L., 1971, Grain-size characteristics of pyroclastic deposits: Journal of Geology, v. 79, p. 696–714.

WALKER, T. R., WAUGH, B., AND CRONE, A. J., 1978, Diagenesis in first-cycle desert alluvium of Cenozoic age, southwestern United States and northwestern Mexico: Geological Society of America Bulletin, v. 89, p. 19–32.

WARESBACK, D. B., AND TURBEVILLE, B. N., 1990, Evolution of a Plio-Pleistocene volcanogenic-alluvial fan: the Puye Formation, Jemez Mountains, New Mexico: Geological Society of America Bulletin, v. 102, p. 298–314.

WEIMER, R. J., 1966, Time-stratigraphic analysis and petroleum accumulations, Patrick Draw field, Sweetwater County, Wyoming: American Association of Petroleum Geologists Bulletin, v. 50, p. 2150–2175.

WRIGHT, J. V., SMITH, A. L., AND SELF, S., 1980, A working terminology of pyroclastic deposits: Journal of Volcanology and Geothermal Research, v. 8, p. 315–336.

WRIGHT, J. V., SELF, S., AND FISHER, R. V., 1981, Towards a facies model for ignimbrite-forming eruptions, *in* Self, S., and Sparks, R. S. J., eds., Tephra Studies: D. Reidel, Dordrecht, Holland, p. 433–439.

YOUNG, R. A., AND LUCAS, J. E., 1988, Exploration beneath volcanics: Snake River plain, Idaho: Geophysics, v. 53, p. 444–452.

ZIMMERLE, W., 1985, New aspects on the formation of hydrocarbon source rocks: Geologische Rundschau, v. 74, p. 385–416.

# PART II
# VOLCANICLASTIC PROCESSES OF TRANSPORT AND DEPOSITION

Volcaniclastic transport and depositional processes caused directly from volcanism differ from normal sedimentary processes in some cases. Part II explores sedimentological processes that result from volcanic eruptions including pyroclastic flows, pyroclastic surges, volcanic debris flows, hyperconcentrated flows and avalanches caused by sector collapse of volcanoes.

# TRANSPORT AND DEPOSITION OF TEPHRA BY PYROCLASTIC FLOWS AND SURGES

STEVEN N. CAREY

*Graduate School of Oceanography, University of Rhode Island, Narragansett, RI 02882*

ABSTRACT: Explosive volcanism at island arcs and convergent continental margins generates pyroclastic flows and surges that transport and deposit large quantities of pyroclastic material in areas adjacent to active volcanic centers. Pyroclastic flows are inferred to be relatively high-concentration, hot, gas-solid dispersions that flow in a nonturbulent or plug-flow manner. Their movement is typically controlled by the underlying topography, as reflected in the common valley-fill relationships of their deposits. Theoretical and experimental studies have confirmed the importance of partial fluidization in the transport of pyroclastic flows. En masse freezing is a common depositional mechanism for relatively low-energy pyroclastic flows that form high-aspect ratio ignimbrites. Highly energetic flows develop a variety of depositional regimes from various portions of the moving flow such as the head, body, and tail. The interaction of these depositional regimes with the topography over which a flow travels leads to marked contrasts in facies and stratigraphic relationships of the resulting deposits.

Pyroclastic surges are low-concentration, gas-solid dispersions that flow in a turbulent fashion. As a result of their more expanded nature, surges are less influenced than flows by underlying topography and produce radially distributed deposits around volcanic centers. Surge deposits are commonly cross-stratified and form dunes, although they can also occur as massive or planar beds. Cross-stratified dune structures indicate that particle support is dominantly by turbulence of the interstitial gases and that grain concentration is relatively low. Particles are also moved along near the base of a surge by saltation and rolling in a traction layer, where high, lateral, shear velocities are developed. Deposition from surge clouds occurs when the velocity in the basal boundary layer or in the overlying turbulent eddies is less than that required to support grains of a certain size. Reduction in the surge velocity will cause material to be transferred from suspension transport to traction transport. The rate at which material is transferred between these regions of the surge cloud versus the rate at which material is deposited from the basal traction flow is critical in determining the facies of the resultant deposits. At low-particle concentrations, the beds should be dominated by cross-stratified units, whereas at higher particle concentrations, more massive types will form. A whole spectrum of facies relationships is likely to exist because of the great differences in the properties of the source materials at different volcanoes and the complicating effect of steam-to-water-phase transitions in "wet" versus "dry" types of surges.

## INTRODUCTION

Explosive volcanic eruptions commonly generate mixtures of juvenile pyroclasts, lithics, magmatic volatiles, and external water, whose bulk density exceeds that of the atmosphere into which they are injected. These mixtures form flows that move downslope under the influence of gravity and thus fall under the general category of pyroclastic density currents. The accumulation of material from these flows can produce thick and widespread deposits that encompass areas up to 20,000 km$^2$ for single eruptive events. These units are stratigraphically important in volcaniclastic sequences because 1) they are formed on relatively short time scales (hours to months) and thus are nearly isochronous horizons, 2) they can be correlated over wide geographic areas, and 3) they are major constructional components of areas with active explosive volcanism, such as island arcs and convergent continental margins. Traditionally, these flows have been discussed in terms of two broad classes: pyroclastic flows and pyroclastic surges. They have been interpreted as high- and low-concentration end members, respectively, of a flow spectrum based on particle concentration (Sparks, 1976). Because of the significant risk to human populations from pyroclastic density currents, recognition of their products on the basis of grain size, lithology and sedimentological features is critical to the establishment of hazard zones around active volcanic centers (e.g., Crandell and others, 1984; Scott, 1989).

It has been over 80 years since the generation of pyroclastic density currents was first recognized during the eruptions of the Soufrière volcano on St. Vincent and Mont Pelée on Martinique in 1902 (Anderson and Flett, 1903; LaCroix, 1904; Perret, 1937). As studies of pyroclastic flows and surges have progressed during the last two decades, it has become clear that there is considerable diversity in the nature of flows at both ends of the particle-concentration spectrum. Some workers are of the opinion that a complete transition occurs between both ends (Cas and Wright, 1987), whereas others argue for distinct gaps (Walker, 1983). Fundamental differences are apparent in the transport and depositional mechanisms at each end of the spectrum, but an increasing number of deposits have been found that cannot be assigned easily to either end member and their interpretation is often controversial. Recent studies, in fact, have emphasized both pyroclastic flows and surges as being complex depositional systems in which the transport and depositional processes are neither spatially or temporally uniform. This is not unexpected as explosive volcanism produces a remarkably wide diversity of flows with contrasting rheological properties, internal energies and volumes. It is therefore unrealistic to hope that all flowage deposits can be conveniently classed into either category and that a single set of transport and depositional processes would apply to each case. The objective of this paper is to present an overview of the morphological, lithological, and sedimentological features of pyroclastic flow and surge deposits that can be used to discriminate them from one another and to relate these characteristics to current theories concerning 1) the physical and rheological properties of the source flows and 2) mechanisms of particle support and deposition. Extensive reviews of these topics can be found in two recent books by Fisher and Schmincke (1984) and Cas and Wright (1987), along with an earlier review by Sheridan (1979).

## PYROCLASTIC FLOW DEPOSITS

Pyroclastic-flow deposits (PFDs) are formed from hot, concentrated flows of juvenile volcanic clasts, crystals and lithics, which travel at velocities of 10 to 300 m/s under

Sedimentation in Volcanic Settings, SEPM Special Publication No. 45

the influence of gravity. Generation of these flows can occur by a variety of mechanisms during explosive volcanic eruptions, such as eruption column collapse (Sparks and Wilson, 1976; Sparks and others, 1978), explosive or gravitational collapse of an active summit dome (Perret, 1937), or low fountaining of material out of a crater such as the 1980 flows of Mount St. Helens (Hoblitt, 1986). PFDs have been recognized at roughly half of the world's volcanoes (Walker, 1983) and are most commonly associated with the eruption of calc-alkaline magmas encompassing the range from basaltic andesite to rhyolite, although alkalic varieties of trachytic to phonolitic composition are also abundant (e.g., Barberi and others, 1978; Mahood, 1984). Table 1 is a recent attempt to classify flows on a genetic basis from Wright and others (1980). In this paper an effort is made to adhere to this scheme, although the use of more specific nomenclature is often necessary to elaborate upon the details of particle transport and deposition.

Pyroclastic flows have been broadly divided into three major categories based on the vesicularity, and to a lesser extent, the composition of their juvenile clasts (Cas and Wright, 1987). Block and ash flows are characterized by the presence of large, dense, poorly vesicular blocks up to several meters in diameter, which are often radially jointed by thermal effects. The 1929 eruption of Mt. Pelée on Martinique produced a thick sequence of block-and-ash-flow deposits in the valley of the Rivière Blanche (Fisher and others, 1980; Fisher and Heiken, 1982). Scoria flows contain abundant scoriaceous material of basaltic andesite or andesite composition, such as the deposits produced by the May 7, 1902, eruption of the Soufrière volcano on St. Vincent (Hay, 1959). Pumice flows are distinguished by an abundance of low-density pumice, usually of high silica composition, and large quantities of fine ash in the flow matrix. If ash is the dominant component (>50%), then the term ash flow is often applied (Smith, 1960a). Examples of deposits from pumice flows include the Valley of Ten Thousand Smokes in Alaska (Curtis, 1968), the Bishop Tuff of Long Valley, California (Bailey and others, 1976), and many of the large ashflow tuffs associated with calderas of the Western Interior United States (Lipman and others, 1984). An important consequence of the variation in magma types that generate pyroclastic flows is that the density of juvenile clasts can be as high as 2500 kg $m^{-3}$ or as low 500 kg $m^{-3}$, depending on the gas content and crystallinity of the source magma. The variance in clast density between all components of a flow thus increases for eruptions that produce highly vesiculated juvenile clasts and will have an important effect on the sedimentological features of the resulting deposits.

### *Volume and Morphology*

Pyroclastic flows travel away from their source vents under the influence of gravity and thus their distribution will be influenced to varying degrees by the topography of the ground surface over which they flow. River valleys channelize small to intermediate flows, whereas ridges act as topographic barriers that can halt or divert a flow's progress. The degree to which the local topography influences the transport and deposition of a flow will depend upon the scale of the flow, i.e., its volume and momentum. These two factors, which are linked to the volume discharge rate of magma and the energetics of flow emplacement, strongly control the geometry, morphology and facies relationships of flow deposits.

One of the most impressive aspects of pyroclastic-flow deposits (PFDs) is the great range in volumes for individual eruptive sequences. Smith (1979) has shown that there is at least a five-order magnitude difference between PFDs associated with small, historic eruptions of stratovolcanos and those derived from large, caldera-forming eruptions. For example, the 1982 explosive eruptions of El Chichón volcano in southeast Mexico produced a small volume (0.04

TABLE 1.—GENETIC CLASSIFICATION OF PYROCLASTIC FLOWS (after Wright and others, 1980)

| Essential fragment | Eruptive mechanism | | Pyroclastic flow | Deposit | Comments |
|---|---|---|---|---|---|
| Vesiculated | Eruption column collapse or "boiling-over" | | Pumice or ash flow | Ignimbrite; pumice and ash flow deposit | Intermediate-volume deposits formed by continuous collapse of a Plinian eruption column as envisaged by Sparks and others (1978), Silicic in composition<br>Small-volume deposits probably formed by interrupted column collapse. Intermediate to silicic in composition |
| Decreasing average density of juvenile clasts | | | Scoria flow | Scoria and ash deposit | Small-volume deposits probably formed by interrupted eruption column collapse produced by short explosions. Basalt to andesite in composition. Some deposits contain large unvesiculated blocks, e.g., those of the Ngauruhoe (1975) eruption. |
| | Lava or dome collapse, or lateral projection | Explosive | Lava debris flow nuée ardente | Block and ash deposits | Small-volume deposits, usually andesitic or dacitic in composition |
| Nonvesiculated | | Gravitational | Lava debris flow nuée ardente | Block and ash deposits | Small-volume deposits usually andesitic or dacitic in composition. These are the hot avalanche deposits of Francis and others (1974) |

km$^3$) of pyroclastic flow material that was largely confined to the major river valleys surrounding the volcano (Sigurdsson and others, 1984). The deposits consist of massive, gray, poorly sorted coarse ash and lapilli with trachyandesite pumice and lithic blocks and extend only 6 km from the summit of the volcano (Fig. 1). Along the Susnubuc River, the total deposit thickness was about 15 m, but because of confinement to the active drainage system of the volcano, the deposits have been eroded quite rapidly. At the other extreme of the flow-volume spectrum are extensive pumice-flow deposits such as the Fish Canyon Tuff, erupted from the La Garita caldera in southwest Colorado (Steven and Lipman, 1976). The total volume of this deposit has been estimated at 3,000 km$^3$ and it covers an area greater than 15,000 km$^2$ (Fig. 2).

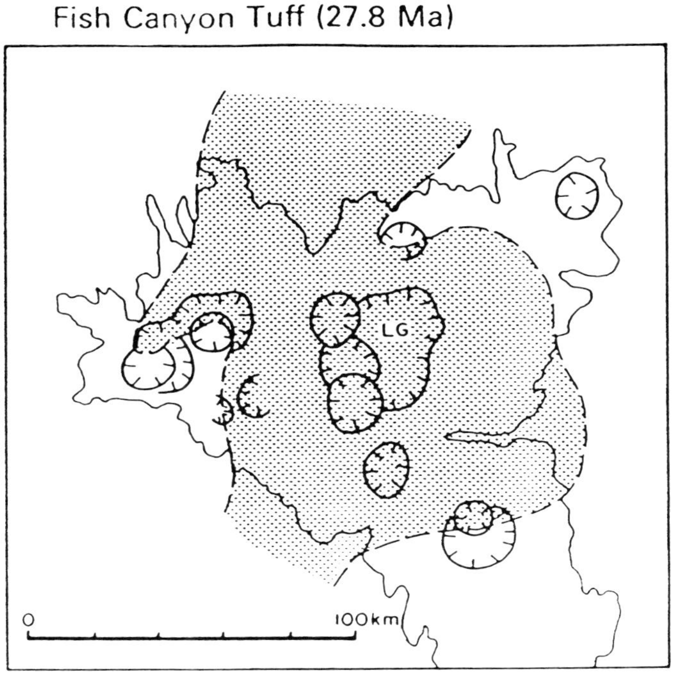


FIG. 2.—Map showing the distribution of the Fish Canyon Tuff (stipled pattern) around the La Garita (LG) caldera in the San Juan Volcanic field. Adapted from Cas and Wright (1987).

The overall geometry of flow deposits can be quite variable depending on the topography surrounding the source vent and the volume of material discharged during an eruption. Small-volume flows tend to fill in valleys and pond in local depressions (e.g., Freundt and Schmincke, 1985). Their direction of transport is strongly controlled by underlying topography, such as the 1982 deposits of El Chichón volcano (Fig. 1). Large-volume flows, on the other hand, form extensive sheet-like deposits, which in some cases are thick enough to overwhelm existing topography. Thus, there are broad differences in the large-scale morphology of PFDs that can be qualitatively related to the total volume of the flow sequence.

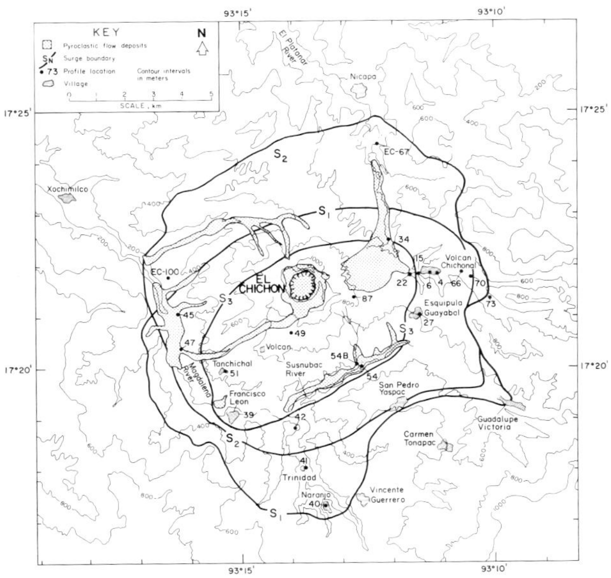


FIG. 1.—Map of the distribution of pyroclastic flows and surges from the March-April 1982 eruptions of El Chichón, from Sigurdsson and others (1984). The heavy lines define the outer limits of the surge zones. Stippled pattern shows the distribution of major pyroclastic flows.

A useful quantitative parameter for characterizing the general form of pyroclastic-flow deposits is the aspect ratio, $V/H$, where $V$ is a measure of flow thickness and $H$ is a horizontal dimension associated with the area covered by the deposits (Walker and others, 1980a). Aspect ratios (thickness to length, $T/L$) have been found to vary from 1/400 to as little as 1/100,000. High-aspect ratio ignimbrites (HARIs), such as the Valley of Ten thousand Smokes in Alaska (1/400), exhibit the classic features of PFDs such as strong topographic control, valley infilling, and a flat, gently sloping upper surface. The deposits are thus generally thick but not widely dispersed.

In contrast, there are other PFDs such as 186 AD Taupo sequence of New Zealand that are very widely dispersed yet relatively thin, yielding low-aspect ratios (e.g., 1/70,000, Walker and others, 1980a). Unlike HARIs, these low-aspect ratio ignimbrites (LARIs) appear to be only minimally affected by the local topography. The Taupo deposit, for example, extends symmetrically to a distance of about 100 km around the source vent of Lake Taupo. Many LARIs show evidence of high emplacement velocities and the ability to surmount topographic barriers that are hundreds of meters in height. The difference in morphology of HARIs and LARIs has thus been interpreted as reflecting emplacement energetics (Walker, 1983). High-aspect ratio ignimbrites are deposited from low-energy flows during eruptions with low magma-discharge rates, whereas low-aspect ratio ignimbrites are deposited from highly energetic flows produced during eruptions of high magma-discharge rate.

### *Sedimentological and Stratigraphic Relations*

Direct observations of pyroclastic flows from historic eruptions have been valuable for understanding their generation, transport velocities, and response to local topography (e.g., Perret, 1937; MacDonald and Alcaraz, 1956;

Moore and Melson, 1969; Davies and others, 1978; Hoblitt, 1986). However, these studies have dealt primarily with small-volume flows derived from eruptions of stratovolcanoes like the Soufrière, Mt. Pelée or Mount St. Helens (<1 $km^3$). There has yet to be direct observation of large-volume pyroclastic flows (>10 $km^3$) similar to ones preserved in the geologic record and typically associated with the formation of calderas (Walker, 1983). Furthermore, because convective, fine-ash clouds typically shield the interior of a moving flow, existing obervations provide little information about the details of particle transport, flow regime or flow structure. The study of these important aspects has relied on the sedimentological characteristics of flow deposits and comparisons with other particulate gravity-flow processes.

Systematic studies of many different PFD sequences have demonstrated that there are distinct lateral and vertical variations in their sedimentologic and stratigraphic characteristics. Sparks and others (1973) proposed a standard "ignimbrite" section, which reflects stratigraphic relations and particle grading patterns commonly found in pumice-bearing pyroclastic flows (Fig. 3). At the base is a stratified and often cross-bedded pyroclastic surge deposit (layer 1, discussed later). Overlying the surge is the main part of the flow deposit, which is divided into a thin, lower basal layer (2a) and an upper, thick layer (2b). Resting on top of layer 2 is a fine ashfall deposit (layer 3) derived from the loss and subsequent fallout of the fine-ash component of the underlying flow (Sparks and Walker, 1977). In this idealized section, layer 2 is interpreted as the deposits from the main body of the pyroclastic flow.

There are several general features of layer 2 deposits that are critical to an assessment of transport and depositional mechanisms. First, they are typically very poorly sorted with respect to grain size and contain clasts that range in size from meter-size blocks to glass shards only a few microns in diameter. Second, they are massive and lack internal stratification, although there are systematic vertical variations in the distribution of large clasts (Sparks, 1976). In layer 2a, both lithics and pumice are often reversely size graded. This grading is not restricted to the lower flow boundary but also can be found along the sides of a deposit, if deposition has been laterally bounded by a steep valley wall. Reverse size grading is also observed in the main flow layer (2b), but is typically restricted to low-density pumice clasts. Dense clasts, such as lithic blocks, grade in the opposite sense, being largest toward the bottom and decreasing in size with height in the deposit (Fig. 3). Detailed grain-size analyses of samples from vertical sections of PFDs have shown that the grading of different clast types is a strong function of their size (Sparks, 1976). The larger the clast, the more susceptible it is to becoming concentrated at the base (lithics) or top (pumice) of a deposit. As particle size decreases, the extent of grading with height in the deposit becomes less and the size distribution of the fine-ash component (<0.5 mm) in the flow becomes similar irrespective of vertical position (Fig. 4). This grading pattern, in which only the coarsest part of the total size distribution changes as a function of position, is known as coarse-tail grading.

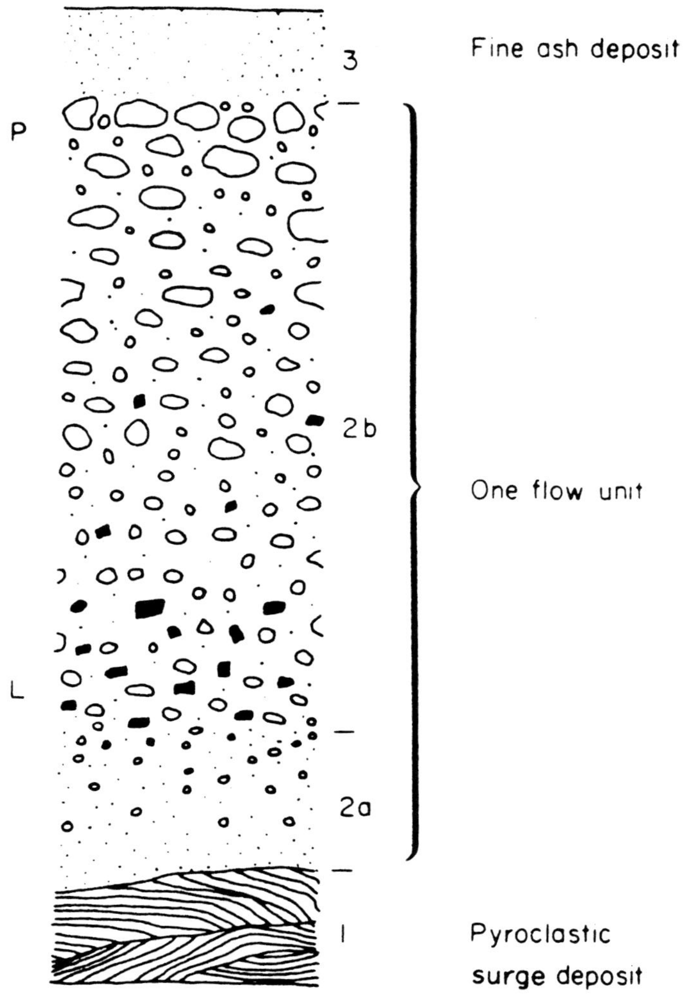


FIG. 3.—Stratigraphic section of the idealized products of an ignimbrite eruption, after Sparks and others (1973). *P* indicates pumice clasts and *L*, lithic clasts. Note concentration of pumice at the top of the flow unit and lithics at the base.

All PFDs do not exhibit the same type of grading patterns. Many deposits show no evidence of grading and others may show normal grading of both pumice and lithics. The appearance of grading patterns will be related strongly to the initial size distribution of the source material and the relative density contrasts between different component types (Sparks, 1976). Block and ash flows consisting of dense, poorly vesiculated clasts are often normally graded with respect to large clasts, as are scoria-flow deposits (Fig. 5).

The relations described above for layer 2 deposits are taken as characteristic of many PFDs of the high-aspect ratio type that are deposited at medial to distal distances from source. Other deposits of lower aspect ratio have been found to have more complex stratigraphic and facies relationships. One of the most thoroughly studied pyroclastic-flow deposits that exhibits marked lateral and vertical facies variations is the Taupo ignimbrite of New Zealand (Walker and others, 1981; Wilson and Walker, 1982). The deposit

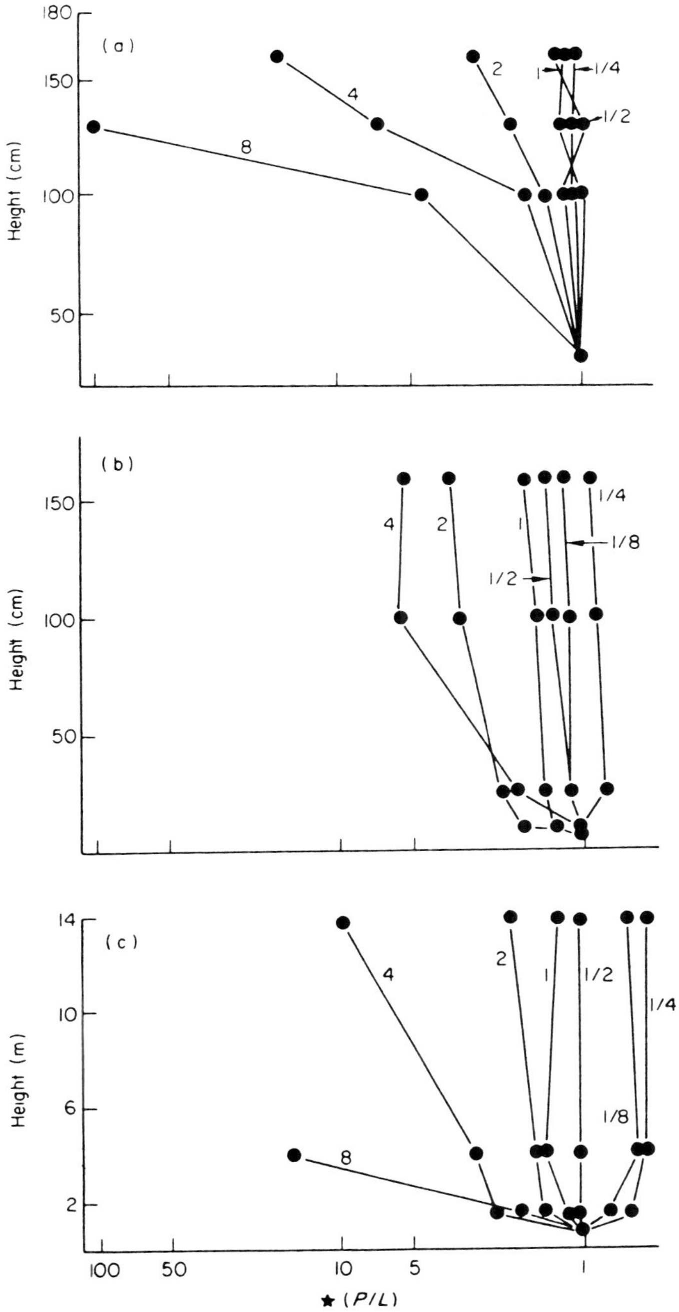


FIG. 4.—Plots of normalized pumice-lithic ratio ($P/L$) as a function of height in pyroclastic flow deposits for grain sizes from 1/8 to 8 mm, from Sparks (1976). a) Minoan ignimbrite, Santorini, b) Ignimbrite D, Vulsini, c) Ignimbrite C, Vulsini. Note the stronger fractionation between pumice and lithics with increasing height as the particle-size class increases.

was emplaced during a 186 AD eruption and covers an area of about 20,000 $km^2$ extending outward in a roughly circular pattern from Lake Taupo in the central North Island of New Zealand. Its stratigraphic relations deviate substantially from the generalized flow stratigraphy of Figure 3 and the deposit is considered an example of a LARI (Fig. 6).

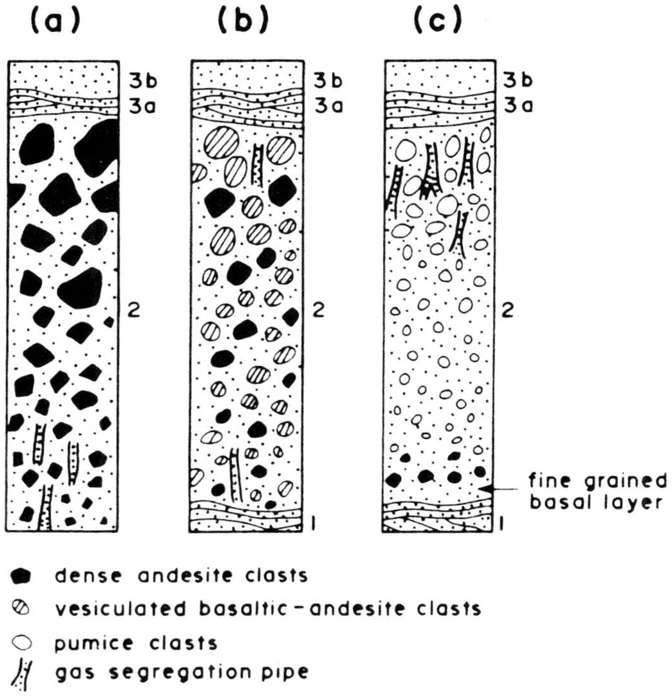


FIG. 5.—Typical patterns of clast type, stratigraphic relations and clast-grading trends in a) block-and-ash flow deposit, b) scoria flow deposit and c) pumice flow deposit, from Cas and Wright, 1987).

Layer 1 deposits of the Taupo consist of two strongly contrasting varieties; one is composed almost exclusively of low-density pumice (1P) and the other (1H) is predominantly dense lithic clasts. Both types are separated from the overlying layer 2 units by a sharp, planar contact. They are often found together with 1P underlying 1H, although the former is sometimes eroded or missing. Grain-size analysis of 1P layers typically shows some degree of fines depletion relative to the layer 2 units (Wilson and Walker, 1982).

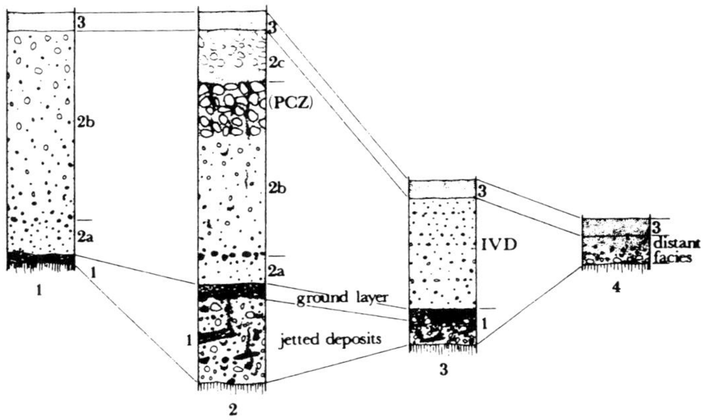


FIG. 6.—Comparison of the stratigraphic relations in the ignimbrite layering model of Sparks and others (1973), column 1, with stratigraphic relations in the valley-pond ignimbrite (column 2), ignimbrite veneer (column 3) and distal facies (column 4) of the Taupo pyroclastic-flow deposits, from Wilson (1985).

Layer 2 deposits of the Taupo also occur as two distinct types. One is similar to the "normal" layer 2 deposits of Figure 3 in that they are found as thick, valley-filling units. These valley-pond ignimbrites (VPIs) vary in thickness from as little as 1 m to as much as 70 m and usually consist of a single flow unit. They often show coarse-tail size grading with low-density pumice concentrated near the flow top and lithics enriched near the base. In some flows the increase in pumice toward the top of the flow is manifested as sharply defined pumice concentration zones (PCZs) that contain the largest pumices in the ignimbrite and are virtually devoid of other components.

A second variety of layer 2 deposits is found as a mantling unit on ridges in between the valley-pond facies and has been termed ignimbrite veneer deposit (IVD; Walker and others, 1981). A characteristic feature of this deposit is that it often shows a layered structure, yet is generally similar in grain size to the valley-pond deposits with the exception of having a smaller maximum pumice size. The layering results from variations in both the content of coarse pumice and slight differences in the bulk grain size. A major difference between the IVD and the VPI is that the latter is usually 5 to 50 times thicker than the former. Field observations also demonstrate that the IVD and VPI are lateral equivalents of each other and were not formed by separate volcanic events.

The Taupo deposits serve to illustrate the contrasting types of facies that can occur in widespread low-aspect ratio ignimbrites. In this case these facies are found to be best developed at distances beyond 13 km from the vent. Recently, there has been considerable progress made on the recognition and description of PFDs close to source. Thick, lithic-rich breccias, termed co-ignimbrite lag breccias, have been recognized as the near-vent facies of some PFDs (Wright and Walker, 1977; Wright and others, 1981; Druitt and Sparks, 1982; Druitt, 1985). These deposits are typically fines-depleted, stratified, 1 to 20 m in thickness and can extend up to 20 km from source (Walker, 1985). A characteristic feature of these deposits is that they contain an abundance a large lithic and poorly vesicular juvenile blocks. Lithic blocks are usually larger than those of associated pyroclastic flows with many exceeding 50 cm in diameter.

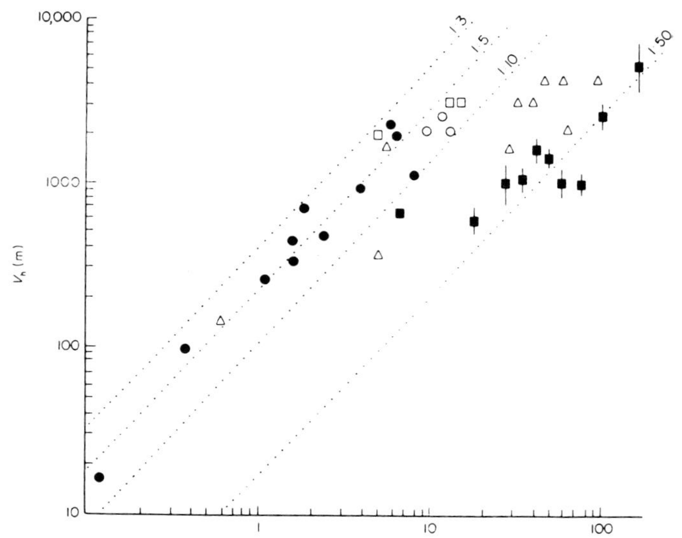


FIG. 7.—Relationship between the vertical height change ($V_h$) and the horizontal distance of travel for different types of particulate flows, from Sparks (1976). $V_h$ is usually taken as the reconstructed height of the caldera prior to collapse. Dottle lines are equal ratio of $V_h/H_h$. Data points are cold-rock avalanches (solid circles), dense pyroclastic flows (open boxes), mud flows (open triangles), and pumice flows (solid boxes).

## *Transport and Mobility*

Studies of large-volume PFDs preserved in the geologic record led to the recognition that these flows are more mobile than other particulate gravity flows, such as cold-rock avalanches (Francis and others, 1974; Sparks, 1976; Francis and Baker; 1977). In particular, large-volume pyroclastic flows with abundant fine ash and pumice show the most pronounced evidence for high mobility (Fig. 7), whereas small-volume flows with dense clasts appear to travel no farther than cold-rock avalanches. However, one of the problems with comparing pyroclastic flows to other types of particulate flows is that it is difficult to define the height at which pyroclastic flows are initiated (Sheridan, 1979). If flows are generated by eruption-column collapse, as suggested by theoretical studies (Sparks and Wilson, 1976; Sparks and others, 1978; Woods, 1988), then the height at which they begin their descent is the elevation of the volcano plus the height of the collapsing fountain above the vent. Numerical models have suggested collapse heights of up to 7 km (Sparks and others, 1978). When these collapse heights are taken into account, the apparent enhanced mobility of large flows is reduced, although large-volume pyroclastic flows are still able to travel large distances (Fig. 7). Support for a strong relationship between collapse height and flow mobility comes from the Taupo pyroclastic flow, which travelled over 100 km from the source at high velocity but was derived from Lake Taupo, an area with virtually no topographic expression. Preceding the generation of the flow, however, was a Plinian phase that produced an inferred eruption column of 55 km height, the greatest known for any explosive eruption (Walker, 1980; Carey and Sigurdsson, 1989).

Field evidence also indicates that some pyroclastic flows are able to surmount topographic barriers up to 600 m high (Yokoyama, 1974; Miller and Smith, 1977). This was originally taken as evidence that flows were highly expanded, but now it is generally accepted that flow momentum determines the extent to which a flow can override topographic obstacles in its path.

Before the grain-size relations of pyroclastic flows were systematically investigated, several workers had proposed that the process of fluidization was important in their transport (Reynolds, 1954; McTaggart, 1962; O'Keefe and Adams, 1965). It was also suggested that the heating and sudden expansion of entrained air provided a way to increase the mobility of flows (McTaggart, 1960). The latter largely can be ruled out because hot, small-volume pyroclastic flows travel no farther than cold-rock avalanches (Fig. 7). Fluidization, on the other hand, is generally regarded as a key factor in the movement of pyroclastic flows (Cas and Wright, 1987).

Fluidization is a process by which a bed of solid particles can acquire the rheological properties of a fluid and flow under the influence of gravity. If gas is passed vertically through a bed of cohesionless particles with a very narrow range in particle size, a series of transformations will occur as the gas velocity is increased. A slight expansion of the bed occurs initially, but at higher velocity, ($U_{mf}$), a point is reached where the drag force exerted by the gas phase balances the weight of the particles and the bed begins to behave as a fluid. A very different behavior has been found for systems with a broad range in particle size, such as pyroclastic flows (Wilson, 1980, 1984). In this situation, it is not possible to fluidize completely all particles simultaneously. At any specified gas velocity, the particles will fall into three groups based on their pneumatic properties (Fig. 8). Large particles with settling velocities ($U_t$) greater than the vertical gas velocity will not be fluidized. Fluidization will only occur for particles that satisfy the condition, $U_t \geq U \geq U_{mf}$. If these particles represent a volumetrically significant part of the total distribution, then a fluid-like matrix will be generated. Very small particles, whose settling velocity ($U_t$) is much less than the gas velocity, will be removed from the bed by the process of elutriation. Thus, in a partially fluidized bed, large clasts are present within a fluidized matrix that is constantly losing fine ash by elutriation.

Sparks (1976) argued that partial fluidization was the likely state for most pyroclastic flows. The reverse grading of low-density pumice fragments and normal grading of dense lithic fragments (Figs. 3, 4) is attributed to the floating of pumice in a fluidized matrix and the sinking of lithic clasts. The important corollary of this explanation is that the density of the fluidized matrix must lie between the density of the pumice and the matrix density of the deposit (0.6 and 1.1 g/cm$^3$, respectively, for the Minoan ignimbrite). In other words, the flows were not highly expanded relative to their deposits when they were in motion and thus were high-particle concentration dispersions. The moving flow is thus envisaged to be a high-concentration dispersion whose enhanced mobility is attributed to a fluidized matrix of fine-ash particles. Large clasts, such as pumices and lithics, are transported within the matrix and are supported by buoyancy forces. Their position within the flow will depend upon whether they are positively or negatively buoyant. At the base of the flow, the observed reverse grading of both lithic and pumice clasts (Fig. 3) can be explained by dispersive forces acting normal to the boundary with the ground surface.

Experimental investigations of the fluidization behavior of pyroclastic flow material reveals that as gas velocity is increased, upward flow is not homogeneously distributed, but becomes preferentially directed along vertical channels (Wilson, 1980, 1984). These channels become depleted in fine ash and enriched in the coarse and dense fragments of the flow, similar to the gas escape pipes observed in many PFDs (e.g., Sheridan, 1970; Walker, 1972). At very high gas velocities, bubbling of the mixture occurs and strongly fines-depleted segregation layers develop at the base and top of the flows. Based on these experiments, Wilson (1980) proposed a three-fold classification scheme of PFDs that related sedimentological characteristics to fluidization state. Type 1 flows lack grading of large clasts, have not suffered extensive loss of fine ash by elutriation, and represent a less expanded flow matrix. Type 2 flows exhibit coarse-tail grading with pumice concentrated near the top and lithics enriched at the base. These flows represent expanded conditions where gas fluxing is sufficient to partially fluidize the matrix and allow for the floating or settling of large clasts. The third type of flow is referred to as segregated, and is characterized by strongly fines-depleted pumice and lithic concentration zones. Such flows represent very high

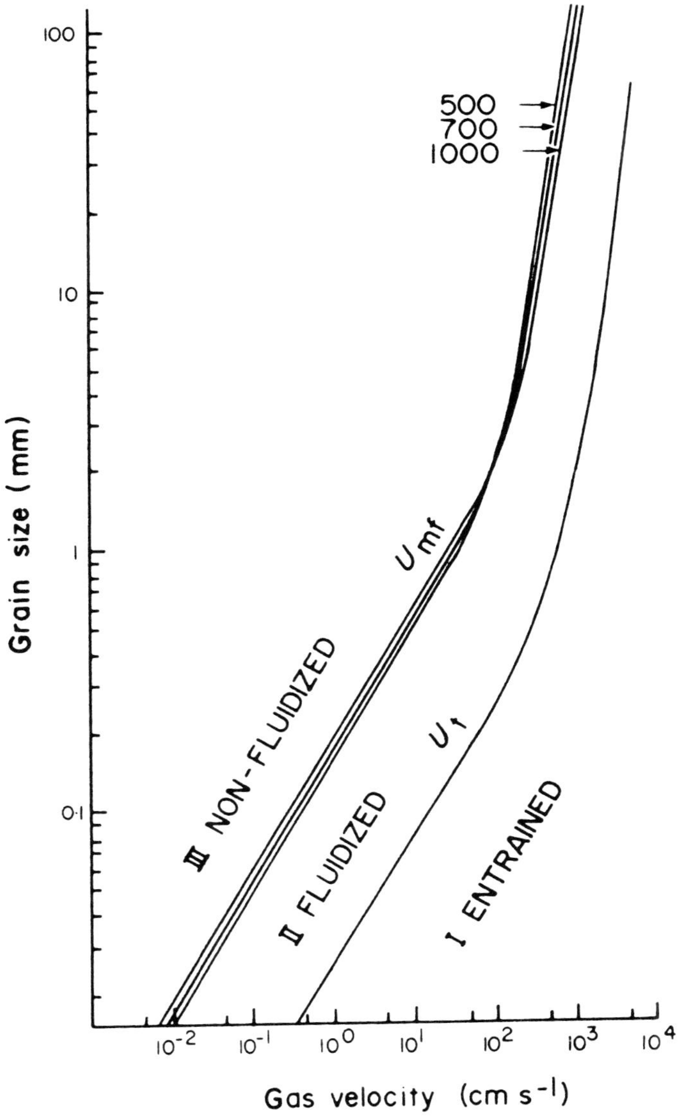


FIG. 8.—Minimum fluidization ($U_{mf}$) and terminal settling velocity ($U_t$) curves for 1.0 g cm$^{-3}$ spheres as a function of grain size and gas velocity, from Sparks (1976). Fluidization curves are calculated at 500, 700 and 1000°C. For any gas velocity, particles will fall into three classes: entrained, fluidized, and non-fluidized.

levels of fluidization and consequently are depleted in the fine ash fraction by elutriation.

An important consideration of the fluidization model is the origin of the gas phase. Sparks (1979) proposed that exsolution of gas from juvenile clasts could provide enough material to fluidize pumice and ash flows. However, Wilson (1985) considers that in the case of the Taupo flow the source of the fluidizing gas was primarily air ingested at the front of the flow and the combustion of entrained vegetation. The great mobility of the Taupo flow is thus viewed not as the consequence of a highly fluidized state but rather a direct result of the high emplacement velocity, which led to enhanced air ingestion and fluidization.

### *Depositional Processes*

It has been shown that PFDs vary greatly in their morphology, volume, and facies relationships. In particular, the facies relationships appear to be related to the energy of flow emplacement, or violence, and the position relative to the vent (Wilson and Walker, 1982). Fluidization is likely to be an important process in many pyroclastic flows and there is good evidence that many flows are high-concentration mixtures of gas and particles that are not greatly expanded relative to the deposits. However, the exact relationship between the moving flow and the deposit that is left on the ground is still subject to much debate (Wilson, 1985).

Sparks (1976) pointed out that many sedimentological features of PFDs are similar to those found in deposits from other high-concentration sediment-gravity flows, such as debris flows and high-concentration turbidity currents. These features include poor sorting, reverse grading of low-density clasts, matrix support of large clasts and the presence of fine-grained basal layers. Deposition in high-concentration sediment gravity flows, such as debris flows, occurs by en masse freezing when the gravitational force can no longer overcome the strength of the material. Sparks (1976) proposed this type of depositional mechanism for pyroclastic flows of the high-aspect ratio type. The deposit is thus viewed as the frozen remnant of the flow itself, much the same way as a lava flow solidifies upon cooling.

The en masse freezing model may be applicable to some PFDs but clearly there are other deposits that require different depositional processes, such as the Taupo ignimbrite. In order to explain the complex facies variations and stratigraphic relations of the Taupo ignimbrite, Wilson and Walker (1982) proposed a model that included different depositional regimes within a moving pyroclastic flow. They argued that, like other types of gravity currents, pyroclastic flows are likely to consist of a head, body and tail from which distinct depositional facies may arise (Fig. 9). In the head, fluidization is enhanced by the entrainment and heating of atmospheric air and the combustion of injested vegetation. High-gas velocities and turbulent conditions favor the segregation of dense clasts to the base of the head and the relative depletion of fine ash. Deposition of the segregated material produces lithic-enriched ground layers, such as 1H, which are then overridden by the main body of the flow (Walker and others, 1981). Another process occurring in the head is the periodic jetting of the flow front caused by the rapid heating and expansion of entrained air. The stratigraphic and grain-size relations of the 1P layer of the Taupo suggest that they were formed by the deposition of

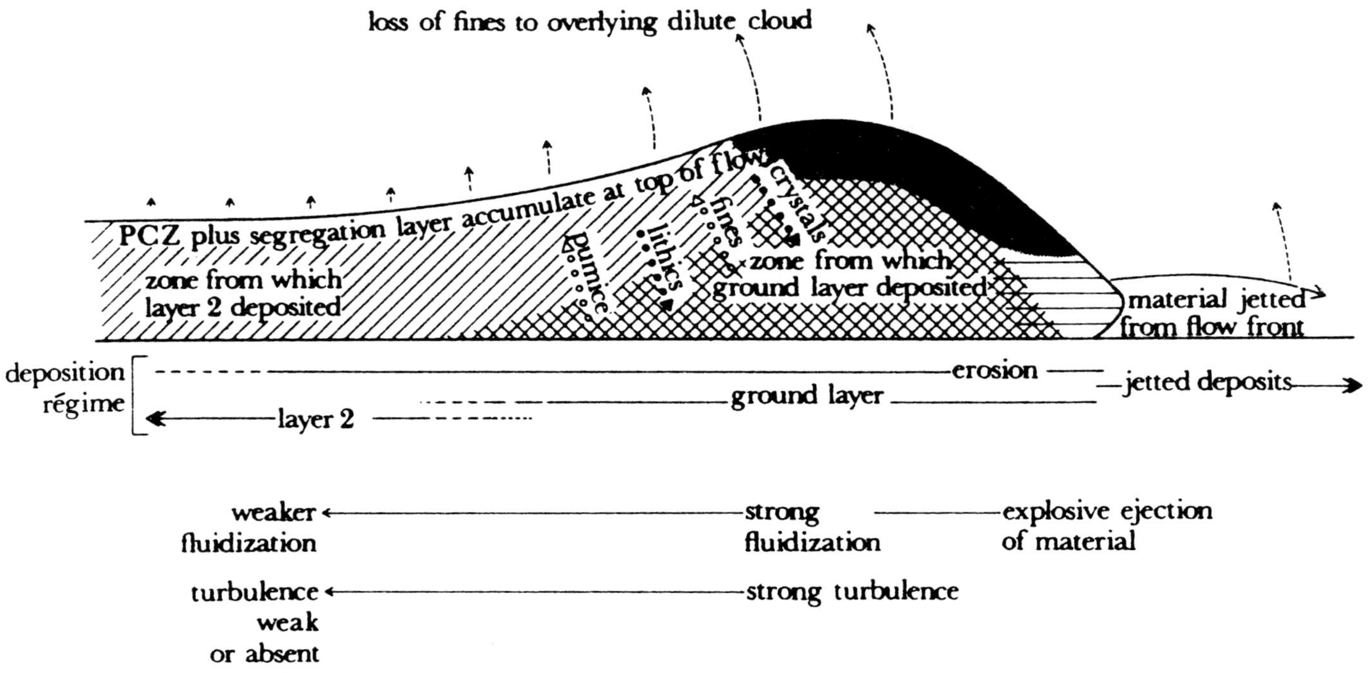


FIG. 9.—Schematic diagram illustrating the development of different depositional regimes at the front of the Taupo pyroclastic flow, from Wilson (1985). See text for discussion of the deposits that result from different parts of the flow.

jetted material that experienced some degree of fines depletion (Walker and others, 1980b). This phenomenon was vividly described by F. Perret during his observations of pyroclastic flows from Mont Pelée in 1929 as a "curious rolling and puffing movement which at the immediate front takes the form of forward-springing jets, suggesting charging lions." Thus, Wilson and Walker (1982) attribute the formation of layer 1 deposits principally to the highly fluidized head region of pyroclastic flows.

Layer 2 deposits of the Taupo, which include both the valley-pond ignimbrite (VPI) and the ignimbrite veneer deposit (IVD), are attributed to deposition from the body/tail of the flow (Fig. 9). The IVD is a somewhat peculiar layer in that it possesses features of all the major pyroclastic-deposit types. It mantles topography like a fall deposit, is stratified like a pyroclastic surge, yet poorly sorted like a pyroclastic flow (Walker and others, 1981). Deposition of an IVD has been interpreted as a layer-by-layer process caused by the variation in flow thickness as the flow passes a single depositional site (Wilson, 1985). As flow thickness increases, high shearing rates at the base result in reduced deposition. Decreasing flow thickness favors enhanced deposition as basal shearing rates are reduced. Variations in the flow thickness are proposed to occur on two scales. One is that the flow consists of separate batches or pulses of material and the second is that waves develop on the upper free surface of the flow. The major difference between this model and that of Sparks (1976) is that IVDs represent only a fraction of the total flow that passes a given depositional site and are not representative of the grain-size characteristics or total thickness of the parental flow. Deposition occurs predominantly from the base of the flow tail, which is slowed down relative to the main flow by its interaction with the ground. Formation of the IVD is thus more similar to the layer-by-layer depositional model of pyroclastic flows proposed by Fisher (1966, 1979).

Characteristics of the valley-pond ignimbrites, such as their massive structure, poor sorting, and pumice concentration zones, indicate that the deposits were derived from only moderately expanded flows, yet there is other evidence to suggest that the flows that passed the areas of VPI deposition were much thicker than the resulting deposits. To resolve this apparent paradox, Wilson and Walker (1982) proposed that the VPIs were derived from the flow tail but that the greater thickness could be related to a combination of a thicker flow tail in the valley floors, drainage of the upper part of the flow tail into the valleys, and the blockage of valley constrictions by logs or large pumices.

The existing depositional models of pyroclastic flows represent two ends of a spectrum in which PFDs are viewed as either frozen remnants of the moving flow or as fractional parts of the flow as it passed any depositional point. In the latter case, the deposits cannot be used to infer the thickness of physical properties of the source flow quantitatively. The difference in the facies relationships of pyroclastic flows, and in particular the formation of layer 1 deposits, has been related to the energy of flow emplacement (Wilson and Walker, 1982). High-velocity flows, such as the Taupo, ingest and heat large amounts of air, which upon expansion lead to strong fluidization in the head region and the production of jetted deposits and lithic-rich ground layers (Fig. 10). As the velocity of the flow front is decreased, the extent of fluidization in the head is diminished and the development of jetting and strong segregation is reduced. At very low velocities, ingestion of air is not significant and layer 1 deposits are formed (Fig. 10). The lack of layer 1 surge deposits (Fig. 3) in the Taupo deposits can be attributed to the high velocity of the flow front and the inability of a dilute surge to move in advance of the flow.

Depositional mechanisms discussed so far strictly apply to pyroclastic flows once they have developed away from the vent. In the near-vent region, the generation of pyro-

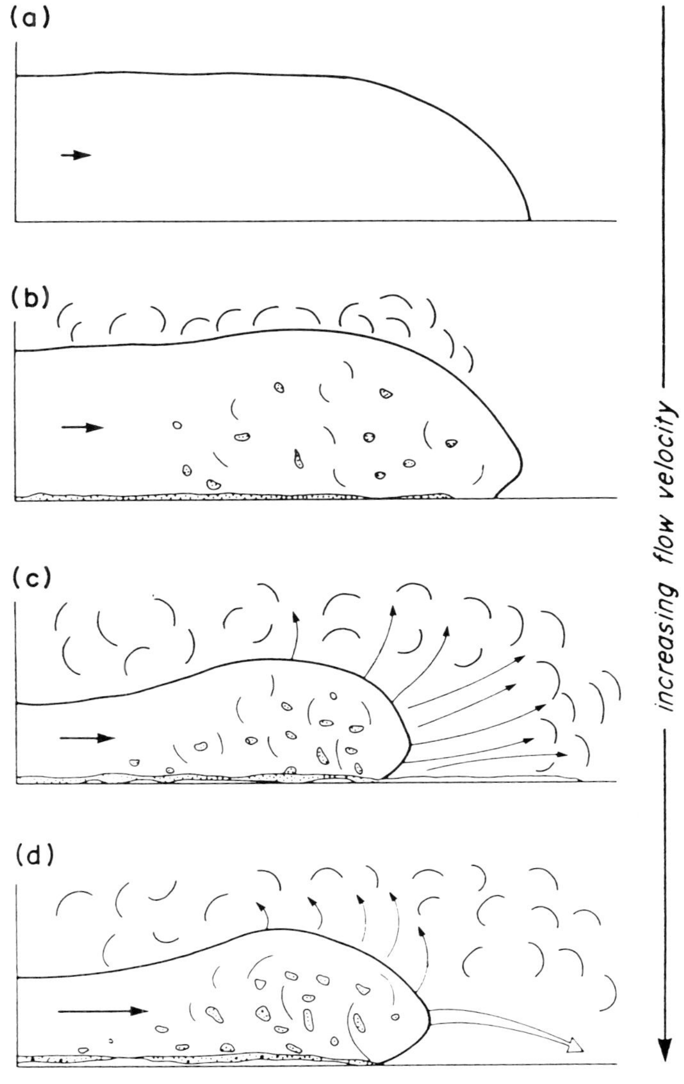


FIG. 10.—Schematic representation of the effect of increasing flow velocity on the extent of fluidization and the development of layer 1 depositional facies, from Wilson and Walker (1982). a) low-flow velocity, no significant air ingestion, no layer deposits, b) limited air ingestion, ground layer formation, c) moderate air ingestion results in formation of ground layers and dilute surges, d) strong air ingestion, highly fluidized head and the production of ground layers and jetted deposits.

clastic flows is thought to take place by deflation of a turbulent, collapsing mixture of gas and particles (Sparks and others, 1978). As deflation occurs, particle concentration increases, fine ash is lost by gas fluxing, and larger clasts settle toward the ground. This region, which Walker (1985) terms the deflation zone, is the area where coarse co-ignimbrite lag deposits are formed (Fig. 11). Several of the features of these breccia provide important constraints on the depositional processes occurring in this region. The breccias typically contain large lithic blocks (>50 cm diameter), are fines-depleted and often stratified. In addition, lateral variations of thickness, especially in the vicinity of pre-existing valleys, confirm a lateral flow component during the formation of the breccias. Walker (1985) suggests that lag breccias are the lateral equivalents of more distal pyroclastic flows. They form by the deposition of large clasts that are too heavy to be transported away from the source region in the deflating pyroclastic flows. Fines depletion is caused by the strong gas fluxing associated with deflation of the collapsing turbulent mixture of gas and particles. As the flow moves away from the source, large clasts are lost from the base (Fig. 11). Trapping of pumice and lithic clasts along the ground favors the production of a deposit in which the particles are not pneumatically equivalent because small pumices can lodge in spaces between large lithics, whereas large pumices will be carried along in the traction load (Walker, 1985).

In some eruptive sequances, lag breccias are interbedded with normal pyroclastic-flow deposits. Such breccias could be interpreted as ground layers originating from the overlying pyroclastic flows. Walker (1985) stresses, however, that lag breccias differ from lithic-enriched ground layers discussed earlier because lag breccias are the lateral equivalents of medial and distal pyroclastic flows and not formed by fluidization processes in the head of a moving flow. The occurrence of breccias interbedded with pyroclastic flows has thus been interpreted as an expansion of the deflation zone caused by an increase in the magma discharge rate at the source during climax of the eruption and may be associated with the onset of caldera collapse (Walker, 1985).

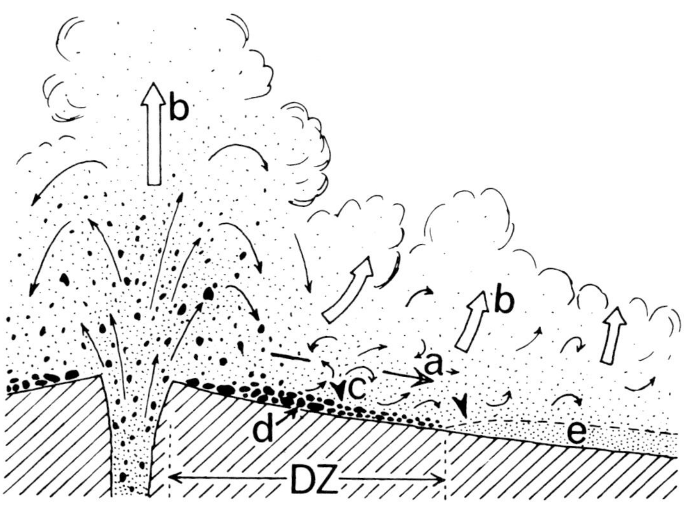


FIG. 11.—Diagram showing the proposed site of lag-breccia formation around a collapsing eruption column, from Walker (1985). *DZ* is the deflation zone where lag breccias accumulate. *a* = lateral motion of turbulent collapsing mixture or gas and pyroclasts, *b* = loss of gas and fine particles, *c* = settling of particles through the mixture, *d* = lag-breccia deposit, *e* = high-particle concentration flow.

### *Welding*

An important post-depositional process affecting some PFDs is welding and compaction (e.g., Ross and Smith, 1961; Smith, 1960b, Ragan and Sheridan, 1972). High temperature, low-clast viscosity, and high lithostatic load are conducive to plastic deformation and sintering together of juvenile material in flows. Pumice clasts can be flattened to form fiamme, imparting an overall foliation or eutaxitic texture to the deposit. In hand specimen, some densely welded flows are difficult to distinguish from glassy lava flows and microscopic examination may be needed to correctly assess the origin. Another feature that welded flows share with lava flows is the development of columnar jointing. Welding results in significant increases in the bulk density and a concomitant decrease in the porosity of a deposit. These parameters change systematically as a function of height in a cooling unit, with the densest zones usually occurring in the lower central parts (Ragan and Sheridan, 1972).

## PYROCLASTIC-SURGE DEPOSITS

It is generally accepted that pyroclastic surges are low-concentration mixtures of volcanic particles, gases and water that are highly expanded and flow in a turbulent fashion as a gravity flow (Cas and Wright, 1987). Field studies and observations of historic eruptions suggest that there are at least three types: base, ground and ash cloud. Base surges are typically associated with phreatomagmatic eruptions and develop as a radially spreading cloud from the base of an eruption column (e.g., Moore, 1967; Waters and Fisher, 1971; Schmincke and others, 1973). Ground surges are found at the base of pyroclastic-flow deposits and have been interpreted as having a genetic relationship to the overlying flow. One interpretation is that they develop by gravitational segregation of an initially homogeneous pyroclastic density current into a denser, slower moving basal pyroclastic flow and an overlying, more dilute turbulent surge, which then moves ahead of the flow and is eventually overridden by it (Sparks and Walker, 1973). Other theories for the formation of ground surges include the initial collapse of fine-grained outer parts of an eruption column (Fisher, 1979), overpressured blasts (Wohletz and others, 1984), turbulence at the base of a pyroclastic flow (Valentine and Fisher, 1986), or turbulence at the head of a pyroclastic flow (Valentine and Wohletz, 1979). Ash-cloud surges are envisaged to form by the elutriation of fine ash from the main body and tail of a pyroclastic flow and move in a turbulent state, but at a lower velocity than its parental flow, so that deposition occurs on top of the sequence (Fisher, 1979).

Pyroclastic-surge deposits (PSDs) are more restricted in their range of volume and areal extent than pyroclastic-flow deposits. Base-surge deposits typically extend to distances of 10 km or less and individual eruptive sequences reach up to several meters close to the source but thin rapidly away (Moore, 1967). They are typically associated with eruptions of basaltic or basaltic-andesite magma and form an important component of volcaniclastic sequences surrounding tuff rings, maars, and some stratovolcanoes that have experienced phreatomagmatic activity. On the other hand, because ground and ash-cloud surges are usually intimately associated with pyroclastic flows, some can extend considerable distances away from the source. For example, Valentine and others (1989) have described ground-surge deposits at the base of the Peach Springs Tuff at distances up to 100 km from the source. Similarly, Fisher (1979) has described ash-cloud surges from the Bandelier Tuff at a distance of 20 km from the source caldera. For the most part, ground and ash-cloud surges rarely exceed more than a few meters in thickness over the majority of their extent.

In contrast to pyroclastic flows, surges are highly expanded, turbulent flows and are therefore less strongly controlled by the topography over which they travel. Individual eruptive sequences thus produce deposits that are radially distributed around the source volcano. For example, during the 1982 eruptions of El Chichón volcano in Mexico, two sets of hot pyroclastic surges were generated and spread out in all directions to a distance of about 7 km from the source (Fig. 1). Strandlines on ridges near the distal end of the surge path show that the moving cloud was at least 300 m thick (Sigurdsson and others, 1984).

The incorporation of water during the formation of base surges has an important effect on the physical properties of the flow. Conversion of water to steam at the source vent utilizes a large fraction of the thermal energy contained in the magma, resulting in mixtures that are relatively cool compared with ground or ash-cloud surges. Temperatures within the surge cloud may be as low as 100°C or less and condensation of steam to water produces an additional component in the turbulent mixture. Thus, surges are often discussed in terms of wet (abundant steam, and water) and dry (hot, water-poor) varieties. The occurrence of accretionary lapilli, soft-sediment deformation structures and the plastering of mud onto objects are taken as diagnostic features of wet surges (Walker, 1984).

### *Sedimentologic and Stratigraphic Relations*

Surge deposits exhibit a wide variety of sedimentary features that reflect the nature of the source eruption, composition of juvenile and lithic material, transport mechanisms, depositional processes, and the influence of local topography. One of the most characteristic features of pyroclastic surge deposits is the occurrence of wave-like or dune structures that show complex internal stratification (e.g., Fisher and Waters, 1970; Schminke and others, 1973; Sheridan and Updike, 1975; Sigurdsson and others, 1987). In addition, however, surges also form at least two other types of bedforms: massive and planar (Wohletz and Sheridan, 1979). These forms can be difficult to identify as surge deposits, especially at distal localities, because other processes such as ash fallout can produce layers with similar appearance.

Dune structures exhibit a complex variety of forms that are influenced both by the moisture content of the surge cloud and its velocity. Base-surge deposits, for example, contain evidence of high-water content in the form of accretionary lapilli and the plastering of cohesive mixtures of wet tephra on objects in the flow path. The eruption of Taal volcano in 1965 produced a radially distributed sequence of surge deposits that is dune bedded within 3 km of the vent (Fig. 12). Dunes are generally aligned perpendicular to the direction of flow and decrease in wavelength from 19 m at the source to 4 m at 2.5 km away. Within this zone, trees and other obstacles were aerodynamically coated with up to 40 cm of thinly stratified mud. Sandblasting of trees also occurred over the entire area affected by the surges, but there was no carbonized vegetation, indicating a relatively low temperature during deposition. In cross section the dunes were asymmetrical with a slightly eroded stoss side steeper than the more gently sloping lee side.

In contrast, the 1982 eruptions of El Chichón volcano in Mexico produced dune-bedded surge deposits under conditions of high temperature, as shown by the carbonization of vegetation throughout their extent (Sigurdsson and others, 1987). The dunes form transverse, sinuous ridges up to 100 m in length with horizontal form index (breadth/wavelength) ranging from 1 to 10. Stoss sides are typically steeper than lee sides and show evidence of erosion. The internal structure consists of multiple wavy bedsets that crosscut one another, indicating that the overall dune form resulted from successive episodes of erosion and deposition. Many of the individual beds consist of distinctive

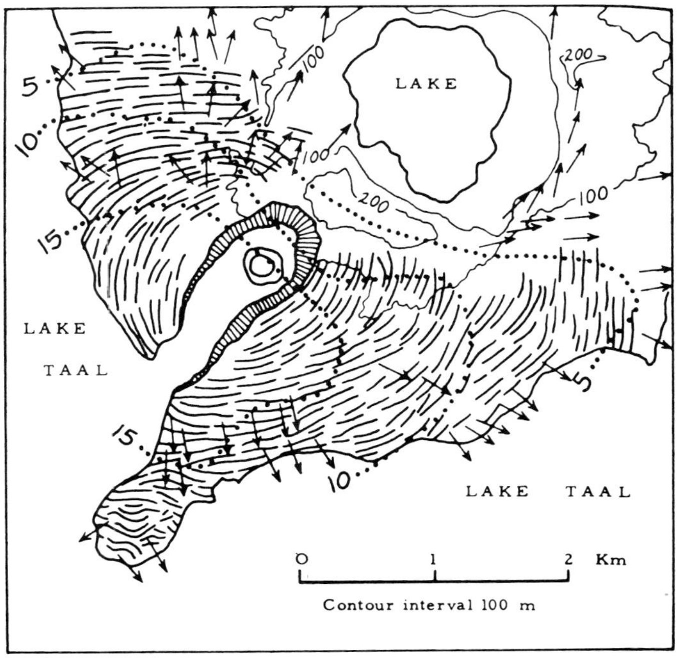


FIG. 12.—Orientation of dune crests in surge deposits from the 1965 eruption of Taal volcano, from Moore (1967). Arrows show the direction of surge movement and the dotted contours are the wave lengths of dunes in meters.

moderately to poorly sorted, reversely graded units with well-rounded pumice clasts.

Allen (1984) has suggested a general classification of dune beds or sand waves from surge deposits based on the migration direction of successive wave crests (Fig. 13). Progressive types migrate away from the source and have cross-stratification that is steeper on the lee side than on the stoss. Regressive types, which migrate toward the source, show the opposite relationship with the steepest cross-stratification on the stoss side. Intermediate between these two forms are stationary types with a symmetrical cross section and similar dip on both the stoss and lee sides.

The grain-size characteristics of pyroclastic-surge deposits are extremely variable as a result of differences in temperature, water content, magma composition, and transport mechanisms of source eruptions. It is often pointed out that surge deposits occupy a field intermediate between pyroclastic flow and fall deposits on a plot of sorting versus median diameter (Fig. 14). Although such a diagram is useful for discriminating some deposits, there is considerable overlap between fields and use of such a diagram in isolation of other criteria may lead to problems. In a stratified-surge sequence, individual beds may be quite well sorted by pneumatic equivalence, but the sequence as a whole could be very poorly sorted owing to the pulsatory nature of particle transport and deposition (Walker, 1984; Cas and Wright, 1987).

A major problem in the characterization of the grain-size relationships of surge deposits is the identification of beds deposited from a single surge event. Walker (1984, p. 282) defines the term bedset as "the assemblage of beds, and the dune structure to which they belong, which can be interpreted to be the product of single surge." The variance in median diameter of individual beds in surge bedsets ranges from 1 to 6 phi and is negatively correlated with the median diameter (Fig. 15, Walker, 1984). Wet deposits are generally finer grained, exhibit less variance, and contain individual beds that are more poorly sorted compared with deposits from hot, dry surges. The poor sorting and reduced variance between beds in wet deposits is attributed to the cohesive effects of water in the surge cloud. However, the high variance in bedsets and coarser grain size of dry surges are interpreted as both a lack of water and a higher velocity of emplacement than in wet surges (Walker, 1984).

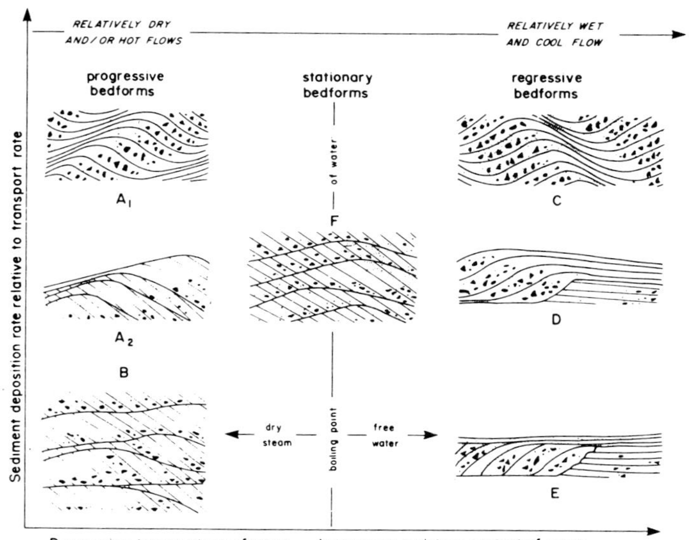


FIG. 13.—Classification scheme of pyroclastic surge bedforms based on sediment deposition rate, temperature, and moisture content, from Allen (1984).

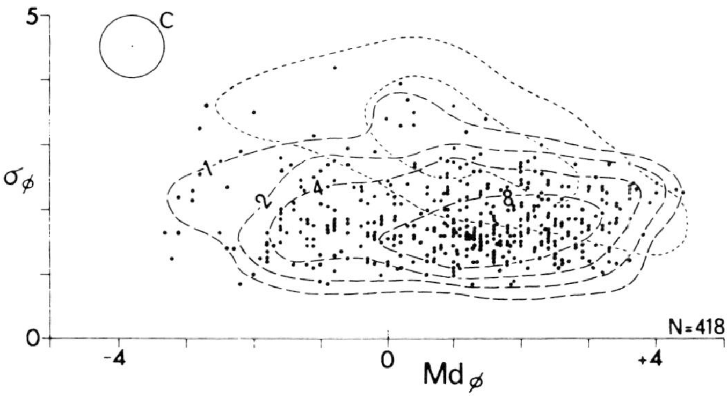


FIG. 14.—Plot of sorting and median grain size for individual beds of dune-bedded surge deposits, from Walker (1984). The field is contoured (dashed lines) as the percentage of the total number lying within a circle of size C located at any point. Dotted lines are the contours for the field of pyroclastic flows for comparison.

## *Transport and Particle-Support Mechanisms*

The sedimentary features of surge deposits provide important constraints on the mechanisms by which particles

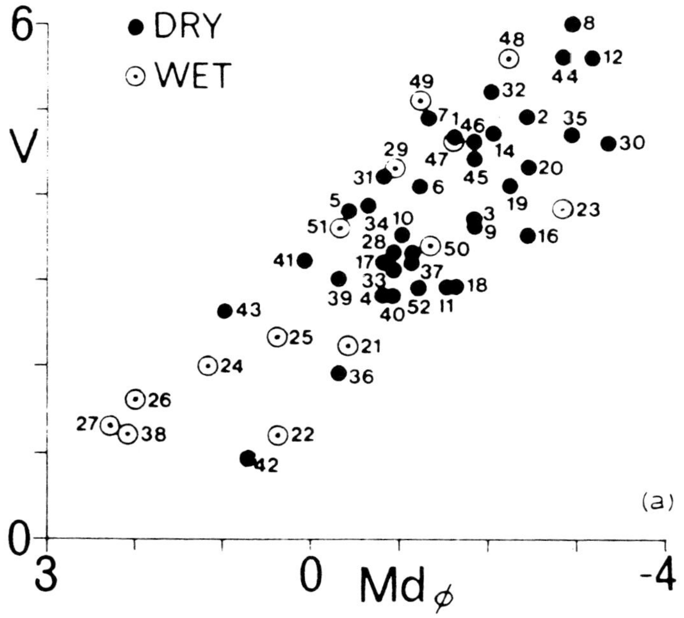


FIG. 15.—Relationship between the variance and median diameter of the coarsest bed in dry- and wet-surge bedsets, from Walker (1984). Variance is the maximum difference in median diameter of individual beds within a bedset sequence. The increase in variance with coarsest bedset is interpreted to reflect increasing velocity of emplacement.

are transported and eventually deposited. By analogy with other types of sediment gravity flows, the cross-stratified dune structures observed in PSDs are good evidence that particle support is dominantly by turbulence of the interstitial gases and that the concentration of grains is relatively low. Low-particle concentration allows for the improved sorting in surge clouds compared to high-concentration pyroclastic flows. Turbulence within surge clouds is maintained by the conversion of gravitational potential energy into kinetic energy during flow downslope and the very low effective viscosities of highly expanded mixtures of gases and particles. In contrast to pyroclastic flows, fluidization is not thought to play an important role in the support of particles because of the high-gas velocities required to fully fluidize material of surge deposits. Such velocities would result in extensive depletion of fines and very good sorting (Sigurdsson and others, 1987).

Gravity plays an important role in driving the flow of a surge cloud by acting on the body much as it would drive a current of dense homogeneous fluid downslope through a less dense fluid (Simpson, 1987). Surges, however, are not single-phase fluids but are mixtures of material with greatly contrasting density. Thus, while the flow as a whole is moving under the influence of gravity, the mixture itself is responding to gravitational forces by internal segregation of particles. As a surge moves away from the source, it will become stratified with respect to particle concentration and is best considered as a stratified flow (Valentine, 1987).

Support of large particles by turbulence in surges is generally argued as an efficient transport mechanism for only limited distances from the source (Walker, 1983; Sparks, 1983; Sparks and others, 1978). In the deposits of the May 18, 1980, eruption of Mount St. Helens and the May 8, 1902, eruption of Mt. Pelée, for example, transport of large clasts by turbulent suspension has been called into question based on calculations of clast settling in turbulent flow (Walker and McBroome, 1983; Walker and Morgan, 1984; Sparks, 1983). Turbulent support of particles is a function of the particle Rouse number (Pn)

$$Pn_i = w/ku^*$$

where $Pn_i$ is the particle Rouse number of a particle of size $i$, $w$ is the particle settling velocity in the flow, $k$ is Von Karman's constant (0.4), and $u^*$ is shear velocity of the flow. For particle support to be effective, $Pn_i$ must be less than 2.5 (Middleton and Southard, 1978). In the case of the Mount St. Helens eruption, measurements of flow velocity (up to 300 m/s) and realistic estimates of surface roughness elements suggest that lithic clasts up to 10 cm in diameter could have been transported by turbulent suspension (Fig. 16). Clasts of this size correspond to the largest ones measured by Moore and Sisson (1981) in the blast deposit and can be found at distances up to 12 km from the volcano. However, the blast cloud of the May 18 eruption was driven both by gravity and the supersonic expansion of the mixture (Keiffer, 1981), which contributed to the energetic transport capabilities of the flow. For the Mont Pelée flow, LaJoie and others (1989) have pointed out that the turbulent suspension of particles could be accounted for by assuming a higher density at the base of the flow.

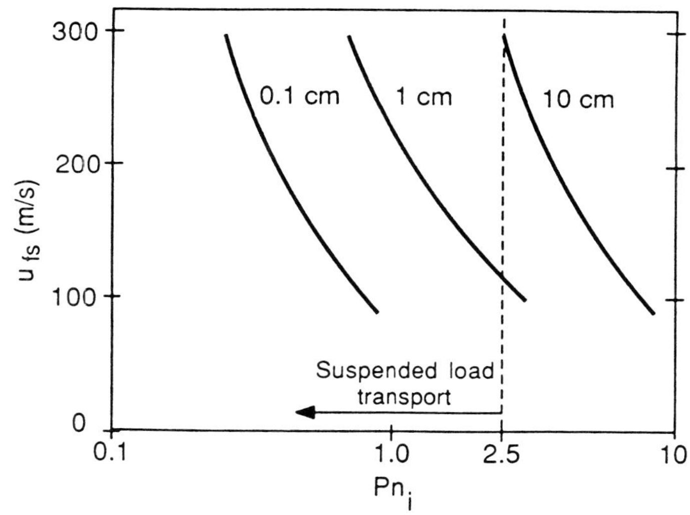


FIG. 16.—Particle Rouse numbers ($Pn_i$) for 0.1-, 1- and 10-cm lithic clasts of the 1980 Mount St. Helens blast as a function of free-stream velocity ($U_{fs}$), from Valentine (1987). Turbulent suspension is effective for values of $Pn_i$ less than 2.5. A 10-cm lithic clast would require a 300-m/s stream velocity for turbulent suspension.

In addition to turbulent suspension, particles in surge clouds are moved along the ground surface by saltation and rolling where high lateral shear velocities are developed (Cas and Wright, 1987). The high degree of grain rounding in the El Chichón surge deposits indicates that rolling, sliding, and saltation of grains along the substrate were particularly important processes (Sigurdsson and others, 1987). Under conditions of high-particle concentration at the base of a surge cloud, transport may also evolve to a series of thin, bed-load grain flows (Fisher and others, 1987; Valentine, 1987; Savage, 1984). There are thus two regions of the surge cloud where different transport mechanisms are likely to operate. At the base, the largest clasts are moved by traction, saltation and thin grain flows, while above, a thick region of turbulent suspension moves the smaller clasts. During transport, there is an exchange of material between both regions as large clasts settle into the traction-saltation domain and re-entrainment of smaller clasts from the base moves them into turbulent suspension.

### *Depositional Processes*

A fundamental difference between pyroclastic flow and surge deposits is that in some instances the density and grain-size distribution of flow deposits correspond closely to the source flow because of their poorly inflated nature, whereas surge deposits represent partial deposition of material from a restricted part of the surge cloud (Fisher, 1986). Deposition from surge clouds will occur when velocities in the basal boundary layer or in the overlying turbulent eddies are less than that required to support particles. During movement away from the source, the velocity of a surge will decrease because of 1) density reduction through sediment loss, 2) reduction of flow thickness, 3) friction at the base and upper surface of the flow, or 4) reduction in the bed slope (Cas and Wright, 1987). Reduction in the surge velocity will cause material to be transferred from suspension transport to traction transport.

The origin of dune-bedded deposits from surge clouds is central to an understanding of the depositional processes of pyroclastic surges. Early studies of surge deposits described some dune-bedded forms using the same terminology that is applied to bedforms developed in fluvial flow over a bed of cohesionless particles, such as antidunes and chute and pool structures (Crowe and Fisher, 1973; Schmincke and others, 1973). Some studies utilized equations developed for aqueous flows in order to calculate the flow regime and velocity of surge clouds. Allen (1984) has questioned the comparison of surge bedforms with fluvial types because of the potential effect of water in wet surges on the interaction between the cloud and the bed surface. The development of regressive structures that migrate toward the source in wet surges may have more to do with the cohesive nature of a water-tephra mixture than whether the flow regime is supercritical or subcritical. Fisher and Schmincke (1984) point out, however, that chute and pool structures in deposits from Laacher See volcano are only found in proximal coarse-grained facies, whereas farther from the source the same beds exhibit progressive bedforms that contain accretionary lapilli. At this time, the relationship between flow velocity and bedforms is not sufficiently known to exclude either view.

Deposition from hot, dry surges may be less complicated because of the lack of abundant water. Several workers have suggested that the development of dune bedforms from dry surges is somewhat analogous to the production of eolian dunes (Allen, 1984; Sigurdsson and others, 1987). The formation of eolian dune forms, such as ballistic ripples, occurs as particles settle out on the lee side of dune crests and then avalanche as a result of oversteepening (Allen, 1984). Avalanching of accumulated material lowers the leeward slope, can result in reverse size grading of beds, as observed in the El Chichón surges (Sigurdsson and others, 1987), but produces cross-bedding with an angle-of-repose slope. Although some dry surges exhibit relatively steep lee side cross-bedding, most are of shallow angle. The comparison should be viewed with caution, however, as there are important differences between the eolian and surge systems that are likely to influence the internal characteristics of dune bedforms in each case. In eolian systems, wind moves highly sorted sediment by a driving force that is primarily a thermally induced pressure difference. Surges, on the other hand, are more concentrated mixtures of volcanic gases, entrained air, and poorly sorted particles that move because of the density contrast of the bulk cloud relative to the surrounding atmosphere. Particle concentration, flow velocity, and variance of particle size are all much greater in surges than in eolian systems. Thus, movement of particles in surges is not just by rolling and saltation of individual grains, but also involves a traction carpet of thin grain flows moving in a pulsatory fashion. Movement of such flows down the lee side of dune crests under conditions of high-shear velocity could produce the lower angle cross-bedding commonly observed in many deposits (Cas and Wright, 1987).

In fluvial systems, dune bedforms are linked to waves that form on a free surface (Allen, 1984). Antidunes, for example, develop in phase with standing waves of supercritical flow (Froude number $> 1$). There is, however, no evidence that a sharp density interface exists on the upper surface of a surge cloude. As previously mentioned, surge clouds may be best viewed as a continuously stratified flow with respect to particle concentration. Valentine (1987) proposed that in stratified surge clouds the development of dune bedforms is related to the propagation of internal waves. The maximum frequency of internal waves in a stratified flow is known as the Brunt-Vaisala frequency ($N$) expressed as

$$N = \frac{1}{2\pi}\left(\frac{-g\, d\rho}{\rho\, dy}\right)^{1/2} \quad (1)$$

where $g$ is the acceleration due to gravity, $\rho$ is bulk density, and $d\rho/dy$ is the vertical density gradient in the flow. Internal waves may have a frequency less than $N$, but cannot exceed it. The wavelength scale of surge bedforms is thought to reflect minimum internal wave propagation in the near-bed region (Valentine, 1987). It can be seen from equation (1) that the frequency is a function of both height in the flow and the particle Rouse number, which governs $d\rho/dy$. During runout of a surge cloud, velocity and turbulence will decrease and result in a higher particle Rouse number and increased density gradient at the base of the flow. These changes would cause a decrease in the maximum wavelength scale of bedforms with increasing distance from source, as documented in many surge deposits of both the wet and dry variety (Moore, 1967; Waters and Fisher, 1971; Sigurdsson and others, 1987).

### *Facies Variations*

Systematic changes in the bedforms of surge deposits as a function of distance from the source have been documented primarily for phreatomagmatic eruptions. Wohletz and Sheridan (1979) defined three facies (sandwave, massive and planar) based on the relative abundance of sandwave, massive, and planar beds. The sandwave facies is found in proximal areas and consists of abundant sandwave beds and minor massive beds. At medial distances all three bedforms occur in the massive facies. Distal localities are characterized by the planar facies, where planar beds are dominant and massive beds are present in minor amounts.

Wohletz and Sheridan (1979) suggest that the observed facies relationships record a progressive change in the nature of particle transport and deposition as a surge cloud deflates by gas loss (Fig. 17). Close to the source, the surge cloud is a highly expanded mixture with a void fraction of 0.95. The wide separation of particles favors transport by saltation and other traction processes (Sheridan and Updike, 1975). Deposition in this regime results in sandwave-type beds with a few massive interbeds originating from local variations in particle concentration. With continued gas loss and settling of particles through the cloud, there is an increase in particle concentration and decrease in void space at the surge base. Particle interaction becomes more

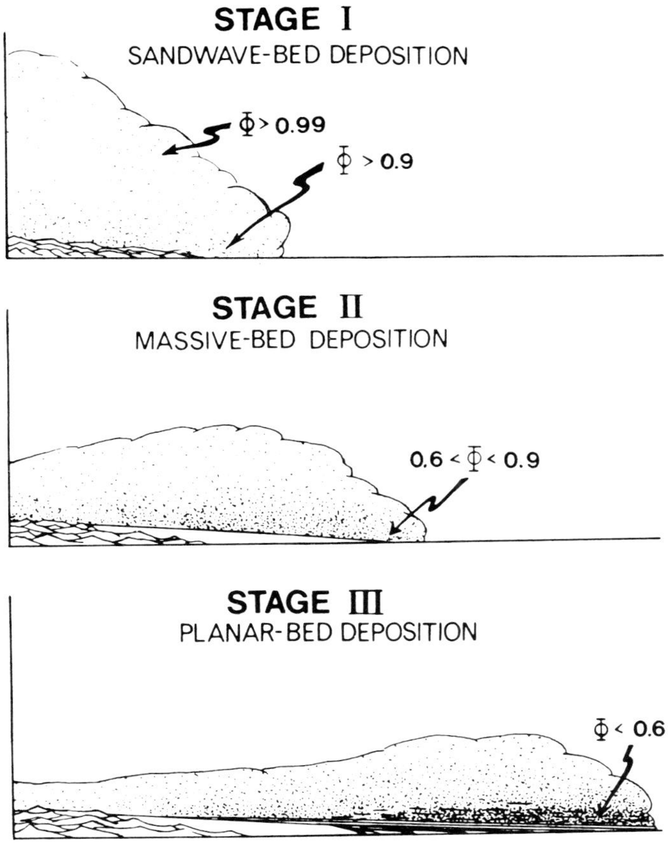


FIG. 17.—Progressive deflation model of surge cloud evolution, from Wohletz and Sheridan (1979). Proximal deposition (stage I) produces sandwave beds from a highly inflated cloud (void fraction, $\Phi = >0.9$). As deflation occurs, particle concentration increases ($0.6 < \Phi < 0.9$) and massive beds are deposited at medial distances. In distal areas, high-particle concentration ($\Phi < 0.6$) leads to the deposition of planar beds.

important and deposition is divided between high- and low-particle concentration regimes. Deposits formed under these conditions are a mixture of sandwave, massive and planar beds. At the distal margin of the surge cloud, deflation leads to a high basal-particle concentration with a void fraction of less than 0.6. Deposition from this regime results in the formation of reversely graded planar beds.

Valentine (1987) presents an alternative view of the facies distribution based on the theory of stratified flow. His model is similar to that of Wohletz and Sheridan (1979) in that the change in bedforms is attributed to an increasing particle concentration at the surge base with greater distance from source. However, the mechanism for increasing particle concentration is proposed to be the reduction in turbulent transport potential of the surge and not the loss of fluidizing gases (deflation). As the surge moves away from the source, turbulence in the basal region is damped out as the concentration of particles increases by gravitational settling.

The facies model of Wohletz and Sheridan (1979) was developed mostly from data on basaltic phreatomagmatic eruptions and does not appear to hold for more widely dispersed hot/dry surges such as those produced at El Chichón in 1982 (Sigurdsson and others, 1987). The El Chichón surges are characterized by sandwave facies throughout their entire extent. Clearly there is much to be learned about the facies relationships of surge deposits. At present there are only a handful of studies that have systematically examined surge bedforms as a function of distance from source. Considering the great differences in the properties of the source materials at different volcanoes and the complicating effect of steam- to water-phase transitions in "wet" versus "dry" types of surges, it is unlikely that a single facies model will apply to surge deposits in general.

## DISCUSSION AND CONCLUSIONS

In the previous sections, the transport and depositional mechanisms have been shown to be quite different at each end of the particle-concentration spectrum. However, deposits are often encountered that bear the characteristics of both high- and low-concentration flows (e.g., Waitt, 1981). The extent to which the processes overlap or act together in flows of intermediate particle concentration is currently a controversial topic in the field of volcanology. A good example of this problem is the analysis of the May 18, 1980, Mount St. Helens blast deposit, where detailed studies have led to two greatly contrasting interpretations of the transport and depositional processes. One group believes that the blast deposit was emplaced by a highly energetic pyroclastic flow (Walker and McBroome, 1983; Waitt, 1984; Walker and Morgan, 1984), whereas another advocates deposition from a turbulent, pyroclastic surge (Fisher and others, 1980; Hoblitt and others, 1981; Moore and Sisson, 1981; Fisher and others, 1987).

Some facies of the May 18 blast deposit clearly have the characteristics of pyroclastic flows. For example, in Coldwater Creek there are deposits of "blast pyroclastic flow" that are up to 6 m thick, contain gas-escape pipes, and have flat tops (Hoblitt and others, 1981; Fisher and others, 1987). However, grain-size characteristics of the flow-like facies indicate more extensive fine-ash loss than typical pyroclastic flows (Walker and McBroome, 1983). A characteristic feature of the 1980 blast deposit is the occurrence of a coarse-grained, relatively fines-depleted basal layer (A1) that is overlain by a finer grained, more poorly sorted and occasionally stratified sequence (A2, Fig. 18). Dune bedforms are also present but are not as extensively developed as in other well-documented surge deposits.

A large part of the St. Helens controversy centers around the ability of pyroclastic density currents to undergo large changes in their particle concentration by flow transformation (Fisher, 1983). Both groups agree that high-concentration pyroclastic flows can transform into surges by expansion of entrained air and primary gases. They differ, however, on whether low-particle-concentration surges can transform into high-concentration pyroclastic flows. Walker and McBroome (1983) argue that surges cannot deflate to form pyroclastic flows because loss of fine ash would be too extensive to maintain transport by partial fluidization. Those who propose a surge origin for the Mount St. Helens blast attribute the pyroclastic flow facies to localized regions of high-particle concentration at the base of the cloud

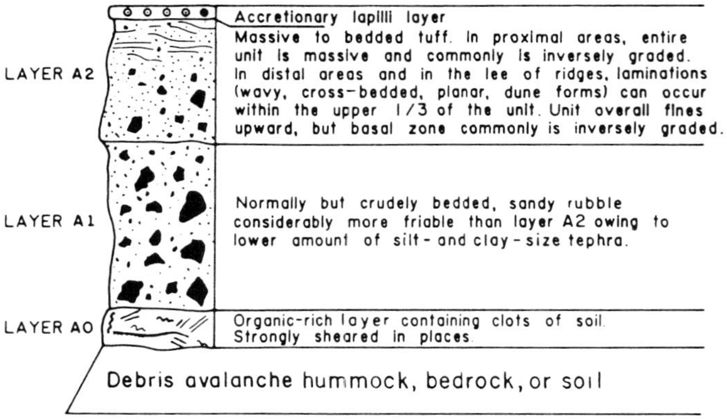


FIG. 18.—Composite stratigraphic section of the May 18, 1980, Mount St. Helens avalanche and blast deposit in the vicinity of South Coldwater Creek, north of the volcano, from Fisher and others (1987).

that evolves by gravitational segregation during flow (e.g., Hoblitt and others, 1981; Fisher and others, 1987; Valentine, 1987).

It may be that highly expanded surge clouds that produce characteristic dune-bedded deposits cannot deflate to high-concentration pyroclastic flows, but that does not necessarily preclude the deflation of moderately inflated flows to particle concentrations close to that of a true pyroclastic flow. A key factor in this argument concerns the state of the initial fragmented mixture as it begins to flow. There is no a priori reason for starting mixtures to be either highly or poorly expanded, i.e., to begin at either end of the spectrum. The degree of expansion will be a complex function of fragmentation processes, dissolved volatile content, lithic abundance, magma discharge rate and the mechanism of flow generation (i.e., column collapse, dome collapse, or Vulcanian explosion). It seems likely, if not inevitable, that there will be a broad range in the initial degree of expansion of pyroclastic density currents. Thus, problems in the transformation to either end member may not be insurmountable for flows with particle concentrations that start out between the two.

In the case of the Mount St. Helens blast deposit, it is possible that elements of both the flow and surge models may apply. It is expected that the initial flow was not a homogeneous mixture but that lateral variations in particle concentration existed as depressurization occurred. The flow expanded rapidly away from the volcano as a ground-hugging cloud at velocities up to 325 m/s (Keiffer, 1981). Its height did not exceed 500 m until it lifted buoyantly off the ground, after travelling 25 km from source (Sparks and others, 1986). The flow consisted mostly of juvenile, poorly vesiculated dacite fragments and lithics from the cone of Mount St. Helens (Moore and Sisson, 1981). Some parts may have been poorly expanded and moved by partial fluidization, resulting in some of the proximal flow-like facies, as suggested by Hoblitt and others (1981). Other, more expanded portions may have initially moved in a turbulent fashion at high velocities but segregation of clasts resulted in increased particle concentration with some loss of fine ash. The grain-size data of Walker and McBroome (1983) indicate that, although some facies of the blast deposits resemble normal pyroclastic flows, there is evidence of more extensive size sorting and fines loss.

Valentine (1987) has proposed that many of the features shown by the May 18 blast deposit can be explained by the interaction of a stratified flow with the rough topography surrounding Mount St. Helens. Deposition of the massive valley-filling facies that resemble pyroclastic flows is attributed to the blocking of the high-density, lower portion of the turbulent cloud by topographic barriers (Fig. 19). Stratified and cross-bedded facies were deposited along ridges where the lower density part of the cloud interacted with the ground. In a recent study of the blast deposit, Druitt (1989) has proposed that there is a completely transitional relationship between proximal deposits that exhibit features typical of deposition from high-energy pyroclastic flows and

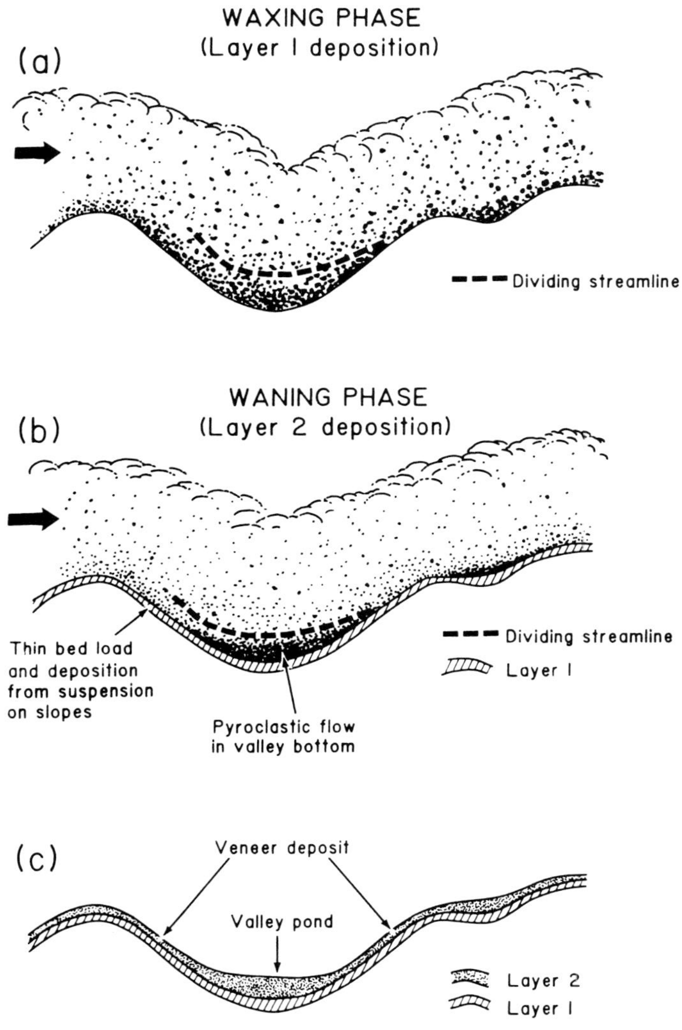


FIG. 19.—Model for the generation of layer A1 and A2 of the Mount St. Helens blast deposit from Valentine (1987). a) Deposition of layer A1 during an initial waxing phase. Blocking of the flow by ridges results in thicker valley deposits. b) deposition of layer A2 results as the flow begins to wane. Pyroclastic flows are formed in some valleys by blocking of the flow. c) Stratigraphic and thickness relationships of A1 and A2 facies between valleys and ridges.

medial/distal deposits that possess the characteristics of low-concentration pyroclastic surges. This gradation is attributed to the progressive dilution of particle concentration in the flow head because of the lack of sufficient recycled fine ash from the body of the flow.

Studies of the Mount St. Helens 1980 blast deposit have shown that it is not always easy, or perhaps even necessary, to characterize a flow as being either poorly or highly inflated. Progress in our understanding of the temporal and spatial evolution of transport and depositional processes in pyroclastic density currents will require a combination of both field and theoretical studies. It is important to continue studies of the facies relationships of deposits from single eruptive sequences using detailed grain size and component abundance measurements. However, because initial conditions of flow and surge generation will remain out of the realm of direct measurement, their early development and movement are best studied through theoretical modeling. The advent of highly sophisticated computer codes to model the movement of particle-gas mixtures (e.g., Valentine and Wohletz, 1989) holds great potential for exploring the wide range of conditions that are likely to occur in natural flow-generating, explosive eruptions.

## ACKNOWLEDGMENTS

Discussions with Steve Sparks on various aspects of pyroclastic flows and surges were most helpful. Careful reviews by Greg Valentine and an anonymous reviewer provided many useful suggestions which led to improvements of the manuscript. J. R. Allen, Ray Cas, Dick Fisher, James Moore, Haraldur Sigurdsson, Steve Sparks, Greg Valentine, George Walker, Colin Wilson and Ken Wohletz are gratefully acknowledged for allowing diagrams from their publications to be used in this review.

## REFERENCES

ALLEN, J. R., 1984, Sedimentary Structures, their Character and Physical Basis: Amsterdam, Elsevier, 663 p.

ANDERSON, T., AND FLETT, J., 1903, Report on the eruption of Soufrière in St. Vincent in 1902 and on a visit to Montague Pelée in Martinique: Philosophical Transactions of the Royal Society of London A, v. 200, p. 353–553.

BAILEY, R. A., DALRYMPLE, G. B., AND LAMPHERE, M. A., 1976, Volcanism, structure, and geochronology of Long Valley Caldera, Mono County, California: Journal of Geophysical Research, v. 81, p. 725–744.

BARBERI, F., INNOCENTI, L. L., MUNRO, T. P. AND SANTACROCE, R., 1978, The Campanian ignimbrite: A major prehistoric eruption in the Neapolitan area (Italy): Bulletin Volcanology, v. 41, p. 10–32.

CAREY, S., AND SIGURDSSON, H., 1989, The intensity of Plinian eruptions: Bulletin of Volcanology, v. 51, p. 28–40.

CAS, R. A., AND WRIGHT, J. V., 1987, Volcanic Succession Modern and Ancient: Allen and Unwin, London, 528 p.

CRANDALL, D. R., BOOTH, B., KUSUMADIVATA, K., SHIMOZUM, D., WALKER, G. P., AND WESTERCAMP, D., 1984, Source Book for Volcanic-Hazard Zonation: UNESCO, Paris, 97 p.

CROWE, B. M., AND FISHER, R. V., 1973, Sedimentary Structures in base-surge deposits with special reference to cross-bedding, Ubehebe Craters, Death Valley, California: Geological Society of America Bulletin, v. 84, p. 663–682.

CURTIS, G., 1968, The stratigraphy of ejecta from the 1912 eruption of Mount Katmai and Novarupta, Alaska, *in* Coats, R., Hay, R., and Anderson, C., eds., Studies in Volcanology: Geological Society of America Memoir 116, p. 153–210.

DAVIES, D. K., QUEARRY, M. W., AND BONIS, S. B., 1978, Glowing avalanches from the 1974 eruption of volcano Fuego, Guatemala: Geological Society of America Bulletin, v. 89, p. 369–384.

DRUITT, T. H., 1989, Emplacement of the May 18, 1980, lateral blast northeast of Mount St. Helens, Washington: New Mexico Bureau of Mines and Mineral Resources Bulletin 131, p. 75.

DRUITT, T. H., 1985, Vent Evolution and lag breccia formation during the Cape Riva eruption of Santorini, Greece: Journal of Geology v. 93, p. 439–454.

DRUITT, T. H., AND SPARKS, R. S., 1982, A proximal ignimbrite breccia facies on Santorini volcano, Greece: Journal of Volcanology and Geothermal Research, v. 13, p. 147–171.

FISHER, R. V., 1966, Mechanism of deposition from pyroclastic flows: American Journal of Science, v. 264, p. 350–363.

FISHER, R. V., 1979, Models for pyroclastic surges and pyroclastic flows: Journal of Volcanology and Geothermal Research, v. 6, p. 305–318.

FISHER, R. V., 1983, Flow transformations in sediment gravity flows: Geology, v. 11, p. 273–274.

FISHER, R. V., 1986, Systems of transport and deposition within pyroclastic surges: Evidence from Mount St. Helens, Washington: EOS Transactions, American Geophysical Union, v. 67, p. 1246.

FISHER, R. V., AND HEIKEN, G., 1982, Mt. Pelée, Martinique: May 8 and May 20, 1902, pyroclastic flows and surges: Journal of Volcanology and Geothermal Research, v. 13, p. 339–371.

FISHER R. V., AND SCHMINCKE, H. U., 1984, Pyroclastic rocks: Springer-Verlag, Berlin, 472 p.

FISHER, R. V., AND WATERS, A. C., 1970, Base surge bedforms in maar volcanoes: American Journal of Science, v. 268, p. 157–180.

FISHER, R. V., GLICKEN, H., AND HOBLITT, R., 1987, May 18, 1980, Mount St. Helens deposits in South Coldwater Creek, Washington: Journal of Geophysical Research, v. 92, p. 10267–10283.

FISHER, R. V., SMITH, A. L., AND ROOBOL, M. J. 1980, Destruction of St. Pierre, Martinique by ash-cloud surges, May 8 and 20, 1902: Geology, v. 8, p. 472–476.

FRANCIS, P. W., AND BAKER, M. C., 1977, Mobility of pyroclastic flows: Nature, v. 270, p. 164–165.

FRANCIS, P. W., ROOBOL, M. J., WALKER, G. P., COLBOLD, P. R., AND COWARD, M. P., 1974, The San Pedro and San Pablo volcanoes of North Chile and their hot avalanche deposits: Geologica Rundscha, v. 63, p. 357–388.

FREUNDT, A., AND SCHMINCKE, H. U., 1985, Hierarchy of facies of pyroclastic flow deposits generated by Laacher See-type eruptions: Geology, v. 13, 278–281.

HAY, R. L., 1959, Formation of the crystal-rich glowing avalanche deposits of St. Vincent, B. W. I.: Journal of Geology, v. 67, p. 540–562.

HOBLITT, R. P., 1986, Observations of the Eruption of July 22 and August 7, 1980, at Mount St. Helens, Washington: U.S. Geological Survey Professional Paper 1335, 44 p.

HOBLITT, R. P., MILLER, C. D., AND VALLANCE, J. W., 1981, Origin and stratigraphy of the deposit produced by the May 18 directed blast. U.S. Geological Survey Professional Paper 1250, p. 401–419.

KEIFFER, S. W., 1981, Fluid dynamics of the May 18 blast at Mount St. Helens: U.S. Geological Survey Professional Paper 1250, p. 379–400.

LACROIX, A., 1904, La Montagne Pelée et ses eruptions: Paris, Masson, 662 p.

LAJOIE, J., BOWDON, G., AND BOURDIER, J. L., 1989, Depositional mechanics of the 1902 pyroclastic nuée-ardente deposits of Mt. Pelée, Martinique: Journal of Volcanology and Geothermal Research, v. 28, p. 131–142.

LIPMAN, P., SELF, S., AND HEIKEN, G., 1984, Calderas and associated igneous rocks: Journal of Geophysical Research, v. 89, p. 8219–8801.

MACDONALD, G. A., AND ALCARAZ, A., 1956, Nuées ardentes of the 1948–1953 eruption of Hibok-Hibok: Bulletin of Volcanology, v. 18, p. 169–178.

MAHOOD G. A., 1984, Pyroclastic rocks and calderas associated with strongly peralkaline magmatism: Journal of Geophysical Research, v. 89, p. 8540–8553.

MCTAGGART, K. C., 1960, The mobility of nuées ardentes: American Journal of Science, v. 258, p. 369–382.

MCTAGGART, K. C., 1962, Nuées ardentes and fluidization—A reply: American Journal of Science, v. 260, p. 470–476.

MIDDLETON, G. V., AND SOUTHARD, J. B., 1978, Mechanics of sediment movement: Society of Economic Paleontologists and Mineralogists Short Course 3, p. 6.37–6.41.

MILLER, T. P., AND SMITH, R. L., 1977, Spectacular mobility of ash flows around Aniakchak and Fisher Calderas, Alaska: Geology, v. 5, p. 173–176.

MOORE, J. G., 1967, Base surge in recent volcanic eruptions: Bulletin of Volcanology, v. 30, p. 337–363.

MOORE, J. G., MELSON, W. G., 1969, Nuées ardentes of the 1968 eruption of Mayon Volcano, Phillippines: Bulletin of Volcanology, v. 33, p. 600–620.

MOORE, J. G., AND SISSON, T. W., 1981, Deposits and effects of the May 18 pyroclastic surge: U.S. Geological Survey Professional Paper 1250, p. 421–438.

O'KEEFE, J. A., AND ADAMS, E. W., 1965, Tektite structure and lunar ash flows: Journal of Geophysical Research, v. 70, p. 3819–3829.

PERRET, F. A., 1937, The eruption of Mt. Pelée 1929–1932: Carnegie Institute of Washington Publication, no. 458, 125 p.

RAGAN, D. H., AND SHERIDAN, M. F., 1972, Compaction of the Bishop Tuff, California: Geological Society of American Bulletin, v. 83, p. 95–106.

REYNOLDS, D. L., 1954, Fluidization as a geological process and its bearing on the problems of intrusive granites: America Journal of Science, v. 252, p. 577–613.

ROSS, G. S., AND SMITH, R. L., 1961, Ash-flow tuffs, their origin, geological relations and identification: U. S. Geological Survey Professional Paper 354F, p. 1–79.

SAVAGE, S. B., 1985, The mechanics of rapid granular flow: Advanced Applied Mechanics, v. 24, p. 289–366.

SCHMINCKE, H. U., FISHER, R. V., AND WATERS, A. C., 1973, Antidune and chute and pool structures in base surge deposits of the Laacher See area, Germany: Sedimentology, v. 20, p. 553–574.

SCOTT, W., 1989, Volcanic and related hazards, *in* Tilling, R. I., ed., Volcanic Hazards: American Geophysical Union Short Course in Geology 1, p. 9–24.

SHERIDAN, M. F., 1970, Fumarolic mounds and ridges of the Bishop Tuff, California: Geological Society American Bulletin, v. 81, p. 851–868.

SHERIDAN, M. F., 1979, Emplacement of pyroclastic flows: A review, *in* Chapin, C. E., and Elston, O. H., eds., Ash Flow Tuffs: Geological Society of America Special Paper No. 180, p. 125–136.

SHERIDAN, M. F., AND UPDIKE, R. G., 1975, Sugarloaf Mountain tephra—A Pleistocene rhyolitic deposit of base-surge origin in Northern Arizona: Geological Society of America Bulletin, v. 86, p. 571–581.

SIGURDSSON, H., CAREY, S., AND ESPINDOLA, J. M., 1984, The 1982 eruption of El Chichón, Mexico: Stratigraphy of pyroclastic deposits: Journal of Volcanology and Geothermal Research, v. 23, p. 11–37.

SIGURDSSON, H., CAREY, S., AND FISHER, R. V., 1987, The 1982 eruption of El Chichón volcano, Mexico (3): Physical properties of pyroclastic surges: Bulletin of Volcanology, v. 49, p. 467–488.

SIMPSON, J. E., 1987, Gravity currents in the environment and the laboratory: Ellis Houwood Limited, Chichester, England, 244 p.

SMITH, R. L., 1960a, Ash-flows: Geological Society of America Bulletin, v. 71, p. 795–842.

SMITH, R. L., 1960b, Zones and zonal variations in welded ash-flows: U.S. Geological Survey Professional Paper 354-F, p. 149–159.

SMITH, R. L., 1979, Ash-flow magmatism, *in* Chapin, C. E., and Elston, W. E., eds, Ash Flow Tuffs: Geological Society of America Special Paper No. 180, p. 5–25.

SPARKS, R. S., 1976, Grain size variations in ignimbrites and implications for the transport of pyroclastic flows: Sedimentology, v. 3, p. 147–188.

SPARKS, R. S., 1979, Gas release rates from pyroclastic flows: An assessment of the role of fluidization in their emplacement: Bulletin of Volcanology, v. 41, p. 1–9.

SPARKS, R. S., 1983, Mont Pelée, Martinique: May 8 and May 20, 1902, pyroclastic flows and surges—Discussion: Journal of Volcanology and Geothermal Research, v. 19, p. 175–184.

SPARKS, R. S., AND WALKER, G. P., 1973, The ground surge deposit: A third type of pyroclastic rock: Nature, v. 241, p. 62–64.

SPARKS, R. S., AND WALKER, G. P., 1977, The significance of vitric-enriched air-fall ashes associated with crystal-enriched ignimbrites: Journal of Volcanology and Geothermal Research, v. 2, p. 329–341.

SPARKS, R. S., AND WILSON, L., 1976, A model for the formation of ignimbrite by gravitational column collapse: Journal of the Geological Society of London, v. 132, p. 441–452.

SPARKS, R. S., MOORE, J. G., and RICE, C. J., 1986, The initial giant umbrella cloud of the May 18, 1980 explosive eruption of Mount St. Helens: Journal of Volcanology and Geothermal Research, v. 28, p. 257–274.

SPARKS, R. S., SELF, S., AND WALKER, G. P., 1973, Products of ignimbrite eruptions: Geology, v. 1, p. 115–118.

SPARKS, R. S., WILSON, L., AND HULME, G., 1978, Theoretical modelling of the generation, movement and emplacement of pyroclastic flows by column collapse: Journal of Geophysical Research, v. 83, p. 1727–1739.

STEVEN, T. A., AND LIPMAN, P. W., 1976, Caldera of the San Juan volcanic field, southwestern Colorado: U.S. Geological Survey Professional Paper 958, 35 p.

VALENTINE, G. A., 1987, Stratified flow in pyroclastic surges: Bulletin of Volcanology, v. 49, p. 616–630.

VALENTINE, G. A., AND FISHER, R. V., 1986, Origin of layer 1 deposits in ignimbrites: Geology, v. 14, p. 146–148.

VALENTINE, G. A., AND WOHLETZ, K. H., 1989, Numerical models of Plinian eruption columns and pyroclastic flows: Journal of Geophysical Research, v. 94, p. 1867–1887.

VALENTINE, G. A., BUESCH, D., AND FISHER, R. V., 1989, Basal layered deposits of the Peach Springs Tuff, northwestern Arizona, USA: Bulletin of Volcanology, v. 51, p. 395–414.

WAITT, R. B., 1981, Devastating pyroclastic density flow and attendant air fall of May 18—stratigraphy and sedimentology of deposits: U.S. Geological Survey Professional Paper 1250, p. 439–458.

WAITT, R. B., 1984, Comment on Walker and McBroome (1983): Geology, v. 13, p. 693.

WALKER, G. P., 1972, Crystal concentration in ignimbrites: Contributions of Mineralogy and Petrology, v. 36, p. 135–146.

WALKER, G. P., 1980, The Taupo pumice: Product of the most powerful known (ultraplinian) eruption?: Journal of Volcanology and Geothermal Research, v. 8, p. 69–94.

WALKER, G. P., 1983, Ignimbrite types and ignimbrite problems: Journal of Volcanology Geothermal Research, v. 17. p. 65–88.

WALKER, G. P., 1984, Characteristics of dune-bedded pyroclastic surge bedsets: Journal of Volcanology and Geothermal Research, v. 20, p. 281–296.

WALKER, G. P., 1985, Origin of coarse lithic breccias near ignimbrite source vents: Journal of Volcanology and Geothermal Research, v. 25, p. 157–171.

WALKER, G. P., AND MCBROOME, L. A., 1983, Mount St. Helens 1980 and Mount Pelée 1902—flow or surge?: Geology, v. 11, p. 571–574.

WALKER, G. P., AND MORGAN, L., 1984, Reply to Hoblitt and Miller (1984) and Waitt (1984): Geology, v. 12, p. 693–695.

WALKER, G. P., HEMING, R. F., AND WILSON, C. J., 1980a, Low-aspect ratio ignimbrites: Nature, v. 283, p. 286–287.

WALKER, G. P., HEMING, R. F., AND FROGGATT, P. C., 1980b, Fines-depleted ignimbrite in New Zealand—The product of a turbulent pyroclastic flow: Geology, v. 8, p. 245–249.

WALKER, G. P., SELF, S., AND FROGATT, P. C., 1981, The ground layer of the Taupo ignimbrite: A striking example of sedimentation from pyroclastic flow: Journal of Volcanology and Geothermal Research, v. 9, p. 409–421.

WATERS, A. C., AND FISHER, R. V., 1971, Base surges and their deposits: Capelinhos and Taal volcanoes: Journal of Geophysical Research, v. 76, p. 5596–5614.

WILSON, C. J., 1980, The role of fluidisation in the emplacement of pyroclastic flows: An experimental approach: Journal of Volcanology and Geothermal Research, v. 8, p. 231–249.

WILSON, C. J., 1984, The role of fluidisation in the emplacement of pyroclastic flows, 2: Experimental results and their interpretaiton: Journal of Volcanology and Geothermal Research, v. 20, p. 55–84.

WILSON, C. J., 1985, The Taupo eruption, New Zealand II. The Taupo ignimbrite: Philosophical Transactions of the Royal Society of London, A. 314, p. 229–310.

WILSON, C. J., AND WALKER, G. P., 1982, Ignimbrite depositional facies: The anatomy of a pyroclastic flow: Journal of the Geological Society of London, v. 139, p. 581–592.

WOHLETZ, K. H., AND SHERIDAN, M. F., 1979, A model of pyroclastic surge, *in* Chapin, C., E., and Elston, O. H., eds., Ash Flow Tuffs: Geological Society of America Special Paper 180, p. 177–194.

WOHLETZ, K. H., MCGRETCHIN, T. R., SANFORD, M. T., AND JONES, E. M., 1984, Hydrodynamic aspects of caldera-forming eruptions: Journal of Geophysical Research, v. 89, p. 8269–8286.

WOODS, A., 1988, The fluid dynamics and thermodynamics of eruption columns: Bulletin of Volcanology, v. 50, p. 169–198.

WRIGHT, J. V., AND WALKER, G. P., 1977, The ignimbrite source problem: Significance of a co-ignimbrite lag-fall deposit: Geology, v. 5, p. 729–732.

WRIGHT, J. V., SMITH, A. L., AND SELF, S., 1980, A working terminology of pyroclastic deposits: Journal of Volcanology and Geothermal Research, v. 8, p. 315–336.

YOKOYAMA, S., 1974, Flow and emplacement mechanism of the Ito pyroclastic flow arom Aira caldera, Japan: Tokyo Kyoihu Daigaku Science Report, Section C, v. 12, p. 17–62.

# LAHARS: VOLCANO-HYDROLOGIC EVENTS AND DEPOSITION IN THE DEBRIS FLOW—HYPERCONCENTRATED FLOW CONTINUUM

GARY A. SMITH
*Department of Geology, University of New Mexico, Albuquerque, NM 87131*
AND
DONALD R. LOWE
*Department of Geology, Stanford University, Stanford, CA 94305*

ABSTRACT: Many volcano-hydrologic events produce high-discharge, sediment-laden flows originating on or near volcanoes, during or following eruptions. Water for some flows is derived from normal precipitation, commonly with enhanced runoff because of reduced infiltration into slopes mantled by pyroclastic debris and on which most vegetation has been destroyed. Flows triggered by eruptions may contain large volumes of water mobilized by eruption-induced snowmelt, discharge of crater lakes, or liquefaction of saturated debris-avalanche material. Sediment is available as unconsolidated pyroclastic or autoclastic material, principally of sand to fine-gravel size, that is thickest on the steep slopes of the volcano itself. Resulting flows generally possess the characteristics of debris flows or hyperconcentrated flows.

Lahars have traditionally been defined as volcanic-debris flows and their deposits. Events and deposits called lahars, however, include a considerably wider range of flow phenomena. More recently, lahar has been applied as a general term for rapidly flowing sediment-laden mixtures of rock debris and water from a volcano. As such, a lahar may include one or more discrete flow processes. This definition emphasizes that volcano-hydrologic events are complex, encompassing a variety of rheological flow types and flow transformations.

The complexity of lahars lies in the common occurrence of flow transformations, especially bulking and dilution. Initially dilute, high-discharge flows readily entrain loose material, particularly by streamed entrenchment, and may evolve into debris flows. Debris flows traveling within stream valleys are subject to dilution by mixing with water along the flow fronts, resulting in progressive downstream transformation into hyperconcentrated and dilute flows.

Characterization of lahar-flow rheology is generally based on study of the resulting deposits. Relative to nonvolcanogenic-debris flows, those developed on volcanoes are typically mud poor and have much greater flow depths and discharges. In most cases, cohesion contributes little to sediment support, although the resulting deposits resemble those of nonvolcanogenic-debris flows. Hyperconcentrated flows, which are transitional between debris flows and more dilute stream flows, have been variously defined on the basis of sediment concentration, rheological properties, and depositional features.

Lahars may be triggered by a variety of processes, including pyroclastic flows and debris avalanches. In some cases, it is possible to infer triggering mechanisms, as well as the extent of flow bulking or dilution, from study of lahar deposits.

## INTRODUCTION

Sediment yields and erosion rates in volcanic areas are among the highest known because of the large volumes of loose, granular sediment generated by volcanic processes, widespread destruction of sediment-stabilizing vegetation, and decreased infiltration on tephra-covered slopes (Segerstrom, 1950; Waldron, 1967; Ollier and Brown, 1971; Vessel and others, 1981; Swanson and others, 1982). In volcanic terranes, enormous volumes of sediment are transported downslope by mass-flow and high-sediment-concentration flow processes, the deposits of which often dominate volcaniclastic sedimentary sequences (Lydon, 1968; Smith, 1986; Palmer and Walton, 1990). Sparked in part by a number of destructive eruptions in the 1980s, there has been a doubling of significant research papers on volcaniclastic sedimentation in the past 5 years (Pierson and Scott, 1985; Smith, 1986; Lowe and others, 1986; Naranjo and others, 1986; Scott, 1988; Walton and Palmer, 1988; Rodolfo, 1989; Arguden and Rodolfo, 1990). This paper considers sediment transport and deposition by high-sediment-load flows in volcanic areas, emphasizing aspects most useful to interpreting continental volcaniclastic sequences.

This treatment will focus on the features of debris flows and more dilute hyperconcentrated flows that permit recognition and discrimination of their deposits, processes of flow generation, and transport and deposition mechanics. It will also be useful to compare and contrast these processes with low-sediment-concentration streamflows and debris avalanches. Perhaps the most widely used term to describe mass flows of volcaniclastic debris is the Indonesian word, lahar. Lahars have been defined both as debris flows composed of volcanic debris and as the deposits of such debris flows (Van Bemmelen, 1949; Crandell, 1971; Neall, 1976; Fisher and Schmincke, 1984). In practice, however, the term has commonly been applied to any poorly sorted volcaniclastic sedimentary deposit, including many lacking definitive evidence for debris-flow deposition. Other investigators have restricted the term lahar to deposits emplaced at higher than ambient temperatures or as a direct result of volcanism (see Neall, 1976; and Smith, 1986; for reviews and citations). Because of this ambiguity, Smith (1986) recommended that the term be abandoned. A broader view, however, has been taken by participants of a recent Geological Society of America Penrose Conference (Smith and Fritz, 1989) and Rodolfo (1989): lahar is "a general term for a rapidly flowing mixture of rock debris and water (other than normal streamflow) from a volcano. A lahar is an event; it can refer to one or more discrete processes, but does not refer to a deposit" (Smith and Fritz, 1989, p. 375). This definition emphasizes the complex behavior of volcano-hydrologic events, involving several different temporally and spatially variable flow processes, and is a useful terminological umbrella for these processes. The definition also has the advantage of recognizing that some aspects of these processes are unique to volcanic areas, but are not always directly related to eruptions, a link that is difficult or impossible to establish in ancient volcaniclastic deposits. A final advantage is maintainance of distinct terms for processes and deposits. Lahars encompass a continuum be-

Sedimentation in Volcanic Settings, SEPM Special Publication No. 45

tween debris-flow and hyperconcentrated-flow processes (Smith and Fritz, 1989; Rodolfo, 1989), although the definition permits inclusion of debris avalanches. In this paper, we focus on debris flows and hyperconcentrated flows, but do recognize the common genetic link of these processes to debris avalanches (e.g., Fairchild, 1987; Crandell, 1989; Glicken, this volume; Palmer and others, this volume).

## TRANSPORT AND DEPOSITIONAL PROCESSES

Flows of volcaniclastic materials, containing little or no magmatic gas, range from dilute streamflow through hyperconcentrated and debris flows to debris avalanche (Fig. 1).

### *Dilute Streamflow*

Dilute streamflow is the flow of a Newtonian fluid. Flow is fully turbulent, except in a thin layer near the bed, and turbulence is the principal sediment-support mechanism (Smith, 1986; Pierson and Costa, 1987; Costa, 1988). Deposition occurs principally by traction, in a "grain-by-grain" fashion.

In general, two types of dilute streamflow deposits can be considered (Fig. 1). The first includes sediments transported over a wide range of discharge conditions in a relatively continuous fashion or flood-deposited material that is redistributed into bedforms at lower stages. The second includes sediments transported only during high-discharge conditions that are not reworked by lower stage flows. Type 1 deposits consist of cross-bedded sands or relatively well-sorted, massive or cross-bedded, well-rounded, imbricated gravel with associated lenses and thin beds of sand. Type 2 deposits tend to be much more poorly sorted, contain more angular grains, and generally lack cross-bedding generated by migrating bedforms, although scour-and-fill structures are common (Fig. 2).

### *Hyperconcentrated Flow*

Turbulent, or transitional-to-laminar, sediment-charged dispersions have long been recognized as important in subaqueous sedimentation (Middleton and Southard, 1984; Lowe, 1979, 1982), but have been generally ignored in studies of terrestrial sedimentation. Recently, however, the concept of hyperconcentrated flow (Beverage and Culbertson, 1964) has been applied by a number of authors to the interpretation of volcaniclastic deposits (Pierson and Scott, 1985; Smith, 1986; Walton and Palmer, 1988; Rodolfo, 1989; Waresback and Turbeville, 1990).

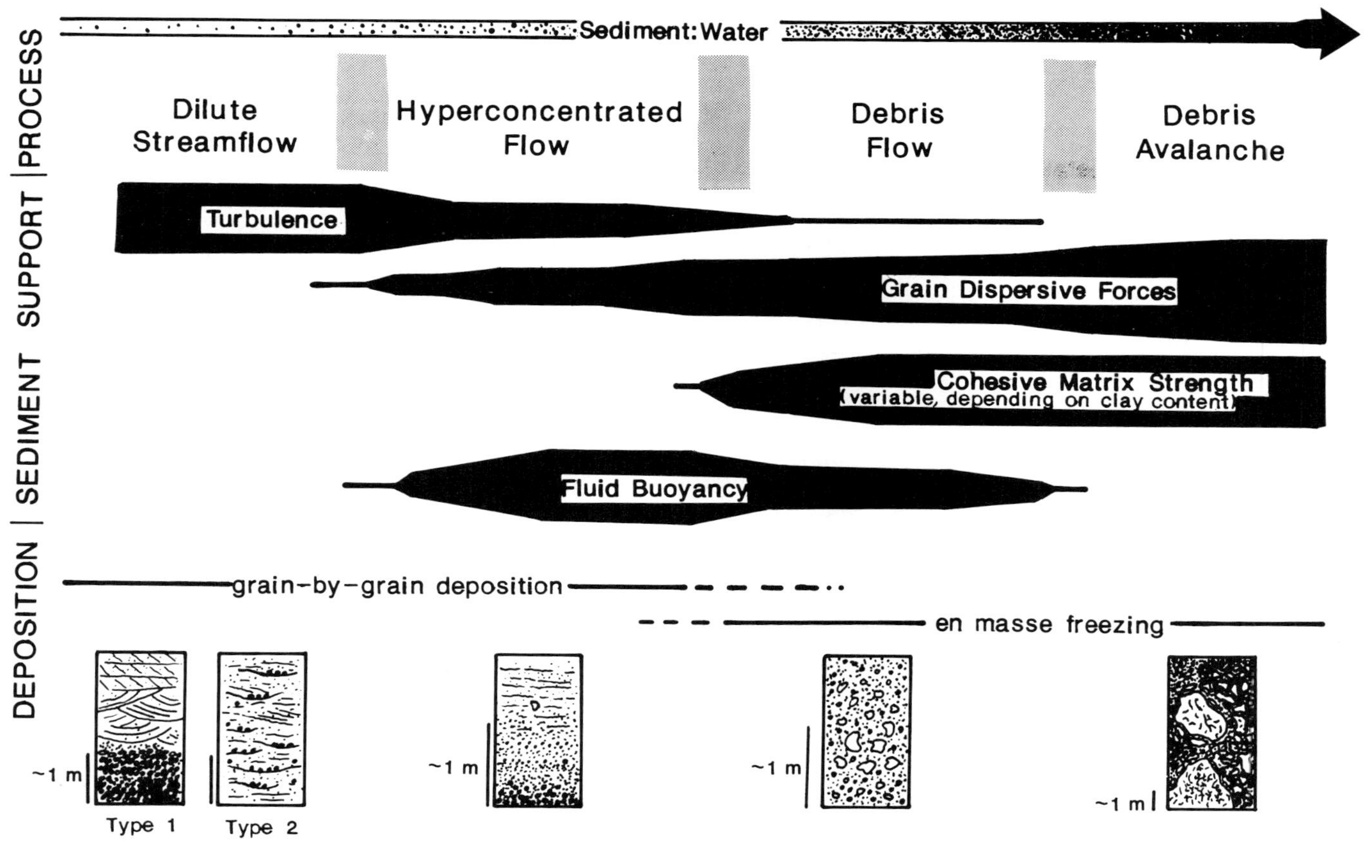


FIG. 1.—Schematic diagram illustrating sediment-support and depositional mechanisms, and ideal examples of deposits for dilute-streamflow, hyperconcentrated-flow, debris-flow, and debris-avalanche processes. Note that each flow type may be characterized by multiple sediment-support mechanisms. Thickness of deposits is variable and scales on deposit drawings are only approximate. Note progressive loss of stratification and grading in deposits as sediment concentrations increase. Refer to text and succeeding figures for details of deposit characteristics.

FIG. 2.—Scour-and-fill bedded, poorly sorted, flood-deposited volcaniclastics deposited in poorly confined, shallow braided channels; late Miocene Ellensburg Formation, Washington.

*Definition.*—

Although defined in a number of different ways, hyperconcentrated flow is considered by all investigators to include flow intermediate in nature between dilute, fully turbulent, normal streamflow and viscous, generally nonturbulent debris flow. The term *hyperconcentrated streamflow* was introduced by Beverage and Culbertson (1964) for flows in which debris was inferred to be supported by a combination of buoyancy, grain interactions, and dampened turbulence (Fig. 1). Based on observations of mud-rich streams in flood in the southwestern United States, they defined hyperconcentrated streamflows as those having sediment concentration between 40 and 80 wt percent; normal streamflow as having less than 40 percent; and debris flow as greater than 80 percent.

Pierson and Costa (1987) applied the term *hyperconcentrated streamflow* to flows having sediment concentrations ranging from that at the onset of measureable yield strength to some threshold concentration above which yield strength is nearly invariant. These definitive sediment concentrations are variable and depend on the grain-size distribution of the mixture and the types and abundance of clay minerals.

Smith (1986) noted that concentration-based definitions are not directly applicable to the geologic record and that there are few rigorous sedimentological descriptions of deposits that can be correlated with flows of specific known sediment concentrations. Smith (1986) defined *hyperconcentrated flood-flow* as high-discharge flow in which turbulence is not the only sediment-support mechanism and in which deposition does not occur en masse, as is commonly assumed for debris flows (Fig. 1). This usage allows recognition of event deposits having characteristics that are neither those of normal streamflow nor debris-flow deposits. Depositional criteria for recognition of hyperconcentrated flood flows (Smith, 1986; Scott, 1988) may not define flow conditions corresponding to the empirical definition of Beverage and Culbertson (1964) or the rheological constraints of Pierson and Costa (1987). Type 2 dilute streamflow deposits (Figs. 1, 2) may also represent hyperconcentrated conditions by the definitions of Beverage and Culbertson (1964) and Pierson and Cosota (1987) despite the precedence of referring to such deposits as "sheetflood" or "shallow braided stream" facies. Until the mechanics of flow and the relationship between flow mechanics and deposit properties are better understood, it seems most prudent to define *hyperconcentrated flows* as non-Newtonian fluid-solid mixtures possessing little or no strength and generating deposits intermediate in nature between those of debris flows and dilute streamflows.

*Deposits.*—

Smith (1986, 1987a,b, 1988) outlined the characteristics of hyperconcentrated flows based on the study of Quaternary and Tertiary volcaniclastic deposits differing from dilute streamflow or debris-flow facies:

1. Sand- and granule-size deposits are massive or exhibit crude horizontal stratification (Figs. 1, 3); stratification is more apparent in pumiceous deposits. Individual strata are ungraded, normal graded, or reverse graded. Cross-bedding or ripple laminations are absent, and scour structures reflecting localized erosion during overall aggradation are rare, even where outsized clasts are deposited with sand (Fig. 3). The deposits generally exhibit distribution-normal grading with stratification better developed in upper, finer grained parts. Outsize cobbles or boulders are common. Low-density particles, such as pumice and organic matter, are concentrated in the upper parts of most deposits, irrespective of grain size.
2. Gravel inferred to have been deposited by hyperconcentrated flows generally occurs in normally graded units, usually shows nonimbricated complex fabric, and lacks intercalated sand lenses or beds typical of dilute streamflow gravel. Similar features were cited as evidence for hyperconcentrated-flow origin of nonvolcaniclastic conglomerates and gravels by Arguden and Rodolfo (1986), Schmitt and Olson (1986), and Wells and Harvey (1987). Unambiguous distinction of hyperconcentrated-flow conglomerates, however, remains hampered by possible similarities to streamflow conglomerates deposited by dilute but high-discharge, high-velocity flood flows.

These observations suggest that hyperconcentrated flows are turbulent enough to allow differential settling and the development of graded beds and, perhaps, the migration of some types of bedforms that account for horizontal stratification (Smith, 1986). Hyperconcentrated-flow deposits closely resemble those of subaqueous, coarse-grained, high-density turbidity currents and probably share inferred transport and depositional characteristics (Lowe, 1982; Smith, 1986).

A somewhat different view of hyperconcentrated flows has been given by Pierson and Scott (1985), Scott (1988), and Major and Scott (1988) based on studies in valleys near Cascade Range volcanoes. These views are summarized by Scott (1988):

FIG. 3.—Graded-stratified hyperconcentrated-flow deposits. Note in each the massive, coarser base and the upward increase in the definition of stratification in finer grained sediment; arrows mark bases and tops of event deposits. Note the lack of cross-stratification or scour structures, even in association with the large outsize clast in (A). Also note, in (A), the inversely graded strata (sub-units of Scott, 1988) near the field notebook. Deposits are (A) Quaternary sediments flanking Ruapehu volcano, Helwyn Camp, New Zealand; (B) late Miocene volcaniclastics, Ellensburg Formation, Washington.

1. Deposits are always in the sand- to pebble-size range, generally massive or rarely crudely stratified, and clast supported. Coarse-tail normal grading may occur and generally includes the concentration of low-density components toward the bases of the deposits. Inversely graded sub-units are defined within the deposits.
2. Hyperconcentrated flows are noncohesive mass flows that deposit sediment in pulses of mass-emplaced materials, thus accounting for absence of stratification, inverse grading, and trapping of low-density material at the bottoms of the deposits.

We feel that, to some extent, both views are correct. The observations and conclusions of Smith (1986) apply most generally to relatively low-concentration hyperconcentrated flows, and are consistent with the observations of turbulent-flow features and deposit stratification associated with an observed hyperconcentrated flow (Pierson and Scott, 1985). The views of Scott (1988) are more applicable to flows that approach debris flows. In view of the rheological spectrum represented by hyperconcentrated flows, a range of characteristics should be expected in the deposits.

### *Debris Flows*

Debris flows are flows of sediment and fluid that exhibit strength or resistance to shear. Strength is typically inferred to relate to a cohesive, clay-rich matrix (Middleton and Southard, 1984; Johnson, 1970). These "ideal" debris flows can be characterized mechanically using a Coulomb-viscous model (Johnson, 1970), but debris flows in volcanic regions commonly differ from those in other areas. Volcanic debris flows are generally clay poor (Fisher and Schmincke, 1984; Smith, 1986) because most form by mobilization of fragmental pyroclastic and autoclastic debris during or soon after eruption. The low-clay content represents an important rheological difference between volcanic and nonvolcanic debris flows, the latter typically forming by failure of soil layers, weathered regolith, or weathered bedrock. Where clays are present in volcanic debris-flow deposits, it has commonly formed diagenetically by alteration of highly reactive volcaniclastic grains. Where primary clay occurs, it is commonly derived from hydrothermally altered source material (Crandell and Waldron, 1956; Scott, 1988) or by the incorporation of soil during transport (Lowe and others, 1986; Palmer and others, this volume). Most volcanic debris flows, therefore, derive their strength or shear resistance from: 1) grain collisions, representing frictional resistance to shear not accounted for by Coulomb-viscous models (Iverson and Denlinger, 1987); and 2) cohesion between silt-size ash particles. Although the terms debris flow and mudflow are often used interchangeably, mudflow is best applied to debris flows that contain appreciable clay and are cohesive (Lowe, 1979). Few volcaniclastic-debris flows are mudflows.

Deposition from ideal, cohesive debris flows occurs en masse for most particles because of cohesion and the development of plugs as the applied shear stress falls below the yield strength of the flowing debris (Johnson, 1970; Middleton and Southard, 1984; Lowe, 1979; Smith, 1986). Resulting deposits, masssive and matrix-supported, are commonly similar in thickness to the moving flow and show consistent trends in cohesion-sensitive relationships, such as bed thickness (BTh) versus maximum particle size (MPS; Bluck, 1967; Nemec and Steel, 1984). The larger volcanic debris flows are relatively noncohesive flows that transport up to $10^7$ $m^3$ of sediment over periods of 3 to 9 hours and to distances exceeding 100 km from source (e.g., Janda and others, 1981; Naranjo and others, 1986; Lowe and others, 1986; Rodolfo, 1989). These high-discharge flows are different from the relatively small-volume, nonvolcanogenic alpine and alluvial-fan debris flows upon which most concepts of debris-flow rheology and deposition are based (Johnson, 1970; Middleton and Southard, 1984). These large-volume, far-traveled flows invariably leave deposits that are significantly thinner than the flows themselves and leave thin veneers of debris-flow matrix with lags of coarse clasts that may mantle, rather than smooth, topography (Kesel, 1973; Lowe and others, 1986; Pierson and Scott, 1985; Rodolfo, 1989). The flows also produce depositional terraces (Rodolfo, 1989) and clast-supported streamlined lags ("whaleback bars," Gilkey, 1983; Scott, 1988). Deposit thickness, therefore, need not correlate to flow thickness and the competency of deep, high-discharge flows to transport large clasts; this may account for the substantial scatter on MPS/BTh plots for volcanic-debris flows (Walton and Palmer, 1988; Arguden and Rodolfo, 1990). Collectively, all of these features suggest that debris-flow deposition is not always en masse and that volcanic-debris flows are not ideal, cohesive debris flows.

In general terms, volcanic-debris flows and their deposits are not radically different from those generated in nonvolcanic settings and can be identified by the same features and studied by the same methods (Smith, 1986; Walton and Palmer, 1988). The two principal differences are clay content and, compared to other terrestrial flows, scale. Most volcanic-debris flows are noncohesive and can be regarded as water-saturated granular flows. Volcanic-debris flows may possess remarkably large flow volumes and discharges that permit travel over many tens of kilometers. The rheological and mechanical consequences of these two distinctions, relative to most nonvolcanic-debris flows, must be considered when examining the record of lahar events.

### *Debris Avalanches*

Debris avalanches are very rapid, inertial, granular flows resulting from large-volume landslides (Pierson and Costa, 1987). They are important in stratovolcano terranes where they result from sector collapse of steep, structurally unstable volcanic cones (Siebert, 1984). For reasons that are as yet unclear, volcanic-debris avalanches are typically more mobile, with longer runout distance for a given height of fall, than slides generated on nonvolcanic slopes (Siebert, 1984). Water in debris avalanches occurs in pores and is carried with the solids rather than contributing to a fluid that supports and transports clasts. In cases where abundant water is incorporated into debris avalanches from ground water, snow or ice on the volcano, or from crater lakes, it may drain away from the coarser constituents of the debris avalanche to generate debris flows.

The nature of the rockslide-debris avalanche process on volcanoes was highlighted by the 1980 failure of the north flank of Mount St. Helens (Voight and others, 1981). Mass failures had been previously observed or inferred on geomorphic evidence, but the deposits were generally termed "lahars" without distinction between debris-flow and debris-avalanche processes and deposits (e.g., Neall, 1976, 1979; Crandell, 1971). Debris avalanches are now recognized to be major contributors to the depositional record of volcanic terranes (Siebert, 1984; Crandell, 1989; Glicken, 1986 and this volume; Palmer and Neall, 1989; Palmer and Walton, 1990; Palmer and others, this volume), and they have also been recognized in sedimentary sequences that accumulated adjacent to high-relief nonvolcanic regions (Yarnold and Lombard, 1989).

Both debris-avalanche and debris-flow deposits are characteristically coarse-grained and poorly sorted, and criteria for distinguishing between these deposit types are not well defined. Morphologically, debris-avalanche deposits are characterized by hummocky upper surfaces, with relief of meters to tens of meters, generated by lateral spreading of the debris and intact transport of large blocks of source material. These hummocks, commonly referred to as "lahar mounds," have been cited as the most distinctive feature of debris-avalanche deposits (Siebert, 1984; Palmer and Walton, 1990). Although obvious in modern surfical deposits, hummocks may be difficult to recognize on the scale of most outcrops or to distinguish from undulating erosional surfaces in ancient sequences. Texturally, debris-avalanche deposits are characterized by an abundance of clasts, ranging in size from centimeters to tens of meters, that are complexly fractured with little or no displacement across fractures (Fig. 4a). Referred to as brecciated clasts and megaclasts (Palmer and others, this volume), debris blocks (Glicken, this volume) or crackle and jigsaw brecciation (Yarnold and Lombard, 1989), this fracturing is generated by dilation at the time of failure (Glicken, 1986) and shattering of the sliding and flowing debris. With further dilation, clasts may disintegrate along these fractures to generate domains in the deposit within which clasts are monolithologic and separated by matrix of the same composition (Fig. 4b). These textural features are diagnostic of debris-avalanche processes, but may also be preserved as inherited features of debris-flow deposits resulting from mobilization of saturated debris-avalanche material, especially if domains of plug flow are established in the debris flow. Clast shattering can also occur within debris flows as a result of collision of clasts. An abundance of brecciated clasts and matrix heterogeneity over large portions of a deposit are, however, the best criteria for identification of debris-avalanche deposits in the stratigraphic record.

FIG. 4.—Sedimentological features of debris-avalanche deposits. (A) Highly brecciated megaclast or crackle breccia; (B) irregular subvertical contact between domains composed of clasts and matrix derived from disintegration of different lithologies (light on left, dark on right) during transport. Both illustrations from the Pleistocene Pungarehu Formation of Neall (1979), near Mount Egmont, New Zealand.

## FLOW GENERATION AND FLOW TRANSFORMATIONS

The complexity of lahar events lies in the multiple origins for sediment-laden flows and the transformations that occur from one flow type to another during movement. These complexities are diagrammatically represented in Figure 5 and are discussed below.

### *Flow Transformations*

Fisher (1983) defined flow transformations as changes in flow behavior between laminar and turbulent states. Surface transformations, resulting from addition of fluid (dilution) or addition of sediment (bulking) are common in lahar events. Two additional types of flow transformation, not necessarily related to changes in Reynolds number, include the generation of debris flows from debris avalanches and the transition from pyroclastic flows and surges, in which gas forms the interstitial fluid, to debris flows and hyperconcentrated flows, in which water is the intragranular fluid.

*Dilution and bulking.*—

Nonvolcanic terrestrial debris flows are best known from studies of flows on arid-climate alluvial fans (Johnson, 1970). Volcanic flows, in contrast, often occur in relatively humid settings or are influenced by snow and ice melting or the release of crater lakes (Major and Newhall, 1989). Water is generally available in large volumes and the rheology of the resulting flow is determined by the manner and degree in which solids and fluids are mixed.

Most nonvolcanic debris flows are triggered by failure of hillslope material that is saturated in situ (Johnson and Hampton, 1969; Johnson, 1970). Entrainment of streambed debris by turbulent, dilute flows to the extent that the rheology could be changed to that of debris flows has been considered unlikely or impossible (Johnson and Hampton, 1969). Observations of recent lahar events, however, indicate that large-volume dilute runoff, generated by eruption-induced snowmelt or heavy rainfall, can bulk up with loose sediment and transform into debris flows and hyperconcentrated flows downstream (Waitt and others, 1983; Pierson and Scott, 1985; Lowe and others, 1986; Rodolfo, 1989). The bulking process offers the best explanation for the abundance of rounded clasts (e.g., Scott, 1988) and downslope increase in the lithologic diversity of clasts (e.g., Crandell and Waldron, 1956; Palmer and Walton, 1990) in many lahar deposits.

Dilution occurs when debris flows mix with water, commonly in perennial stream channels draining volcanic centers or small ponds or lakes. Dilution of debris flows to form hyperconcentrated flows was documented during a 1982 eruption-related lahar event at Mount St. Helens (Pierson and Scott, 1985). Mixing of turbulent streamwater with the moving front of the debris flow diluted the flow front to form a hyperconcentrated flow. Downstream, passage of the hyperconcentrated flow was followed by the remaining undiluted part of the debris flow; progressive dilution over a 24-km reach eventually transformed the entire debris flow to a hyperconcentrated flow (Pierson and Scott, 1985). The depositional record of dilution, therefore, is a bipartite de-

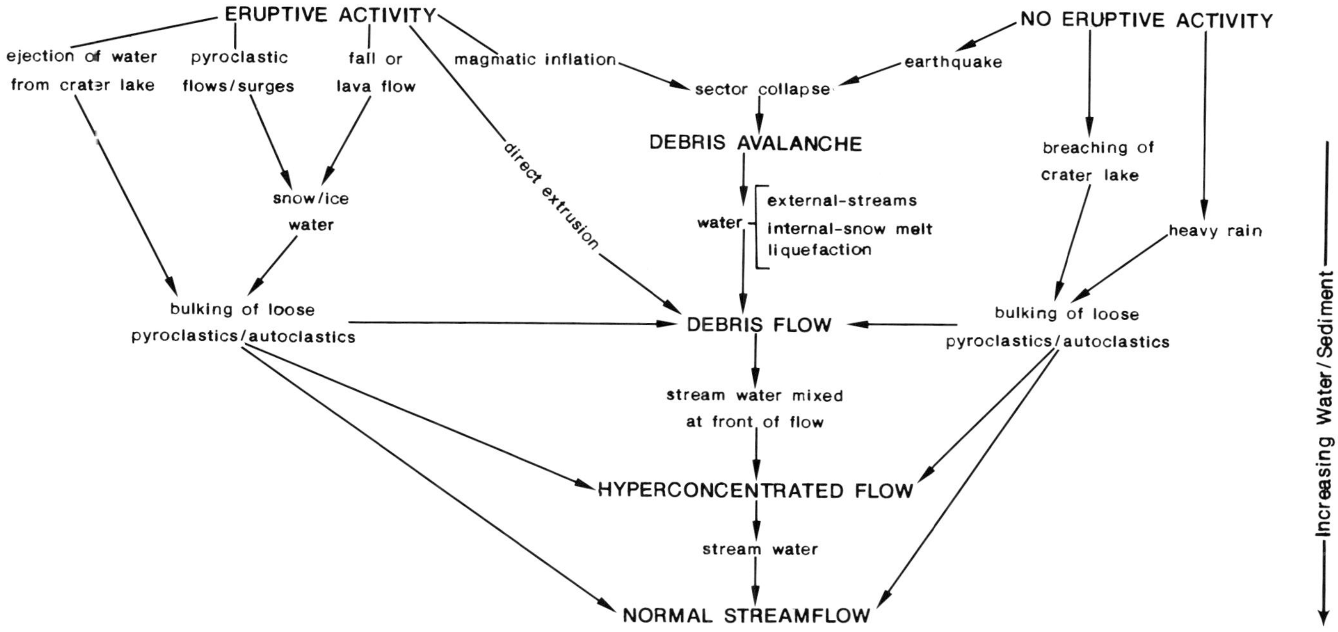


FIG. 5.—Schematic diagram illustrating genetic relationships of volcanic phenomena and the generation of debris avalanches, debris flows, and hyperconcentrated flows. Principal processes of sediment transport and deposition are vertically arranged in the center according to the ratio of sediment to water in the moving flow; they can be related to one another by dilution or bulking processes. Processes that may release large volumes of water, or sediment, or both, are listed near the top of the diagram. Paths drawn along the margins of the chart indicate the types of flow phenomena that may be expected to result depending on the degree of dilution or bulking.

positional unit composed of a lower part displaying characteristics of hyperconcentrated-flow deposits grading upward into an upper debris-flow facies (Pierson and Scott, 1985; Smith, 1986; Fig. 6a). The hyperconcentrated-flow facies thicken downslope as the debris-flow deposited material diminishes. Smith (1986, 1987a,b, 1988) has demonstrated that similar features can be identified in ancient volcaniclastic deposits (Fig. 6b).

Pierson and Scott (1985), Scott (1988), and Major and Scott (1988) consider all hyperconcentrated flows to be generated by the dilution of debris flows, necessitating that each hyperconcentrated-flow deposit, which they term the

FIG. 6.—Single-event deposits in which massive, debris-flow deposits overlie basal, stratified, hyperconcentrated-flow units produced by more watery flows that were probably generated by dilution of the debris-flow front. (A) Deposit of May 1984 lahar, North Fork Toutle River, near Mount St. Helens, Washington; lens cap for scale; (B) late Miocene laharic deposit, Jemez Mountains, New Mexico.

"lahar runout facies," correlates with a debris-flow event upslope. We see no reason why this has to be the case. It seems plausible that some flows originate as hyperconcentrated flows or as streamflows that bulk up to sediment concentrations typical of hyperconcentrated flow (Fig. 5). Nonetheless, dilution is an important process for producing hyperconcentrated flows, as indicated by observations of Mount St. Helens flows (Pierson and Scott, 1985; Scott, 1988) and downslope outcrop transitions from debris-flow-dominated to hyperconcentrated-flow-dominated facies (Smith, 1987b, 1988). Further downslope dilution of hyperconcentrated flows to dilute, normal streamflows has also been documented (Scott, 1988). This transition undoubtedly results not only from continued mixing of water with the flow but decrease in sediment concentration because of deposition.

Bulking and dilution can occur in the same lahar event. The 1982 eruption of Mount St. Helens (Waitt and others, 1983; Pierson and Scott, 1985) released a relatively dilute, snowmelt flood that bulked up to become a debris flow and then was diluted downslope to hyperconcentrated and, eventually, dilute streamflow. Appreciation of these complex transformations, in which more than one process is active at a fixed point through time, and at different points at the same time, is fundamental to understanding the complexities of the resulting deposits.

*Debris flows derived from debris avalanches.—*

Debris-flow deposits are commonly associated with those of debris avalanches. In the case of valley-confined avalanches, the related debris-flow facies typically occur downslope of the avalanche materials (Fairchild, 1987; Glicken, 1986 and this volume), whereas debris-flow deposits form a laterally adjacent, peripheral apron to unconfined avalanche deposits (Neall, 1979; Palmer and others, this volume). Clearly, the wetter debris flows are derived from the debris avalanches, although the mechanics of debris-flow generation are not always evident. A large, destructive debris flow during the 1980 eruption of Mount St. Helens was caused by liquefaction of debris-avalanche material (Fairchild, 1987). In other cases, it seems likely that water-rich portions of the avalanche separate from the avalanche and extend onward as debris flows (Crandell, 1989; Palmer and others, this volume).

*Transformations from pyroclastic flows and surges to aqueous flows.—*

Pyroclastic flows and surges (Carey, this volume) are primary volcanic phenomena in which hot pyroclastic fragments are transported by a high-temperature fluid of primarily magmatic gases and fine ash. The close spatial and temporal occurrence of pyroclastic flows or surges and lahars has suggested possible genetic relationships between these flow processes (see Neall, 1976, and Major and Newhall, 1989, for reviews). This relationship has been most recently evaluated at Mount St. Helens (Janda and others, 1981; Pierson, 1985; Brantley and Waitt, 1988; Waitt, 1989) and at Nevado del Ruiz (Naranjo and others, 1986; Lowe and others, 1986). It remains unclear, however, whether direct transition from gaseous transport to aqueous transport occurs by mixing of moving pyroclastic flows and surges with stream or snowmelt water (Pierson, 1985) or whether the debris flows are generated by bulking of snowmelt or streams with pyroclastic material after the pyroclastic flow or surge has come to rest (Brantley and Waitt, 1988; Waitt, 1989). Only the first mechanism would represent a direct transformation between the two processes.

## *Origins of Lahars and Resulting Deposits*

The most exhaustive and complete classification of lahars is that of Neall (1976). The flow chart in Figure 5 is an effort not so much to classify lahars as to portray the complex origins and transformations between processes comprising lahar events.

In general, lahars can be divided into those that are a direct result of eruptive activity and those that are not temporally related to eruptions. The latter may still be indirectly related to previous eruptions if heavy rains or crater-lake overflows mobilize recently erupted, loose pyroclastic and autoclastic debris. Lahars and debris avalanches unrelated to eruptions can be as voluminous and destructive as those triggered by eruptions (e.g., Healy, 1954; Richardson, 1968; Crandell, 1971; Williams, 1952; Siebert, 1984) and the deposits of the two types can be indistinguishable.

With the exception of lahars related to debris avalanches, most fragments in the resulting deposits are of pyroclastic or autoclastic origin (Fisher and Schmincke, 1984). Large volumes of such debris blanket the steep slopes of most volcanoes. Epiclastic materials, generated by weathering and erosion of pre-existing rocks, are less abundant and are generally incorporated into flows by bulking of alluvium. The composition and grain size of clasts within debris flows are, therefore, strongly dependent upon the nature of eruptive activity at the source.

Although the end products of the various evolutionary pathways (Fig. 5) are commonly similar in composition, texture, and structure, it may be possible, in some cases, to deduce specific triggering mechanisms from the resulting deposits.

*Debris flows related to debris avalanches.—*

The existence of large-volume debris-flow deposits, commonly 10 m or more thick and containing clasts meters in diameter that have moved tens of kilometers, implies mobilization of enormous volumes of material at the source. The simplest explanation for this phenomenon would call upon mass failure of a large part of the source volcano. If large volumes of water were liberated at the site of the landslide, the transport process may have been a debris flow from the outset (Crandell and Waldron, 1956; Crandell, 1989); otherwise a transition from debris avalanche to debris flow by one of the mechanisms described in the previous section may occur. Although suggestive, large volume or long runout distance of a debris-flow deposit are not, by themselves, adequate indicators of a genetic relationship to debris avalanches. There are other features that may strengthen the association. Debris-flow deposits related to debris avalanches in the Cascade Range tend to contain a relatively high abundance of clay (3 to 5 percent) largely related to hydrothermally altered or weathered avalanche material (Crandell and Waldron, 1956; Scott, 1988). Large brecciated megaclasts, as much as 10 m across, generated by debris avalanches, may be transported by derivative debris flows for ten of kilometers (Scott, 1988; Smith, unpub. data). The occurrence of these large shattered clasts in thick debris-flow deposits that are tens of kilometers from their source strongly suggest a relationship to rockslide-debris avalanche processes (Fig. 7a).

*Debris flows and floods related to lake release or ejection.—*

The type lahars of Indonesia are generated by eruptions through crater lakes (Neall, 1976). Crater-lake water may also be incorporated into flows by collapse of walls of craters containing lakes (e.g., Healey, 1954) and nonexplosive displacement of lake water by extrusions in the crater. Large debris flows or floods can also be generated by the release of lakes formed when lava flows, debris avalanches, or other accumulations of volcaniclastic debris dam valleys draining the volcanic slopes (e.g., Scott, 1988). Most flows generated by lake breakouts are probably impoverished in sediment, but may bulk up as they move downslope. If the volume of water is large, the final flow will probably be a hyperconcentrated flow. Flow deposits related to lake breakouts can be expected to contain heterolithic clast assemblages, a high content of rounded clasts bulked up from alluvium, and outsize clasts of the natural-dam material. Scott (1988) has documented a series of mid-Holocene debris-flow deposits near Mount St. Helens that are composed almost entirely of rounded cobbles and boulders and scat-

FIG. 7.—Clast and matrix features of debris-flow deposits. (A) Close view of 8-m-diameter debris-flow clast showing crackle brecciation suggestive of debris-avalanche origin (compare with Fig. 4a); early Pleistocene (?) Tivives Formation, Bajamar, Costa Rica. (B) Radial prismatic joints characteristic of *in situ* cooling and thermal contraction of hot clast; clast in lithic pyroclastic-flow deposit, late Miocene Ellensburg Formation, Washington. (C) Thick debris-flow deposit containing a prominent clast, about 1 m in diameter, with radial cooling joints; Oligocene Espinaso Formation, New Mexico. (D) Pumiceous debris-flow deposit illustrating vesicular matrix; Holocene, near Taupo, New Zealand.

tered angular megaclasts of altered, shattered volcanic rock. This association suggests that the debris flows originated as lake-breakout flood surges resulting from failure of a debris-avalanche dam (Scott, 1988). Also, because hydrothermal circulation is associated with most crater lakes, fragments derived from floors or walls of crater lakes are often intensely altered.

*"Hot" lahars and their deposits.—*

Many lahars involve mobilization of hot volcanic materials. These may be derived from 1) transformation of pyroclastic flows and surges to aqueous flows, and 2) erosion of hot pyroclastic debris or fragmented lava flows by snowmelt or subsequent heavy rain. The resulting deposits tend to be dominated by pyroclastic or autoclastic fragments and may be nearly monolithologic. Where composed largely of pyroclastic material, massive debris-flow deposits can closely resemble unwelded ignimbrites or the deposits of lithic pyroclastic flows (Smith, 1986). In such cases, paleomagnetic methods can be used to estimate the emplacement temperature and, possibly, the mode of deposition. Debris flows derived from hot pyroclastic deposits may transport clasts having temperatures up to several hundred degrees Celsius. These can be recognized from rock-magnetism studies because all such clasts will yield magnetic-pole orientations consistent with the field orientation at the time of deposition rather than random orientations expected from cold deposition (Hoblitt and Kellogg, 1979; Crandell, 1987; Arguden and Rodolfo, 1990). Depending on thermal conductivity and cooling rates, some lithic clasts in pyroclastic or debris flows develop radial, thermal contraction joints as they cool (Fig. 7b,c). Later movement of the cooled clasts generally causes disintegration along these fractures. Hence, the occurrence of radially jointed clasts in a lahar deposit suggests that the clasts were hot when emplaced (Hoblitt and Kellogg, 1979; Crandell, 1980).

The rheological behavior of debris flows and floods transporting large volumes of hot fragments could be expected to differ from that of flows transporting material at ambient temperature. Locally, the interstial fluid might consist of water and vapor rather than water alone (Arguden and Rodolfo, 1990). More watery flows would tend to have lower kinematic viscosities at higher temperatures and rising viscosities as the flows cool. Hence, transport mechanics would vary as the flows cool downslope (Smith, 1986).

Arguden and Rodolfo (1990) have attempted to evaluate these differences in observed hot and cold debris-flow deposits on the flanks of Mayon volcano in the Philippines. At the time of deposition, the two types of flows are easily distinguished on the basis of related thermal phenomena. After cooling and burial, however, the deposit differences are subtle. Inversely graded basal zones tend to be thicker in hot debris-flow deposits, emplaced during or immediately after eruptions, than in cold debris-flow deposits, triggered by monsoons 2 years after the eruptions. The hot debris-flow deposits also tend to be more poorly sorted, contain more fine ash, and are distributed to greater distances. Arguden and Rodolfo (1990) argue, using BTh/MPS plots, that cold flows have greater cohesive strength.

Care is necessary in interpreting features that may be used as evidence for the presence of a vapor phase in volcanic debris flows. Debris-flow matrices, modern and ancient, are commonly vesicular (Fig. 7d), a result of entrapped air, or, in some cases, voids left following decay of plant material (Johnson, 1970; Crandell, 1971). This phenomenon is well known not only from volcanic debris-flow deposits (Crandell, 1971; Fisher and Schmincke, 1984; Major and Voight, 1986; Arguden and Rodolfo, 1990) but also from nonvolcanic deposits of varied origins (Johnson, 1970; Sharp and Nobles, 1953; Bull, 1964; Wasson, 1978; Allen, 1982). The presence of vesicles should not, therefore, be taken as evidence for the presence of steam (cf. Arguden and Rodolfo, 1990). In rare instances, debris-flow or hyperconcentrated-flow deposits have been observed to contain narrow, tabular or cylindrical, vertical fines-depleted zones that resemble vertical structures attributed to volatile escape in ignimbrites (Crandell, 1987; Arguden and Rodolfo, 1990; Smith, unpub. data). Similar structures, however, can be produced by expulsion of water through rapidly deposited sediment (Johnson, 1986; Collinson and Thompson, 1982; Nocita, 1989), indicating that they are not diagnostic of gas escape.

*Direct extrusion of debris-flows.—*

In relatively low-temperature hydrovolcanic eruptions, water may exit a volcanic vent in a liquid, rather than vapor, state. Slurries of tephra have been observed to exit vents (Wohletz and Crowe, 1978; Smith, 1980) and the occurrence of units resembling debris-flow deposits interbedded with other hydrovolcanic products, such as base-surge deposits, suggests that debris flows can be a primary eruptive phenomenon (Sheridan and Wohletz, 1983). The importance of directly extruded debris flows in the sedimentological record of volcanic terranes is rarely considered. Lydon (1968) noted voluminous late Tertiary volcanic debris-flow aprons in northern California that are associated with breccia-filled dikes having textures like those of the debris-flow deposits. He suggested direct extrusion of hundreds to thousands of cubic kilometers of debris-flow sediments with water derived from both the phreatic zone and exsolved magmatic volatiles. To our knowledge, Lydon's interpretation has yet to be evaluated critically by petrologists. Conventional wisdom suggests that external water sources are adequate to account for large-volume debris-flow accumulations, but perhaps this is not always the case.

## CONCLUSIONS

Volcano-hydrologic events commonly generate high-discharge flow events referred to as lahars. Lahar flow processes vary spatially and temporally between hyperconcentrated flow and debris flow. Deposits of these processes may be found with those of dilute streamflows and debris avalanches. None of these processes is unique to volcanic regions, but several significant differences should be noted when comparing with nonvolcanic examples.

Most volcanic debris flows have a larger discharge and volume than those in alpine and alluvial-fan settings. They rarely contain significant clay, suggesting that resistance to shear is more a function of high-grain concentration than a result of cohesion. These two characteristics contribute to the long runout distances of many volcanic-debris flows and their tendency to veneer landscapes with deposits whose thickness may not be related to the depth of the moving flow.

Hyperconcentrated flows are frequently generated by dilution of debris flows by streamwater. These flows transport sediment by a combination of dampened turbulence, grain-dispersive pressure, and fluid buoyancy. Deposits are recognized by features intermediate between anud gradational to those ascribed to dilute streamflows and debris flows.

The complexity of many volcaniclastic facies sequences rests in the variety of ways in which flowing sediment/water mixtures are triggered, and genetic relationships between flow types. Sediment concentration, flow rheology, sediment transport, and depositional mechanics may vary spatially, or temporally at a fixed depositional site, as sediment is bulked into the flow, water is added, or sediment is deposited. These changes produce complex lateral and vertical facies sequences produced by single events. Textural and compositional characteristics of deposits may yield interpretation of flow-generation mechanisms and relationships to primary pyroclastic or debris-avalanche deposits.

## ACKNOWLEDGMENTS

Ongoing research on volcaniclastic sedimentation by Smith is sponsored by the donors to the Petroleum Research Fund of the American Chemical Society. Smith acknowledges the Caswell Silver Foundation at the University of New Mexico for overseas research. Smith also appreciates the many on-the-outcrop discussions with K. M. Scott, V. E. Neall, B. A. Palmer, B. Alloway, and S. Donoghue that helped to clarify some of the concepts presented here. We appreciate the thoughtful reviews of W. J. Fritz and R. V. Fisher.

## REFERENCES

Allen, J. R. L., 1982, Sedimentary Structures, their Character and Physical Basis: Amsterdam, Elsevier, v. 2, 663 p.

Arguden, A. T. and Rodolfo, K. S., 1986, Sedimentary facies and tectonic implications of lower Mesozoic alluvial-fan conglomerates of the Newark basin, northeastern United States: Sedimentary Geology, v. 51, p. 97–118.

Arguden, A. T. and Rodolfo, K. S., 1990, Sedimentologic and dynamic differences between hot and cold laharic debris flows of Mayon Volcano, Philippines: Geological Society of America Bulletin, v. 102, p. 865–876.

Beverage, J. P., and Culbertson, J. K., 1964, Hyperconcentrations of suspended sediment: American Society of Civil Engineers, Proceedings, Hydraulics Division Journal, v. 90, no. HY6, p. 117–128.

Bluck, B. J., 1967, Deposition of some Upper Old Red Sandstone conglomerates in the Clyde area: A study in the significance of bedding: Scottish Journal of Geology, v. 3, p. 139–167.

Brantley, S. R., and Waitt, R. B., 1988, Interrelationships among pyroclastic surge, pyroclastic flow, and lahars in Smith Creek valley during the first minutes of 18 May 1980 eruption of Mount St. Helens, U.S.A.: Bulletin of Volcanology, v. 50, p. 304–326.

Bull, W. B., 1964, Alluvial fans and near-surface subsidence in western Fresno County, California: U.S. Geological Survey Professional Paper 437-A, 71 p.

Collinson, J. D., and Thompson, D. B., 1982, Sedimentary structures: London, Allen and Unwin, 194 p.

Costa, J. E., 1988, Rheologic, geomorphic, and sedimentologic differentiation of water floods, hyperconcentrated flows, and debris flows, *in* Baker, V. R., Kochel, R. C., and Patton, P. C., eds., Flood Geomorphology: New York, J. Wiley and Sons, p. 113–122.

Crandell, D. R., 1971, Post-glacial lahars from Mt. Rainier volcano, Washington: U.S. Geological Survey Professional Paper 677, 75 p.

Crandell, D. R., 1980, Recent eruptive history of Mount Hood, Oregon, and potential hazards from future eruptions: U.S. Geological Survey Bulletin 1492, 81 p.

Crandell, D. R., 1987, Deposits of pre-1980 pyroclastic flows and lahars from Mount St. Helens volcano, Washington: U.S. Geological Survey Professional Paper 1444, 91 p.

Crandell, D. R., 1989, Gigantic debris avalanche of Pleistocene age from ancestral Mount Shasta volcano, California, and debris-avalanche hazard zonation: U.S. Geological Survey Bulletin 1861, 32 p.

Crandell, D. R., and Waldron, H. H., 1956, A recent volcanic mudflow of exceptional dimensions from Mt. Rainier, Washington: American Journal of Science, v. 254, p. 349–362.

Fairchild, L. H., 1987, The importance of lahar initiation processes, *in* Costa, J. E., and Wieczorek, G. F., eds., Debris Flows/Avalanches: Process, Recognition, and Mitigation: Geological Society of America Reviews in Engineering Geology, v. 7, p. 51–62.

Fisher, R. V., 1983, Flow transformations in sediment gravity flows: Geology, v. 11, p. 273–274.

Fisher, R. V., and Schmincke, H.-U., 1984, Pyroclastic Rocks: Berlin, Springer-Verlag, 472 p.

Gilkey, K. E., 1983, Sedimentology of the North Fork and South Fork Toutle River mudflows generated during the 1980 eruption of Mount St. Helens: Santa Barbara, University of California, Unpub. M.S. Thesis, 226 p.

Glicken, H., 1986, Rockslide-debris avalanche of May 18, 1980, Mount St. Helens volcano, Washington: Santa Barbara, University of California, Unpub. Ph.D. Dissertation, 303 p.

Healy, J., 1954, Origin of flood and Ruapehu lahars, *in* Tangiway Railway Disaster, Report of Board of Inquiry: Wellington, Government Printer, p. 6–8, 28–31.

Hoblitt, R. P., and Kellogg, K. S., 1979, Emplacement temperature of unsorted and unstratified deposits of volcanic rock debris as determined by paleomagnetic techniques: Geological Society of America Bulletin, v. 90, pt. 1, p. 633–642.

Iverson, R. M., and Denlinger, R. P., 1987, The physics of debris flows–A Conceptual assessment, *in* Beschta, R. L., Blinn, T., Grant, G. E., Swanson, F. J., and Ice, G. G., eds., Erosion and Sedimentation in the Pacific Rim: International Association of Hydrological Sciences Publication 165, p. 155–165.

Janda, R. D., Scott K. M., Nolan, K. M., and Martinson, H. A., 1981, Lahar movement, effects, and deposits, *in* Lipman, P. W., and Mullineaux, D. R., eds., The 1980 Eruptions of Mount St. Helens, Washington: U.S. Geological Survey Professional Paper 1250, p. 461–478.

Johnson, A. M., 1970, Physical Processes in Geology: San Francisco, Freeman, Cooper and Co., 577 p.

Johnson, A. M., and Hampton, M. A., 1969, Subaerial and Subaqueous Flow of Slurries: Final report to U.S. Geological Survey, Branner Library, Stanford University, California (unpaginated).

Johnson, S. Y., 1986, Water-escape structures in coarse-grained, volcaniclastic, fluvial deposits of the Ellensburg Formation, south central Washington: Journal of Sedimentary Petrology, v. 56, p. 905–910.

Kesel, R. H., 1973, Notes on the lahar landforms of Costa Rica: Zeitschrift fur Geomorphologie, v. 18, p. 78–91.

Lowe, D. R., 1979, Sediment gravity flows: Their classification and some problems of application to natural flows and deposits, *in* Doyle, L. J., and Pilkey, O. H. Jr., eds., Geology of Continental Slopes: Society of Economic Paleontologists and Mineralogists Special Publication 27, p. 75–82.

Lowe, D. R., 1982, Sediment gravity flows II. Depositional models with special reference to the deposits of high-density turbidity currents: Journal of Sedimentary Petrology, v. 52, p. 279–297.

Lowe, D. R., Williams, S. N., Leigh, H., Connor, C. B., Gemmell, J. B., and Stoiber, R. E., 1986, Lahars initiated by the 13 November 1985 eruption of Nevado del Ruiz, Colombia: Nature, v. 324, p. 51–53.

Lydon, P. A., 1968, Geology and lahars of the Tuscan Formation, northern California: Geological Society of America Memoir 116, p. 441–475.

MAJOR, J. J., AND NEWHALL, C. G., 1989, Snow and ice perturbations during historical volcanic eruptions and the formation of lahars and floods: Bulletin of Volcanology, v. 52, p. 1–27.

MAJOR, J. J., AND SCOTT, K. M., 1988, Volcaniclastic sedimentation in the Lewis River Valley, Mount St. Helens, Washington—Processes, Extent, and Hazards: U.S. Geological Survey Bulletin 1383-D, 38 p.

MAJOR, J. J., AND VOIGHT, B., 1986, Sedimentology and clast orientations of the 18 May 1980 southwest-flank lahars, Mount St. Helens, Washington: Journal of Sedimental Petrology, v. 56, p. 691–705.

MIDDLETON, G. V., AND SOUTHARD, J. B., 1984, Mechanics of sediment movement: Society of Economic Paleontologists and Mineralogists, Eastern Section, Short Course 3, 401 p.

NARANJO, J. L., SIGURDSSON, H., CAREY, S. N., AND FRITZ, W. J., 1986, Eruption of Nevado del Ruiz volcano, Colombia, on 13 November 1985: Tephra fall and lahars: Science, v. 233, p. 941–963.

NEALL, V. E., 1976, Lahars–Global occurrence and annotated bibliography: Victoria University of Wellington, New Zealand, Publication 5, 18 p.

NEALL, V. E., 1979, Geologic map of New Zealand (New Plymouth, Egmont, and Manaia): New Zealand Geological Survey Sheets P19, P20, P21, scale 1:50,000 plus 35 p. text.

NEMEC, W., AND STEEL, R. J., 1984, Alluvial and coastal conglomerates: Their significant features and some comments on gravelly mass-flow deposits, *in* Koster, E. H., and Steel, R. J., eds., Sedimentology of Gravels and Conglomerates: Canadian Society of Petroleum Geologists Memoir 10, p. 1–31.

NOCITA, B. W., 1989, Soft-sediment deformation (fluid escape) features in a coarse-grained pyroclastic-surge deposit, north-central New Mexico, Reply: Sedimentology, v. 36, p. 947–949.

OLLIER, C. D., AND BROWN, M. J. F., 1971, Erosion of a young volcano in New Guinea: Zeitschrift fur Geomorphologie, v. 15, p. 12–28.

PALMER, B. A., AND NEALL, V. E., 1989, The Murimotu Formation–9500 year old deposits of a debris avalanche and associated lahars, Mount Ruapehu, North Island, New Zealand: New Zealand Journal of Geology and Geophysics, v. 32, p. 477–486.

PALMER, B. A., AND WALTON, A. W., 1990, Accumulation of volcaniclastic aprons in the Mount Dutton Formation (Oligocene-Miocene), Marysvale volcanic field, Utah: Geological Society of America Bulletin, v. 102, p. 734–748.

PIERSON, T. C., 1985, Initiation and flow behavior of the 1980 Pine Creek and Muddy River lahars, Mount St. Helens, Washington: Geological Society of America Bulletin, v. 96, p. 1056–1069.

PIERSON, T. C., AND COSTA J. E., 1987, A rheologic classification of subaerial sediment-water flows: Geological Society of America Reviews in Engineering Geology, v. 7, p. 1–12.

PIERSON, T. C., AND SCOTT K. M., 1985, Downstream dilution of a lahar: Transition from debris flow to hyperconcentrated streamflow: Water Resources Research, v. 21, p. 1511–1524.

RICHARDSON, D., 1968, Glacier outburst flood in the Pacific Northwest: U.S. Geological Survey Professional paper 600-D, p. 79–86.

RODOLFO, K. S., 1989, Origin and early evolution of lahar channel at Mabinit, Mayon Volcano, Philippines: Geological Society of America Bulletin, v. 101, p. 414–426.

SCOTT, K. M., 1988, Origins, behavior, and sedimentology of lahars and lahar-runout flows in the Toutle-Cowlitz River system: U.S. Geological Survey Professional Paper 1447-A, 74 p.

SCHMITT, J. G., AND OLSON, T. J., 1986, Hyperconcentrated flood-flow deposition of coarse-grained sediment on temperate, semi-arid to humid alluvial fans: Geological Society of America Abstracts with Programs, v. 18, p. 741.

SEGERSTROM, K., 1950, Erosion studies at Paricutin, State of Michoacan, Mexico: U.S. Geological Survey Bulletin 965-A, 165 p.

SHARP, R. P., AND NOBLES, L. H., 1953, Mudflow of 1941 at Wrightwood, southern California: Geological Society of America Bulletin, v. 64, p. 547–560.

SHERIDAN, M. F., AND WOHLETZ, K. H., 1983, Hydrovolcanism: Basic considerations and review: Journal of Volcanology and Geothermal Research, v. 17, p. 1–29.

SIEBERT, L., 1984, Large volcanic debris avalanches: Characteristics of source areas, deposits, and associated eruptions: Journal of Volcanology and Geothermal Research, v. 22, p. 163–197.

SMITH, G. A., 1980, Non-magmatic historic eruptions and the origin of volcanic mudflows and floods of Mt. Pelée, Martinique: Coompass, v. 57, p. 33–38.

SMITH, G. A., 1986, Coarse-grained nonmarine volcaniclastic sediment: Terminology and depositional process: Geological Society of America Bulletin, v. 97, p. 1–10.

SMITH, G. A., 1987a, Sedimentology of volcanism-induced aggradation in fluvial basins: Examples from the Pacific Northwest, U.S.A., *in* Ethridge, F. G., Flores, R. M., and Harvey, M. G., eds., Recent Developments in Fluvial Sedimentology: Society of Economic Paleontologists and Mineralogists Special Publication 39, p. 217–228.

SMITH, G. A., 1987b, The influence of explosive volcanism on fluvial sedimentation: The Deschutes Formation (Neogene) in central Oregon: Journal of Sedimentary Petrology, v. 57, p. 613–629.

SMITH, G. A., 1988, Sedimentology of proximal to distal volcaniclastics dispersed across an active foldbelt: Ellensburg Formation (late Miocene, central Washington): Sedimentology, v. 35, p. 953–977.

SMITH, G. A., AND FRITZ, W. J., 1989, Volcanic influences on terrestrial sedimentation: Geology, v. 17, p. 375–376.

SWANSON, F. J., COLLINS, B., DUNNE, T., AND WICHERSKI, B. P., 1982, Erosion of tephra from hillslopes near Mt. St. Helens and other volcanoes: Public Works Research Institute, Japan, Technical Memorandum 1908, p. 183–221.

VAN BEMMELEN, R. W., Geology of Indonesia: The Hague, The Netherlands, Government Printing Office, 732 p.

VESSEL, R. K., AND DAVIES, D. K., 1981, Nonmarine sedimentation in an active fore arc basin, *in* Ethridge, F. G. and Flores, R. M., eds., Recent and Ancient Nonmarine Depositional Environments: Society of Economics Paleontologists and Mineralogists Special Publication 31, p. 31–45.

VOIGHT, B., GLICKEN, H. X., JANDA R. H., AND DOUGLASS, P. M., 1981, Catastrophic rockslide avalanche of May 18, *in* Lipman, P. W., and Mullineaux, D. R., eds., The 1980 Eruptions of Mount St. Helens: U.S. Geological Survey Professional Paper 1250, p. 347–377.

WAITT, R. B. Jr., 1989, Swift snowmelt and floods (lahars) caused by great pyroclastic surge at Mount St. Helens volcano, Washington, 18 May 1980: Bulletin of Volcanology, v. 52, p. 138–157.

WAITT, R. B. Jr., PIERSON, T. C., MACLEOD, N. S., JANDA, R. J., VOIGHT, B., AND HOLCOMB, R. T., 1983, Eruption-triggered avalanche, flood, and lahar at Mount St. Helens—Effects of winter snowpack: Science, v. 221, p. 1394–1397.

WALDRON, H. H., 1967, Debris flow erosion control problems caused by the ash eruptions of Irazu Volcano, Costa Rica: U.S. Geological Survey Bulletin 1241-I, 37 p.

WALTON, A. W., AND PALMER, B. A., 1988, Lahar facies of the Mount Dutton Formation (Oligocene-Miocene) in the Marysvale volcanic field, southwestern Utah: Geological Society of America Bulletin, v. 100, p. 1078–1091.

WARESBACK, D. B., AND TURBEVILLE, B. N., 1990, Evolution of a Plio-Pleistocene volcanogenic alluvial fan: The Puye Formation, Jemez Mountains, New Mexico: Geological Society of America, v. 102, p. 298–314.

WASSON, R. J., 1978, A debris flow at Reshun, Pakistan Hindu Kush: Geografiska Annaler, v. 60, p. 151–159.

WELLS, S. G., AND HARVEY, A. M., 1987, Sedimentologic and geomorphic variations in storm-generated alluvial fans, Howgill Fells, northwest England: Geological Society of America Bulletin, v. 98, p. 182–198.

WILLIAMS, H., 1952, Volcanic history of the Meseta Central Occidental, Costa Rica: University of California Publications in the Geological Sciences, v. 29, p. 145-180.

WOHLETZ, K. H., AND CROWE, B. M., 1978, Development of lahars during the 1976 activity of La Soufriere de Guadeloupe, F.W.I.: Geological Society of America Abstracts with Programs, v. 10, p. 154.

YARNOLD, J. C., AND LOMBARD, J. P., 1989, A facies model for large rock-avalanche deposits formed in dry climates, *in* Colburn, I. P., Abbott, P. L., and Minch, J., eds., Conglomerates in Basin Analysis: A Symposium Dedicated to A.O. Woodford: Pacific Section, Society of Economic Paleontologists and Mineralogists, v. 62, p. 9–32.

# RAIN-LAHAR GENERATION AND SEDIMENT-DELIVERY SYSTEMS AT MAYON VOLCANO, PHILIPPINES

KELVIN S. RODOLFO AND A. TEVFIK ARGUDEN
*Department of Geological Sciences M/C 186 University of Illinois at Chicago, P.O. Box 4348, Chicago, Illinois 60680*

ABSTRACT: At least 59 rain lahars have occurred on Mayon Volcano, on the Philippine island of Luzon, since its last eruption in 1984. In the southeastern sector of the volcano, where lahar activity has been greatest, we have evaluated the generation of these flows and their effects by measuring lahar-generating rainfall on the slopes during each main typhoon and rain season (September-December) since 1986 and by mapping the adjacent Mabinit and Matanag channels in detail since 1985. Sixteen debris flows occurred during the monitoring period. Each was triggered by a rainfall that lasted at least 1.4 hours, delivered a minimum of 40 mm of rain at an overall rate of 11 mm/h or more, and included at least one 10-min interval during which at least 10 mm fell. The empirical relationship between the threshold values of lahar-triggering rainfall duration (D) and intensity (I) is the power function $I = 27.3D^{-0.38}$, a substantially higher threshold than one determined for debris flows world-wide. This higher threshold is due to the coarse, granular, and very porous volcaniclastic surface materials, which also render the role of antecedent rain insignificant in generating lahars.

Comparable sediment-delivery systems are situated along the ESE (Basud) and SSE (Bonga) radii of the volcano. Each system is composed of a summit ravine with a fan of pyroclastic deposits at its base, and a pair of channels, one along each side of the fan. Only one of each channel pair continues to be a major lahar conduit. Basud Channel, the principal recipient of runoff and sediment from its ravine and fan of the Basud system, experienced the most frequent lahars in the first year after the eruption; however, its principal source of lahar sediment, an ash capping on Basud pyroclastic fan, was depleted very quickly, thus it has experienced debris flows only twice since 1986. Over the same period, 14 debris flows have occurred along Mabinit Channel because its delivery system has the largest catchment area in its sector and includes the largest ravine of the edifice. Bonga Ravine is deeply incised into a composite lava-tephra sequence; strata cropping out in the steep ravine sides avalanche frequently. Debris that collects along the axis of the ravine during the relatively dry months is mobilized into lahars by the first large storms of the typhoon season.

Erosion and deposition by lahars keep the active portion of Mabinit Channel narrow and deep. The debris-flow phases of a lahar spread out upon entering a widening stretch of channel, and the thinner lateral portions stop and aggrade while the thicker central portions continue moving downchannel, leaving a narrowed channel. Waning-stage or subsequent hyperconcentrated and flood flows cut down through the new debris-flow deposits. A narrow, deep channel results, constricted between the vertical walls of new debris-flow terraces. The terraces also store material for subsequent lahars to erode and incorporate.

A debris-flow fan at the end of Mabinit Channel, initially produced by unconfined debris flows during the 1984 eruption, has continued to evolve. The apex of the fan has twice been extended headward by avulsing debris flows, during Typhoons Saling in 1985 and Unsang in 1989. These events have added about 9 percent of area to the east side of the fan.

When debris flows stop occurring along a channel such as Matanag Channel, it is quickly widened by laterally eroding floods and hyperconcentrated flows, which also deposit and thus aggrade the channel floor.

## INTRODUCTION

### *Definition of Lahar*

One of the most destructive phenomena associated with composite volcanoes is the *lahar*. The term, of Indonesian origin, has a confusing and interesting history of usage in geology. Scrivenor introduced it in 1929, in a report on diamicts that were produced in 1919 when an eruption ejected large volumes of water from Gunong (Mount) Kelut's crater lake, generating catastrophic flows of volcanic debris mixed with water. Scrivenor translated lahar as "mudstream," but Indonesian dictionaries define it simply as "lava"; the root word, manipulated into verb form, means "to erupt." Lay inhabitants of volcano slopes and aprons are not likely to distinguish between flows of lava and flows of debris variably admixed with water; all of these more or less fluid downslope movements are hazards from which to run away, not to examine closely. In 1949, van Bemmelen (p. 191) expanded the definition: ". . . *volcanic breccias, transported by water . . . a mudflow, containing debris and angular blocks of chiefly volcanic origin*." In a strict sense, "*mudflow*" and Scrivenor's "*mudstream*" are sedimentologically incorrect for most lahars, which contain very little silt and clay. More seriously, van Bemmelen's designation of *lahar* for the phenomenon as well as its deposit probably encouraged even looser usage. Then, shortly after the pioneering work on debris flows by Yano and Daido (1965) and by Johnson (1970), some workers began to use the term in too *narrow* a way, as a synonym for both "volcanic debris flow" and the deposits left by such flows. The compounded confusion eventually led Smith (1986) to recommend that the term be discarded entirely. The term, however, usefully denotes a *specifically volcanogenic, hazardous* phenomenon, a *natural unit* that is commonly *complex in flow behavior*. We have found, as have workers on Mount St. Helens (Pierson and Scott, 1985), that a single lahar can change in character from debris flow to hyperconcentrated streamflow, and *vice versa*. Debris flows, with water contents ranging from as little as 10 weight percent up to no more than about 25 weight percent (Pierson, 1986; Pierson and Costa, 1987), are non-Newtonian fluids that move as fairly coherent masses in what is thought to be predominantly laminar fashion, although the relative importance of laminar and turbulent flow has not been established and is debatable. Hyperconcentrated streamflows contain 25- to about 40-weight-percent water (Beverage and Culbertson, 1964); thus, they possess some yield strength, but these flows are characteristically turbulent. At Mayon Volcano on the Philippine island of Luzon (Figs. 1, 2), it is not unusual for a single lahar to involve more than one debris-flow phase, with transitional as well as precursor and waning-stage hyperconcentrated-streamflow phases. Given appropriate circumstances, either flow type can erode or deposit along any reach of its channel, and the morphologic and sedimentologic effects of such a lahar can be very complex.

Sedimentation in Volcanic Settings, SEPM Special Publication No. 45

Our usage of *lahar* in this report is consonant with the definition agreed upon at an international conference of volcaniclastic sedimentologists in 1988: ". . . *a rapidly flowing mixture of rock debris and water (other than normal stream flow) from a volcano. A lahar is an event; it can refer to one or more discrete processes* [such as debris flow and hyperconcentrated streamflow], *but does not refer to a deposit*" (Smith and Fritz, 1989, p. 375).

### *Rain Lahars*

Generation of lahars by crater-lake floods, such as at Gunong Kelut, is relatively rare. Lahars can form in a variety of other ways, as summarized by Neall (1976). The most intensely studied ones are those that occurred at Mount St. Helens during and after its 1980 eruption (Janda and others, 1981; Pierson 1985, 1986; Fairchild, 1987; Scott, 1988; Waitt, 1989). Of these, the largest were formed when debris avalanches and pyroclastic surges entered streams. Others, and also the lethal flows at Nevado del Ruiz in Colombia in 1985 (Lowe and others, 1986; Naranjo and others, 1986), were generated when hot ejecta melted and eroded summit snow and ice. In a recent global review, Major and Newhall (1989) have documented more than 40 volcanoes at which lahars have been triggered by the melting of snow and ice during eruptions.

Intense rainfall and the resulting runoff on volcano slopes trigger "*rain lahars,*" which include not only hot flows that occur during eruptions, but also "cold" flows at ambient temperatures during repose periods, months or even years after an eruption. About 600 (46 percent) of the world's active volcanoes are situated in the humid tropics, and most of these lack ice caps. Thus, rain lahars probably are common, and their deposits must also be abundant in the stratigraphic record. But they are also the most poorly documented and the least understood, because they occur mainly in developing countries where scientifically trained personnel are scarce. Cold-rain lahars are less likely to be scientifically evaluated than those that occur during eruptions, which tend to draw scientists from other countries. The hazards posed by cold-rain lahars may also be greater because the threatened populations have not been alerted by the spectacle and noise that accompany an eruption. This is exemplified by the single most lethal event in the 374-year recorded history of Mayon Volcano: the cold-rain lahar that killed 1,500 people in November, 1875, 3 years after a major eruption (Ramos-Villarta and others, 1985).

We have been studying rain lahars at Mayon Volcano since 1985. Our main interest has been their debris-flow phases, but in 1988 we initiated a program addressed specifically to hyperconcentrated streamflow (Ledda and Rodolfo, 1990). The sedimentology and rheology of both hot- and cold-debris flows at Mayon, reported previously (Umbal, 1986; Arguden and others, 1989; Rodolfo and others, 1989; Arguden and Rodolfo, 1990), will not be discussed here. Instead we will address these questions: What parameters determine whether or not a rainstorm will trigger a laharic debris flow? Are these parameters unique and site-specific, and are they different for nonvolcanic debris flows? Given the same meteorologic setting and slope characteristics, why do some channels on an edifice experience debris flows at times when only hyperconcentrated-stream flows or flash floods occur in others nearby?

### *Mayon Volcano*

Mayon Volcano is a classic composite cone composed of andesite and basaltic-andesite lavas and the deposits of pyroclastic flows and lahars (Newhall, 1977, 1979; Punongbayan and Ruelo, 1985). It is very active, having erupted at least 44 times since 1616 (Ramos-Villarta and others, 1985). During a third of Mayon's eruptions, lahars are generated on the volcano slopes by heavy rains from storms, or from moist surface air that has been uplifted by eruption updrafts (Rodolfo, 1989). Intervals between the 20 largest eruptions, including the last three (1968, 1978 and 1984) average about 10.5 years and range from 2.7 to 28 years (Punongbayan, 1985). During these intervals, debris is mobilized into lahars by monsoonal and typhoon rains of sufficient intensity and duration. A typical major rain lahar, such as the one triggered by Typhoon Rosing 3 years after the latest eruption, lasts from several hours to 2 days and is mainly hyperconcentrated streamflow in character, with interspersed pulses of debris flow that each typically lasts several minutes to 1 hour and has a velocity of 3 to 6 m/s (Arguden and others, 1989; Rodolfo and others, 1989; Rodolfo and others, 1989; Ledda and Rodolfo, 1990).

The extraordinary symmetry of Mayon (Fig. 1) signifies that all radii are equally affected by aggradational and degradational processes in the long term. In the short term, however, activity is far from even. For example, the limited volumes of lava extruded during a typical effusive eruption can contribute to the growth of the edifice only as one or two narrow, radial tongues, and pyroclastic flows may leave large, discrete deposits such as those labelled Bonga Fan and Basud Fan in the southeastern sector of the volcano (Fig. 1). During periods between eruptions, erosional and depositional processes vary significantly between areas, and even between closely spaced channels. Since 1986, we have monitored rainfall and lahar activity at Basud and Bonga fans. We will describe the characteristics of lahar-generating rainfall at those sites, and compare it to published thresholds for debris flows that have occurred elsewhere in the world (Caine, 1980).

We will also show how rainfall of similar intensity and duration causes greatly different lahar activity at two comparable geomorphic complexes, each situated near one of the two sites, that continue to serve as sediment-delivery systems for cold-rain lahars, and explain how these contrasts arise. Then, concentrating on the most active of the two systems, we will describe the differences in geomorphic evolution and sedimentologic activity along two channels that belong to it, Mabinit and Matanag channels. Since 1985, we have mapped both channels repeatedly by planetable and alidade at a scale of 1:1,000. At present, Mabinit Channel is by far the most active conduit for lahars (Fig. 2). Its origin during the 1984 eruption, and the changes it underwent during its first year of existence, have already been described in detail (Rodolfo, 1986, 1989). Here, we will discuss the continued evolution of the channel and will

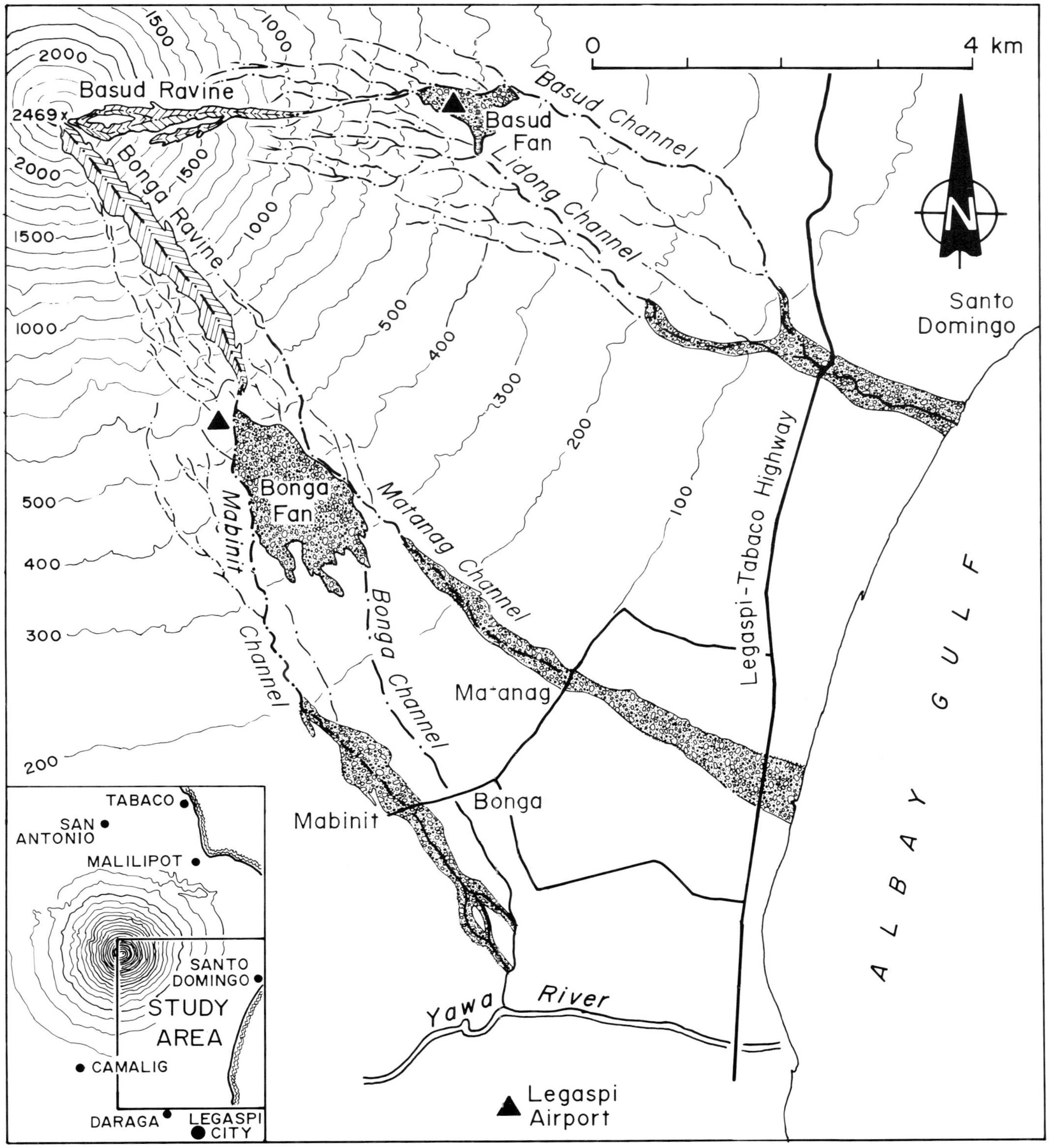


FIG. 1.—The southeastern sector of Mayon Volcano. To facilitate comparison of the drainage systems above Bonga and Santo Domingo, details of the channels between Lidong and Matanag channels have been omitted. The solid triangles denote rain-gauging stations. Contour interval is 100 m.

FIG. 2.—Northward view of Mayon Volcano, showing Bonga Ravine and part of Bonga Fan (barren area above and to the right of the treeline). Composite stratigraphy is exposed at the sides of the ravine.

also describe the continued growth of its debris-flow fan. This treatment may find application to other fans, both volcanic and nonvolcanic.

### *The Record of Lahar Activity at Mayon Volcano*

From 1616 to 1981, 29 lahars were recorded at Mayon, 21 during eruptions (Umbal, 1986; Rodolfo, 1989). Only eight that were triggered by typhoons during repose periods were either lethal or destructive enough to be recorded, but many more doubtless occurred, judging from the 138 rain lahars that we have documented during the 5 years following the most recent eruption.

The 1984 eruption, which lasted from 10 September to 6 October, produced at least 12 hot lahars, the first occurring on 14 September (Corpuz, 1985; Umbal, 1986). In the remaining 3 months of 1984 after the eruption, abundant loose ash and coarser ejecta blanketing the upper and middle slopes were mobilized by intense rainfall into 67 other lahars, which occurred along 32 gullies and channels in all sectors of the volcano. Since the end of 1984, heavy typhoon or monsoonal downpours have continued to generate lahars, mainly in the southeastern sector. Umbal (1986) documented 32 dates in 1985, when at least 44 individual lahars occurred, along only five channels. As the supply of loose volcaniclastic debris has been depleted, the frequency of lahars has declined steadily. Ten events were recorded in 1986, four in 1987, ten in 1988 (an unusually wet and stormy year at Mayon), and only three in 1989. Eight lahars are of greatest interest for this study, having had major debris-flow phases that caused large-scale channel modifications. Five of these were generated by typhoons: Saling I in 1985, Rosing in 1987, Unsang and Yoning in 1988, and Saling II in 1989. The other three resulted from exceptionally heavy monsoonal downpours on 20 November 1987, 20 November 1988, and 14 February 1989. Typhoons Saling I and Unsang have special importance, because they generated debris flows that avulsed from the middle stretch of Mabinit Channel and buried large areas that had been unaffected by lahars since well before the 1984 eruption. We will describe all of these changes in Mabinit Channel except those produced during Typhoon Saling II, which have not yet been completely mapped. Except for Saling I, we have detailed rainfall information for all these precipitation events, data that bear on how lahars are generated.

## LAHAR-TRIGGERING RAINFALL

The climate of the Mayon region has been described by Rodolfo and others (1989) and Arguden and Rodolfo (1990). Rain falls throughout the year, but half of it comes during a distinct wet season from October to January. Much of it is brought by typhoons and weaker storms that cross southern Luzon, most frequently between October and December. Most lahars occur in October and November, because by December the supply of loose debris has been largely depleted by the earlier flows of the season. Since the 1984 eruption, however, rainstorms of exceptional intensity have triggered major lahars as late as March.

The long-term (30-year) average annual rainfall of the region around Mayon computed from the official records of the Philippine government is 3,300 mm (Rodolfo and others, 1989), but this figure and all the official precipitation data are of little value in analyzing how lahars are generated. The records are gathered at the Legaspi City airport, which, although only 12 km from the Mayon summit, is less than 10 m above sea level. Thus, rain falls at the airport and on the slopes in significantly different periods and quantities, even during regional storms. We have gathered continuous rainfall data during the typhoon season from September to mid-December at two stations 600 m above sea level: on Basud Fan (Fig. 1) since 1986, and near Bonga Fan since 1988. Figure 3 illustrates the daily rainfall at Legaspi airport and at the slope sites during a recent, particularly stormy and wet 31-day period in which several major lahars occurred. Orographic effects enhanced the rain accumulation on the slopes commonly to more than twice, and on one date to more than five times the amount recorded at the airport. Even if the amounts that fall on the slopes and at the airport were similar, the airport data consist of daily totals only, and thus are not detailed enough to use in reconstructing how and when lahars were trig-

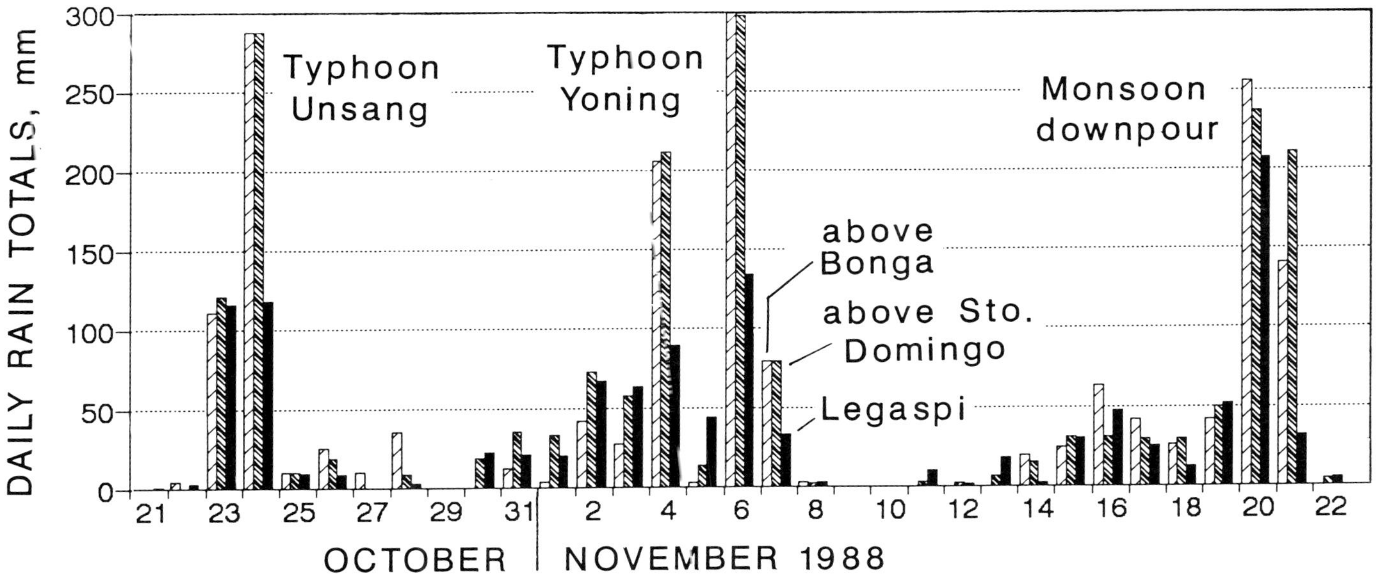


FIG. 3.—Total daily rainfall above the apex of Bonga Fan, on Basud Fan above Santo Domingo, and at Legaspi airport, from October 22 to November 22, 1988. For the location of these stations, see Figure 1.

gered. Detailed analyses of our rainfall data from the sites on the slope are the subject of another report in preparation. The relations between significant parameters are embodied in Figure 4.

### *Parameters of Triggering Rainfall*

Caine (1980) has synthesized published data from all over the world for 73 rainfall events that caused debris flows and described a generalized threshold for such events in terms of their intensities I (in mm/hr) and durations D (in hr). His threshold, which is plotted as the dotted curve in the lower right panel of Figure 4, is the simple power function

$$I = 14.82\ D^{-0.39} \qquad (1)$$

Researchers at Sakurajima Volcano (Ministry of Construction, Japan, 1988) have introduced an important refinement to Caine's approach that we have adopted in modified form. We define a "*triggering rain*" as a rainfall event that includes no pauses longer than an hour, and results in flow of water and debris in any proportion, of any magnitude and duration, in Basud or Mabinit channels (Fig. 1). The amount and duration of the triggering rain refers only to the accumulation and the time elapsed up until a flow starts. The additional rain that falls during the flow we treat separately as *sustaining rainfall.*

The four most important parameters that determine whether or not a rainfall event will trigger a debris flow at Mayon are plotted as the four axes of Figure 4. The *quantity of triggering rainfall* and its *duration* (upper and right axes; upper right panel) together determine the *intensity* averaged over the entire triggering event. Intensity is treated as a parameter in its own right in Figure 4 (the lower vertical axis). As at Sakurajima, a fourth critical determinant of the capacity of a rain event to trigger a flow is the short-term intensity, defined as the *quantity of rain that accumulates during the most intense 10 minutes,* the left axis of Figure 4. The lower left panel of this figure shows, for example, that even if a rain event has the high overall intensity of 40 mm/h, it will not cause a debris flow unless it includes a 10-min interval during which a threshold for short-term intensity is exceeded.

### *Results*

Figure 4 presents data from 78 precipitation events during the 4 years of our rainfall study: 23 measured near Bonga Fan in 1986 and 1987, and 56 measured at both rain stations in 1988 and 1989. Some events lacked the intensity and duration to generate streamflow of any kind. Other events triggered normal or hyperconcentrated streamflows; the distinctions between these flow types are still poorly understood, and thus they are undifferentiated and plotted with identical symbols in Figure 4. The plot does distinguish those events that generated debris flows, and between the events that produced flows of any type in Mabinit and Basud channels. Each debris flow was initiated by a rainfall that lasted 1.4 hours or longer and delivered a minimum of 40 mm of rain at an overall rate of at least 11 mm/h, with 10 mm or more falling during the most intense 10 min.

Our data clearly show that debris-flow initiation at Mayon requires substantially more intense rainfall for any given duration (and *vice versa*) than Caine's (1980) threshold values (Eq. 1). Only thirteen of the normal- and hyperconcentrated-streamflow events we observed plot below or at Caine's threshold. Thirty-nine events that exceeded Caine's threshold triggered only normal or hyperconcentrated streamflows, and nine other events that plot at or above his threshold did not result in channelized flow of any sort. During our monitoring period, 14 rainfalls triggered 16 debris flows; all of these plot well above Caine's threshold. Modifying Caine's approach because our data are less nu-

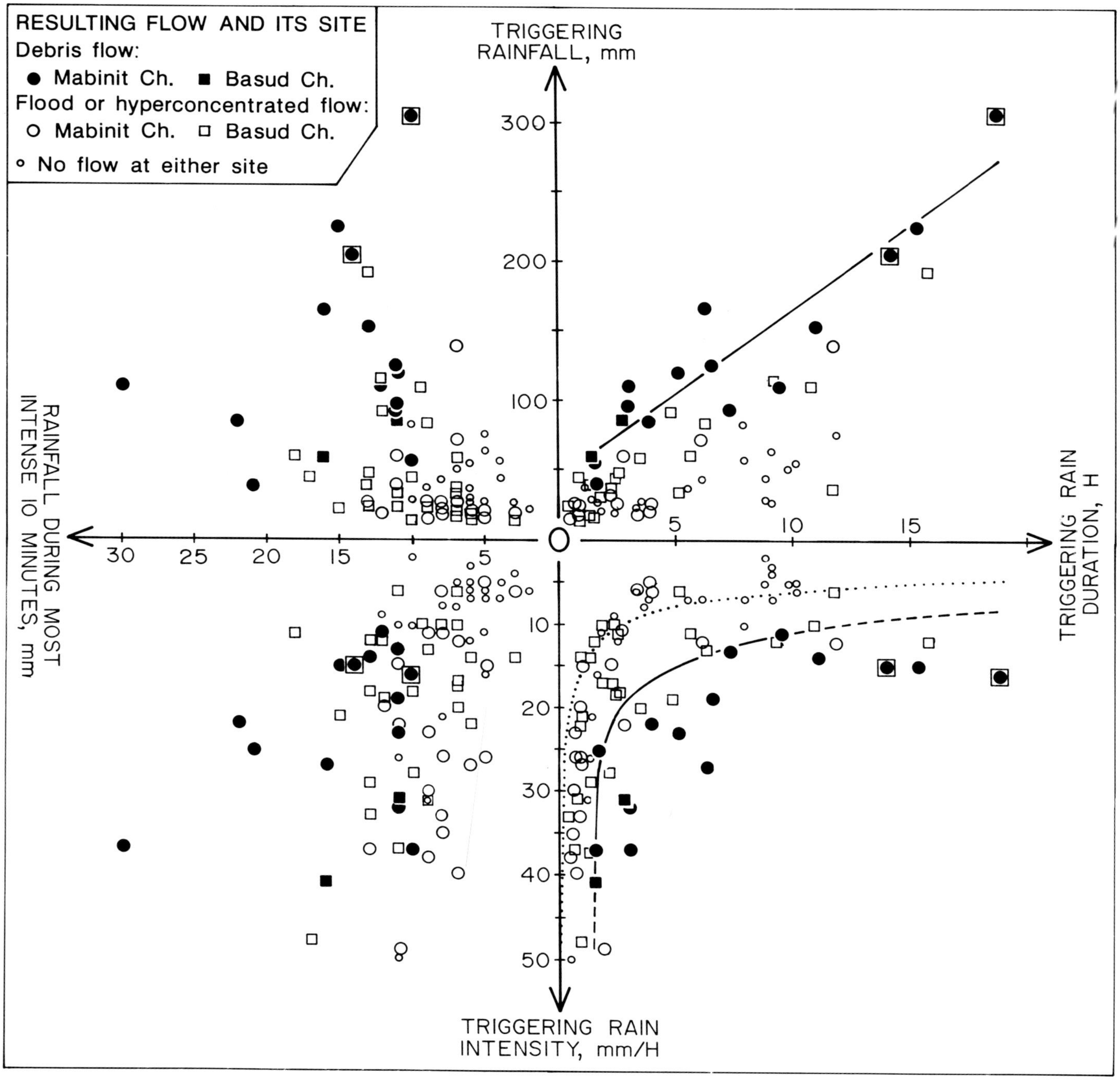


FIG. 4.—The parameters of lahar-triggering rainfall, and the sites and types of resulting flows. The two solid circles in open squares in each panel represent two rain events that caused debris flow in Mabinit Channel and only floods or hyperconcentrated flows in Basud Channel, regional storms during which precipitation, although measured only above Santo Domingo, is presumed to have been equal at both sites. Upper right panel: Triggering rainfall versus duration. The line is a least-squares fit ($r^2 = 0.68$) for the 16 rain events that triggered debris flows, and indicates an average threshold intensity of 14.3 mm/h. Lower right panel: Duration versus intensity. The dotted curve represents Caine's (1980) threshold equation $I = 14.82\ D^{-0.39}$; the solid curve is the empirical equivalent for Mayon, determined from data gathered through 1989: $I = 27.3D^{-0.38}$. The lower left panel plots the average intensity for the entire triggering event against the maximum accumulation over a 10-min period, which is a measure of maximum short-term intensity. The upper left panel plots total rainfall against short-term maximum intensity and illustrates the two thresholds of 40-mm total rainfall and 10 mm/10 min.

merous, we determined an empirical intensity-duration threshold for the Mayon debris-flow events by plotting all those data on a log-log graph, determining a least-squares regression line for the 14 events, and drawing a parallel to it through the five minima. The result, translated to the lower panel of Figure 4 as the solid curve with a dashed extrapolation, represents the threshold equation

$$I = 27.3D^{-0.38} \qquad (2)$$

Ten rainfalls that exceeded this threshold produced only hyperconcentrated or normal streamflows, three along Mabinit Channel and seven along Basud Channel.

Rainfall at both our monitoring stations on the slopes are generally quite similar in quantity and timing, especially during the regional storms that cause rain lahars. The disparity in debris-flow occurrence along Basud and Mabinit channels is due to differences in geomorphology and in the quality and quantity of volcaniclastic material available to Mabinit and Basud channels.

TABLE 1.—BONGA AND BASUD SEDIMENT DELIVERY SYSTEMS

| Dimensions | Summit Ravines | | | |
|---|---|---|---|---|
| | Bonga | Basud | | |
| Length, m | 3950 | 3500 | | |
| Maximum width, m | 400 | 225 | | |
| Average width, m | 290 | 75 | | |
| Area, km$^2$ | 1.14 | 0.53 | | |
| Axial depth, m | 125 | 30 | | |
| | Pyroclastic Fans | | | |
| | Bonga | Basud | | |
| Length, m | 1900 | 1000 | | |
| Maximum width, m | 1100 | 675 | | |
| Average width, m | 750 | 300 | | |
| Area, km$^2$ | 1.58 | 0.32 | | |
| Apex elevation, m | 540 | 600 | | |
| Distal elevation, m | 300 | 460 | | |
| | Channels | | | |
| | Bonga Area | | Basud Area | |
| | Mabinit | Matanag | Lidong | Basud |
| Upper catchment, km$^2$ | 2.52 | 1.87 | — | 3.62 |
| Lower catchment, km$^2$ | 3.38 | 3.55 | 5.32 | 4.79 |
| Total catchment, km$^2$ | 6.90 | 5.42 | 5.32 | 8.41 |

SEDIMENT-DELIVERY SYSTEMS

The 1984 eruption left two comparable geomorphic complexes, upslope of Barangay (village) Bonga and the town of Santo Domingo, that continue to serve as sediment-delivery systems for lahars (Fig. 1; Table 1). Generalized in Figure 5, each system consists of a ravine, a fan-shaped pyroclastic-flow deposit at the mouth of the ravine, and a pair of channels, one along each lateral margin of the fan.

### *Summit Ravines*

The highest element of each sediment-delivery system is a prominent ravine that was eroded and enlarged by pyroclastic flows during the 1984 eruption. Bonga Ravine (Fig. 2) is by far the largest such feature of Mayon Volcano, and no earlier ravines of comparable size are mentioned in the recorded history of the volcano. Its area is about 1.8 times that of Basud Ravine, which is also much shallower, its axis and sides cut smoothly into a single lava flow (Fig. 6). Bonga Ravine (Fig. 2) is much more deeply incised, and its steep walls expose eight layers of fairly massive lava, intercalated with friable layers of tephra. This composite stratigraphy also characterizes the sides of the contiguous upper reaches of Mabinit Channel (Fig. 7).

### *Pyroclastic Fans*

At the mouth of each ravine, the pyroclastic flows left large, fan-shaped accumulations, consisting mainly of coarse, poorly sorted volcaniclastic debris, that are informally referred to as Bonga and Basud pyroclastic fans. (In 1985, Okkerman and others called the Basud feature the Sta. Misericordia Fan, after the *barangay* in which a PHIVOLCS field station is located, but we use Basud to be brief and consistent.) Both fan deposits have proximal apices close to the 600-m elevation, where radial profiles change in slope from 15° to 12°; otherwise, they are significiantly different. Bonga Fan extends farther downslope, much of its distal edge below the 300-m elevation, whereas that of Basud Fan is between 400 and 500 m. Bonga Fan, with an area of 1.6 km$^2$, is almost five times larger than Basud Fan, and its effect on pre-eruption drainage patterns was much more drastic. Before the eruption, in the southeast sector of the

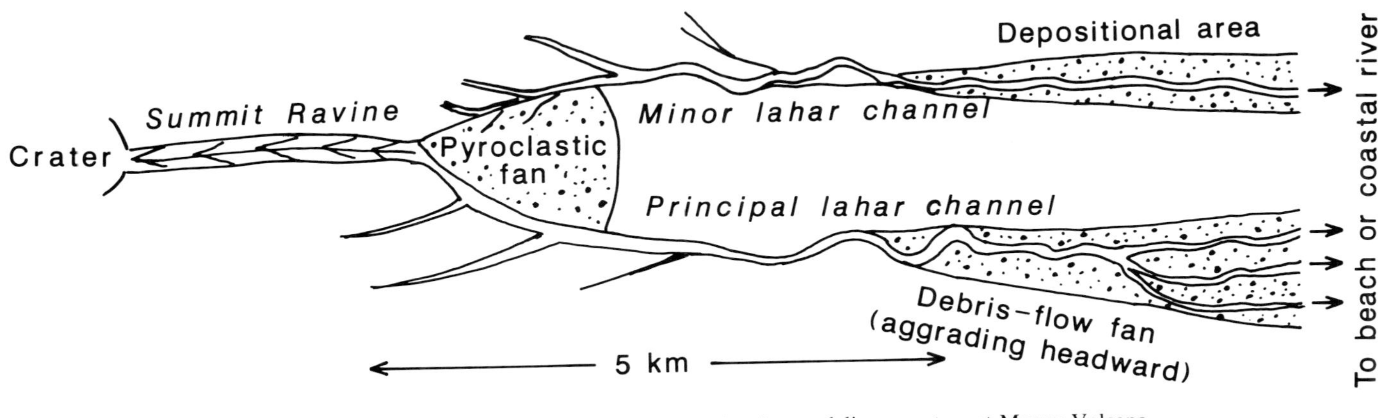


FIG. 5.—The geomorphic elements of a generalized sediment-delivery system at Mayon Volcano.

volcano, Bonga Channel was the largest and best developed channel, with a head at the highest elevation of 1 km, in the site of the future Bonga Ravine. Bonga Fan completely buried the stretch of Bonga Channel between 280- to 580-m elevations; the surviving portion of the channel above the fan is now part of Mabinit Channel, and the portion below the fan, formerly a site for frequent and lethal cold-rain lahars, is now largely inactive (Rodolfo, 1989). Avalanches that occurred late in the eruption buried the pyroclastic-flow deposits at the apex of Bonga Fan. These deposits contain only traces of clay and less than 4 percent silt; at least 50 percent is coarser than sand and includes numerous blocks and boulders derived mainly from the older lava flows that crop out on the walls of Bonga Ravine (Arguden and Rodolfo, 1990). Similar avalanche deposits are missing on Basud Fan, which, however, received a 10- to 15-cm-thick cover of fine ash that is missing on Bonga Fan (Okkerman and others, 1985). Deposited during the later part of the 1984 eruption, when vulcanian activity was most pronounced and the prevailing winds were westerly (Corpuz, 1985), this ash carapace on Basud Fan is now largely eroded away.

### *Lahar Channels*

Since the latest eruption, remnants of pre-eruption channels and gullies along the lateral margins of both pyroclastic fans have been reintegrated into new drainages: the Mabinit-Matanag and Basud-Lidong channel pairs. Only one of each pair has experienced large lahars since 1986: Basud and Mabinit channels; each of these is the downslope extension of the summit ravine of its system. Of the four channels, Basud Channel was the most active conduit for cold-rain lahars until 1986. Since that time, Mabinit Channel in the Bonga system, which also experienced many lahars in the early post-eruptive period, has been equally active, and has been the only channel that has experienced lahars with debris-flow phases.

All the channels have widths that range from 15 to over 100 m, and the depths of 1 to 23 m. Along most of their 6- to 8-km lengths, excepting portions of their upper reaches where old lava flows may be exposed, the channels are incised into coarse, poorly sorted deposits of pyroclastic flows and lahars. Channel walls, which are frequently modified by slumping and lateral undercutting by lahars, are steep and commonly vertical, despite the friable nature of the volcaniclastic sediments, owing to the angularity of the intermediate and finer grained components. The channels are dry except during and shortly after intense rain because the sediments are very porous and permeable.

FIG. 6.—View toward the summit along Basud Ravine, which is clean of debris and smoothly floored with massive lava.

FIG. 7.—Upper Mabinit Channel, at the base of Bonga Ravine, showing the composite layering that provides debris-flow materials. The channel walls here are 40 to 50 m high. The uppermost 3 m is a debris-flow deposit, underlain by a 7-m-thick lava flow, which in turn overlies two pyroclastic-flow deposits. Note the accumulation of coarse talus in the channel.

*Mabinit and Matanag Channels (Bonga System).—*

Before the 1984 eruption, the only portions of what is now Mabinit Channel were the 300-m-long segment of the pre-eruption Bonga Channel from 600- down to 540-m elevations, and an 800-m-long reach of a minor gully from 300 down to 240 m. The rudiments of its present course were shaped by erosion and deposition by lahars during the eruption (Rodolfo, 1989).

At present, Mabinit Channel is the most active lahar conduit, not only in its own sediment-delivery system, but for the entire volcano. Its upper catchment area is larger than that of Matanag Channel; more importantly, it includes Bonga Ravine. The strata cropping out in the steep ravine sides avalanche frequently throughout the year. Debris that collects along the axis of the ravine during the relatively dry months is mobilized into lahars by the first large storms of the typhoon season. Once a lahar has been initiated along the upper reaches of Mabinit Channel, it bulks up to as much as twice its initial volume by eroding the channel (Rodolfo, 1989). Mabinit Channel follows the western periphery of the fan but does not dissect it, and thus receives some runoff but little sediment from the fan.

Matanag Channel, which existed before the eruption, was lined with boulder-studded levees during the eruption, when the first hot debris flows filled it to overflowing, stranding the boulders along the channel rims. These boulder levees were 50 to 80 m long, 5 to 15 m wide, and up to 2 m high (Fig. 8, sections D, E, F). Mabinit Channel did not exist to receive similar levees (Rodolfo, 1989). Immediately after the eruption, the floor of Matanag Channel was discontinuously lined with small terraces, produced when channelized debris flows stopped moving and left deposits that were then dissected by hyperconcentrated or normal flows, either during the waning stages of the depositing lahars, or during later runoff events (Section B-B', Fig. 8). Standing no more than 2 m above the channel floor, they were smaller than similar features in Mabinit Channel (Rodolfo, 1989); only one was as long as 70 m. Matanag Channel was not seriously modified by the deposition of Bonga Fan and has been the main recipient of sediment from the Fan. Bonga Fan, however, did not have an ash cover of low permeability like the capping on Basud Fan. Consequently, all but the most intense rains percolate into the fan instead of running overland, and thus the fan has provided little sediment for lahars. Debris flows have not occurred along Matanag Channel since 1985.

*Basud vs. Lidong Channels (Basud System).—*

The deposition of Basud Fan by pyroclastic flows in 1984 did not greatly change the pre-eruption configurations of Basud and Lidong channels. Both served to channelize pyroclastic flows during the eruption, down to elevations of 300 and 170 m along Lidong and Basud channels, respectively. Below these elevations, hot lahars left debris-flow and hyperconcentrated-streamflow deposits in the channels. Since the eruption, Basud Channel has been a much more active lahar conduit than Lidong Channel because its catchment area is more than twice as large, and, more importantly, because the upper third of its drainage area includes Basud Ravine and most of Basud Fan, the major local sources of sediment. Lidong Channel receives most of its water below Basud Fan, from tributaries that run alongside the fan.

Okkerman and others (1985) have reported that debris flows were exceptionally common along Basud Channel in the first year following the 1984 eruption because the new ash cover of Basud Fan had a low infiltration capacity that was easily exceeded by monsoonal and typhoon downpours, and thus was mobilized frequently by overland flow. These authors also observed that the entire surface of the fan was deeply incised with a system of gullies that was integrated by micropiracy to feed only four channels by the end of March 1985. By the end of the 1985–1986 typhoon season, this erosional work was essentially complete, and the supply of easily eroded ash had been largely exhausted.

The Legaspi-Tabaco highway at Basud Channel was frequently blocked and damaged by debris flows in 1984 and 1985, and the local public works office dredged the channel many times, using the dredgings to build large spur dikes. Even after Basud Channel ceased to experience large debris flows, massive earth-moving efforts continued along it because hyperconcentrated flows and flash floods still damage the dikes and the highway, disrupting traffic. Thus, it has not been possible to monitor the natural evolution of Basud Channel, as we have done for Matanag and Mabinit channels, where artificial modifications have not been as drastic. Ignoring the effects of engineering, Basud Channel has tended to widen and fill over time, evolving in a manner similar to that of Matanag Channel, which is described below.

POST-ERUPTION CHANGES IN CHANNELS OF THE BONGA SYSTEM

Figures 8 and 9, respectively, illustrate how Matanag and Mabinit channels have evolved since 1984. The data shown in these figures are summarized and compared in Figure 10, to show the great differences in how the two channels have developed.

*Matanag Channel*

The adjustment of Matanag Channel to the eruption consisted simply of widening and aggradation that were essentially complete after the 1986–1987 typhoon season. Changes were greatest below the 250-m elevation. In 1985, the channel was distinctly sinuous, and its average width was only 28 m. By 1986, its average width had increased by over 41 percent, to 40 m, and a major consequence was to greatly reduce the sinuosity (compare the two maps of Fig. 8). The increase in width was compensated for by a 58 percent decrease in average depth of about 3 m; the channel cross-section areas remained fairly constant. In the following year, the channel again widened, but not so drastically, and depths remained fairly constant.

A major consequence of the post-eruption widening of Matanag Channel was to obliterate all the terraces and boulder levees left by the eruption. Unlike at Mabinit Channel, new channel terraces did not form, and large-scale avulsions and overbank deposition have not happened in subsequent years, simply because large debris flows no longer occurred.

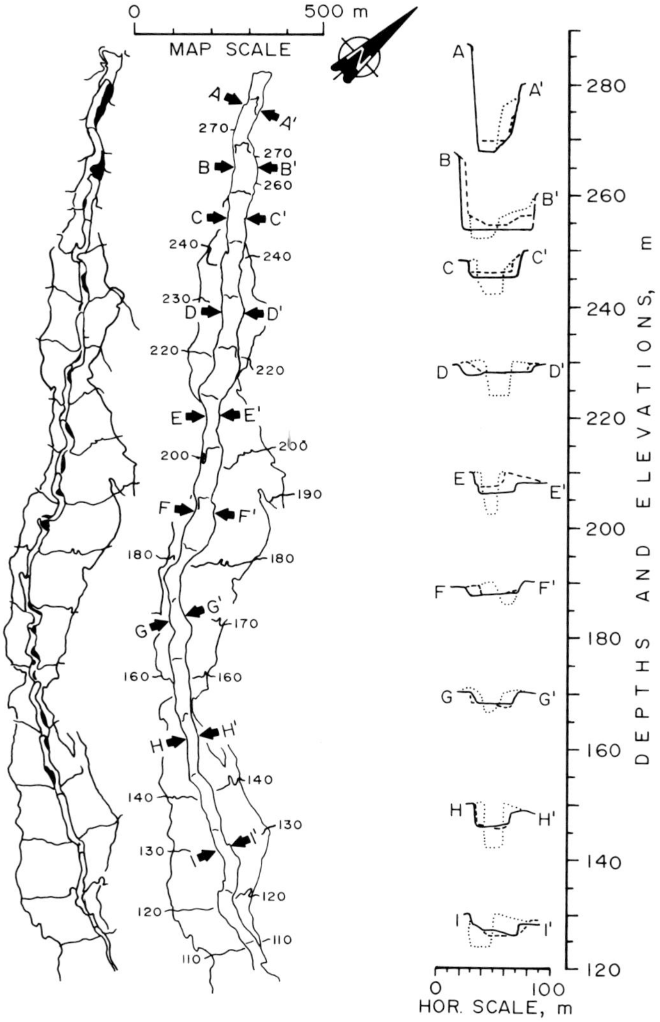


FIG. 8.—Evolution of Matanag Channel. Left map is the 1985 survey; note narrowness and sinuosity of channel, and small terraces. Right map is the last (1988) survey. Dotted, dashed, and solid cross-section lines respectively show the channel configurations in 1985, 1986, and 1988.

### *Mabinit Channel*

Like Matanag Channel, the overall width of Mabinit Channel has consistently increased over time. Shortly after the eruption, its average width was about 36 m, and its was remarkably sinuous for an ephemeral channel on such a steep slope, particularly along its middle stretch betwen the elevations of 178 and 240 m (Rodolfo, 1989). As Figures 10 and 11 show, the average bank-to-bank width had increased to 54 m in 1986, then to 67 m in 1987, and to 80 m in

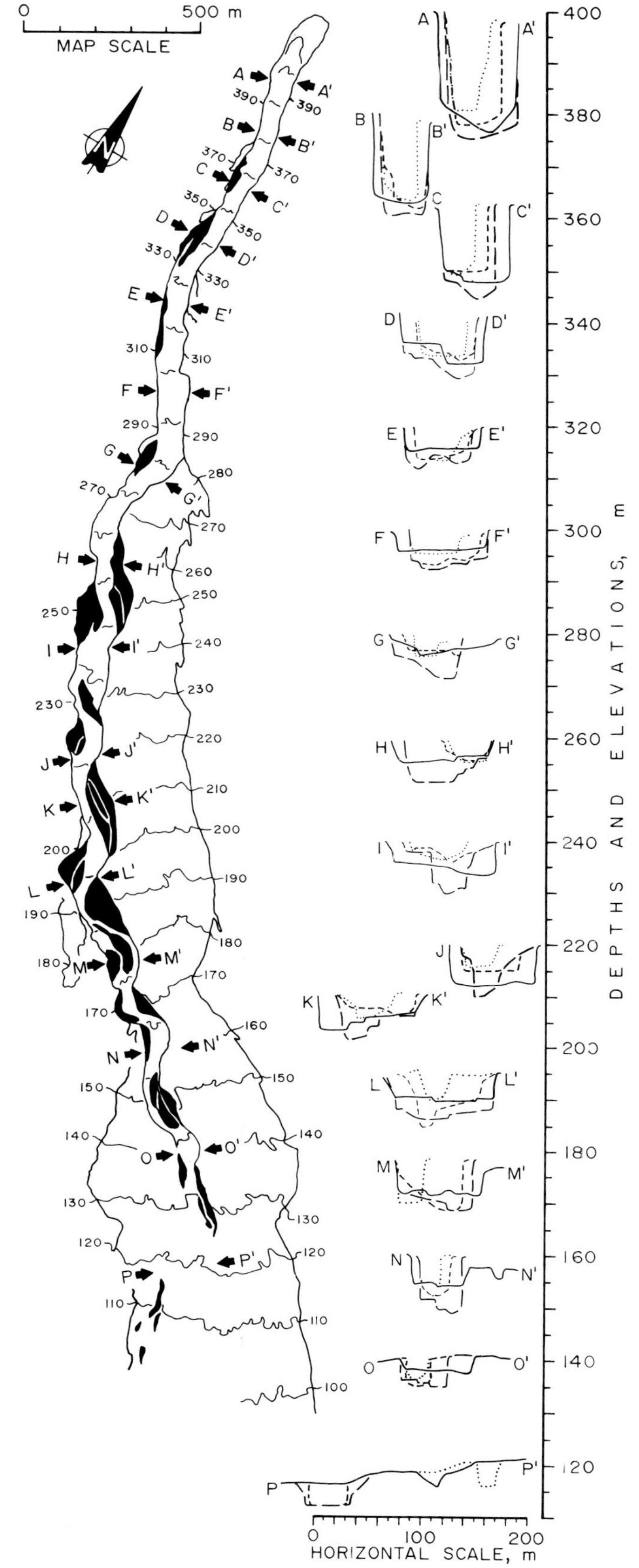


FIG. 9.—Evolution of Mabinit Channel since 1985. The map is the survey in 1989, after Typhoon Unsang. Dark areas are terraces; white bands in them denote multiple levels. In the cross sections, the dots, short dashes, long dashes, and solid lines respectively represent the channel configurations in 1985, 1986, 1988, and 1989.

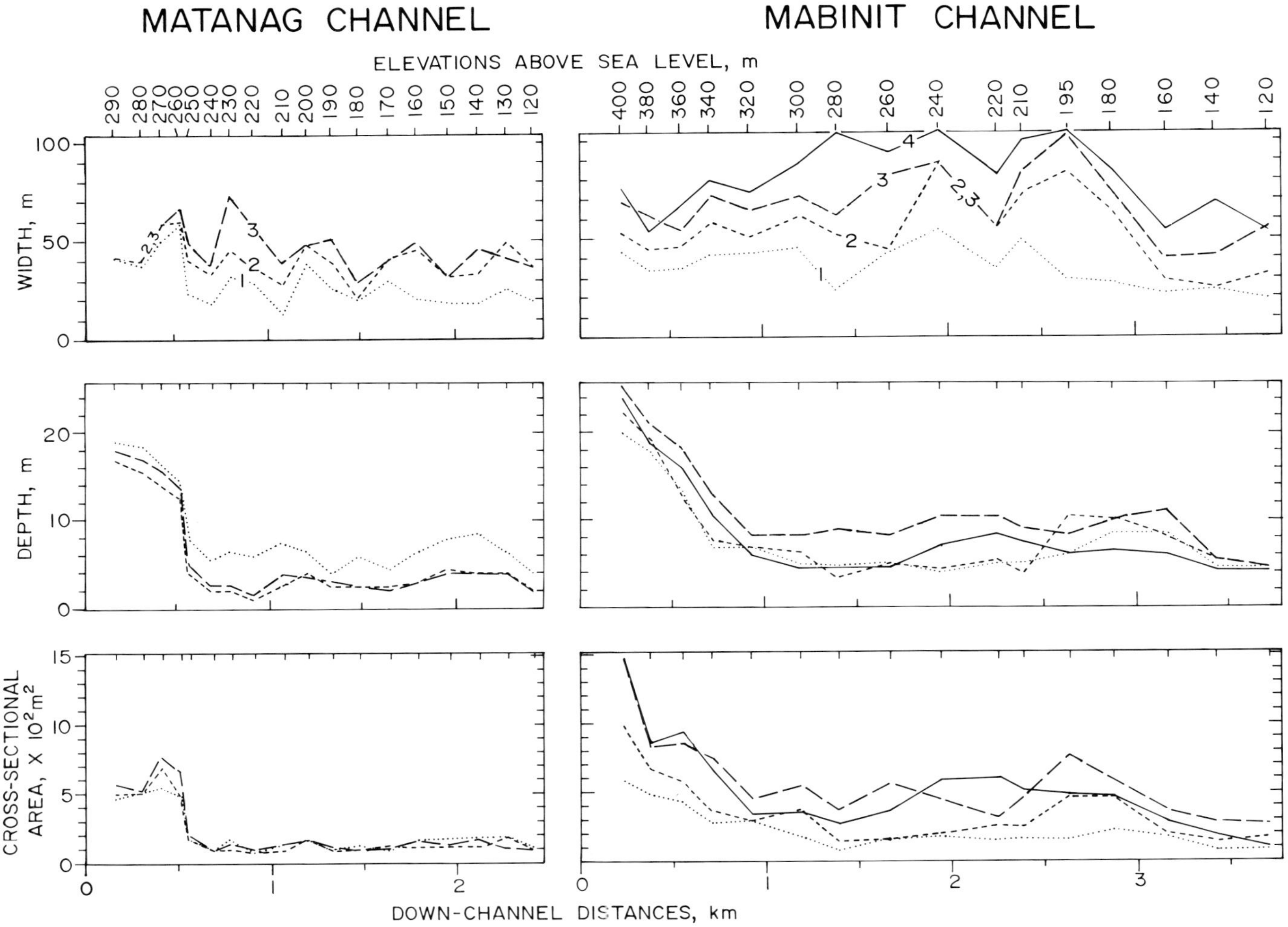


FIG. 10.—Summary and comparison of the changes in Matanag and Mabinit channels. The dots, short dashes, long dashes, and solid lines connect the data for 1985, 1986, 1988, and 1989, respectively, and the numbers mark the lines with the ordinal sequence of surveys.

1988, an overall increase of 126 percent. Over the same period, depths increased by 11 percent, but this increase was not continuous. The lahars of Typhoon Rosing in 1987 were exceptionally erosive, scouring out the channel along most of its length. Before we could remap the entire channel, a strong monsoonal downpour 5 days later restored some of the channel fill; nevertheless, the 1987 season deepened the channel by an average 3.5 m (Arguden and others, 1989). Additional filling during the 1987–1988 typhoon season, however, restored the average depth to 1986 values, and much of the channel became even shallower than it was in 1985. The overall widening and deepening increased the average cross-sectional area by 135 percent over the 1985 value.

It is important to note that individual lahars do not occupy the entire channel, but are confined to a smaller active channel that is frequently modified but has tended to remain narrow and sinuous (Figs. 9, 10, 11). Only rarely and locally is this active channel overtopped; the flats bounded by the main channel wall and the rim of the active channel are analogous to the floodplains of more normal streams.

Three striking features that are not shared by Matanag Channel mark how lahar activity has shaped the evolution of Mabinit Channel (Fig. 11). Terraces of dissected debris-flow deposits, which were initially quite small, have been eroded away and replaced with larger features at least twice. Major debris flows have twice overtopped the left (east) bank along its middle stretches and left large sheet deposits. Finally, the channel has been repeatedly filled and replaced below the 125-m elevation.

*Terraces.*—

When first mapped after the eruption, the channel contained more than 30 terraces that stood no higher than 2 m above the channel floor, the largest only 30 m wide and 150 m long. The lahars of Typhoon Saling on 18 October 1985 replaced them with terraces commonly 40 m wide, and one almost 280 m long (Rodolfo, 1989). In 1987, the lahars of Typhoon Rosing scoured virtually all of these features away, leaving the channel wide, box-shaped, and 6 m deeper in the 1.7-km stretch between the 300- and 200-m elevations. Five days later, a heavy monsoon rain gen-

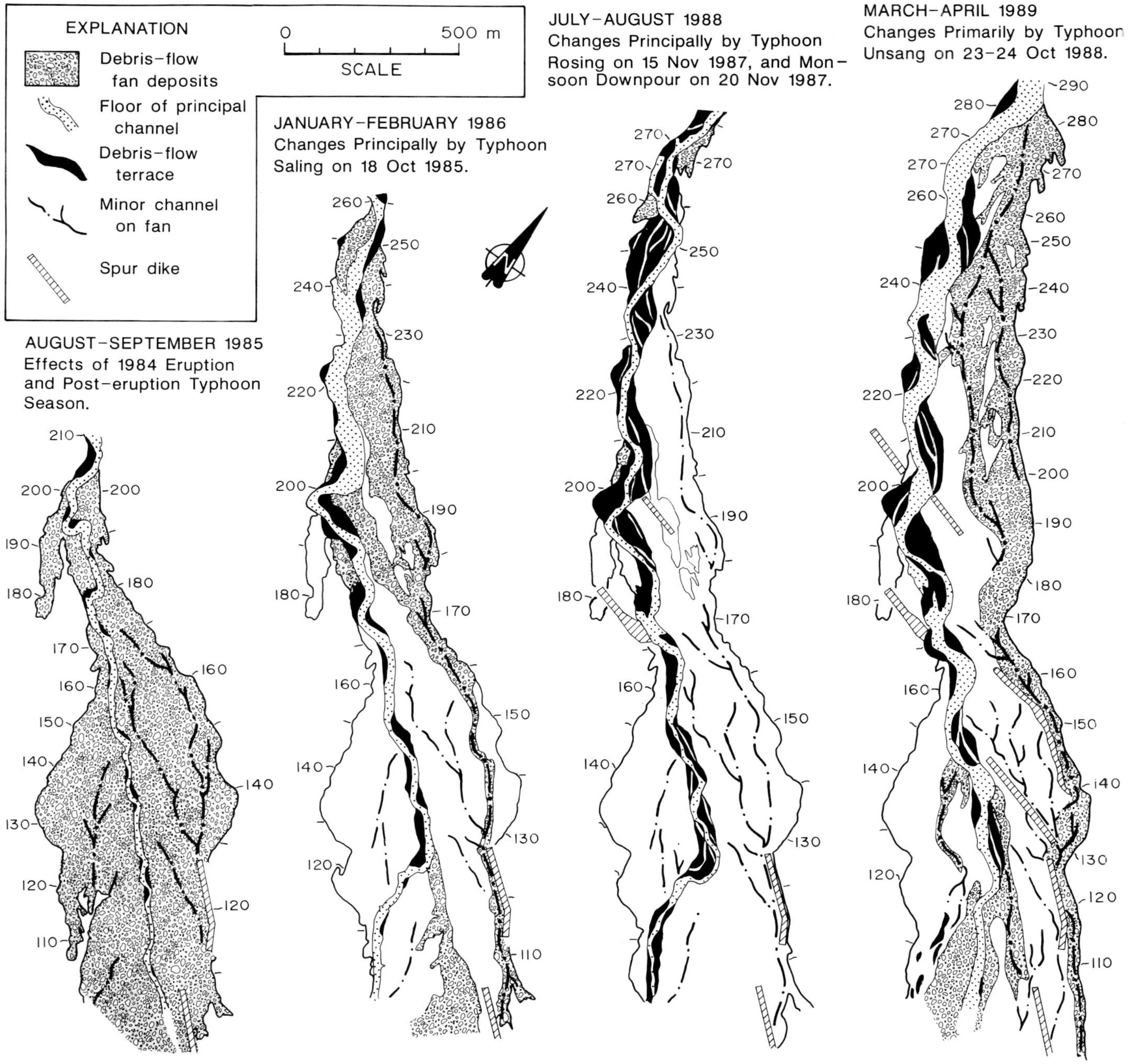


FIG. 11.— Evolution of lower Mabinit Channel and its debris-flow fan, as mapped by planetable and alidade at an original scale of 1:1,000 during the indicated periods. White bands in the debris-flow terraces indicate multiple levels 1 to 2 m above each other. The terraces in the 1988 map were left by lahars of the 20 November 1987 monsoon, after the channel was scoured clean of terraces by the Typhoon Rosing lahar. In each panel, only the latest overbank debris-flow deposits are shown with the gravelly pattern. Those mapped in 1989 were left exclusively by the lahars of Typhoon Unsang; however, the channel had been modified after this typhoon, by lahars generated during Typhoon Yoning on 4–6 November 1988, and during monsoonal downpours on 20 November 1988 and 14 February 1989.

erated a lahar that did not enlarge the channel further. Instead, several debris-flow surges filled the channel almost to bankful. These deposits were then dissected by waning-stage hyperconcentrated streamflows and flood flows, leaving multistoried terraces up to 70 m wide and 360 m long; between them, the new channel was as narrow and sinuous as it had been in 1986. Remnants of some of these terraces survived major lahars of the next season. At the end of the 1988–1989 lahar season, these remnants, together with terraces that were generated during Typhoons Unsang and

Yoning and subsequent monsoonal downpours, comprised terrace complexes comparable in size to those of the previous season.

*Avulsions.—*

Typhoon-triggered debris flows have twice overtopped Mabinit Channel at elevations well above the initial field of debris-flow deposits left by the 1984 eruption. The first event occurred during Typhoon Saling in 1985. The principal avulsion site was the stretch between 224-and 226-m elevations, with minor overtopping also at the 252 to 248- and 200 to 204-m elevations. Coalescing, these flows blanketed an area of $1.6 \times 10^5 m^2$ with a 1-m thickness of bouldery sediment. This deposit was 150 m wide and stretched 1.4 km downslope to the 156-m elevation (Rodolfo, 1989).

Also during Typhoon Saling, a minor overtopping occurred at 275 to 268 m. During the monsoon downpour of 20 November 1988, the same site again experienced a minor avulsion. Finally, during Typhoon Unsang in October 1988, a major overtopping here produced a sheet debris flow 200 m wide extending 3 km to the Mabinit-Bonga road, covering a $2.8 \times 10^5 m^2$ area with depositis that thinned both laterally and distally from maximum thickness of about 1 m near the channel.

*Abandonment and replacement of distal end of the channel.—*

Following the 1984 eruption, the 1.2-km stretch of Mabinit Channel below the 175-m elevation was remarkably straight (Fig. 11). In 1985, debris flows of Typhoon Saling plugged the channel below 130 m. Deflected to a minor swale, the flows carved a new channel of comparable size (Rodolfo and others, 1989). This new channel segment was enlarged and lined with debris-flow terraces by the lahars of 1987. In 1988 the new channel was plugged in turn, again below 130 m, by debris flows of Typhoon Unsang. The new channel replacing it was carved in a medial position between it and the initial post-eruption channel.

## DISCUSSION AND CONCLUSIONS

### *Rainfall Threshold for Mayon Debris Flows*

The virtual identity of the exponents in Caine's (1980) and our equations (Eqs. 1, 2), which govern the shapes of the curves, is to be expected. It is the twofold difference between the coefficients 14.82 and 27.3 that result in the substantial separation between the two curves plotted in the lower right panel of Figure 4. The differences in the two thresholds would have been even greater had Caine's data distinguished, as ours did, between the portion of each rainfall event that triggered a debris flow and the portion that fell after the flow commenced.

One possible reason why our results deviate from Caine's is the exceedingly heterogeneous quality of his data, as he acknowledged. Another possible explanantion is that our monitoring period may simply not have been long enough to provide a statistically valid number of debris flows, although the large proportion of our data that fall between the two thresholds, particularly in the short-duration portion of the field, indicates otherwise. In any case, our threshold could well change as we acquire additional data. Other reasons lie in the quality of the available sediment, and in geomorphic details.

### *Sedimentologic and Geomorphic Controls for Lahar Activity*

That the volcanogenic qualities of the available sediment supply and landforms play major roles in generating rain lahars at Mayon is illustrated by the pronounced differences in lahar frequency and character, not only between, but also within the two sediment delivery systems, despite the fairly uniform nature of rainfall at the same elevations on the slopes (Fig. 3). Debris flows were more frequent in Basud Channel early in the post-eruptive period but no longer occur there, although hyperconcentrated and normal streamflows are still very common; few lahars have occurred along either Matanag or Lidong channels; and present debris-flow activity is now restricted to Mabinit Channel. Geomorphologic and sedimentologic controls operate as two sets of factors that behave both independently and synergistically.

A volcanogenic sedimentologic factor that operates fairly independently is one reason why the rainfall intensity-duration threshold for debris-flow initiation is so high at Mayon. The materials that underlie the middle and lower slopes and constitute channel boundaries are pyroclastic and laharic sediments that, by virtue of their granular, fines-depleted nature (Arguden and Rodolfo, 1990), are very permeable. A debris flow may be initiated either if overland flow occurs and bulks up by erosion, or if porewater pressures build up sufficiently to cause slope failure (Fig. 12). The second case is rare at Mayon because of the high permeability. Either circumstance can occur only if the rainfall exceeds the high infiltration capacities by being exceptionally intense, although its duration may be relatively short. In contrast, all but two of the debris flows in Caine's data set occurred in nonvolcanic settings underlain by soils containing substantially more clay and with low permeabilities. For debris flows to occur in such areas, threshold amounts of rain must fall over antecedent periods that are measured in days or even months (Wieczorek, 1987). At Mayon, we have found that the amounts of rain that fall during the 24, 12, or even only the 6 hours that precede a triggering rainfall event have little bearing on whether or not the event causes a flow of any type. This is consistent with the common observation that channels are dry less than 3 hours after a major rainfall or after the passage of a major lahar. The presence of even a few percent clay, as is generally the case with nonvolcanic-debris flows, would also have enhanced their mobility and runout distances (Costa, 1984).

The intensity-duration threshold and other rainfall parameters that govern the initiation of debris flows at Mayon are by no means universally applicable to other volcanoes, one reason being their greatly variable rates of sediment supply. Nojiri River, the most active lahar conduit on Sakurajima Volcano in Japan, has experienced 254 debris flows in the period from 1976 to 1987 alone. There, debris flows are triggered if a rainfall exceeds 10 mm, falls continuously for at least 1 hour, has an intensity of more than 5 mm/h, and

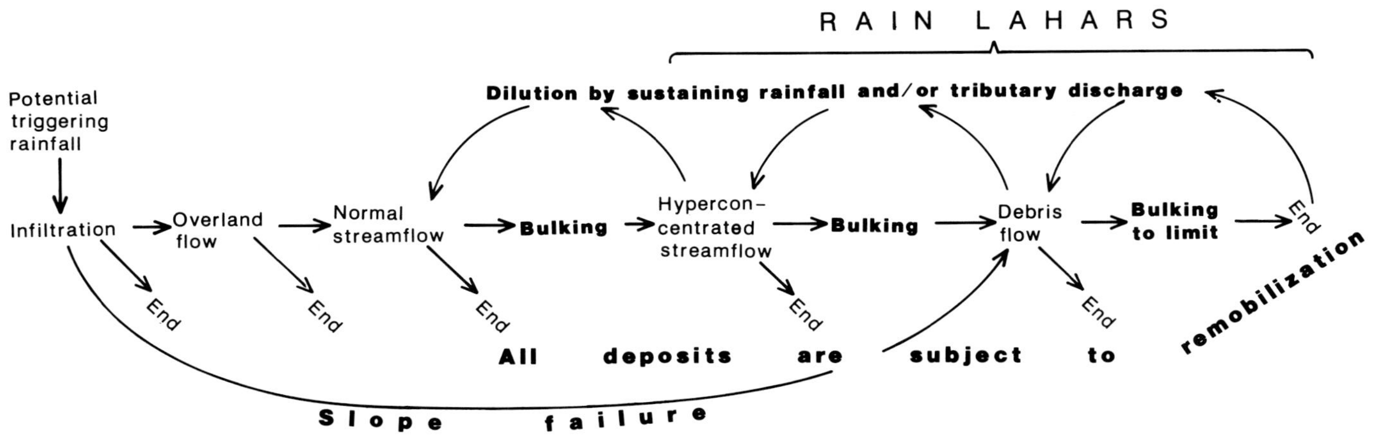


FIG. 12.—Schematic diagram for the generation of rain lahars, and for all the possible changes in state that a lahar can undergo. The process can terminate anywhere along the sequence; it is driven farther along by the magnitude and intensities of the rainfall event until a debris flow attains the maximum solid content possible (about 90 wt%). Dilution can reverse the sequence, and it is possible for a single lahar to go through the debris-flow—hyperconcentrated-streamflow—debris-flow loop more than once.

if more than 5 mm accumulate during the most intense 10-min interval (Ministry of Construction, Japan, 1988). All these values are lower than those for Mayon and demonstrate how important the availability of erodible sediment is for debris flows to occur. Sakurajima has provided abundant ash for debris flows by erupting more than 5,400 times since October 1955.

For debris flows to occur at Mayon, an independent geomorphic factor, high slopes, is essential to counteract the coarseness and high permeability of the sediment supply. Not only does the scarceness of clay and silt (Arguden and Rodolfo, 1990) make it difficult for pore pressures to build up, but also these sediments cannot possibly remain mobile on the very low slopes over which nonvolcanic-debris flows have been known to flow for long distances.

A geomorphic response to a particular sediment-supply characteristic, and a feedback from this response in turn influencing the sediment supply, are exemplified by the changing nature of post-eruption rain lahars in Basud Channel. Lahars were frequent immediately after the 1984 eruption, owing to the thin, relatively impermeable capping of ash on Basud Fan. When gullying breached this ash layer so water could percolate into the permeable underlying pyroclastic-flow deposits, the numbers and volumes of lahars passing through the Basud system were drastically reduced (Okkermann and others, 1985; Umbal, 1986). A similar history occurred at Mount St. Helens after its 1980 eruption, in areas blanketed by an ash fall of similar thickness. There, only 3 years were sufficient for a stable rill network to develop and cut through the ash, exposing more permeable material, and for sheetwash and rill erosion to decline by one to two orders of magnitude (Collins and Dunne, 1986).

The complex interplay between geomorphology and sediment availability is perhaps best illustrated by the Bonga sediment-delivery system. The apex of Bonga Fan (also that of Basud Fan) is at the 600-m elevation, which is marked by a fairly pronounced break in slope of 3°. The cause-and-effect relationship of pyroclastic-fan deposition and this slope break is still unresolved. In this system, Bonga Channel was rendered inactive when its upper reaches were buried by the emplacement of Bonga Fan. The course of the new Mabinit Channel, the principal heir to Bonga Channel's function, was dictated by the west margin of the fan, where gully remnants that survived the emplacement of the fan were integrated into a single drainage. The great differences in lahar activity along Matanag and Mabinit channels are due to their respective placements with respect to the fan and the summit ravine, and their different sources are also reflected in subtle differences between their debris-flow deposits (Arguden and Rodolfo, 1990). Once initiated at the base of the ravine and the upper reaches of Mabinit Channel, a lahar is nourished by water from tributaries and sediment in the main channel walls and stored in terraces.

### *Rain Lahars as Agents of Erosion and Deposition*

As a sedimentologic environment, an active composite volcano like Mayon is unique in the sporadic, catastrophic way that the sediment supply is renewed and drainages are drastically modified or even obliterated by an eruption. Both these changes can occur penecontemporaneously when summit ravines and pyroclastic deposits like the Bonga and Basud features are formed. A radial channel can adjust with remarkable rapidity to the large changes suddenly imposed on it, owing to the steep slopes, and to the large changes in sediment supply, both of which are much higher than those of more normal fluvial systems. In the tropics, the speed of these adjustments is compounded by high rainfall, especially when much of it comes during a major season of intense storms, as at Mayon. A single lahar during Typhoon Saling I in 1985 lasted 9 hours, widened its channel by an average of 25 m, caused up to 66 m of lateral erosion, and deposited an estimated at 2.5-3 × $10^5 m^2$ meters of sediment (Rodolfo, 1989). Depths can be increased by 6 m, as during the Rosing lahar of 1988.

Were lahars not so complicated, it would be easier to catalog their various sedimentary and landform products as caused by erosion and deposition. As Figure 12 shows, however, rain lahars can be very complex. Initially, they may originate directly from failures of saturated slope soils or bank material, or when streamflow undercuts a channel bank. Alternatively, a lahar can evolve along the "mainstream" line in Figure 12 from infiltration to normal streamflow and thence, by bulking—incorporating additional sediment directly by eroding the channel boundary, or causing walls to slump into the flow by undercutting them—until it becomes the most dilute end member of the lahar continuum, hyperconcentrated streamflow with the minimal sediment contents (which was set arbitrarily at about 40 weight percent by Beverage and Culbertson in 1964). At this juncture the new lahar follows one of three paths. It may simply end by depositing its load and waning into normal streamflow; it may be driven farther to the right in the depicted sequence, bulking into a debris flow by eroding and incorporating additional channel material; or it may be diluted back into normal streamflow by sustaining rainfall or tributary discharge.

Once a lahar attains debris-flow state, it does not immediately cease being turbulent, as was noted of flows along the South Toutle River during the 1980 Mount St. Helens eruption (Janda and others, 1981). One of its options, besides ending or becoming diluted back into hyperconcentrated state, is to continue eroding its channel until it attains the maximum possible solid content, about 88 weight percent in Mayon flows (Rodolfo and others, 1989). Even if it "freezes" and blocks its channel, it can be remobilized if hyperconcentrated or normal streamflow backs up and breaches the temporary dam. Furthermore, the surges that are characteristic of debris flow can remobilize debris at one site while depositing debris at another (Fisher, 1971).

The net result of all this complexity is that at any one site a single lahar often leaves a stack of very diverse deposits, from both hyperconcentrated streamflows and debris flows in any of their various states (Rodolfo, 1989; Rodolfo and others, 1989; Arguden and Rodolfo, 1986, 1990). This stratigraphy most often is incomplete because one or more stage of that one lahar may erode, leaving only single erosional horizons in place of earlier deposits of the same lahar, that themselves could well have been marked by other hiatuses. If these deposits were as well exposed as the Paleogene rocks at Bridge Point, New Zealand, they would probably also show up as parts of a complex association of multiple channels and ". . . wedges of asymptotic layers of fine and coarse debris . . .," so well illustrated and described by Cas and Landis (1987, p. 901), notwithstanding the marine setting of that surtseyan volcanic pile. With every additional experience of viewing debris flows in motion, however, we have become more reluctant than these workers to interpret coarse, chaotic channel deposits contained by fine-grained, reversely graded boundary deposits as being due to plug flow. The mid-channel portions of debris flows we have watched always displayed some turbulence, the boulders bobbing and saltating instead of being more sedately rafted along, as we would expect in plug flows. Turbulence and mixing in the center of a channelized debris flow would still result in a chaotic arrangement of its clasts after a debris flow freezes.

We should again note that some of the most notable features left by a lahar, the channel terraces described earlier, commonly are the result of deposition and erosion by both hyperconcentrated and debris flow.

Whether debris flows are erosional or depositional has been a question that has livened gatherings of volcaniclastic geologists. Students of active channelized flows, and many who have described their deposits and morphological effects, generally emphasize their erosiveness. The abnormally high depths and densities of the flows enable them to exert large shear stresses on their boundaries (Costa, 1984), and they have been demonstrated to contain large proportions of accidental clasts, plucked along their channels (Scott, 1988). On the other hand, many students of ancient volcaniclastic successions point out that many of their deposits have non-erosional bases. Historically, the proponents of both sides in a geological debate generally turn out to be correct, as we believe is the case here. The tradition of referring to both the phenomenon and its deposit as a debris flow is unfortunate and may have helped to generate the debate. A debris-flow deposit is no more a debris flow than a corpse is a living person. The deposit is a "dead" flow; deceleration ("dying") is a loss of competence. Overtopping debris flows at Mayon invariably have nonerosional bases, and buried vegetation again exposed in a channel wall (likely by a succeeding debris flow) commonly revives and flourishes. The problem arises largely from a failure to distinguish between channelized and sheet debris flows.

A much more meaningful question regarding the relative abilities of debris-flow and hyperconcentrated-streamflow phases to erode during a lahar, before emplacement of the deposits we see after the event, may be impossible to answer from direct observation, given the complex modes of flow and of the transitions between them. We have observed indirectly, however, that hyperconcentrated streamflow, although capable of some vertical erosion, is a much more effective agent of lateral erosion, and that debris flows are more effective in eroding vertically. On 10 October 1989, the active portion of Mabinit Channel was considerably widened but its floor lowered less than a meter by two typhoon-triggered lahars in rapid succession that were exclusively hyperconcentrated streamflows. Each was less than 2 m deep, lasted less than 2 hours, and ended in waning-stage normal streamflow (Ledda, 1989, unpubl. fieldnotes; Ledda and Rodolfo, 1990; Rodolfo and Ledda, 1990). These were followed a few hours later by a lahar with a debris-flow head that had a snout 4 m high and a velocity of 6 m/s and lasted only 11 minutes. This debris flow, followed by less than 2 hours of waning-stage hyperconcentrated and normal flow, left the channel 4 m deeper in its narrow thalweg.

Our long-term monitoring indicates clearly that debris flows tend to keep a channel narrow and deep, whereas hyperconcentrated streamflows widen a channel by lateral erosion and decrease its depth by aggradation. The debris flows of the 1984 Mayon eruption left the lahar channels with width-to-depth ratios that were quite low compared to those of typical stream channels. The data summarized in

Figures 8, 9, 10, and 11 clearly document that the active portion of a channel is maintained in this narrow, deep configuration only if the lahars occurring in it include debris-flow phases. Thus, although the bank-to-bank width of Mabinit Channel has continued to increase, its active channel is persistently narrow relative to its depth because Bonga Ravine continues to provide the debris and runoff that are repeatedly mobilized into debris flows. This configuration, by maximizing flow depth and minimizing basal flow surface and friction (Costa, 1984), increases the efficiency of the channel as a conduit for debris flows. By concentrating the weight of the flow, it also enhances the ability of a debris flow to erode and further deepen the channel. Together, the debris-flow, hyperconcentrated-flow, and flood-flow stages of the Mabinit lahars also collaborate to maintain this shape by creating terraces that constrict the channel. Entering a wide stretch of channel, a debris flow spreads out. Its lateral portions are those most likely to slow down or stop, due to decreased depth and increased basal area and friction. Its thickest portion, which tends to be axial along straight reaches and displaced toward the outsides of bends, is the most likely to continue moving down-channel. This may leave a more or less pronounced thalweg. Hyperconcentrated flow and flood flow, either during waning stages (the "lahar runout" of Scott, 1988), or caused by subsequent rainfalls, cut through the new deposits, following whatever thalweg is available. The result is a narrowed channel bounded by the vertical walls of the new terraces. The terraces play an additional role, storing erodible material with which subsequent lahars can bulk up along their route. During periods when only normal or hyperconcentrated streamflow occur, the active channel again tends to widen.

It is difficult to say whether the narrow and deep configuration represents an equilibrium shape for channels that funnel laharic debris flows; indeed, the idea that any landform can attain equilibrium on the slope of a volcano as active as Mayon is not easy to accept. On the other hand, it is clear that, in the absence of subsequent debris flows, Matanag Channel was quickly widened and shoaled by stream floods and hyperconcentrated streams, and in a very short time attained much larger width/depth ratios. Schumm (1960) has shown that the width/depth ratio of fluvial channels is governed primarily by the coarseness of its perimeter materials; Matanag Channel appears to be evolving in the direction indicated by this relationship. This adjustment has already removed the boulder levees and terraces that originally characterized it as a debris-flow channel.

### *Evolution of the Mabinit Debris-Flow Fan*

Since 1985, our maps have monitored the continued growth of the Mabinit debris-flow fan. The fan was initiated during the 1984 eruption when sheets of hot debris flows, lacking a channel, spread out over a $5.1 \times 10^5 m^2$ area. Higher shear stress in the thicker central portions of the flows may have left a poorly defined initial channel, which was enlarged by lahar runout and by cold lahars after the eruption (Rodolfo, 1989). The apex of the fan was initially at the 200-m elevation, but the avulsions during Typhoons Saling in 1985 and Unsang in 1989 progressively extended it upslope, to the 250-m, then to the 280-m elevation. These events increased the fan area by about 9 percent. Each time, the location of the egress site was controlled largely by the presence of channel reaches trending orthogonally across the radial-slope direction. Whether fortuitously or not, these sites are also marked by small (0.8°) breaks in slope along the profile from the summit to Mabinit. The avulsions may have been facilitated directly by these slope breaks; alternatively, the channel had responded previously to the presence of the slope breaks by deviating from the principal downslope direction, thereby creating the overbanking sites. Similarly, a change in regional slope from 2.8° at the 130-m elevation may have governed the repeated channel plugging and replacement at that point.

ACKNOWLEDGMENTS

We are very grateful for the cooperation, ideas, and many kindnesses extended to us by all the members of the Philippine Institute of Volcanology and Seismology (PHIVOLCS)—University of Illinois Department of Geological Sciences (UICDOGS) Lahar Study Group, especially Jesse Umbal, Rosalito Alonso, Jessie Daligdig, Gerald Ledda, Hernulfo Ruelo, Rene Solidum, and Raymundo Punongbayan, PHIVOLCS Director. Karen Prestegaard and SEPM reviewers Tom Pierson and Anonymous made many invaluable suggestions and kept us honest in general. The fieldwork was supported by a grant to the PHIVOLCS-UICDOGS Cooperative Program in Volcanology and Related Fields from the Philippine National Science and Technology Authority, Grants EAR-8515304 and EAR-8720819 from the U.S. National Science Foundation, and a Geological Society of America Travel Grant to Arguden.

REFERENCES

Arguden, A. T., and Rodolfo, K. S., 1986, Sedimentology of hot and cold lahar deposits produced by the 1984 eruption of Mayon Volcano, Philippines: Philippines Journal of Volcanology, v. 3, p. 60–78.

Arguden, A. T., and Rodolfo, K. S., 1990, Sedimentologic and dynamic differences between hot and cold laharic debris flows of Mayon Volcano, Philippines: Geological Society of America Bulletin, v. 102, p. 865–876.

Arguden, A. T., Umbal, J. V., and Rodolfo, K. S., 1989, 17 October 1985 typhoon-triggered lahar on Mayon Volcano, Philippines: Proceedings of the International Conference on Volcanoes, Kagoshima, Japan, July 19–23, 1988, p. 654–657.

Beverage, J. P., and Culbertson, J. K., 1964, Hyperconcentrations of suspended sediment: American Society of Civil Engineers, Journal of the Hydrology Division, v. 90, no. HY6, p. 117–126.

Caine, N., 1980, The rainfall intensity-duration control of shallow landslides and debris flows: Geografiska Annaler, v. 62, p. 23–27.

Cas, R. A. F., and Landis, C. A., 1987, A debris-flow deposit with multiple plug-flow channels and associated side accretion deposits: Sedimentology, v. 34, p. 901–910.

Collins, B. D., and Dunne, T., 1986, Erosion of tephra from the 1980 eruption of Mount St. Helens: Geological Society of America Bulletin, v. 97, p. 896–905.

Corpuz, E. G., 1985, Chronology of the September-October 1984 eruption of Mayon Volcano, Philippines: Philippine Journal of Volcanology, v. 2, p. 36–51.

Costa, J. E., 1984, Physical geomorphology of debris flows, *in* Costa, J. E., and Fleischer, P. J., eds., Developments and Applications of Geomorphology: Berlin, Springer-Verlag, p. 268–317.

FAIRCHILD, L. H., 1987, The importance of lahar initiation processes, *in* Costa, J. E., and Wieczorek, G. F., eds., Debris Flows/Avalanches: Process, Recognition, and Mitigation: Geological Society of America Reviews in Engineering Geology, v. VII, p. 51–62.

FISHER, R. V., 1971, Features of coarse-grained, high-concentration fluids and their deposits: Journal of Sedimentary Petrology, v. 41, p. 916–927.

JANDA, R. J., SCOTT, K. M., NOLAN, K. M., AND MARTINSON, H. A., 1981, Lahar movements, effects, and deposits, *in* Lipman, P. W., and Mullineaux, eds., The 1980 eruptions of Mount St. Helens, Washington: U.S. Geological Survey Professional Paper 1250, p. 461–478.

JOHNSON, A. M., 1970, Physical Processes in Geology: San Francisco, Freeman, Cooper, 577 p.

LEDDA, P. G., AND RODOLFO, K. S. 1990, Flow processes and deposits of Lahars along Mabinit Channel, Philippines: EOS, Transactions, American Geophysical Union, v. 71, no. 7, p. 513.

LOWE, D. R., WILLIAMS, S. N., LEIGH, H., CONNOR, C. B., GAMMEL, J. B., AND STOIBER, R. E., 1986, Lahars initiated by the 13 November, 1985 eruption of Nevado del Ruiz, Colombia: Nature, v. 324, p. 51–53.

MAJOR, J. J., AND NEWHALL, C. G., 1989, Snow and ice perturbation during historical volcanic eruptions and the formation of lahars and floods: Bulletin of Volcanology, v. 52, p. 1–27.

MINISTRY OF CONSTRUCTION, JAPAN, 1988, Debris flows at Sakurajima: Ohsumi Work Office, Kyushu Regional Construction Bureau, 65 p.

NARANJO, J. L., SIGURDSSON, H., CAREY, S. N., AND FRITZ, W., 1986, Eruption of the Nevado del Ruiz Volcano, Columbia on 13 November, 1985: Tephra falls and lahars: Science, v. 233, p. 961–967.

NEALL, V. E., 1976, Lahars as major geological hazards: Bulletin of the International Association of Engineering Geology, v. 14, p. 233–240

NEWHALL, C. G., 1977, Geology and petrology of Mayon Volcano: Davis, California, University of California, unpubl. M.S. Thesis, 292 p.

NEWHALL, C. G., 1979, Temporal evolution of the lavas of Mayon Volcano, southeastern Luzon, Philippines: Journal of Volcanological and Geothermal Research, v. 6, p. 61–83.

OKKERMAN, J. A., GERONIMO, S. G., UMBAL, J. V., AND PALAD, J. N., 1985, Study into the mechanisms of debris flow during/after the 1984 eruption of Mayon Volcano: Philippine Journal of Volcanology, v. 2, p. 94–142.

PIERSON, T. C., 1985, Initiation and flow behavior of the 1980 Pine Creek and Muddy River lahars, Mount St. Helens, Washington: Geological Society of America Bulletin, v. 96, p. 1056–1069.

PIERSON, T. C., 1986, Flow behavior of channelized debris flows, Mount St. Helens, Washington, *in* Abrahams, A. D., ed., Hillside Processes: Boston, Massachusetts, Allen & Unwin, p. 269–296.

PIERSON, T. C., AND COSTA, J. E., 1987, A rheologic classification of subaerial sediment-water flows *in* Costa, J. E., and Wieczorek, G. F., eds., Debris Flows/Avalanches: Process, Recognition, and Mitigation: Geological Society of America Reviews in Engineering Geology, v. 7, p. 93–105.

PIERSON, T. C., AND SCOTT, K. M., 1985, Downstream dilution of a lahar: Transition from debris flow to hyperconcentrated streamflow: Water Resources Research, v. 21, p. 1511–1524.

PUNONGBAYAN, R. S., 1985, An approach for estimating ages of active volcanoes: Philippine Journal of Volcanology, v. 2, p. 191–205.

PUNONGBAYAN, R. S. AND RUELO, H. B., 1985, Profile morphology and internal structure of Mayon Volcano: Philippine Journal of Volcanology, v. 2, p. 160–171.

RAMOS-VILLARTA, S. C., CORPUZ, E. G., AND NEWHALL, C. G., 1985, Eruptive history of Mayon Volcano, Philippines: Philippine Journal of Volcanology, v. 2, p. 1–35.

RODOLFO, K. S., 1986, Mabinit lahar channel, Mayon Volcano, Philippines: Philippine Journal of Volcanology, v. 3, p. 73–89.

RODOLFO, K. S., 1989, Origin and early evolution of lahar channel at Mabinit, Mayon Volcano, Philippines: Geological Society of America Bulletin, v. 101, p. 414–426.

RODOLFO, K. S., ARGUDEN, A. T., SOLIDUM, R. U., AND UMBAL, J. V., 1989, Anatomy and behavior of a post-eruptive rain lahar triggered by a typhoon on Mayon Volcano, Philippines: International Association of Engineering Geologists Bulletin, v. 40, p. 79–90.

RODOLFO, K. S., AND LEDDA, P. G., 1990, Scouring and sedimentation by lahars along Mabinit Channel, Mayon Volcano, Philippines: Geological Society of America Abstracts with Programs, v. 22, no. 7, p. 455.

SCHUMM, S. A., 1960, The shape of alluvial channels in relation to sediment type: U.S. Geological Survey Professional Paper 352-B, p. 17–30.

SCOTT, K. M., 1988, Origins, behavior, and sedimentology of lahars and lahar-runout flows in the Toutle-Cowlitz river system: U.S. Geological Survey Professional Paper 1447-A, 74 p.

SCRIVENOR, J. B., 1929, The mudstreams ('lahars') of Gunong Keleot in Java: Geological Magazine, v. 66, p. 433–434.

SMITH, G. A., 1986, Coarse-grained nonmarine volcaniclastic sediment: Terminology and depositional processes: Geological Society of America Bulletin, v. 97, p. 1–10.

SMITH, G. A., AND FRITZ, W. J., 1989, Penrose Conference report: Volcanic influences on terrestrial sedimentation: Geology, v. 17, p. 376.

UMBAL, J. V., 1986, Mayon lahars during and after the 1984 eruption: Philippine Journal of Volcanology, v. 3, p. 38–59.

VAN BEMMELEN, R. W., 1949, The Geology of Indonesia: Government Printing Office, The Hague, The Netherlands, 1A:732 pp.

WAITT, R. B., 1989, Swift snowmelt and floods (lahars) caused by great pyroclastic surge at Mount St. Helens Volcano, Washington, 18 May 1980: Bulletin of Volcanology, v. 52, p. 138–157.

WIECZOREK, G. F., 1987, Effect of rainfall intensity and duration on debris flows in central Santa Cruz Mountains, California *in* Costa, J. E., and Wieczorek, G. F., eds., Debris Flows/Avalanches: Process, Recognition, and Mitigation: Geological Society of America Reviews in Engineering Geology, v. 7, p. 93–105.

YANO, K., AND DAIDO, A., 1965, A fundamental study on mudflow: Bulletin of Disaster Prevention Research Institute Publication, Kyoto, Japan, v. 14, p. 69–83.

# VOLCANIC-DEBRIS-AVALANCHE DEPOSITS IN NEW ZEALAND—LITHOFACIES ORGANIZATION IN UNCONFINED, WET-AVALANCHE FLOWS

BETH A. PALMER,[1] BRENT V. ALLOWAY,[2] AND VINCENT E. NEALL

*Department of Soil Science, Massey University, Palmerston North, New Zealand*

ABSTRACT: In New Zealand, volcanic-debris avalanches have originated mostly in water-rich source volcanoes. Most wet-debris avalanches produced three distinctive mappable lithofacies that occupy adjacent areas. The axial-A lithofacies is dominated by brecciated and stratified megaclasts that form a self-supporting framework. Relict stratigraphy from the source area is preserved locally both within and among megaclasts. Interclast matrix in this lithofacies is silty sand, and the associated surface of the deposit is characterized by closely spaced, large mounds (hummocks). The axial-B lithofacies, in contrast, is interclast-matrix rich, so that at many exposures brecciated clasts and megaclasts are surrounded by clasts and an interclast matrix of silty sand. Mounds on the surface of this lithofacies are lower and more widely spaced than on the surface of axial-A. The marginal lithofacies is distinguished by an allophane-rich, muddy or silty interclast matrix and a relatively planar upper surface. Clasts are derived from disaggregation of brecciated megaclasts and sediment eroded from underlying deposits.

The lithofacies represent division of the debris-avalanche deposits into megaclast-rich (axial-A) and interclast matrix-rich (axial-B) parts and transformation of the avalanches into lahars (marginal lithofacies). Megaclasts preserving original volcanic stratigraphy in axial-A represent collapse and transport of coherent pieces of the edifice by the debris avalanches. Transformation into lahars occurred as the coarse avalanche debris came to rest and the finer portion continued to flow, enhanced by a secondary component eroded from underlying deposits. In general, flow across relatively featureless ring plains allowed multidirectional-lithofacies development, so that lobes of axial-A are surrounded by axial-B and marginal lithofacies both laterally and, where visible, distally.

Variations in avalanche size and ring-plain physiography produced different lithofacies architectures in the New Zealand deposits. Large debris avalanches were little affected by physiography and spread out in fan-shaped sheets comprising near-concentric zones of each lithofacies. Smaller avalanches were confined at some point during flow, so that central parts of the deposit are shoestring diamictons of axial-B and the marginal lithofacies. The tripartite axial-A through axial B and marginal lithofacies associations are present in proximal areas of these deposits. Distal areas are usually axial-B and marginal lithofacies.

## INTRODUCTION

The eruption-triggered collapse of the north flank of Mount St. Helens on 18 May 1980 (Voight and others, 1981) focused global scientific attention on the role of debris avalanches in the evolution of volcanoes. Such collapses have been recognized, and the resultant deposits mapped, for almost a century, but only since 1980 has attention been concentrated on this phenomenon. In New Zealand, early workers recognized that an extensive apron of volcaniclastics surrounded the two major andesitic stratovolcanoes of Egmont and Ruapehu. The aprons were called ring plains (Gibson and Morgan, 1927) and the distinctive mounds (hummocks) that characterize parts of the ring-plain landscape were recognized as having been emplaced by a gravity-flow mechanism (Grange, 1931), considered at the time to be lahars (see Smith and Lowe, this volume). The 1980 Mount St. Helens event led to reinterpretation of the New Zealand ring-plain deposits (Neall and Alloway, 1986) and to recognition of a genetic relationship between debris avalanches and lahars in a collapse event.

The cause of mass failures of volcanic edifices has been the subject of much debate in New Zealand. Grant-Taylor (1964) proposed that because ring-plain deposits of Egmont Volcano overlie a sequence of uplifted Quaternary marine benches, they must have been deposited during glacial periods that alternated with interglacial bench-cutting episodes. This required that large volumes of ice, melted during eruptions, provide the water necessary for lahars to cover wide areas of the ring plain. Later work demonstrated that some of these deposits originated as volcanic-debris avalanches. Some debris-avalanche deposits occurred in postglacial time when ice sheets were not present at Egmont Volcano (Neall, 1979). Furthermore, volcanic-debris avalanches have occurred at volcanoes where snow and ice are absent; e.g., White Island in 1914. Thus, the eruption-triggered flank failure at Mount St. Helens that created a 2.8-$km^3$ debris avalanche provided a documented mechanism that supported New Zealand work on the origin of volcanic-ring plains.

A debris-avalanche interpretation of ring-plain deposits is made still more plausible by the recognition that some volcanoes exhibit a tendency for repetitive collapses. In New Zealand, collapses occur in different directions, resulting in different-age segments within each ring plain. Between periods of ring-plain accumulation, fluvial erosion incises valleys that form new paths to guide younger flows. Each segment is recognized as a *lahar planeze* (Neall, 1982) and named separately (Grant-Taylor, 1964; Neall, 1979).

The relatively physiographically unconfined terrain of the Egmont and Ruapehu ring plains allowed the debris avalanche/lahar events to spread radially outward in an unrestricted fashion. This resulted in the development of lithofacies distributions that are different from previously described debris-avalanche deposits that were generally deposited in valleys (Crandell and others, 1984; Glicken, 1986). In this paper we concentrate on the lithofacies architecture of unconfined debris-avalanche deposits. This is of considerable importance to those deciphering volcaniclastic sedimentation in the pre-Quaternary record, as well as to those interpreting potential volcanic hazards at current sites of active volcanism.

Present addresses: [1]Department of Geology and Geological Engineering, University of Idaho, Moscow, Idaho, USA 83843

[2]Department of Geology, University of Toronto, Scarborough Campus, Ontario, Canada.

Sedimentation in Volcanic Settings, SEPM Special Publication No. 45

### *Terminology*

Different terminologies have been used in the description of volcanic debris-avalanche deposits, many of which are confusing because of their ambiguity. For example, the terms "block" and "matrix" have been used to describe individual components of a deposit as well as named facies (Crandell and others, 1984). We define the terms used in this paper as follows:

Matrix—All components of the deposit comprising material <2 mm in diameter. The matrix may be **interclast matrix** present between clasts or megaclasts, or **intraclast matrix** present within clasts or megaclasts.

Clast—All components of the deposit between 2 mm and 1 m maximum size that are bounded by an outer surface and show an internal lithological homogeneity. This definition is intended to include all portions of the flow that acted as coherent units. Clasts may be fragments of lava, a crystal, a diamicton, or any pre-existing rock type incorporated into the flow. The term is also intended to include a single rock type that has become extensively brecciated but retained its relative shape during transport, here referred to as a **brecciated clast.**

Megaclast—All components of the deposit >1 m maximum size that are bounded by an outer surface and show an internal lithological homogeneity. Single rock megaclasts, **brecciated megaclasts,** and stratified megaclasts are recognized.

Lahar—Rapidly flowing mixture of rock debris and water (excluding streamflow) from a volcano (Smith and Fritz, 1989). Lahars generated from debris avalanches in New Zealand are debris flows.

All of the matrix, clasts, and megaclasts that are interpreted as the original constituents derived from collapse of the volcanic edifice are called the **primary component.** All rip-up clasts, usually comprising volcaniclastic sediment, tuff, peat, lignite, and wood, that were incorporated in the flow during its transport are called the **secondary component.**

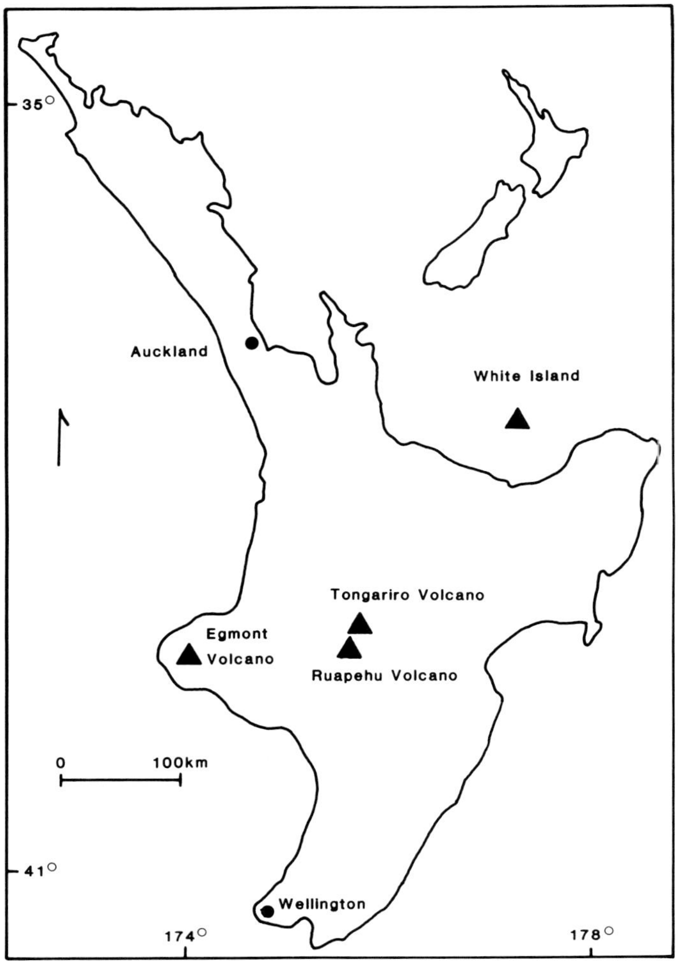


FIG. 1.—Andesitic stratovolcanoes (triangles) of the central North Island, New Zealand.

## VOLCANIC-DEBRIS-AVALANCHE OCCURRENCE IN NEW ZEALAND

All of the Quaternary volcanic-debris-avalanche deposits in New Zealand are confined to four andesitic stratovolcanoes. One avalanche deposit is recognized at White Island and one at Tongariro Volcano (Fig. 1); two are currently recognized at Ruapehu Volcano. Ten debris-avalanche deposits are recorded in the 130-ka history of Egmont Volcano. Here, collapses occurred radially in every direction except to the north, where an old volcanic massif buttresses the lower flank. The repeated collapses at Egmont result from the structure of the volcano. The steep (average 32°) volcanic cone consists of lava flows that rarely exceed 4 km in length, which overlie unconsolidated pyroclastic-flow deposits. Deep fluvial erosion within these deposits seems to have weakened the entire structure. In contrast, at Ruapehu Volcano the cone comprises lavas that extend up to 14 km from the source, constructing a more stable edifice with a smaller proportion of unconsolidated volcaniclastic deposits.

Summary data for each volcanic-debris-avalanche deposit currently recognized in New Zealand are presented in Table 1. The largest is the Pungarehu Formation that involved a large-scale, westward-directed collapse (c. 7.5 $km^3$) of Egmont Volcano. A short time earlier, an eastward collapse, resulting in the Ngaere formation, produced the most areally extensive avalanche deposit at Egmont Volcano that covers over 500 $km.^2$ On average, there has been one debris avalanche every 13 ka from Egmont Volcano, but these events have not been evenly spaced over time.

## DEBRIS-AVALANCHE LITHOFACIES ON UNCONFINED APRONS

Description of debris-avalanche deposits has focused on subdivision of the deposit into brecciated clasts or megaclasts (block facies) and fine-grained material (matrix or

TABLE 1.—DEBRIS-AVALANCHE/LAHAR DEPOSITS IN NEW ZEALAND

| Location and Deposit* | Age Years BP | Maximum Observed Thickness m | Maximum Width km | Runout km | Area $km^2$ | Volume $km^3$ |
|---|---|---|---|---|---|---|
| **Egmont** | | | | | | |
| Opua Fm | 7,000 | 6 | 20.4 | 25+ | 120 | 0.4 |
| Warea Fm | 12–14,000 | 5 | 5.1 | 31+ | | 1.2 |
| Pungarehu Fm | 20,000 | 60+ | 20.2 | 26+ | 250+ | 7.5+ |
| Ngaere fm | 23,000 | 25 | — | 39+ | 500 | 5.8 |
| Opunake Fm | 33,000 | 2.5 | 1.8 | 32+ | 45 | 0.6 |
| Stratford Fm | | | | | | |
| 1 | 48,000 | 17 | 32.1 | 32+ | 510 | 2.6 |
| 2 | | 1.5 | 10.7 | 39+ | 90 | 0.1 |
| 3 | 80,000 | 6 | 32.1 | 29+ | 255 | 1.0 |
| Okawa fm | 100,000 | 25 | 8.0 | 39+ | 255 | 3.6 |
| Motunui lahar | 115,000 | 4.3 | 8.0 | 39+ | 30 | 0.1 |
| **Ruapehu** | | | | | | |
| Murimotu Fm | 9,500 | 14 | 5.0 | 15+ | 23 | 0.2 |
| Unnamed | >50,000 | 30+ | — | 40+ | — | — |
| **Tongariro** | | | | | | |
| Te Whaiau fm | >20,000 | — | — | 8+ | 8 | 0.1 |
| **White Island** | 75 | 20 | 0.5 | 1+ | 0.1 | 0.005 |

*Information on unnamed Ruapehu deposit from K. A. Hodgson (ipers. commun.); Te Whaiau formation from W. Prebble, (pers. commun.).

mixed facies; Crandell and others, 1984; Glicken, 1986; Ui and Glicken, 1986). This classification means that an individual exposure can contain both block and mixed facies.

New Zealand debris-avalanche/lahar deposits, in contrast, can be divided into three lithofacies that occupy distinct positions with respect to each other. These are mappable units, so that most individual exposures comprise only one lithofacies. The lithofacies represent changes in the nature of flow within the landslide mass as it travelled away from the source. Two lithofacies, axial-A and axial B, are attributed to the debris-avalanche phase of the landslide. A third, the marginal lithofacies, represents deposition during a debris-flow phase. These lithofacies are distinguished on the basis of fabric, abundance of brecciated megaclasts, texture of the interclast matrix, and surface morphology.

### *Axial-A Lithofacies*

The axial-A lithofacies is dominated by megaclasts, with <30 percent sandy interclast matrix (Fig. 2A). Closely spaced, large mounds dominate the surface of this lithofacies (Table 2). The megaclasts are intensely brecciated and self-supporting, composing >70 percent of individual exposures.

Some megaclasts show relict stratification that was inherited from the source area and retained during emplacement. Various stages of disruption of the stratification are observed and in places the shattered lava and volcaniclastic deposits form a chaotic ensemble of juxtaposed clast- and matrix-supported diamicton (Fig. 2B).

### *Axial-B Lithofacies*

In contrast to axial-A, the axial-B lithofacies is interclast-matrix rich (30–90 percent; Fig. 3). Surface mounds are lower and more widely spaced than on the axial-A lithofacies (Table 2). Megaclasts are usually brecciated and locally preserve relict stratigraphy. Abundance and size of primary megaclasts decrease with increasing distance from axial-A (Alloway, 1989). This trend is associated with corresponding increase in the abundance of secondary component. Rip-up clasts are usually masses of fluvial and debris-flow deposits, lignite, or fine-grained primary or reworked andesitic ash. Subrounded to rounded secondary clasts are common in the distal parts of axial-B.

### *Marginal Lithofacies*

The marginal lithofacies contains >90 percent interclast matrix of allophane-rich sandy silt to mud. Allophane is a short-range-order aluminosilicate produced by the weathering of volcanic ash (Yoshinaga, 1986). The surface of the marginal lithofacies is predominantly planar with scattered low mounds (<2 m high) in some areas. The contact between axial-B and the marginal lithofacies is gradational and marked by a progressive decrease in the amount of primary component, which in some deposits is subordinate to the secondary component. The most distinctive feature of the marginal lithofacies is secondary clasts and megaclasts of stratified fluvial and debris-flow deposits (Fig. 4A). Some of these are as large as 100 m across and are most common at or near the base of the deposit. Other clasts include masses of volcanic ash and loess, pumiceous tephra, lignite, wood, and fluvial mud and sand (Fig. 4B). Abundance and size of the rip-up clasts decreases to the distal reaches of the deposits, where the marginal lithofacies is a homogeneous mixture of gravelly, sandy mud.

## LITHOFACIES ARCHITECTURE AND EVOLUTION ON UNCONFINED RING PLAINS

### *The General Case*

The three lithofacies described above occupy consistently distinct positions within individual debris-avalanche deposits (Figs. 5, 6). In general, cross sections through the deposits, both parallel and transverse to major flow direction,

FIG. 2.—Megaclasts in the axial-A lithofacies. (A) Shattered fabric typical of brecciated megaclasts (Pungarehu Formation, Egmont Volcano). Note lack of much rotation among fragments. Scale is 12 cm long. (B) Brecciated megaclast with stratification in the Murimotu Formation, Ruapehu Volcano. Arrows point to the deformed contacts between different layers. Hammer is 32 cm long.

TABLE 2.—LITHOFACIES IN NEW ZEALAND DEBRIS-AVALANCHE/LAHAR DEPOSITS

| Lithofacies | Description | Basal and Upper Surfaces | Geometry | Interpretation |
|---|---|---|---|---|
| Axial-A | Framework-supported fabric. Tabular brecciated megaclasts dominate, >10 m. These megaclasts are monolithologic with silty sand intraclast matrix of same lithology as megaclast fragments. Primary megaclasts of stratified deposits and hydrothermally altered material to 100 m long. Scarce secondary material. Silty sand interclast matrix. Mounds cored by single and multiple brecciated clasts. | Basal surface not exposed. Surface mounds to 50 m high and basal diameters as much as 500 m. Closely spaced mounds, 150–300/km.$^2$ | Sheet or lobate | Clast and megaclast-rich debris avalanche |
| Axial-B | Tabular to equant brecciated megaclasts are dominant clast type; stratified megaclasts less common. Secondary megaclasts of stratified sedimentary deposits, and clasts of lignite and tephra. Angular to subrounded clasts, some with rock-flour rims. Silty sand interclast matrix. Mounds cored by single megaclasts. | Planar to irregular, non-scoured to scoured base. Surface mounds <10 m high with basal diameters <25 m. Widely spaced mounds, 20–150/km$^2$ | Sheet or shoestring | Interclast matrix-rich debris avalanche. |
| Marginal | Matrix-supported fabric. Brecciated clasts rare, no stratified primary megaclasts. Secondary component dominates. Stratified secondary clasts and megaclasts common, as much as 100 m long. Secondary clasts of lignite, tephric loess, tephra, and diamicton abundant locally. Subrounded to rounded clasts dominate, some subangular clasts. Allophane-rich sandy mud to sandy, silt interclast matrix. | Planar to irregular, non-scoured to scoured base. Upper surface mostly planar, with scarce mounds, <25/km$^2$ | Sheet or shoestring | Lahar (debris flow) |

FIG. 3.—Brecciated megaclast (arrows mark boundaries) surrounded by sandy interclast matrix and clasts from the axial-B lithofacies in the Murimotu Formation, Ruapehu Volcano. Measuring tape is 1.5 m long.

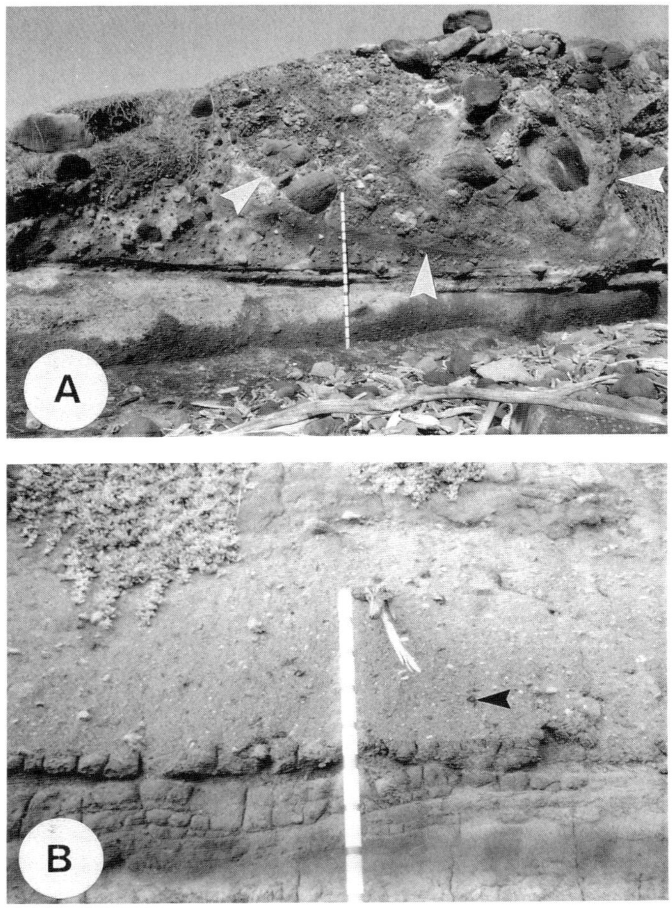


4.—(A) Stratified secondary megaclast of fluvial and debris-flow deposits from the marginal lithofacies of the Opua Formation, Egmont Volcano. Arrows point to the boundary of the megaclast. Jacob staff is 1.5 m long. (B) Wood (top of staff) and lignite (arrow) in distal parts of the marginal lithofacies from a deposit in the Stratford Formation, Egmont Volcano. Jacob staff is marked in a 0.1-m intervals.

show gradation of axial-A and axial-B to the marginal lithofacies (Fig. 7). The distal and lateral changes in lithofacies are characterized by a decrease in size and abundance of primary component and an increase in abundance of secondary component. The grain size of the interclast matrix also changes, from silty sand in axial-A and axial-B to sandy mud or silt in the marginal lithofacies. These trends represent the natural partitioning of the debris-avalanche deposits into clast- and megaclast-rich (axial-A) and interclast matrix-rich (axial-B) parts near the source, and transition into a debris-flow deposit (marginal lithofacies) distally. This occurs by incomplete mixing of relatively coarse and fine primary material at the source, progressive disaggregation and deposition of primary megaclasts, and erosion and incorporation of ring-plain deposits into the flow.

### *Origin of Primary Interclast Matrix*

In some debris-avalanche deposits, axial-B and the marginal lithofacies are present on the volcanic cone (Figs. 5, 6, 7). These lithofacies are not a series of overlapping deposits from successive slides in the source area. Rather, they are either a thin veneer or ponded deposits left behind as the debris avalanche overtopped ridges and entered different catchments (Palmer and Neall, 1989).

The identification of lithofacies on a volcanic cone is important because these deposits record early division of the debris avalanches into megaclast-rich and matrix-rich flow. Large amounts of fine-grained material must have been in these debris avalanches during this early phase. In coeval axial-A deposits on the ring plain, fragments are not rotated within megaclasts and original stratigraphic relationships are preserved. This indicates that little interclast matrix could have been derived from mechanical abrasion and disaggregation during early phases of flow. Rather, two sources of primary fine-grained material are identified in the source areas. At Egmont Volcano, sections of tephra up to 5 m thick mantle the upper flanks of the cone, providing a ready supply of matrix-size material for incorporation into the debris avalanche (e.g., on a 90° sector of Egmont cone, as much as 0.4 km$^3$ of tephra mantles the surface). At Rua-

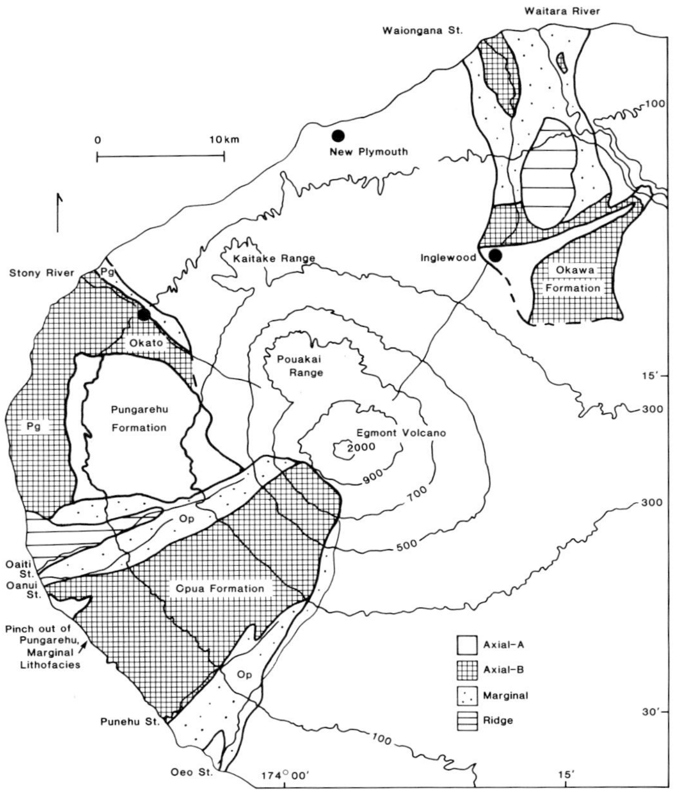


FIG. 5.—Lithofacies relationships in some debris-avalanche deposits on the Egmont ring plain. Lateral lithofacies relationships in the Okawa formation are typical (Alloway, 1989) of those developed in avalanches that were channelized at some point during flow. The Pungarehu Formation (Neall, 1979) was deposited by the largest debris avalanche known on the Egmont ring plain. The axial-B lithofacies is exposed along the coastline and the change from axial-A through marginal lithofacies is present laterally. The southern margin of the Pungarehu Formation is exposed below the Opua Formation, in southern Taranaki. In the Opua Formation (Neall, 1979), axial-A lithofacies, if present, is buried by younger lava flows closer to the source. Contours are in meters.

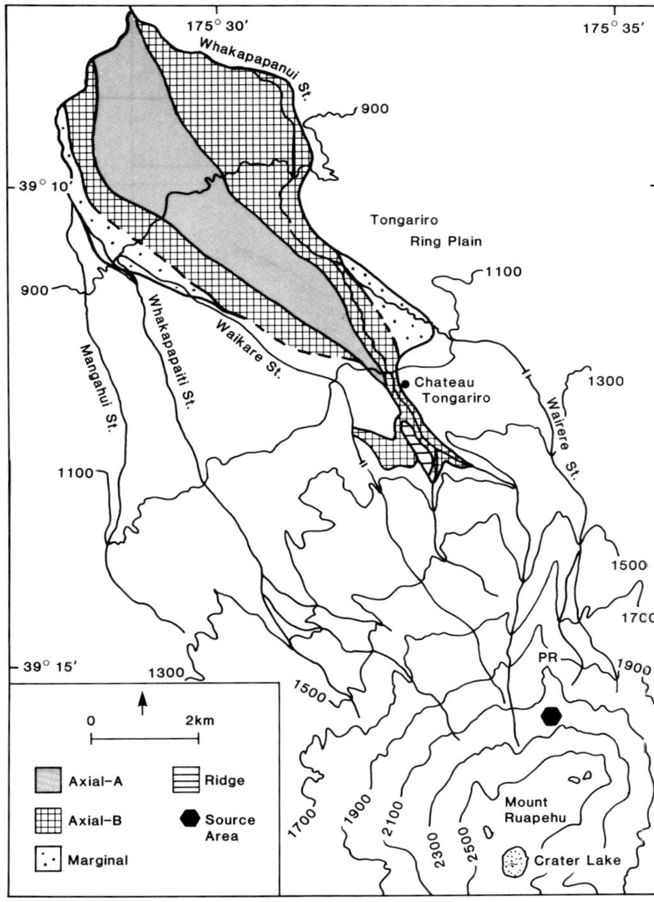


FIG. 6.—Lithofacies in the 9.5-ka Murimotu Formation, Ruapehu Volcano (Palmer and Neall, 1989). Contour intervals are in meters.

pehu and Tongariro volcanoes, hydrothermal alteration of the source rocks (Hackett, 1985; Hackett and Houghton, 1989) may have contributed much of the matrix-size material (Palmer and Neall, 1989; W. Prebble, pers. commun., 1989).

### *Lithofacies Development*

On the ring plains, the matrix-rich flow drained both laterally and distally away from axial-A, leaving behind a landscape dominated by closely spaced mounds. Deposition of axial-B in these areas occurred between and around the lobes of axial-A (Figs. 5, 6).

The reason why parts of the avalanche continue to be mobile while other parts remain immobile has never been fully explained. As the avalanche spreads and thins, settling of megaclasts may explain part of the phenomenon, but many megaclasts do not occur within mounds. Melting blocks of ice to produce the mobile parts of the flow is a slow process that does not occur in the time span of debris avalanche/lahar emplacement. It is discounted as a mechanism for both marginal-facies development and mound formation. An historical example that supports this conclusion is the 1914 White Island volcanic landslide, where mounds formed in areas where snow and ice are absent.

From the evidence in New Zealand, it appears that once a collapse of the volcanic edifice occurred, lavas, volcaniclastic deposits, and water were not completely mixed in the flow so that two quasi-separate phases were present at initiation. Once the flow achieved maximum velocity on the lower flanks, it decelerated and the megaclasts lost their momentum, coming to rest throughout the axial part of the deposit. The interclast matrix-rich part of the flow, which may have had a considerable water content, remained mobile. As this mobile portion continued to flow, the initial surface of the axial-A deposit subsided and the mounds began to protrude through the gradually lowering surface level. As the finer material continued to flow outward, the piles of clasts and megaclasts became exposed with slopes reflecting the angle of repose of the component materials. Progressive disaggregation of megaclasts decreased the megaclast content of the flow with respect to the matrix component, resulting in gradually fewer mounds and the de-

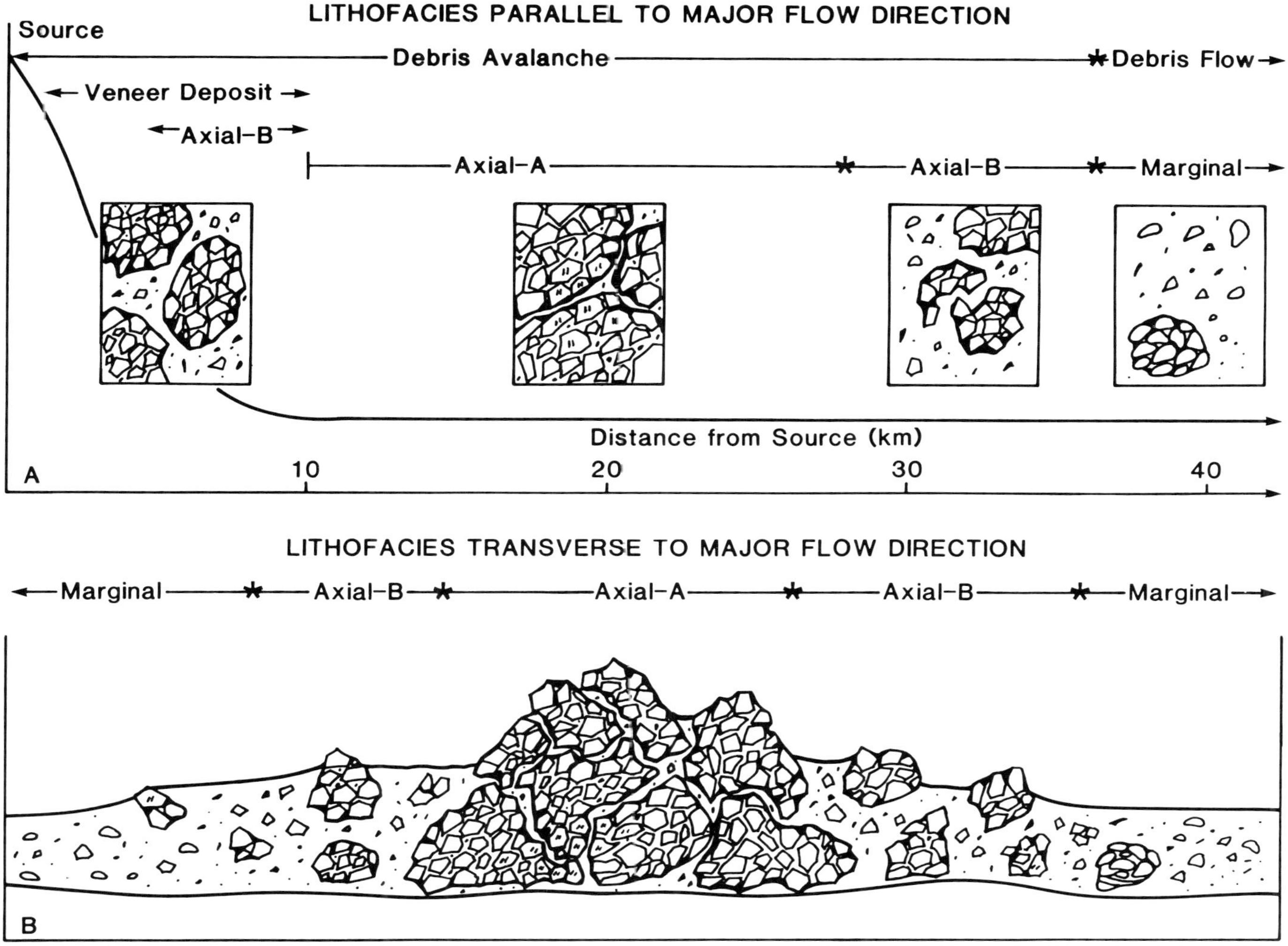


FIG. 7.—Characteristic lithofacies changes in New Zealand debris-avalanche/lahar deposits. Distance from source is estimated from average-size deposits on the Egmont ring plain.

position of axial-B lithofacies. In the Pungarehu Formation, a reduction in height of the flow from at least 50 m at 13 km from the source to at least 15 m at 26 km from the source occurred.

Although unconfined, the interclast matrix-rich part of the debris avalanches was powerfully erosive in places, stripping tephra, fluvial, and debris-flow deposits during flow across the ring plain. Progressive incorporation of weathered volcanic ash enhanced the transformation to debris flows both parallel and oblique to major flow direction. This multidirectional-flow transformation resulted in deposition of the marginal lithofacies in areas surrounding axial-B. The debris flows were extremely mobile. In some cases, they continued downstream for as much as 60 km away from the ring plains.

A cemented cloak of matrix and clasts coats many mounds in the Pungarehu Formation. The cloak is thin or absent on the tops of the mounds and thickens downslope. Named locally a "shell," this layer is extremely hard and resistant to earth-moving machinery. It appears that analagous "cloaks" formed on the 1980 Mount St. Helens deposit where rain, wind, ice and snow effectively built a scree slope around each mound following deposition and prior to establishment of a stabilizing vegetative cover. We interpret the "shells" on the mounds of New Zealand deposits as colluvial matrix-rich deposits that formed in the 5- to 10-year period following landslide emplacement. Long-term post-depositional weathering of the fine material present in the "shell" has led to siliceous cementation.

### *Recognition of Debris-Avalanche-Generated Debris-Flow Deposits*

Deposits of debris flows generated from debris avalanches differ in both geometry and texture from those initiated by other mechanisms, such as collapse of crater lakes or transformation from pyroclastic flows. The marginal lithofacies typically forms laterally extensive diamicton sheets as much as 17 m thick that can be correlated over 30 km transverse to major flow direction. Deposits are tabular, except where confined to channels and where thickness changes in response to underlying topography. Additionally, matrix in marginal lithofacies of Quaternary deposits is dominated by clay and silt that has a high allophane content in the humid climate of New Zealand.

In contrast, debris flows initiated from other mechanisms were mostly restricted to channels on the ring plain. Deposits of these flows are lenticular. In places where these debris flows were large enough to overtop channels, deposits extend less than 10 m transverse to flow. Matrix in these deposits is sand, with only a few percent silt.

Textural and geometic differences between debris flows related to debris avalanches and those initiated through other mechanisms reflect differences in initiation mechanism, grain size of the primary component, and magnitude of the event. The New Zealand debris avalanches were large events formed from collapse of sections of the cone where the presence of either hydrothermally altered material or thick tephra cover provided clay to the matrix of the resulting deposit. Debris flows generated from large debris avalanches were also large enough to erode the thick tephra cover that mantled the landscape prior to collapse.

Debris flows not generated from debris avalanches were smaller and predominantly confined to channels, where mostly sandy sediment was available for erosion and incorporation into the flows. In interchannel areas, these debris flows were generally non-erosive and did not incorporate large volumes of a secondary weathered tephra component, resulting in deposits with a sandy rather than a muddy matrix.

### *Lithofacies Architecture*

Lithofacies architecture varies among New Zealand debris-avalanche deposits, mostly as a function of the size of the event and physiography of the ring plain in the flow path. Size and physiography influenced both the proximal-to-distal position of individual lithofacies and the geometry of the deposits. Physiography also controlled flow direction of some debris avalanches, affecting the lateral development of lithofacies.

Large debris avalanches surmounted physiographic barriers as much as 50 m high and deposited widespread sheets of debris (Fig. 5, Pungarehu Formation). The spatial arrangement of lithofacies in these deposits is a simple pattern of central lobes of axial-A surrounded by axial-B and the marginal lithofacies, respectively. Debris-flow deposits related to large avalanches are also present in river valleys downstream of the volcanoes. Marginal lithofacies in areas up to 60 km downstream record the long runout associated with these events.

On the ring plain, deposits of all lithofacies are sheet-like diamictons that thicken and thin in response to underlying physiography. In the marginal lithofacies, thickness changes perpendicular to the major flow direction are abrupt at the most distal edge of the deposit, reflecting channelization of the debris flow as the flow left the ring plain. Beyond the ring plain, the marginal lithofacies either underlies terraces or completely fills paleovalleys.

Smaller events were confined or unconfined along different portions of their flow paths. In the proximal portion of the Okawa debris avalanche, pre-existing physiography did not affect distribution of axial-A, axial-B, or the marginal lithofacies (Fig. 5). At 22 km from the source, a steep escarpment directed flow into steep-sided channels. Beyond 30 km from the source, where physiography was subdued, the flows spread out to encompass large areas. In deposits such as these, axial-A forms a lobe near the base of the volcano, surrounded by axial-B. The axial-B and marginal lithofacies comprise proximal to distal parts, respectively, of shoestring deposits in areas where flow was channelized. The marginal lithofacies dominates the distal ring-plain portion of the deposit, forming a laterally extensive diamicton sheet.

Deposits of small debris avalanches that crossed relatively featureless ring plains display similar architecture to large events. With the small avalanches, however, relatively small physiographic barriers can strongly influence lithofacies arrangement. For example, a ridge of only 4 m relief on the southwestern margin of the Tongariro ring plain prevented the Murimotu debris avalanche from spreading northward (Fig. 6). Consequently, axial-B lithofacies composes the north margin of the deposit. In the south, physiography was more subdued and a complete lateral gradation from axial-A through marginal lithofacies is present.

## CONCLUSIONS

Volcanic-debris-avalanche deposits are characterized by a distinctive assemblage of lithofacies. Several factors control the development and spatial arrangement of lithofacies. These include size of the debris avalanche, geology of the source area, availability of water, and physiography of areas surrounding the volcanoes (Neall, 1979; Crandell and Waldron, 1956; Crandell and others, 1974; Nakamura, 1978; Kawachi, 1983; Crandell and others, 1984; Seibert, 1984; Glicken, 1986; Naranjo and Francis, 1987; Crandell, 1989; Palmer and Neall, 1989).

In New Zealand, volcanic-debris avalanches originated predominantly from collapse of andesitic stratovolcanoes. The source areas contained enough water from meteoric and hydrothermal sources, plus the possible addition of water from melting snow and glaciers, to produce wet-debris avalanches.

Wet-avalanche flow produced distinctive, mappable lithofacies. Incomplete mixing of the source materials led to early separation of megaclasts and interclast matrix, allowing the wet, interclast matrix-rich flow to drain away from the axial-A lithofacies deposited by the megaclast-rich part of the flow. As matrix-rich debris-avalanche flow transformed to debris flows, sediment was eroded from the underlying deposits. The debris flows were extremely mobile and were able to flow >40 km away from the volcanoes. These extremely mobile debris flows, depositing muddy marginal lithofacies, are a distinctive feature of wet-debris avalanches (Siebert and others, 1987; Crandell, 1989).

The marginal lithofacies represents a mobile phase of the flow and is an integral and synchronous part of the avalanche event. Evidence in the New Zealand deposits indicates that three factors contribute to the development and mobility of marginal lithofacies. (1) Hydrothermal alteration of the source area. At Ruapehu and Tongariro volcanoes, collapse in hydrothermally altered source areas supplied primary fine-grained material to the landslide. (2) Water. Both Egmont and Ruapehu volcanoes are the source of numerous rivers draining the region. Fed from springs around the base of the upper lava cones, considerable volumes of water seep from the volcanic pile and the associated ring-plain deposits, indicating that the coarser volcaniclastic deposits act as aquifers. The mean discharge of all the rivers draining Egmont Volcano to within a 12-km radius of the summit totals 28 million $m^3$ ($14 \times 10^9$ l) per week (Taranaki Catchment Commission 1984). After one week without rain, half this value is likely to be the amount of water stored in the volcanic edifice. Thus, there is evidence for large volumes of water present in the volcanic piles, which will be a primary component in a collapse. In some situations, additional sources of water may be provided to an incipient landslide mass by rain, eruption-induced melting of snow and ice, or by water in crater lakes or hot springs. (3) Allophane and other short-range-order clays. Under the humid-temperate climatic regime of western North Island, abundant andesitic ash (mainly the glassy mesostasis component) rapidly weathers to short-range-order clay materials such as allophane. Allophane is renowned for its low-yield strength. Thus, during collapse events, ash within the volcanic pile contributes allophane as a primary component to the flow. Even greater volumes of allphanic tephra, which may have totalled 30 to 60 $km^3$, now mantle older remnant landscapes to >30 in thickness. As unconfined flows reach the ring plain, they erode this allophanic cover, enhancing the mobility of the marginal lithofacies that then spreads as a thin (usually >5 m) veneer across low-gradient surfaces.

The influence of volume and physiography on lithofacies distribution of New Zealand volcanic-debris avalanches is difficult to evaluate. A feature that is a small physiographic hurdle for one debris avalanche could be a major obstacle for another. In general, most of the New Zealand debris avalanches were unconfined, so that they fanned out as they crossed the ring plains, producing characteristic, multidirectional changes in lithofacies.

Deposits of volcanic-debris avalanches in New Zealand differ from those reported from other areas of the world, where physiographic barriers generally confine flow (Fig. 8). In these areas, the major flow direction is downstream, and the dominant direction of lithofacies development is parallel to flow (e.g., Glicken, 1986; Crandell, 1989). Spatial arrangement of lithofacies in the resulting deposit is a proximal-to-distal arrangement of axial-A, axial-B, and marginal lithofacies, if the avalanche was wet enough to produce a debris-flow phase. In some cases, valleys may be large enough, or debris avalanches small enough, to allow development of a minor-flow component oblique to dominant-flow direction. At Mount Shasta, California, the 300-ka debris-avalanche deposit displays a tripartite lithofacies succession in a downstream direction (Crandell, 1989). Laterally, however, only the axial-A and axial-B portion of the succession is present. The megaclast-rich portion of the

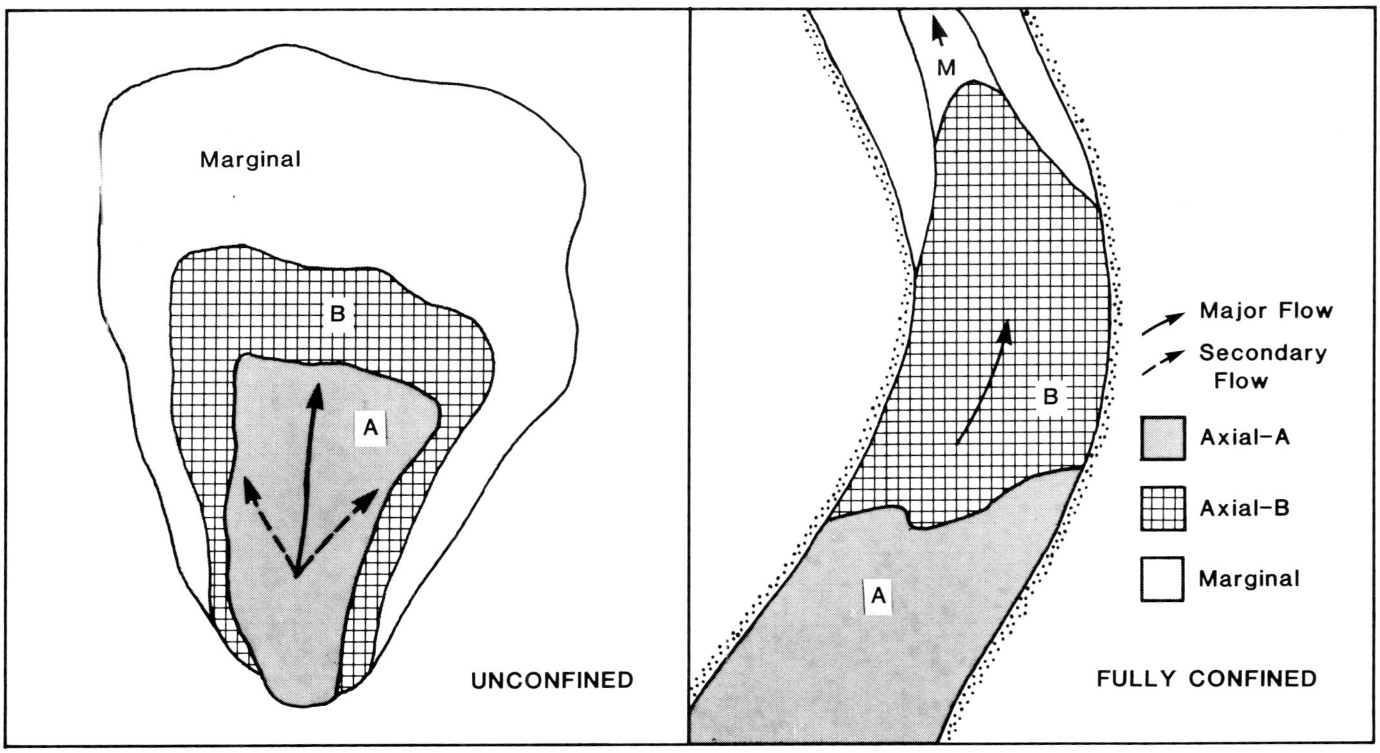


FIG. 8.—Generalized lithofacies relationships in confined and unconfined debris-avalanche/lahar deposits.

avalanche was not large enough to fill the valley completely, allowing the interclast matrix-rich part of the avalanche to drain obliquely away from the axial-A lithofacies. Together, the megaclast-rich and interclast matrix-rich portions of the avalanche, however, did fill the valley completely, preventing development of a debris-flow phase oblique to major flow direction.

The recognition of the relationship between the three lithofacies we describe and unconfined volcanic-ring plains is important in paleogeographic and provenance studies. The relationships described provide another example of the genesis and consequent spatial organization of lithofacies in volcanic-debris-avalanche deposits, which may be used as a tool for deciphering deposits in ancient volcanic successions.

## REFERENCES


ALLOWAY, B. V., 1989, The Late Quaternary cover bed stratigraphy and tephrochronology of central Taranaki, New Zealand. Massey University, Palmerston North, New Zealand, Unpubl. Ph.D. Dissertation 406 p.

ALLOWAY, B. V., 1989, Gigantic debris avalanche of Pleistocene age from ancestral Mount Shasta Volcano, California, and debris-avalanche hazard zonation: U.S. Geological Survey Bulletin 1861, 32 p.

ALLOWAY, B. V., MILLER, C. D., GLICKEN, H. X., CHRISTIANSEN, R. L., AND NEWHALL, C. G., 1984, Catastrophic debris avalanche from ancestral Mount Shasta Volcano, California: Geology, v. 12, p. 143–146.

ALLOWAY, B. V., MULLINEAUX, D. R., SIGAFOOS, R. S., AND RUBIN, M., 1974. Chaos Craigs eruptions and rock-fall avalanches, Lassen Volcanic National Park, California: U.S. Geological Survey, Journal of Research, v. 2, p. 45–59.

ALLOWAY, B. V., AND WALDRON, H. H., 1956, A recent volcanic mudflow of exceptional dimensions from Mount Rainier, Washington: American Journal of Science, v. 254, p. 349–362.

CRANDELL, D. R., 1989, Gigantic debris avalanche of Pleistocene age from ancestral Mount Shasta Volcano, California, and debris-avalanche hazard zonation: U.S. Geological Survey Bulletin 1861, 32 p.

CRANDELL, D. R., MILLER, C. D., GLICKEN, H. X., CHRISTIANSEN, R. L., AND NEWHALL, C. G., 1984, Catastrophic debris avalanche from ancestral Mount Shasta Volcano, California: Geology, v. 12, p. 143–146.

CRANDELL, D. R., MULLINEAUX, D. R., SIGAFOOS, R. S., AND RUBIN, M., 1974, Chaos Crags eruptions and rock-fall avalanches, Lassen Volcanic National Park, California: U.S. Geological Survey, Journal of Research, v. 2, p. 45–59.

CRANDELL, D. R., AND WALDRON, H. H., 1956, A recent vocanic mudflow of exceptional dimensions from Mount Rainier, Washington: American Journal of Science, v. 254, p. 349–362.

GIBSON, W., AND MORGAN, P. G., 1927, The Geology of Egmont subdivision. New Zealand Geological Survey Bulletin, no. 29, 92 p.

GLICKEN, H., 1986, Rockslide-debris avalanche of May 18, 1980, Mount St. Helens Volcano, Washington. The University of California, Santa Barbara, Unpubl. Ph.D. Dissertation 303 p.

GRANGE, L. I., 1931, Conical hills on Egmont and Ruapehu Volcanoes. New Zealand Journal of Science and Technology v. 12, p. 376–384.

GRANT-TAYLOR, T. L., 1964, Geology of Egmont National Park, *in* Scanlan, A. B., ed., Egmont National Park: Egmont National Park Board, New Plymouth, p. 13–26.

HACKETT, W. R., 1985, Geology and petrology of Ruapehu Volcano and related vents. Victoria University, Wellington, New Zealand Unpubl. Ph.D. Dissertation, 407 p.

HACKETT, W. R., AND HOUGHTON, B. F., 1989, A facies model for a Quarternary andesitic composite volcano: Ruapehu, New Zealand: Bulletin of Volcanology, v. 51, p. 51–68.

KAWACHI, S., 1983, Ohtsukigawa debris avalanche, Yatsugatake volcanic chain, central Japan: Geological Society of Japan Journal, v. 89, p. 173–182.

NAKAMURA, Y. 1978, Geology and petrology of Bandai and Nekoma volcanoes: Sendai, Japan, Tohoku University Science Reports, Series III, v. 14, p. 67–119.

NARANJO, J. A., AND FRANCIS, P., 1987, High velocity debris avalanche at Lastarria Volcano in the north Chilean Andes: Bulletin of Volcanology, v. 49, p. 509–514.

NEALL, V. E., 1979. Geologic map of New Zealand, (New Plymouth, Egmont and Manaia: New Zealand Geological Survey, Sheets P19, P20, P21, plus 35 p. text. scale 1:50,000.

NEALL, V. E., 1982, Land forms of Taranaki and the Wanganui Lowlands, *in* Soons, J. M., and Selby, M. J., eds., Landforms of New Zealand: Longman Paul, Auckland, p. 193–212.

NEALL, V. E., AND ALLOWAY, B. V., 1986, Quaternary volcaniclastics and volcanic hazards in Taranaki. New Zealand Geological Survey Record 12, p. 101–137.

PALMER, B. A., AND NEALL, V. E., 1989, The Murimotu Formation, 9500-year old deposits of a debris avalanche and associated lahars, Mount Ruapehu, North Island, New Zealand: New Zealand Journal of Geology and Geophysics, v. 32, p. 477–486.

SIEBERT, L., 1984, Large volcanic debris avalanches: Characteristics of source areas, deposits, and associated eruptions: Journal of Volcanology and Geothermal Research, v. 22, p. 163–197.

SIEBERT, L., GLICKEN, H. X., and UI, T., 1987, Volcanic hazards from Bezmianny- and Bandai-type eruptions: Bulletin of Volcanology, v. 49, p. 435–459.

SMITH, G. A., AND FRITZ, W. J., 1989, Penrose Conference Report: Volcanic influences on terrestrial sedimentation: Geology, v. 17, p. 375–376.

TARANAKI CATCHMENT COMMISSION, 1984, Surface Hydrology, Taranaki Ring Plain Water Resources Survey. Taranaki Catchment Commission, Stratford, New Zealand, 123 p. plus Appendices.

UI, T., AND GLICKEN, X. H., 1986, Internal structural variations in a debris-avalanche deposit from ancestral Mount Shasta, California: Bulletin of Volcanology, v. 48, p. 189–194.

VOIGHT, B., GLICKEN, H., JANDA, R. J., AND DOUGLASS, P. M., 1981, Catastrophic rockslide avalanche of May 18, *in* Lipman, P. W., and Mullineauy, D. R., eds. The 1980 Eruptions of Mount St. Helens, Washington: U.S. Geological Survey Professional Paper 1250, p. 347–377.

YOSHINGA, N., 1986, Mineralogical characteristics II. Clay minerals, *in* Wada, K., ed., Ando Soils in Japan: Kyushu University Press, p. 41–56.

# SEDIMENTARY ARCHITECTURE OF LARGE VOLCANIC-DEBRIS AVALANCHES

HARRY GLICKEN
*Department of Geological Sciences, University of California, Santa Barbara, California 93106*

ABSTRACT: Volcanic-debris-avalanche and related deposits form important parts of the sedimentary record in volcanic regions. A rigorously defined terminology and various sedimentary elements are used to describe the deposits. Particles are clasts (hard rocks) and debris-avalanche blocks (soft pieces of debris). Facies, which may be observed on the map scale or exposure scale, are block facies, consisting of debris-avalanche blocks, and mixed facies, a mixture of rock types. The hummocks of debris-avalanche deposits are classified into three types: type A, block facies hummocks with no mixed facies; type B, predominantly mixed facies hummocks; and type C, hummocks composed of debris-avalanche blocks resting in mixed facies. Fractures are common in the block facies on both macroscopic and microscopic scales. The fractures form from propagation of waves of compression and rarefaction that develop as the mass moves over rough topography during the initial sliding phase of the events.

Lahars (volcanic mudflows and debris flows) are commonly associated with volcanic-debris avalanches. Lahars may form from the debris avalanche during emplacement or minutes to hours after emplacement, or from breakout of debris-avalanche-dammed lakes hours to years after emplacement.

## INTRODUCTION

Volcanic-debris avalanches and related deposits are an important part of the sedimentary record in volcanic regions. Recent work (Ui and others, 1986; Francis and Wells, 1986; Boudon and others, 1987; Seibert and others, 1987; Palmer and others, this volume) has shown that debris-avalanche and related deposits are common around modern stratocones, and similar deposits have been found in the pre-Quaternary volcanostratigraphic record as well (e.g., Sundell, 1985). Around some stratovolcanoes (e.g., Augustine Volcano, Alaska, Siebert and others, 1989; Mount Egmont, New Zealand, Palmer and others, this volume), the deposits compose the bulk of the volcaniclastic rocks.

"Sedimentary architecture" is the term used here to describe how the deposits are put together by various sedimentary elements. It is somewhat similar to architectural-element analysis, which is a way to understand facies in sedimentary rocks by dividing features into basic architectural elements (Miall, 1985).

## TERMINOLOGY

In order to study debris-avalanche deposits, it has been necessary to define rigorously and to use consistently (Glicken, 1986; Glicken and others, 1989b; Siebert and others, 1987, 1989) a single set of definitions. Some of this terminology is adapted from previously published studies of large volcanic-debris-avalanche deposits, and is drawn from the literature on sedimentology and on volcaniclastic rocks. The usage of terms in this work is summarized in Table 1. Some of the terms have rough equivalents to those used by Palmer and others (this volume) to describe debris-avalanche deposits in New Zealand that are not confined to valleys.

The texture of volcaniclastic deposits commonly refers to the grain-size distribution of the material (e.g., Crandell, 1971; Murai, 1961). Here, it is used in the more general sense (e.g., Fisher and Schmincke, 1984) to refer to the size, shape, and fabric (pattern of arrangement) of the particles that form the deposit. A "particle" is usually considered to be a "separable or distinct unit in a rock" (Bates and Jackson, 1980, p. 457). The word "structure" is applied to features visible on the scale of exposures (e.g., a degassing or dewatering structure). The distinction between textures and structures is somewhat arbitrary because exposure-scale features also represent the pattern of arrangement of particles. "Texture" is used in this work to refer to these patterns on any scale.

Two different kinds of particles are defined to describe debris-avalanche deposits. A "clast" is defined as a rock of any size from microns to meters in diameter that would not break if passed through a sieve or immersed in water: a hard rock. Each clast can be considered one particle. A "debris-avalanche block" (or, for brevity, "debris block"; Waitt and others, 1989) is defined as a coherent, unconsolidated or poorly consolidated piece of the volcano that was transported to its place of deposition relatively intact: a soft piece of debris. Each debris-avalanche block may be thought of as one particle that contains many smaller particles (clasts). The usage of the term "debris-avalanche block" is similar to that of "megablock" of Mimura and Kawachi (1981) and Ui (1983). "Megablock" is not appropriate because the debris blocks can be any size. This distinction between consolidated clasts and unconsolidated or poorly consolidated debris-avalanche blocks is essential for the analysis of exposures.

Several terms are used in this and related work to describe the disintegration of material. The breaking of individual clasts is "fracturing." Pervasive fracturing of clasts is called "shattering." When material expands from its original density on the mountain (at least in part by shattering of clasts), it is said to "dilate"; when it breaks apart into its constituent clasts that then move apart, it "disaggregates."

Two end-member facies are used to describe the texture of debris-avalanche deposits, block facies and mixed facies. This terminology follows the precedent of Mimura and Kawachi (1981), Crandell and others (1984), Glicken and others (1989b), Siebert and others (1989), Ui (1983), Ui and Glicken (1986), and Glicken (1986), except that the previously used "matrix facies" is renamed "mixed facies" to avoid confusion with the sedimentological definition of the word "matrix." Both facies can be present at the same exposure; mixed facies may surround block facies. Both facies would be classically described as angular, unstratified, unsorted rubble consisting of material (clasts) that ranges

Sedimentation in Volcanic Settings, SEPM Special Publication No. 45
 ISBN 0-918985-89-7

TABLE 1.—TERMINOLOGY OF DEBRIS AVALANCHE DEPOSITS

Equivalent terms from Palmer and others (this volume) in *italics*.

| | |
|---|---|
| Texture | Size, shape, and fabric (pattern of arrangement) of particles that form the deposit. |
| Particle<br>*Clasts or megaclasts* | A separable or distinct unit in the deposit or the moving material. |
| Clast<br>*Single rock clasts* | A rock of any size that would not break if passed through a sieve or immersed in water. |
| Debris-avalanche block (debris block) | An unconsolidated (or poorly consolidated) piece of the old mountain transported to its place of deposition |
| *Brecciated megaclasts and stratified megaclasts* | An unconsolidated (or poorly consolidated) piece of the old mountain transported to its place of deposition relatively intact. Contains many clasts, which may range in size from microns to meters. |
| Fracturing | Breaking of individual clasts. |
| Shattering | Thorough fracturing of clasts. |
| Dilation | Volume expansion of a mass of material. |
| Disaggregation | Breaking apart of unconsolidated or poorly consolidated material into its constituent clasts. |
| Block facies | Part of the debris-avalanche deposit that consists of debris-avalanche blocks. |
| Mixed facies<br>*Interclast matrix and clasts* | Part of debris-avalanche deposit that is completely mixed. May contain clasts of all rock types and all sizes from microns to meters. |

in size from microns to meters in diameter. This usage of the word "facies" is consistent with the definition of facies in the Glossary of Geology (Bates and Jackson, 1980, p. 220).

The block facies consists entirely of debris-avalanche blocks, coherent unconsolidated or poorly consolidated pieces of the old mountain that were transported relatively intact (Fig. 1A). Debris blocks range in size from a few centimeters to more than a hundred meters wide. Commonly, the debris blocks are deformed to different degrees.

Debris-avalanche blocks may be partially or completely fractured (shattered) from their origin as pieces of the old mountain, so that although the original volcanic stratigraphy or structure is locally preserved (the "mutual arrangement of separate fragments"; Gorshkov and Dubik, 1970, p. 275) few clasts on the order of meters wide from the old mountain remain. Shreve (1968) observed the same texture in the nonvolcanic Blackhawk slide and named it the "three-dimensional jigsaw puzzle effect." In debris-avalanche deposits, the shattering has produced unconsolidated or poorly consolidated rubble that was cohesive enough upon deposition to form hummocks (mounds of Palmer and others, this volume).

The term mixed facies is used here in the sense of Crandell and others (1984), Ui (1983) and Ui and others (1986) to refer to the completely mixed parts of the debris avalanche deposit. The mixed facies may contain clasts representing many rock types from the mountain in an ungraded, unsorted and unstratified mixture, and may contain juvenile material (rock solidified from magma of a concurrent eruption). Locally, fragments of wood and bits of soil and rock from the terrain underlying the deposit are present in the mixed facies.

## FACIES ELEMENTS OF VARIOUS SCALES

The deposits can be considered on map, exposure, and microscopic scales. On the map scale, deposits can be divided into areas that contain only block facies, areas that contain only mixed facies, and areas that contain both (Fig. 2). Block facies may be further divided into lithologic types that may be correlated with rock types on the source volcano (Fig. 2).

On the scale of exposures, deposits may also be divided into areas that contain only block facies and areas that con-

FIG. 1.—(A) Debris-avalanche block. In May 18, 1980, debris-avalanche deposit at Mount St. Helens preserved but shattered and faulted lava-flow stratigraphy in canyon of 1984. Exposure is 50 m deep. (B) Mixed facies. Shasta Valley debris-avalanche deposit near Mount Shasta, California. Overlain by soil. Dark patches surrounded by mixed facies are remnant debris-avalanche blocks. Vertical stripe under man is a rill.

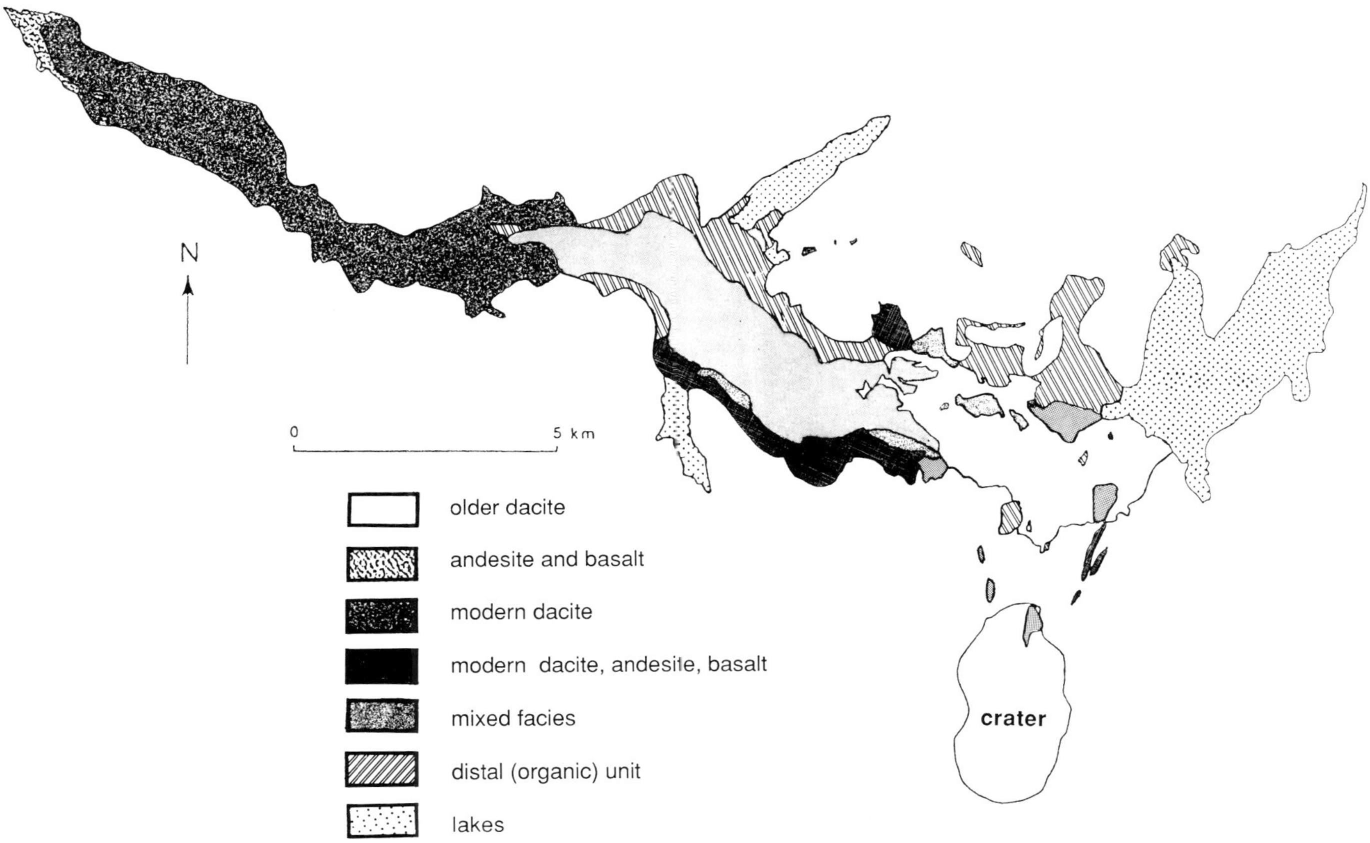


FIG. 2.—Generalized lithologic map of May 18, 1980, debris-avalanche deposit at Mount St. Helens. Generalized from 1:2,400 geologic map of Glicken (1986, pl. 4).

tain only mixed facies (Fig. 3A, B). Debris-avalanche blocks may be fractured and full of jigsaw cracks (Fig. 4) or thoroughly shattered with no jigsaw cracks preserved. Exposures may include structures or stratigraphy preserved from the source volcano. These features may be difficult to distinguish from features created by the transport and flow of the debris avalanche, but they may locally display clues such as chilled margins on dikes (indicating igneous dike rather than clastic dike) or highly contorted beds (indicating deformation during transport; Fig. 5).

## FACIES AND MORPHOLOGY

The morphology of the surface of the deposit can be related to facies as seen in exposures. The surface of volcanic-debris-avalanche deposits generally consists of hummocks that may be classified into a number of different types based on the relation of block facies to mixed facies (Fig. 6A–C).

*Type A, block facies hummocks with no mixed facies.—*

One or more debris blocks extend from hummock to hummock. The debris blocks may be of any size, ranging up to several hundred meters across. There is no mixed facies in the hummocks or in the interhummock areas. These hummocks occur primarily in proximal parts of debris-avalanche deposits.

*Type B, predominantly mixed-facies hummocks.—*

These hummocks are composed almost entirely of mixed facies, and are generally found in distal areas of channelized debris-avalanche deposits. Vertical exposures show that there is no debris block at the cores of the hummocks, but there may be small debris blocks scattered throughout the hummocks. These hummocks are less than a few meters high, generally much smaller than type A hummocks. Because of the relatively well-mixed and uniform nature of the mixed facies, mixed-facies hummocks are not present in many debris-avalanche deposits.

*Type C, hummocks of debris blocks resting in mixed facies.—*

These hummocks are made entirely of large debris blocks of the block facies that rest in, and likely were carried by, the mixed facies. In many volcanic-debris-avalanche deposits (e.g., the deposit north of Mount Shasta, California; Crandell and others, 1984), most of the hummocks are interpreted to be type C hummocks; at other deposits, type C hummocks are rare. In prehistoric deposits, it is common that exposures are only present inside hummocks and not

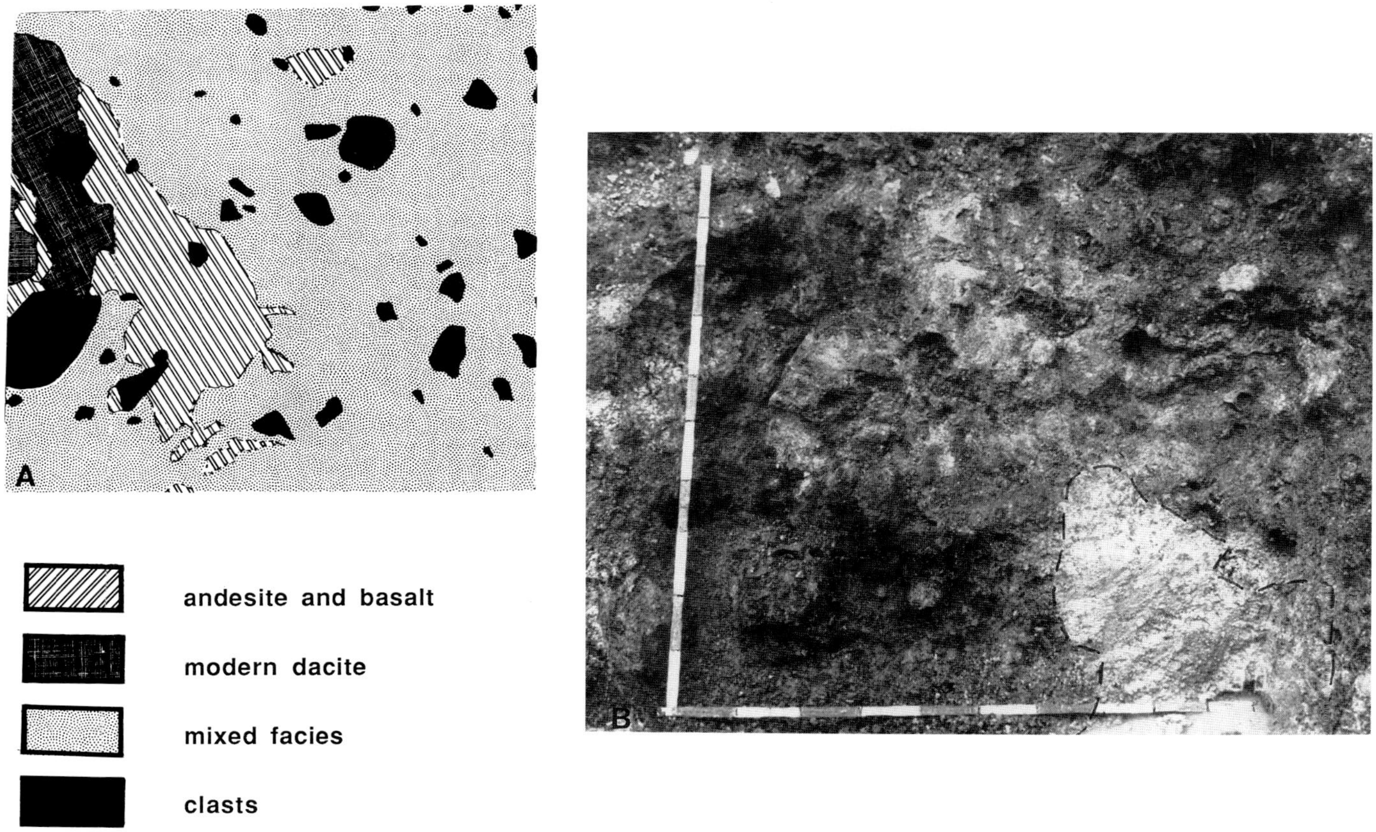


FIG. 3.—(A) Line drawing of 1-m$^2$ exposure of 1980 Mount St. Helens debris avalanche showing both block and mixed facies. Unit designations same as on Figure 2. Dashed lines are boundaries around debris-avalanche blocks. Dotted lines are indistinct boundaries around blocks. (B) A 1-m$^2$ exposure of block facies of debris-avalanche deposit from Papandayan Volcano, Indonesia (Glicken and others, 1987). Scale shows ties at 10-cm intervals. Area in bottom right corner outlined with dashed line is one debris-avalanche block of soft hydrothermally produced clay. Patchwork pattern in exposure indicates various degrees of hydrothermal alteration in debris-avalanche block.

FIG. 4.—Jigsaw cracks in boulder in Mount St. Helens 1980 debris-avalanche deposit. Cracks like this are rare in the Mount St. Helens deposit, but common in others. Rock hammer for scale.

in interhummock areas, occasionally leading to a misinterpretation that all hummocks are type C hummocks.

Debris-avalanche hummocks are interpreted to form in three ways. Many hummocks probably formed from a combination of two or more of the proposed mechanisms.

The hummocks of type A commonly represent the horsts of a simple horst-and-graben system (Voight and others, 1981, 1983). Contacts from the volcano preserved intact during transport are faulted down between the hummocks. The horsts and grabens may form in response to spreading of the debris avalanche.

Some type A and type C hummocks represent the surface topography of debris-avalanche blocks (pieces of the old mountain). Type C hummocks result from debris-avalanche blocks suspended in mixed facies.

Some hummocks of type A and essentially all of type B hummocks form as material was decelerated by basal or lateral shear. Debris avalanches commonly consist of a combination of many different materials and are characterized by strength inhomogeneity. The parts of the moving debris avalanche that are strong enough to resist basal shear slow down, while adjacent parts of the material characterized by less strength flow past, sculpting the slower moving material into hummocks.

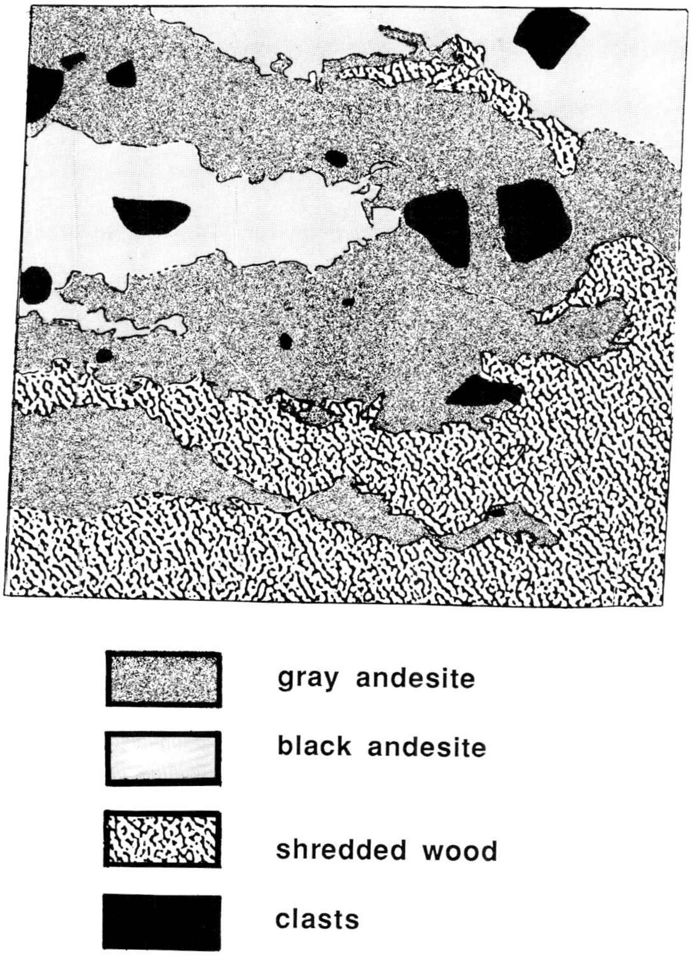


FIG. 5.—A 1-m$^2$ exposure of the May 18, 1980, debris-avalanche deposit at Mount St. Helens near Castle Lake. Gray andesite breccia layer shows disaggregation and rolling of strata during debris-avalanche transport.

## FRACTURES

Fractures are present to varying degrees in debris-avalanche deposits. In some debris-avalanche deposits, meter-size clasts with pervasive jigsaw cracks are common. However, in other deposits, the material is so thoroughly shattered that few jigsaw cracks are recognized. This may result from pre-eruption deformation or pre-eruption brecciation of the source material by earlier mass movements (an interpretation of Hausbeck and Swanson, 1989, for the fractured walls of the 1980 crater at Mount St. Helens).

Fractures are common features of the material of the block facies. Fractures on an exposure scale (Fig. 4; jigsaw cracks as defined by Ui, 1985) are commonly mimicked by fractures, showing clear displacement, on a microscopic scale seen in scanning electron microscope (SEM) images (Fig. 7; Glicken and others, 1989a). Disaggregation and mixing commonly break apart fractured material on the microscopic scale in mixed facies, but SEM images show that the fractures are present in pyrogenic crystals as well as in glass of both the block facies and the mixed facies.

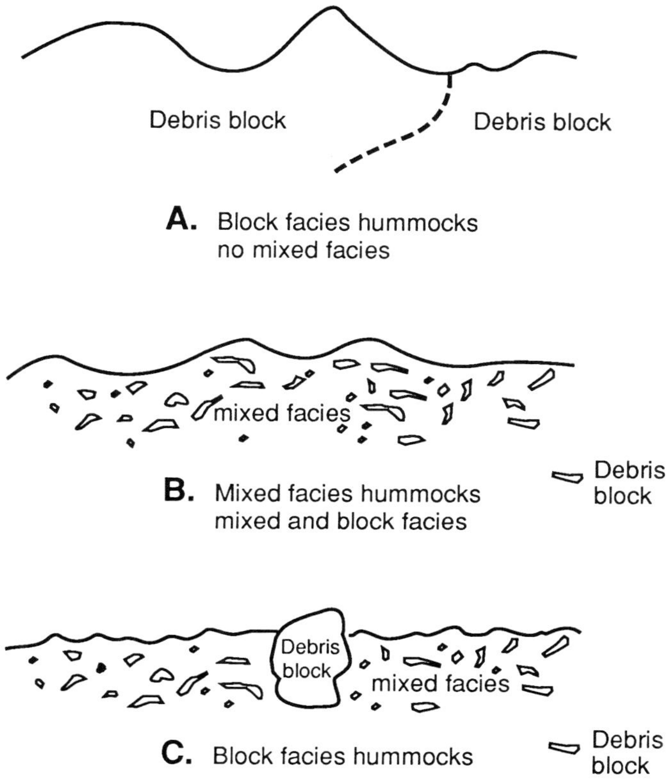


FIG. 6.—Diagram of hummock types. Type A, block facies with no mixed facies. Type B, predominantly mixed facies. Type C, debris-avalanche block suspended in mixed facies; mixed facies probably carried debris-avalanche block.

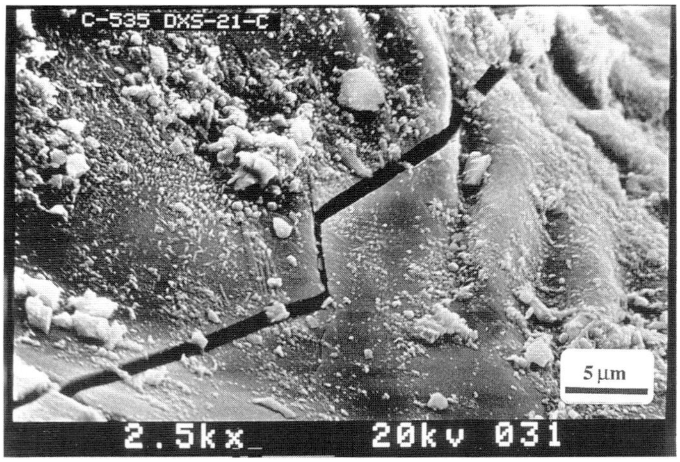


FIG. 7.—SEM photomicrograph of debris-avalanche sample from 1980 deposit of Mount St. Helens, showing characteristic jigsaw-type crack. From Komorowski and others (in prep.).

Pervasive shattering of clasts in the Mount St. Helens debris avalanche probably produced much of the dilation measured from Mount St. Helens to the debris-avalanche deposit. The mass movement began as a retrogressive rockslide, clearly visible on eyewitness photographs (Voight and others, 1983) and then broke up into smaller pieces (debris blocks). Density measurements from the Mount St. Helens deposit and from the same *in situ* rock types at Mount St. Helens suggest that debris blocks were dilated to their maximum extent of about 20 percent during the sliding of the initial rockslide (at the mountain), and were not dilated further during avalanche flow (Glicken, 1986,). Dilation, then, was created by sliding and resulted in profound loss of strength (Voight and others, 1983), facilitating continued sliding and the transition to avalanche flow, but continued expansion of the material was not required for the avalanche to flow. Theoretical mechanics indicate that this is a general case for mass movements that begin as a slide and transform into flow (Savage, 1984).

Fractures form and clasts shatter under the tensile stress (rarefaction) that follows the compressive stress resulting from a collision. The compressive stress also generates a tensile stress approximately normal to the compressive stress during a collision (Jaeger and Cook, 1979).

In sliding mass movements, the shear generated as the sliding material moves over rough topography likely results in repeated propagation of waves of compression and rarefaction. Irregularities in topography compress parts of the base of the moving material and the rarefaction that follows compression results in fracturing of individual clasts. Neighboring fragments remained in contact with one another, so the structures and stratigraphy of the volcano can be preserved relatively intact.

## RELATIONSHIP TO LAHARS

Lahars are commonly associated with debris avalanches (Scott and Janda, 1987). Lahars are masses of flowing volcanic debris intimately mixed with water (Fisher and Schmincke, 1984) and include mudflows and debris flows (terminology of Pierson and Scott, 1985). Lahars form from: 1) transformation of water-saturated parts of mixed facies to lahars during emplacement, 2) slumping and flowing of water-saturated parts of the debris avalanche minutes to hours after emplacement, or 3) breakout of debris-avalanche-dammed lakes hours to years after emplacement (Siebert and others, 1987). Scott and Janda (1987) have noted that origins (1) and (2) commonly produce lahars with clay contents of more than 3 percent.

The first mechanism is illustrated by relationships around Mount Egmont Volcano, as are well described by Palmer and others (this volume) and Neall (1979). In exposures perpendicular to the flow direction of the debris avalanche (Palmer and others, their fig. 7), the hummocky mixed facies of the debris avalanche (the mixed parts of Palmer and others' axial facies) is observed to grade laterally into flat-surfaced, mixed lahar deposits (parts of Palmer and others' marginal facies). As Palmer and others suggest, this is probably a common feature of unconfined, wet-avalanche flows. The first mechanism is also illustrated by the clay-rich, semi-confined Papandayan debris-avalanche deposit in Indonesia, which grades distally into a flat-surfaced lahar (Glicken and others, 1987), by the Osceola mudflow at Mount Rainier, Washington, which is a clay-rich, flat-surfaced lahar interpreted to have formed from proximal debris avalanches of wet, hydrothermally altered sediments (Crandell, 1971; Scott and Janda, 1987), and by the relatively small 1984 debris-avalanche deposit of Ontake Volcano, Central Japan (Endo and others, 1989).

The second mechanism is best illustrated by the emplacement of the 1980 debris avalanche at Mount St. Helens. Within hours after emplacement, lahars formed from slumping and flowing of water-saturated debris-avalanche material. Eyewitness observations by the author indicated that the moving lahars ponded between hummocks, broke out of the ponds, coalesced with other lahars, and then flowed downstream (Fig. 8; Glicken, 1986). The water content of the Mount St. Helens debris avalanche was calculated to be about 15 percent by volume; 20 percent of this water was removed in lahars that contained 35 to 50 percent water (Glicken, 1986; Fairchild, 1985). A debris-avalanche water content much higher than 15 percent by volume is probably necessary for lahars to be formed by the first mechanism.

The third mechanism is illustrated by lahars that formed from floods that developed from the breakout of debris-avalanche-dammed lakes within years after the emplacement of prehistoric debris avalanches at Mount St. Helens. These lahars can only develop from debris avalanches that are deposited in confined channels. Huge lahars, 2.5 to 3 ka, correspond with the Pine Creek eruptive period at Mount St. Helens and are interpreted to have formed by breakout of a lake dammed by a debris avalanche at that time (Scott, 1989). This interpretation is supported by evidence from mapping of the crater walls and deep canyons on the North Flank of Mount St. Helens (Hausbeck and Swanson, 1989) that show fracturing interpreted to be evidence for debris avalanches in the Pine Creek eruptive period. Similar huge lahars were forecast to occur from breakout of three large lakes dammed by the 1980 debris avalanche of Mount St. Helens (Fig. 9; Glicken and others, 1989b; Glicken and Voight, in prep). The lahars did not occur because outlet channels were constructed by the U.S. Army Corps of Engineers (Glicken, 1989b).

FIG. 8.—Moving lahars (dark material) on lighter surface of Mount St. Helens debris-avalanche deposit. Photograph, about 150 m across, by the author from a moving helicopter on afternoon of May 18, 1980.

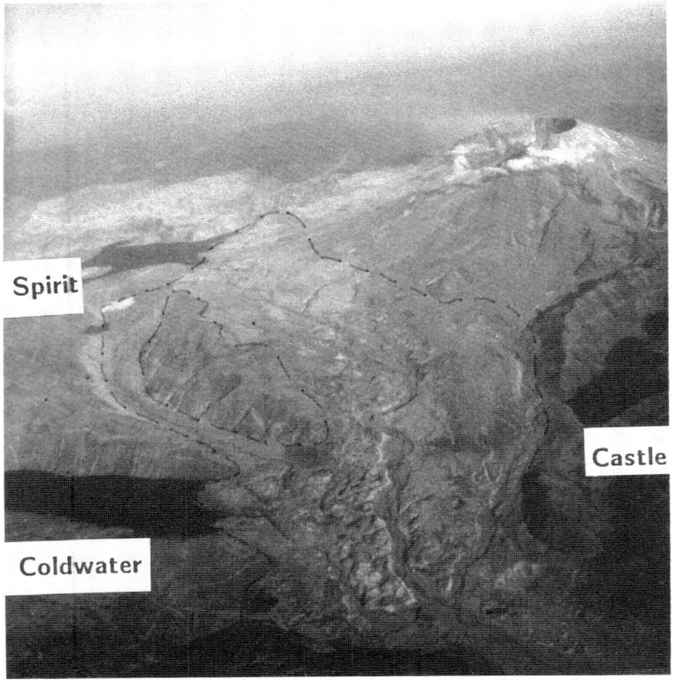


FIG. 9.—Oblique aerial view from the West. Mount St. Helens and three lakes (Spirit, Castle, and Coldwater) dammed by the May 18, 1980, debris-avalanche deposit are shown. Outlets for each of the lakes were constructed by the U.S. Army Corps of Engineers to avoid the formation of large lahars. Dashed line around debris avalanche. Photograph by the author, summer 1985.

CONCLUSIONS

Volcanic-debris avalanches are common in volcanic regions. When well exposed, a detailed analysis of sedimentary features can be undertaken. A terminology has been developed to understand and analyze the deposits. Individual particles are called clasts if they are hard rocks and debris-avalanche blocks (or debris blocks, for brevity) if they are pieces of the volcano transported relatively intact (soft pieces of debris).

Facies types may be differentiated in exposure scale or in map scale. The commonly hummocky morphology of debris avalanches may or may not reflect the facies that underlie it, as hummocks form by different mechanisms.

Fractures are commonly seen in debris-avalanche deposits, either in the field or in samples observed under the scanning electron microscope. The fractures are thought to originate as the material dilates during the initial sliding phase of the events. As sliding material moves over irregularities in the topography, the base is compressed and the consequent rarefaction fractures rocks.

Lahars are commonly associated with debris avalanches. Water-saturated parts of the mixed facies of a debris avalanche may form lahars if the parts continue to flow after the block facies stops (generally in unconfined, wet flows; Palmer and others, this volume), or lahars may form from slumping and flowing of water-saturated parts of the debris-avalanche deposit minutes to hours after emplacement, as at the 1980 Mount St. Helens debris avalanche. In the case of channelized debris-avalanche deposits, lahars may form from breakout of debris-avalanche-dammed lakes hours to years after emplacement.

ACKNOWLEDGMENTS

The ideas in this paper have evolved through discussions with many people over the last decade. Particularly important were discussions with Rocky Crandell, Richard Fisher, Lee Siebert, Tadahide Ui, and Barry Voight.

REFERENCES

BATES R. L., AND JACKSON, J. A., eds., 1980, Glossary of Geology, 2nd edition: Falls Church, Virginia, American Geological Institute, 751 pages.

BOUDON, G., SEMET, M. P., AND VINCENT, P. M., 1987, Magma and hydrothermally-driven sector collapses: The 3100 and 11,000 year b.p. eruptions of la Grande Decouverte (la Soufrière) volcano, Guadeloupe, French West Indies: Journal of Volcanology and Geothermal Research, v. 33, p. 317–323.

CRANDELL, D. R., 1971, Postglacial lahars from Mount Rainier Volcano, Washington: U.S. Geological Survey Professional Paper 677, 73 p.

CRANDELL, D. R., MILLER, C. D., GLICKEN, H. X., CHRISTIANSEN, R. L., AND NEWHALL, C. G., 1984, Catastrophic debris avalanche from ancestral Mount Shasta, California: Geology, v. 12, p. 143–146.

ENDO, K., SUMITA, M., MACHIDA, M., AND FURUICHI, M., 1989, The 1984 collapse and debris-avalanche deposits of Ontake Volcano, Central Japan, *in* Latter, J. H., ed., Volcanic Hazards: IAVCEI Proceedings in Volcanology 1: Springer-Verlag, Berlin Hiedelberg, p. 210–229.

FAIRCHILD, L. H., 1985, Lahars at Mount St. Helens, Washington: Seattle, Washington, University of Washington, Unpubl. Ph.D. Dissertation 374 p.

FISHER, R. V., AND SCHMINCKE, H. U., 1984, Pyroclastic Rocks: New York, Springer-Verlag, 472 p.

FRANCIS, P. W., AND WELLS, G. L., 1986, Landsat thematic mapper observations of debris-avalanche deposits in the Central Andes: Bulletin of Volcanology, p. 258–278.

GLICKEN, H., 1986, Rockslide-debris avalanche of May 18, 1980, Mount St. Helens Volcano, Washington: Santa Barbara, University of California, Unpubl. Ph.D. Dissertation, 303 p.

GLICKEN H., ASMORO, P., LUBIS, H., FRANK, D., AND CASADEVALL, T., 1987, The 1772 debris avalanche and eruption at Papandayan Volcano, West Java, Indonesia, and hazards from future similar events: Abstract Volume, Hawaii Symposium on How Volcanoes Work, Hilo, Hawaii p. 91.

GLICKEN, H., KOMOROWSKI, J. C., AND SHERIDAN, M., 1989a, Particle-particle interaction in the Mount St. Helens debris avalanche: Scanning electron microscope approach: EOS, Transactions, American Geophysical Union, v. 70, p. 1423.

GLICKEN, H., MEYER, W., AND SABOL, M. A., 1989b, Geology and groundwater hydrology of Spirit Lake Blockage, Mount St. Helens, Washington, with implications for lake retention: U.S. Geological Survey Bulletin 1789, 33 pages, 3 plates.

GORSHKOV, G. S., AND DUBIK, Y. M., 1970, Gigantic directed blast at Shiveluch Volcano (Kamchatka): Bulletin Volcanologique, v. 34, p. 262–288.

HAUSBECK, B. P., AND SWANSON, D. A., 1989, Structure and avalanche history of the north flank of Mount St. Helens, Washington: EOS, Transactions, American Geophysical Union, v. 70, p. 1422.

JAEGER, J. C., AND COOK, N. G. W., 1979, Fundamentals of rock mechanics, 3rd edition: London, Chapman and Hall, p. 169.

MIALL, A. D., 1985, Architectural-element analysis: A new method of facies analysis applied to fluvial deposits: Earth-Science Reviews, v. 22, p. 261–308.

MIMURA, K. AND KAWACHI, S., 1981, Nirasaki debris avalanche, a catastrophic event at the Yatsugatake volcanic chain, central Japan: Abstracts, 1981 International Association of Volcanology and Chemistry of the Earth's Interior, Symposium on Arc Volcanism, p. 237.

MURAI, I., 1961, A study of the textural characteristics of pyroclastic flow deposits in Japan: Tokyo University Earthquake Research Institute Bulletin, v. 39, p. 133–248.

NEALL, V. E., 1979, Geologic map of New Zealand (New Plymouth,

Egmont, and Manaia): New Zealand Geological Survey, Sheets P19, P20, P21, scale 1:50,000 plus 35 p. text.

PIERSON, T. C., AND SCOTT, K. M., 1985, Downstream dilution of a lahar: Transition from debris flow to hyperconcentrated streamflow: Water Resources Research, v. 21, p. 1511–1524.

SAVAGE, S. B., 1984, The mechanics of rapid granular flows: Advances in Applied Mechanics, v. 24, p. 289–366.

SCOTT, K. M., 1988, Origins, behavior, and sedimentology of lahars and lahar-runout flows in the Toutle-Cowlitz River system: U.S. Geological Survey Professional Paper 1447-A.

SCOTT, K. M., 1989, Magnitude and frequency of lahars and lahar-runout flows in the Toutle-Cowlitz River system: U.S. Geological Survey Professional Paper 1447-B.

SCOTT, K. M., AND JANDA, R. J., 1987, Multiple origins of catastrophic lahars at Cascade Range volcanoes—Probabilities of precursor events, and hazards in downstream areas: Abstract Volume, Hawaii Symposium on How Volcanoes Work, Hilo, Hawaii, p. 229.

SHREVE, R. L., 1968, The Blackhawk Landslide: Geological Society of America Special Paper, No. 108, p. 1–47.

SIEBERT, L., GLICKEN, H., AND KIENLE, J., 1989, Debris avalanches and lateral blasts at Mount St. Augustine Volcano, Alaska: National Geographic Research, v. 5, p. 232–249.

SUNDELL, K. A., 1985, The Castle Rocks Chaos: A gigantic Eocene landslide-debris flow within the southeastern Absaroka Range, Wyoming: University of California, Santa Barbara, Unpubl. Ph.D. Dissertation 286 p.

UI, T., 1983, Volcanic debris-avalanche deposits—Identification and comparison with non-volcanic debris stream deposits: Journal of Volcanology and Geothermal Research, v. 18, p. 135–150.

UI, T., 1985, Debris avalanche deposits associated with volcanic activity: Proceedings, IVth International Conference and Field Workshop on Landslides, Tokyo, Japan, p. 405–410.

UI, T., AND GLICKEN, H., 1986, Internal structural characteristics of a debris avalanche from Mount Shasta, California, U.S.A.: Bulletin of Volcanology v. 48, p. 189–194.

UI, T., YAMAMOTO, H., SUZUKI-KAMATA, K., 1986, Characterization of debris-avalanche deposits in Japan: Journal of Volcanology and Geothermal Research, v. 27, p. 255–264.

VOIGHT, B., GLICKEN, H., JANDA, R. J., AND DOUGLASS, P. M., 1981, Catastrophic rockslide-avalanche of May 18, in Lipman, P. W., and Mullineaux, D. R., eds., The 1980 eruptions of Mount St. Helens, Washington: U.S. Geological Survey Professional Paper 1250, p. 347–378.

VOIGHT, B., JANDA, R. J., GLICKEN, H., AND DOUGLASS, P. M., 1983, Nature and mechanics of the Mount St. Helens rockslide—Avalanche of 18 May 1980: Geotechnique, v. 33, p. 243–273.

WAITT, R. B., HOBLITT, R. P., CRISWELL, C. W., SCOTT, K. M., GLICKEN, H., AND BRANTLEY, S. R., 1989, Excursion 2A: Recent volcaniclastic deposits and processes at Mount St. Helens volcano, Washington, *in* Chapin, C. E., and Zidek, J., eds., International Association of Volcanology and Chemistry of the Earth's Interior General Assembly, Field Excursions to Volcanic Terranes in the Western United States, Volume II: Cascades and Intermountain West: New Mexico Bureau of Mines and Mineral Resources Memoir 47, p. 51–88.

# PART III
# VOLCANICLASTIC SEDIMENTATION IN CONTINENTAL SETTINGS

Nonmarine sedimentation in volcanic regions may largely occur in response to volcanism, as indicated by study of especially well-preserved Cenozoic sequences. Part III explores the facies associations, and the impact that is made upon sedimentary and physiographic environments by single and multiple explosive volcanic events.

# FACIES SEQUENCES AND GEOMETRIES IN CONTINENTAL VOLCANICLASTIC SEDIMENTS

GARY A. SMITH
*Department of Geology, University of New Mexico, Albuquerque, New Mexico 87131*

ABSTRACT: Studies of the response of fluvial systems to volcanic eruptions generating voluminous pyroclastic material point to two fundamental conditions of landscape and sedimentation. The *syneruption* period is characterized by geologically instantaneous production of large volumes of volcaniclastic sediment, combined with enhanced and more variable runoff leading to sedimentation principally by high-sediment-load flood and debris-flow processes. The periods of high-volume syneruption sedimentation are short and separated by relatively longer inter-eruption periods when volcanism has little or no impact on the fluvial systems. During the inter-eruption periods, sediment delivery is greatly diminished and normal streamflow processes are dominant. Incision of deposits aggraded during the syneruption period is typical.

The syneruption and inter-eruption deposits can often be distinguished on the basis of sediment composition (syneruption deposits show less lithologic diversity), sediment grain size (syneruption deposits tend to be rich in sand-size pyroclasts, whereas gravel-bedload facies are more abundant in inter-eruption deposits), and the distinctive large lateral extent of volcanism-related debris-flow and flood deposits. On broad alluvial-plain aprons (syn. ring plains) associated with continental-margin arcs, distinctive facies geometries exist. Syneruption deposits are laterally extensive sheets, whereas inter-eruption facies, if present, are typically much thinner and confined to valley systems incised into syneruption sheets. Aggradation of syneruption deposits is driven by volcanism. Aggradation of inter-eruption deposits is independent of volcanism and requires other allogenic driving forces, of which basin subsidence is probably most effective, to induce deposition of preservable thicknesses of material. The facies geometry, therefore, is variable depending on the extent to which sedimentation is in response to extrabasinal volcanism or intrabasinal subsidence.

## INTRODUCTION

There have been two principal objectives for most recent studies of ancient continental volcaniclastic sedimentary sequences. Several studies examined the sedimentary record to determine the volcanological and petrological nature of the source area (Smith and others, 1987, 1988; Parker and others, 1988; Turbeville and others, 1989). The second motivation has been understanding volcanic influences on sedimentation and how volcanism-induced deposition is recognized (Walton, 1986; Smith, 1987a, 1987b, 1988; Waresback and Turbeville, 1990). This understanding is fundamental to reconstructing depositional patterns and analyzing basin-fill sequences in volcanic areas. Only then can paleogeographic and paleotectonic reconstructions be made with confidence, and other controls on depositional character, including tectonics and climate change, be assessed. This paper focuses on the second objective and proposes a conceptual model of facies sequences and geometries in volcaniclastic alluvium. The model is based on a number of studies around modern, active volcanoes and a synthesis of observations in ancient volcaniclastic sequences.

## SEDIMENT PRODUCTION AND EROSION IN VOLCANIC AREAS

A unique and important aspect of volcaniclastic sediments, when compared to conventional siliciclastics, is that they are principally composed of fragments generated by volcanic processes rather than by weathering. Pyroclastic and autoclastic fragments (classifications of Fisher, 1961; Schmid, 1981; Fisher and Schmincke, 1984) are essentially instantaneous sediments produced at short-term rates that are much higher than those at which epiclastic material is generated by weathering of pre-existing rocks. In addition, this debris is loosely consolidated, or unconsolidated, and emplaced as thick, landscape-mantling layers or valley fills that are susceptible to rapid erosion by fluvial processes.

Sediment yields in watersheds draining recently active volcanoes reach the $10^5$- to $10^6$-$m^3$/yr range (Vessel and Davies, 1981; Swanson and others, 1982). The impact of the 1980 eruption of Mount St. Helens, Washington, on the North Fork Toutle River is illustrative of volcanism-induced sedimentation (Lehre and others, 1983). The eruption resulted in the deposition of $68 \times 10^6$ $m^3$ of pyroclastic material and emplaced an additional $2{,}500 \times 10^6$ $m^3$ of debris-avalanche material within the drainage basin. Approximately $12 \times 10^6$ $m^3$ of sediment was remobilized as debris flows at the time of the eruption. During the next year, $56 \times 10^6$ $m^3$ of sediment was eroded, about half of which was redeposited within the Toutle River valley (Lehre and others, 1983). These sediment yields are comparable to the annual suspended sediment load of the Mississippi River (Milliman and Meade, 1983), but were derived from a drainage basin with a modest area of about 750 $km^2$. Four years after the eruption, suspended-sediment yield at the mouth of the Columbia River, 90 km from the volcano, remained at levels four times higher than pre-eruption conditions (Meade and Parker, 1985) and were about three orders of magnitude higher in the Toutle River (Pringle, 1990). Perturbations of this nature may persist for many years or decades (Vessel and Davies, 1981).

Erosion rates are high not only because of the large volumes of unconsolidated material, but also because sediment-stabilizing vegetation is often destroyed over large areas by the eruption processes. Measured denudation rates for hillslope-mantling tephras by sheet and rill erosion, in humid to tropical climates, range from 25 to 105 mm/yr (Segerstrom, 1950; Waldron, 1967; Ollier and Brown, 1971; Swanson and others, 1982). These values are three to four orders of magnitude greater than estimated long-term denudation rates for high-relief, nonvolcanic areas (Judson and Ritter, 1964).

Another important aspect of pyroclastic sediments is the dominance of coarse-silt to fine-gravel particle sizes (e.g., Davies and others, 1978; Vessel and Davies, 1981; Hume and others, 1975; Lehre and others, 1983). The dominance of sand may bring about significant changes in river morphology and depositional processes because streams with mountainous headwater regions are normally characterized

Sedimentation in Volcanic Settings, SEPM Special Publication No. 45

by gravel-bedload regimes. This response is generally one of channel widening and decrease in sinuosity (e.g., Kuenzi and others, 1979; Vessel and Davies, 1981).

The loss of vegetation and the relative impermeability of fine-grained or poorly sorted tephra, compared to soils (Swanson and others, 1982), cause increased and more variable runoff. These changes in runoff conditions, combined with increased sediment load, increase the volume and flashiness of the total discharge. Storm-generated water discharges in the Toutle River following the 1980 eruption of Mount St. Helens were about an order of magnitude greater than pre-eruption discharges; sediment discharge, for a given water discharge, increased by two to three orders of magnitude, reaching $10^7$ tons per day (U.S. Army Corps of Engineers, 1982).

## DEPOSITIONAL SITES FOR VOLCANICLASTIC SEDIMENTS

Continental volcaniclastic sediments are generally deposited in one of three sites. Debris-avalanche, debris-flow and flood-deposited sediment may be stored locally in volcano-flank fans or immediately adjacent narrow valleys. These depositional sites are most important for volcanoes that are developed in maturely dissected terrains, such as the Quaternary Cascade Range in the northwestern United States (e.g., Smith and others, 1987; Crandell, 1988).

The second depositional site is broadly dispersed sedimentation on "ring plains" or "aprons," built by alluvial streams, extending for tens of kilometers away from volcanic centers. The term "ring plain" was introduced in New Zealand by Morgan and Gibson (1927, p. 14) for "flat or nearly flat land in a nearly circular or annular area surrounding Mount Egmont." This geomorphic feature is underlain by volcaniclastic material deposited by debris avalanches, debris flows, and fluvial processes, with lesser eolian contributions. The term ring plain has subsequently been applied more generally in New Zealand to low-relief alluvial plains and alluvial fans adjacent to stratovolcanoes (Fig. 1A). Ring plains are only truly circular where developed around isolated volcanoes; more commonly they are linear bodies of fragmental material, with wedge-shaped cross sections, that flank linear volcanic-vent complexes, extending 10 to 30 km from the bases of the volcanoes (e.g., Neall, 1979; Hackett and Houghton, 1989).

Smith (1987a, b) used the term "apron" to describe features similar to ring plains. Volcaniclastic aprons are characterized as relatively thin (<0.5 km) accumulations of volcaniclastic material adjacent to volcanoes that possess a wedge-shaped cross section thinning away from source. Examples in the Pacific Northwest extend about 20 to 40 km from the volcanoes (Smith, 1987a; Fig. 1B).

While not rigorously defined, the term "apron" was used by Smith (1987a, b) to emphasize that although these sequences were dominated by debris-flow and flood-deposited sediments, the depositional morphologies were relatively low-gradient alluvial plains, with parallel or contributary drainage networks, rather than alluvial fans. Sediment is, however, commonly distributed over fan-shaped segments of the apron. This occurs when aggrading rivers exit laterally restricted valleys onto the apron. Aggradation causes channel avulsion in the lower, unconfined reach with channel position pivoting about a fixed point at the mouth of the confined reach (nodal avulsion of Leeder, 1978). The result is a fan-shaped volume of sediment deposited by a shifting river channel (Fig. 2). Unlike alluvial fans, which develop because of transition from confined channelized flow to unconfined sheet flow, these fan-shaped accumulations of sediment preserve relatively low fluvial gradients and lack abrupt downslope decreases in sediment size, characteristic of loss of flow competency due to shallowing of flow on alluvial fans (Schofield, 1965; Hume and others, 1975; Smith, 1987b, 1988).

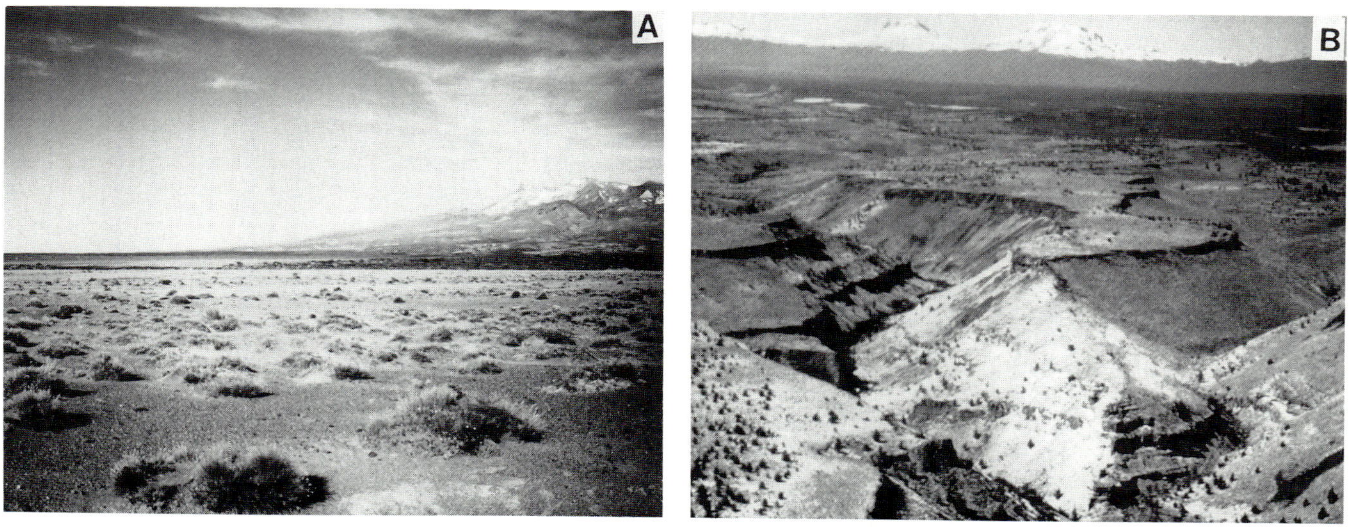


FIG. 1.—Views of volcaniclastic aprons. (A) A part of the low-relief alluvial-plain ring plain of Ruapehu volcano, at right, in New Zealand. (B) Oblique aerial view of the dissected, low-relief, late Miocene alluvial-plain apron constructed adjacent to the Oregon Cascade Range, on skyline and about 50 km distant. Apron sediments and volcanic rocks of the Deschutes Formation are exposed in canyons. Photo by P.O. Hales.

FIG. 2.—View of Ruapehu ring plain from near volcano summit. Note dark, fan-shaped area of laharic deposits along the Whangaehu River in background. Small, 1989 debris-flow deposit in foreground.

Aprons, or ring plains, of volcaniclastic debris are recognized as important components of volcanic terrains (Hume and others, 1975; Nemec and Muszyinski, 1984; Walton, 1979, 1986; Smith, 1987a, b; Hackett and Houghton, 1989; Guest and others, 1988; Palmer and Walton, 1990). In the cited examples, deposition occurred primarily in response to volcanism in structurally stable areas. Where subsiding basins occur adjacent to the volcanic-source region, volcaniclastic deposition may be greatly enhanced by tectonism to produce thicker sedimentary sequences (e.g., Van Houten, 1976; Dill, 1987; Smith, 1988; Waresback and Turbeville, 1990).

The third category of depositional localities is distal sites, of two types. They may be downstream equivalents of volcano-flanking fan or apron facies. With increasing distance from the source, sedimentation is primarily by normal fluvial processes because of limited runout distances of debris avalanches and debris flows and moderation of flood discharges with distance (Smith, 1988). Other distal sites receive large volumes of fallout tephra but do not directly drain the volcanic-source terrane. Though these situations have not been well studied, long-term, high-frequency delivery of large volumes of fine ash to watersheds could have considerable impact on fluvial or eolian processes, especially in dry climates (Vicars and Breyer, 1981).

## CONTROLS ON THE STRATIGRAPHY OF VOLCANICLASTIC-APRON SEQUENCES

Before considering facies sequences of volcaniclastic aprons in detail, it is necessary to consider the controls on apron stratigraphy. Style of volcanism is of fundamental importance. The volume of pyroclastic and autoclastic materials generated dictates the extent of eruption influences on sedimentation. Eruptive style also determines the constructional topography and relief of the source area. The hydraulic head for transport is greatest where the source has high relief, as in the case of steep stratovolcanoes. In contrast, some rhyolitic caldera centers have very low relief. Walton (1986) argues that the paucity of debris-flow deposits in the Tertiary rhyolitic volcaniclastics of West Texas, despite the characteristic occurrence of these facies in andesitic-stratovolcano terranes, resulted from low-relief source areas. Rhyolitic centers in New Zealand typically have negative relief (Walker, 1984), precluding the development of flanking aprons but favoring deposition by fluvial systems draining toward the volcano-tectonic depression (R. Smith, this volume).

Tectonics are another important control. Subsidence within the source area may favor storage in basins along the axis of the volcanic chain and starve adjacent apron sites from sediment (Smith and others, 1987). Structural activity at the depositional site determines the extent of volcanism-induced versus subsidence-induced sedimentation (Smith, 1988; Smith and Vincent, 1987). Intra-basinal structures may also be important for determining sediment dispersal (Smith, 1988).

Climate is an important variable, but is difficult to resolve. The availability of water will determine the ability to transport eruption-generated sediment to adjacent aprons. Climate and vegetation will also determine the degree of change on the hydrologic and sediment supply systems that is brought about by an eruption. It is also important to consider that the climate of mountainous source areas may be very different from adjacent low-lying depositional sites. Most continental-margin volcanic arcs experience moist, maritime climates generating perennial streams that flow across semi-arid to arid basins that lie in the rain shadow of the arc (Smith, 1987b). Care must be exercised when interpreting climate from sedimentary facies. Debris-flow deposits and associated poorly sorted shallow-braided-stream deposits, typically related to arid-region alluvial fans, are widespread products of eruption-induced deposition in even the most humid of climates (c.f., Fritz, 1982).

## FACIES GEOMETRY AND SEQUENCE MODEL FOR VOLCANICLASTIC APRONS

Following is a conceptual model for facies geometry and sequence in volcaniclastic aprons. It is developed from consideration of a number of studies of both Quaternary and pre-Quaternary deposits in valley-fill and apron settings. Principal sources of observational data are Campbell (1973a, 1973b), Van Houten (1976), Davies and others, (1978), Kuenzi and others, (1979), Vessel (1979), Vessel and Davies (1981); Kesel and Lowe (1987); R. Smith (1989 and this volume), G. Smith (1987a, 1987b, 1988, and in prep.).

### *Defining Conditions*

The basic model will consider volcaniclastic aprons adjacent to relatively high-standing volcanic terranes. The climate is relatively moist in the source areas and perennial streams flow across the aprons.

Three essential characteristics of volcanism-induced sedimentation are inferred. First is the geologically instantaneous production of large volumes of sediment, combined with enhanced and more variable runoff, leading to sedimentation principally by high-sediment-load flood and debris-flow processes (see Smith and Lowe, this volume, for a review of these processes). Secondly, these transport and depositional processes are fundamentally similar to those in nonvolcanic areas, though there are some important distinctions in detail. The scale at which these processes operate, and of the deposits generated, is typically much larger in volcanic areas, where factors discussed earlier lead to higher flow volumes, higher discharges, greater runout distances, and thicker and more extensive deposits (Smith, 1986, 1987a). Thirdly, the periods of *syneruption,* high-volume sedimentation, are short and temporally separated by relatively longer *inter-eruption* periods when volcanism has little or no impact on the fluvial systems (Smith and Vincent, 1987). During the inter-eruption periods, sediment delivery is greatly diminished and normal streamflow processes are dominant.

### *Syneruption Versus Inter-eruption Periods*

Studies of Holocene eruptions and related sedimentation indicate important geomorphic and sedimentologic changes (Kuenzi and others, 1979; Vessel, 1979; Vessel and Davies, 1981; R. Smith, 1989 and this volume). When explosive eruptions produce pyroclastic material in volumes exceeding geomorphic thresholds for aggradation, river systems are abruptly transformed to wider, flashy discharge, sand-bedload, often braided, aggrading streams that are impacted by debris flows and hyperconcentrated-flood flows (Smith, 1986; Smith and Lowe, this volume). Aggradational conditions may continue for years or decades following the cessation of eruptive activity until loose debris is completely removed from hillslopes or vegetation is adequately re-established to stabilize the debris. Eventually, preeruption conditions are restored, aggradation ceases and stream beds are incised to form narrower, typically more sinous, gravel-bedload channels.

For the model presented here, two general conditions of landscape and sedimentation are recognized (Fig. 3). The inter-eruption condition is considered the norm, without significant influence of volcanic activity. The syneruption condition is a perturbation brought about by eruptive activity and includes not only the time interval of eruption at the volcano, but also the relatively short period (years to decades) following cessation of activity, while hydrologic and hillslope-sediment-supply systems remain disturbed. The transition from inter-eruption to syneruption conditions is very abrupt. The transition from syneruption to inter-eruption conditions proceeds through a series of adjustments in the geomorphic systems as sediment load and discharge variability diminish (Vessel, 1979; Vessel and Davies, 1981; R. Smith, 1989 and this volume).

### *Syneruption Versus Inter-eruption Deposits*

Three criteria are used to distinguish syneruption deposits from inter-eruption facies—sediment grain size, sediment composition, and nature and extent of facies. Syneruption deposits are likely to be composed primarily of monolithologic pyroclastic sediments (Hume and others, 1975; Smith, 1988). These are produced in large volumes that considerably dilute the contribution of "background" sediment components that are more lithologically diverse and include a larger proportion of epiclastic fragments, possibly of nonvolcanic composition. Syneruption sediments may be intercalated with pyroclastic-fall or flow deposits.

Deposition in response to pyroclastic eruptions will generally provide sand-rich facies. Inter-eruption conditions in the larger streams draining the volcanic highlands are more likely to be of gravel-bedload character (Vessel, 1979; Smith, 1988).

Syneruption periods are characterized by flood and debris-flow deposition extending tens of kilometers from the source (Smith, 1987a). High aggradation rates and channel avulsion favor development of laterally widespread sheets of detritus. Inter-eruption deposits are dominated by normal, dilute streamflow facies whose lateral extent will be largely governed by rates of lateral migration of channels rather than by vertical aggradation (Smith, 1988). An important exception to these facies distinctions are large-volume depositional events near volcanoes that are not responses to eruptions. Earthquake-generated debris-avalanches, and derivative-debris flows, and large-debris flows or floods caused by failure of crater-lake impoundments generate deposits that are difficult to distinguish from syneruption facies.

### *Stages of Geomorphic and Sedimentologic Change*

The above discussion leads to a three-stage model of geomorphic and sedimentologic changes (Fig. 3) that is supported by study of Holocene (Vessel and Davies, 1981; R. Smith, this volume) and older (Smith, 1987a, 1987b, 1988) deposits. The onset of a major pyroclastic eruption initiates widespread syneruption aggradation of nearly monolithologic, pebbly-sand, flood and debris-flow deposits, in sheets that are meters to tens of meters thick. As hydrologic and sediment-supply systems approach pre-eruption conditions, channels are incised into the syneruption sheet facies, with concommitant delivery of sediments to more distal depositional sites. The third stage is the reestablishment of inter-eruption, gravel-bedload conditions with transport of polylithologic detritus. Streams at this stage are confined to varying extents within channels incised into the syneruption sheets, whose upper aggradational surfaces are abandoned, stabilized by vegetation, and are sites for soil development.

This model describes an aggradation-degradation cycle controlled by volcanism-related sediment input. Rates of aggradation and degradation in response to changes in load are among the highest known, with meters of change recognizeable in periods of a few years (Gessler, 1971). Theoretical considerations suggest that rates of degradation and aggradation should differ, with approximately three times more time required to complete 95 percent of the degradation process than for an equivalent amount of aggradation because of temporal differences in the rate of sediment supply for each process (Gessler, 1971). An example of these rate differences, though not as great as predicted by Gessler's calculations, is aggradation in response to hydraulic mining in the Yuba River drainage in California (Gilbert, 1917). Mining occurred between 1850 and 1880. Increased load induced aggradation, not only during the period of

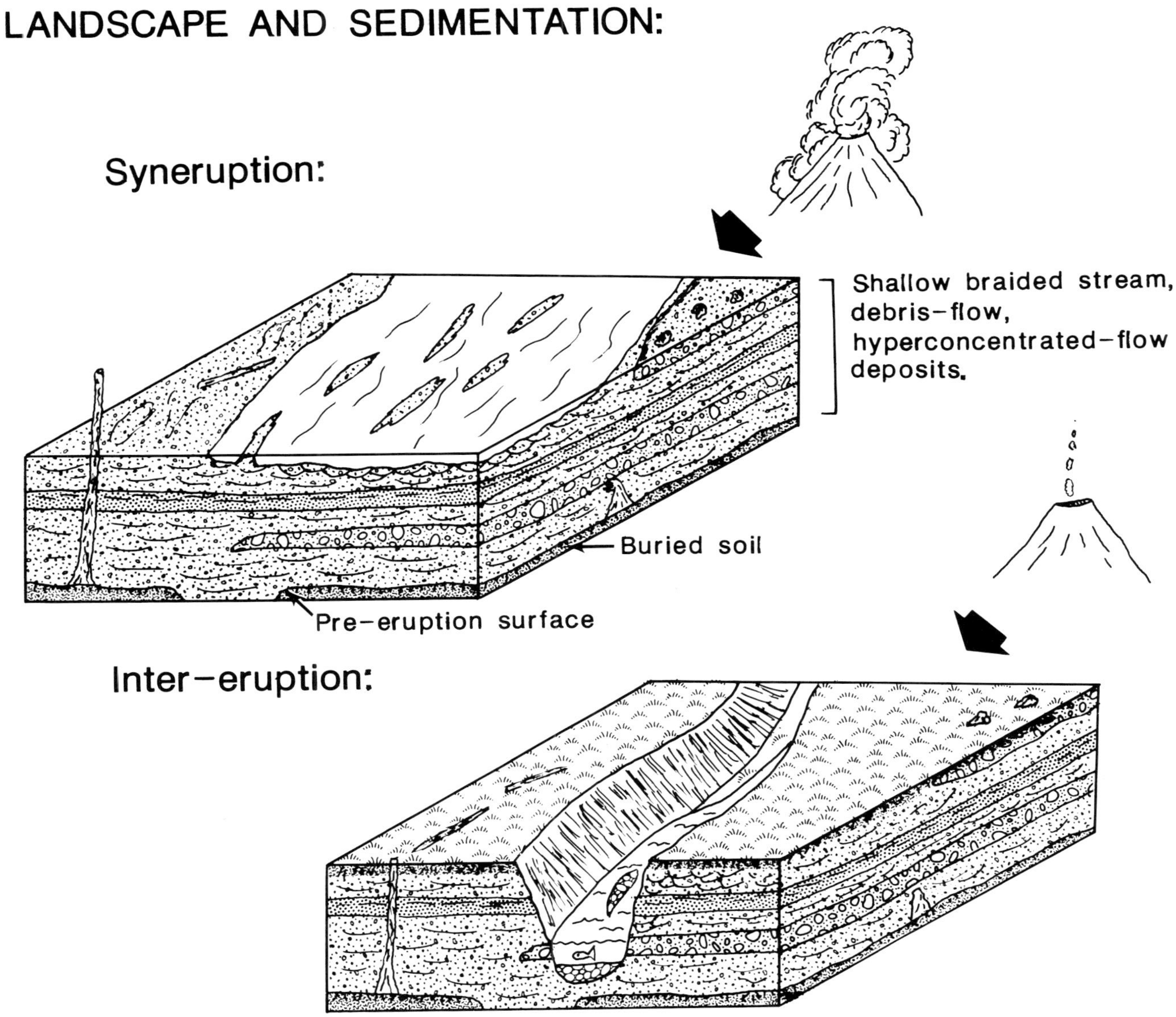


FIG. 3.—Schematic representation of geomorphic, sedimentologic, and stratigraphic characteristics of syneruption and inter-eruption conditions. See text for discussion.

mining but continuing to about 1903. During that interval, the river bed was raised by 26 m for an average aggradation rate of about 0.5 m/yr. Incision over the ensuing 10-year period was at an average of 0.3 m/yr. This example serves as a man-induced sedimentation analog for the volcanism-induced scenario developed here.

Aggradation-degradation cycles in response to fluctuating sediment loads in streams draining volcanoes are not only indicated by the studies of Holocene eruption and sedimentation products (Kuenzi and others, 1979; Vessel and Davies, 1981; R. Smith, this volume), but also by the common observation that narrow, valley-confined rivers draining active volcanoes contain aggradational terraces that correlate not to climatically induced sediment-load/discharge variations, but to volcanic eruptive periods that inundated valleys with material that was then partly excavated during inter-eruption periods of diminished load (Campbell, 1973a, b; Van Houten, 1976; Vessel and Davies, 1979; Crandell, 1987; Smith, 1987a). Because Quaternary climatic fluctuations are known to have produced similar aggradation-degradation cycles, one must endeavor to separate such potential influences in the sedimentary record from those related to syneruption and inter-eruption landscape conditions; the previously discussed criteria for recognizing syneruption deposits must be rigorously applied.

### *Predictive Facies Sequences and Geometries*

To evaluate the stages of landscape and sedimentation change in terms of a sedimentary rock unit, it is necessary to determine the degree of preservation of distinctive characteristics. Syneruption deposits have a high preservation potential, because they are relatively thick and laterally extensive sheets. Inter-eruption deposition, however, is independent of volcanism and requires base-level adjustments within the basin in order to generate preserveable deposits.

Two types of facies geometry are produced (Fig. 4). Each type features syneruption sheets of ash-rich, sandy, flood and debris-flow facies, bounded by erosion surfaces and paleosols. The models differ on the extent of preservation of normal fluvial, largely gravel-bedload, inter-eruption facies, which is presumed, in this case, to be controlled by basin subsidence. If deposition occurs on an apron that occupies a tectonically stable site or one that is subsiding very slowly at long-term rates lower than volcanic-sediment production, there will be little or no depositional record of inter-eruption periods. If the subsidence rate matches or exceeds the volcanism-related sediment input, a more complete record results, represented by both syneruption facies, emplaced in response to eruptions, and inter-eruption facies, deposited in response to subsidence. Two variations on the second type occur (Fig. 4). For type 2a, the inter-eruption facies are largely, if not completely, restricted to channels incised within the syneruption sheets. In type 2b, inter-eruption facies exceed the confines of the channel cut during the incision interval and extend laterally over the syneruptive sheets.

The facies models emphasize the punctuated, episodic nature of sediment production in most volcanic areas. If applied to sedimentary sequences, they have important implications. Type 1 facies sequences are very biased depositional records that provide insight into fluvial style and paleohydraulics only for the relatively brief syneruption aggradational pulses and provide little or no record of the normal fluvial conditions (Smith, 1987a). Type 2a geometry is favored by relatively low-subsidence rates, high-eruption frequencies, or both, whereas type 2b is related to relatively higher subsidence rates and/or lower eruption frequencies. These models also illustrate the importance of analyzing lateral facies relationships in areas of good exposure. Widely spaced vertical sections may not allow for distinction between the various facies geometries shown.

There are problems in applying these models to outcrops. First is the problem of recognizing paleosols that separate syneruption sheets. These facies are generally well-drained, sandy sediments that are not ideal parent materials for developing distinctive soils in short time periods, especially in dry climates. Although such soils are commonly bioturbated, even where poorly developed, the author has found oxidation colors associated with cambic horizons to be very important for locating relatively thin (20 cm to 1.5 m) paleosols in thick volcaniclastic sequences (Fig. 5; Smith, 1987b). In many pre-Quaternary volcaniclastic sequences, however, rock color is largely determined by diagenetic alteration of volcanic grains to zeolite, carbonate, and clay, making recognition of paleosols very difficult. With type 2b facies geometry, there may be inadequate time for soil development before burial, or soils may be eroded by laterally migrating inter-eruption chaannels.

Incision following syneruption episodes is an integral aspect of the presented models. Resulting erosion surfaces may be difficult to recognize from examination of widely spaced vertical sections in areas of limited, laterally continuous exposure, especially if the erosion surfaces are broad features with low relief or paleochannels are very widely spaced. In these cases, the alternation of syneruption and inter-eruption facies types in vertical section may be taken as evidence of type 2 facies geometry, although distinction of types 2a and 2b might not be possible (Fig. 4). Erosion surfaces almost certainly do develop; it is unlikely that subsidence rates could accommodate the short-term, syneruption aggradation rates on the order of meters per year. The

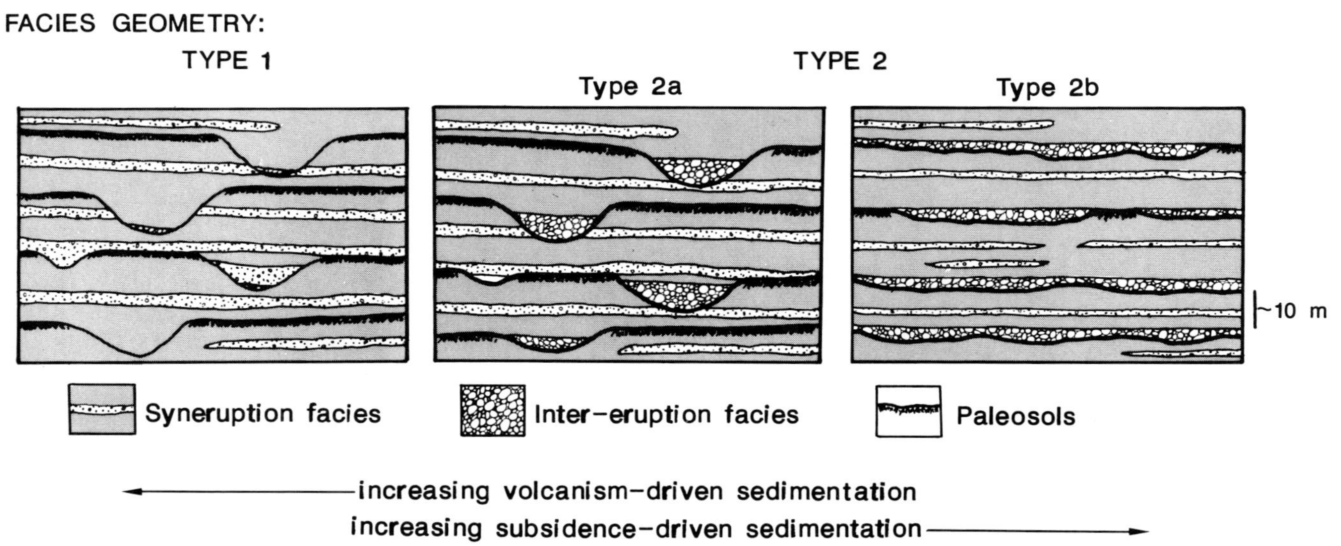


FIG. 4.—Two dimensional facies geometries in volcaniclastic-apron sequences. Syneruption sheets are bounded by erosion surfaces and paleosols. Significant preservation of inter-eruption, gravel-bedload fluvial facies is diagnostic of type 2 geometry. Type 2a is distinguished from type 2b by the lateral extent of inter-eruption conglomerates and depth of stream incision.

FIG. 5.—Paleosol at top of syneruption aggradational sequence in late Miocene Ellensburg Formation, Washington. Note pedogenic obliteration of sedimentary structures and white, silica-lined root traces at level indicated by figure. Upward darkening is due to outcrop reddening that defines a cambic-B soil horizon.

surfaces may have low relief, however, especially if inter-eruption periods are relatively short and the incision process is not completed (Gessler, 1971), or channels become armored against further incision during degradation.

Not all erosional surfaces are bounding surfaces for syneruption sheets. Channels and complex erosion surfaces also occur *within* syneruptive sheets. Aggradation during syneruption periods that last years or decades is not continuous, and incision and reworking of event deposits is common (Lisle and others, 1982; U.S. Army Corps of Engineers, 1982). Surfaces bounding syneruption sheets may be distinguished from those within the sheets by correlation to bounding paleosols or, somewhat arbitrarily, defining an erosion-surface hierarchy based on magnitude of incision, spacing of erosion surfaces, or both.

## FIELD EXAMPLES OF FACIES-MODEL APPLICATIONS

### *Ellensburg Formation, Miocene, Washington*

Dacitic lithic and vitric pyroclastic sediments were shed eastward from the late Miocene Washington Cascade Range. These sediments, named the Ellensburg Formation, are preserved within synchronously formed basins. The resulting facies sequences, exposed over a paleoreach of about 15 to 120 km from the source, are described by Smith (1988). The best exposures, in a small basin about 50 km from presumed source vents, are composed of vertically stacked, tabular sheets of nearly monolithologic, highly pumiceous, pebbly volcanic sandstones and intercalated ashfall tuffs separated by thoroughly bioturbated and oxidized paleosols (Figs. 5, 6). Deposition is inferred to have occurred in broad, shallow braided streams that were occasionally overrun by hyperconcentrated flows and less commonly by debris flows. Continuity of distinctive ledge-forming units over distances of 5 to 10 km, parallel to depositional strike, indicates depositional surfaces of minimal relief. Only near the basin axis is there any variation in bed continuity. There, the sandy, pumiceous sheets are interrupted by channels and small paleovalleys similar in depth to the thickness of the sheets. Above the erosion surfaces are polylithologic, well-rounded gravels with associated, relatively pumice-poor sandstones and overbank mudstones. This sequence (Fig. 6) illustrates type 2a facies geometry and is the predicted result of episodic deposition of broad syneruption sheets of pumiceous sandy alluvium in a slowly subsiding basin, where inter-eruption sedimentation, in response to subsidence, was related to a gravel-bedload river flowing along the basin axis. The importance of laterally continuous exposure is illustrated well in this example; erosion surfaces and inter-eruption facies are only preserved in a very small part of the outcrop belt (Fig. 6).

Outcrops of the Ellensburg Formation in a basin farther downstream, about 80 km from volcanic vents, contain only rare debris-flow and flood deposits (Fig. 7; Smith, 1988). Where these event beds occur, they are associated with trough-cross-bedded, fluvial, pumiceous dacitic sandstones. These sandstones are 5 to 15 m thick and are erosively overlain by polylithologic conglomerates or are separated by paleosols with bioturbation extending 2 to 3 m into underlying sandstones. Quality and orientation of outcrops hinder recognition of deep channels. Although normal fluvial lithofacies dominate the section, the alternation of monolithologic sand-bedload deposits with less common polylithologic gravel-bedload facies is a conspicuous feature of this relatively distal section. Smith (1988) interpreted these relationships to suggest what is herein termed type 2 facies geometry, and although the outcrop data do not allow for easy distinction between types 2a and 2b, two observations suggest that type 2a geometry occurs here. Conglomerates are less common in vertical section than fluvial sandstones, which would be expected if the conglomerates were in relatively confined paleovalleys and sandstones represented the deposits of a broader, sand-bedload stream. Secondly, thorough, deep bioturbation of the sandstones suggests long intervals of stasis and pedogenesis without periodic floodplain accretion; this would be favored by pedogenesis and bioturbation that occurred contemporaneously with deposition of gravel-bedload facies in confined paleovalleys incised into the sandy facies.

### *Deschutes Formation, Miocene, Oregon*

Volcaniclastic strata of the late Miocene Deschutes Formation form an alluvial apron that extends about 40 km

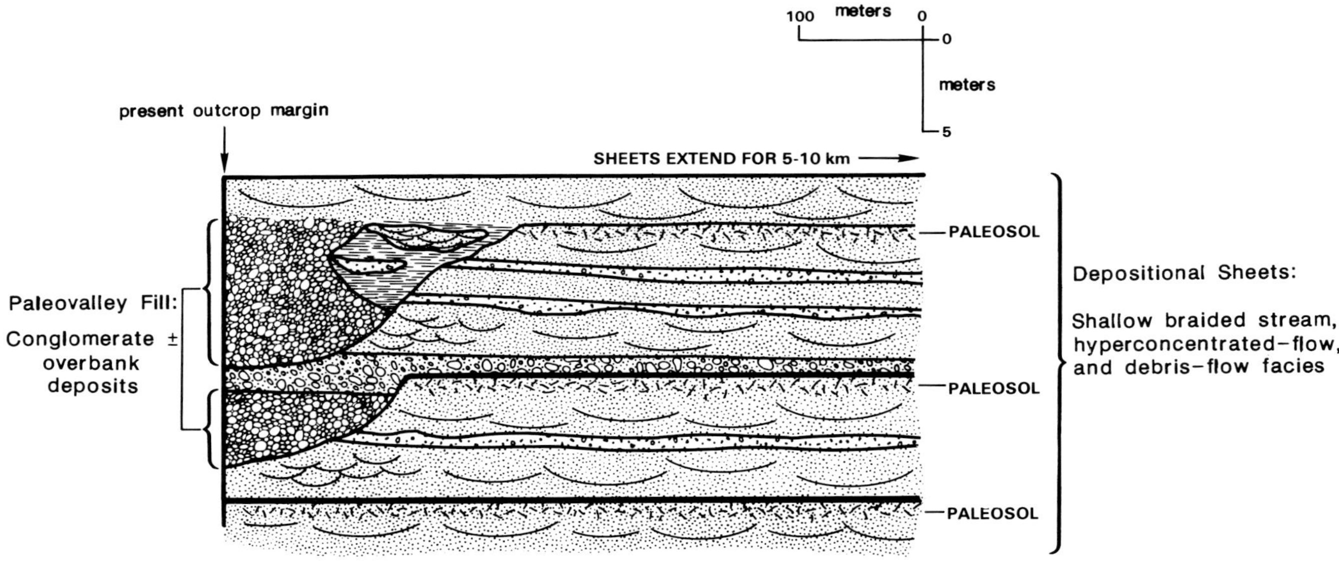


FIG. 6.—Schematic illustration of outcrop relationships in the Ellensburg Formation near Naches, Washington. Laterally extensive, sandy, flood and debris-flow facies, separated by paleosols (Fig. 5) are truncated near the basin axis by erosional valley margins. Paleonvalleys were subsequently filled with channel gravels and related flood-plain deposits (after Smith, 1988).

eastward from a 100-km-long segment of the Oregon Cascade Range (Fig. 1B; Smith, 1987b). Deposition occurred during a 3.5-ma period of large-volume extrusive magmatism, with most of the about 250-m section emplaced in less than 300 ka (Smith and others, 1987). The depositional site was a lowland between pre-existing structural and volcanic constructional high-lands; there are no clearly defined structural basin margins.

Excellent exposures 35 to 45 km from the Cascades illustrate facies that compose the eastward-sloping alluvial plain. Two facies geometries are represented. In the lower 100 m of the section, polylithologic conglomerates up to 30 m thick occupy broad valleys, up to 1 km across. The valleys are incised into more laterally extensive sheets of pebbly sandstones, composed mostly of vitric pyroclastic material, that were deposited in shallow, very broad (>1 km wide) braidplains in association with debris-flow and hyperconcentrated-flow deposits and ignimbrites. In the remainder of the section, relatively well-sorted and well-rounded cobble conglomerates are absent or restricted to laterally discontinuous units only 1 to 2 m thick. Shallow-braided-stream, sheetflood, debris-flow, and hyperconcentrated-flow deposits compose virtually all of the sedimentary component of this upper part of the section and occur in sequences 10 to 50 m thick that are bounded by paleosols and erosion surfaces with up to 70 m relief (Fig. 8). Smaller, steep-sided paleochannels with depths of 2 to 5 m occur within these sequences (Smith, 1987a).

The lower part of the section represents type 2a facies geometry, whereas the remaining section is typical of type 1. The models suggest that the difference in facies geometry reflects change in the relative rates of subsidence and sedimentation, variables that can be semi-quantitatively evaluated for the Deschutes Formation. Figure 9 shows stratigraphic accumulation rates for volcanic and sedimentary lithologies from measured sections with chronostratigraphic control provided by isotopic dates of lava flows (Smith, 1986; Smith and others, 1987). Inflection points define a period of rapid accumulation that corresponds to the interval of type 1 facies geometry. The earlier time segment is represented by type 2a geometry, and the later period of diminished accumulation is a paleosol-dominated section related to sediment starvation following structural isolation of the Deschutes apron from its volcanic source area (Smith, 1987b; Smith and others, 1987). Also shown on Figure 9 is an approximation of cumulative subsidence estimated by reconstructing profiles of paleochannels, whose morphologies are preserved by channel-filling lava flows that were tilted during Deschutes deposition so that the profiles were inclined opposite to the paleoslope. Correction of this tilt to assumed paleogradients for paleochannels at different stratigraphic levels provides an estimate of the relative subsidence in the basin (Smith and others, 1989). During type 2a deposition, both stratigraphic accumulation and subsidence rates were comparable and of low magnitude. During the bulk of Deshutes Formation deposition, stratigraphic accumulation rapidly outstripped the slow rate of subsidence, which is probably an isostatic adjustment to the development of the apron. Subsidence was inadequate to accommodate the rapidly aggrading section, leading to inter-eruption periods dominated by incision and reworking with little or no net deposition. The Deschutes Formation illustrates the importance of basin subsidence as a control on facies geometry.

## COMPARISON TO VOLCANICLASTIC ALLUVIAL FANS

Alluvial fans are semiconical accumulations of sediment, generally less than 10 km in radius, with depositional slopes of 1° to 5°, that form at the base of a mountain front or escarpment as a result of loss in flow competency where

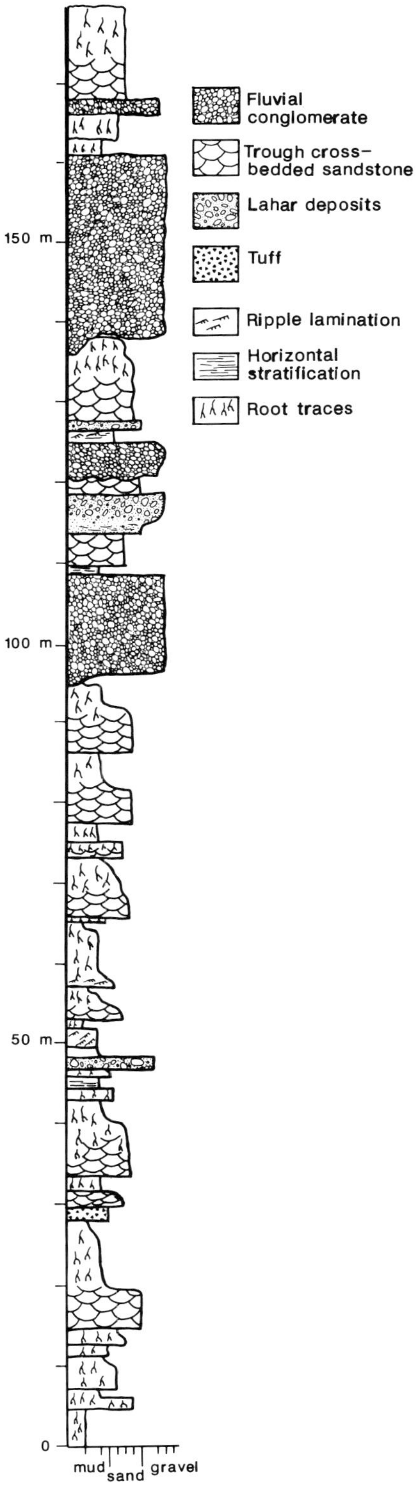


FIG. 7.—Graphic measured section of Ellensburg Formation, south of Yakima, Washington. Pumiceous, monolithologic, fluvial sandstones and siltstones, with related tuffs and lahar deposits, dominate the section, and typically form 5- to 10-m-thick sequences that are pedogenically modified over 30 to 50 percent of their thickness. Polylithologic conglomerates, with erosional bases, are less common, but typically thicker.

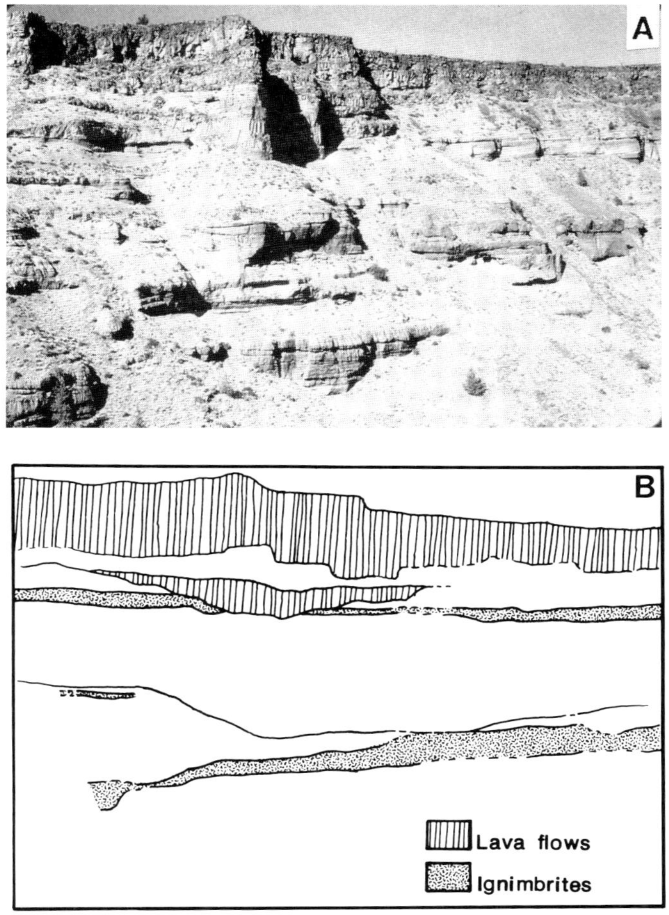


FIG. 8.—(A) Outcrop view and (B) interpretive diagram for a part of the upper Deschutes Formation. Height of canyon wall is approximately 225 m. Note the magnitude of erosion surfaces, as much as 40 m of relief, separating delineated depositional packages of shallow-braided-stream and debris-flow facies. Some channel bases are marked by channel-filling lava flows and ignimbrites.

channelized flow spreads laterally (Bull, 1972; Fraser and Suttner, 1986; McPherson and others, 1987). Volcaniclastic alluvial fans form on the flanks of volcanoes (Kesel and Lowe, 1987; Crandell, 1988) or along basin margins developed adjacent to volcanic centers (Waresback and Turbeville, 1990). Ancient alluvial-fan sequences are typically recognized by their abundance of debris-flow, shallow-braided-stream, and sheetflood facies (Bull, 1972; Nilsen, 1982; Rust and Koster, 1984; McPherson and others, 1987). These criteria are not adequate in volcanic settings, however, because identical facies may extend to distances of many tens of kilometers across low-gradient alluvial-plain surfaces that constitute volcaniclastic aprons (Smith, 1987a, b). Volcaniclastic alluvial fans, however, do retain the characteristic features of rapid downslope facies changes from channelized to unchannelized deposits and concomitant rapid clast-size decrease (Rust and Koster, 1984; Waresback and Turbeville, 1990).

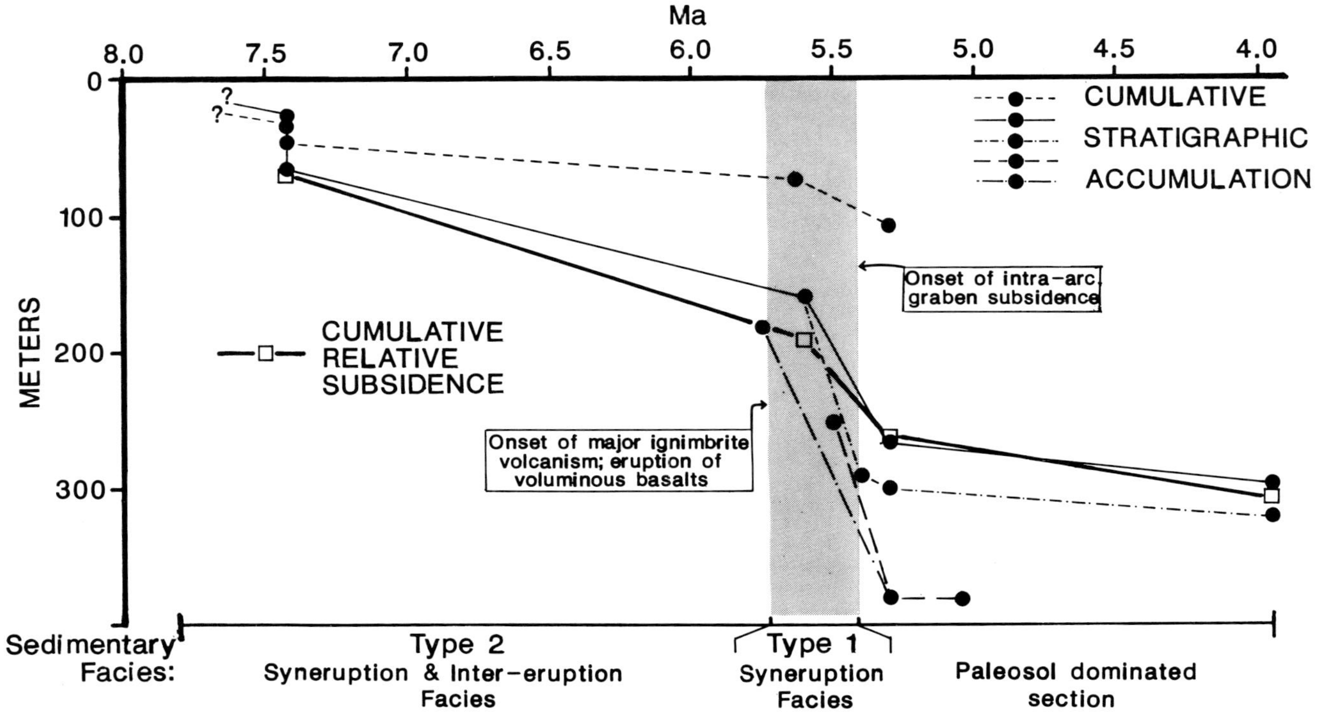


FIG. 9.—Cumulative relative subsidence in the northern Deschutes basin compared to cumulative stratigraphic accumulation calculated for five different sections (modified from Smith and others, 1989). Pyroclastic volcanism occurred throughout the time interval shown, but was greatest following about 5.7 Ma. Intra-arc subsidence, beginning at about 5.4 Ma, starved the Deschutes basin of Cascade-derived sediment, leading to slow accretion of a paleosol-dominated section. Note the high accumulation rate, synchronous with large-volume extrusion of magmas, during the shaded period. Subsidence was inadequate to accommodate this rapidly aggraded section, leading to the development of type 1 facies geometry.

There are two reasons why volcaniclastic alluvial-fan sequences may differ from the apron facies models. First, the morphology of alluvial fans is generated by progradation and aggradation on a portion of the fan surface, followed by avulsion of the feeder channel as gradient favors deposition on other segments of the fan. Shifting of depositional lobes may not lead to incision of a channel following syneruption aggradation on a fan lobe. The second potential difference, in the case of arid climates, is that the eruption produces little change in the sediment supply and hydrologic systems. Loose sediment and absence of significant sediment-stabilizing vegetation characterizes both syneruption and inter-eruption periods, making it difficult, if not impossible, to relate parts of the depositional record to these different periods.

The Rio Toro Amarillo alluvial fan, on the flank of Irazu volcano in Costa Rica, was strongly affected by the 1963–1965 eruptions of the volcano (Fig. 10; Kesel and Lowe, 1987). Prior to the eruption, only a small part of the central fan surface was occupied by active channels. Eruption-induced sedimentation caused continued aggradation of the central part of the fan, followed by avulsion of the channel system to a new position on the fan (Fig. 10). There is no indication of subsequent incision of any part of the fan surface. In this case of a tropical fan, it may still be possible to distinguish syneruption and inter-eruption deposits in the absence of a separating surface with erosional relief. Debris-flow and flood deposits may be more numerous and widespread in the syneruption facies. On the Rio Toro Amarillo fan, the inter-eruption streams have anastomosing morphology, whereas braided patterns dominate when eruption-induced sedimentation occurs (Kesel and Lowe, 1987). These differences may be preserved in the stratigraphic record.

Subarid volcaniclastic alluvial-fan deposits are associated with Cenozoic volcanic fields in the southwestern United States. The Oligocene Espinaso Formation near Cerrillos, New Mexico, is a >450-m thick alluvial-fan sequence composed almost entirely of latite-clast debris-flow deposits, with minor sheetflood deposits (Disbrow and Stoll, 1957; Taylor and others, in prep.). Tracing of lithologic markers defines depositional gradients of about 5°, and very rapid downfan decrease in maximum clast size (from 8 m to 0.5 m over a distance of 5 km). Except for some evidence of fan-head entrenchment, there are no obvious erosion surfaces within the sequence. There are no recognizable vertical variations in facies types or abundance or sediment composition that allow distinction of inter-eruptive and syneruptive deposits.

Late Miocene volcaniclastic alluvial-fan sediments in the Peralta Tuff Member of the Bearhead Rhyolite, in the Jemez Mountains, New Mexico, also lack definitive evidence for distinct inter-eruption and syneruption depositional intervals (Abitz and Smith, 1989; Smith, 1989, and in prep.).

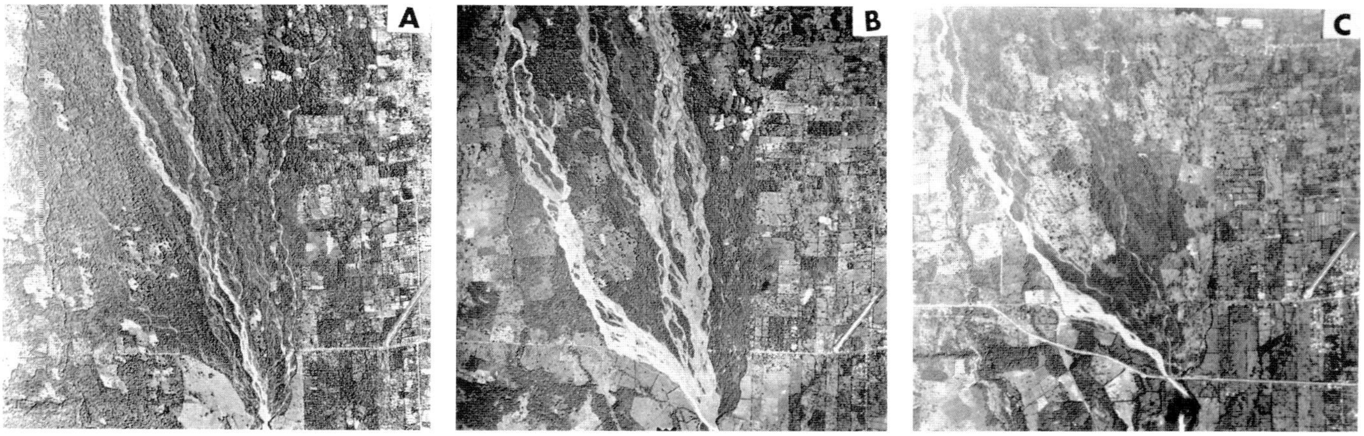


FIG. 10.—Aerial photographs, approximately 5 km across, of the Rio Toro Amarillo alluvial fan on the north slope of Irazu volcano, Costa Rica. Photographs were taken (A) 3 years before; (B) 7 years after; and (C) 20 years after, the 1963–1965 eruption of Irazu. Note the change in stream morphology, channel shifting, and sediment deposition resulting from the eruption. Used with permission from Kesel and Lowe (1987). Volcano is out of view to bottom of photos.

These sediments record a shallow-braided-stream- and sheetflood-dominated fan on which depositional lobes are separated by paleosols, but lacking erosion surfaces with greater than 1 to 2 m of relief. Sediments in the 275-m-thick section are interbedded with pyroclastic-fall, flow, and surge deposits and include local eolian tuffs derived from ashfall deposits. The Espinaso and Peralta alluvial-fan sequences represent aggradation in response to high-discharge events initiated by storms affecting ephemeral drainages in a dry climate. Although the sediment was largely generated by volcanic eruptive activity, there is no evidence for volcanism causing episodic variations in the tempo and style of deposition.

Waresback and Turbeville (1990) describe the Pliocene Puye Formation, which accumulated adjacent to the Jemez Mountains volcanic field and prograded into the Rio Grande rift basin. Rapid downstream facies and clast-size changes, and the marked paucity of channelized deposits beyond 16 km from the source, define a large coalesced alluvial-fan complex, about 25 km across from apex to toe. Better sorted streamflow deposits separating debris-flow and sheetflood facies sequences may represent inter-eruption reworking of event deposits related to source-area volcanism (Waresback and Turbeville, 1990), although these possible inter-eruption units are rare in the section compared to paleosols that bound intervals of debris-flow and flood deposits (Smith, unpubl. data).

The large Puye alluvial fans, fed by relatively large drainage basins at high elevation may represent an intermediate condition between the alluvial fans described above and volcaniclastic aprons. Perennial streams may have crossed the Puye fans, and distinct perturbations in hillslope-sediment supply and hydrologic systems may have accompanied eruptions. Distinctions between inter-eruption and syneruption conditions are not as apparent on the small Espinaso and Peralta fan sequences, which were fed from small, low-elevation ephemeral drainage basins. The lack of erosion surfaces in the Puye fan sequences, and rapid lateral facies and clast-size changes still permit its distinction from the alluvial-plain aprons discussed previously.

## CONCLUSIONS

The most widespread sedimentary records of volcanism are volcaniclastic aprons (syn. ring plains) that develop adjacent to volcanic chains or fields. They are typically the result of deposition by perennial streams, and comprise sedimentary sequences that are several hundred meters thick and extend as much as 50 to 100 km from the volcanic source area. The most conspicuous aspect of sedimentation in this setting is the alternation between two landscape and sedimentation conditions, syneruption and inter-eruption. Studies of sedimentation in response to Holocene eruptions indicate that distinctive facies are associated with each condition. Rapid and widespread aggradation of debris-flow and sandy-flood deposits in response to eruptions is followed by incision and re-establishment of inter-eruption gravel-bedload streams. In the stratigraphic record of ancient volcaniclastic aprons, these two conditions are recognized by alternations of the two facies associations and, where exposure is adequate, erosion surfaces can be defined that separate them.

Aggradation of syneruption sheets is a response to extrabasinal volcanism, whereas inter-eruption sedimentation is independent of volcanism and must be a response to basinal base-level adjustments, frequently brought about by subsidence. Therefore, the facies sequences and geometries are potentially complex, depending on the relative importance of eruption-induced and subsidence-induced deposition. The episodic nature of pyroclastic sediment input is reflected in facies sequences and styles that depend on eruptive style, eruption frequency, and proximity to source.

Once the eruption-induced deposition is recognized, other controls on sedimentation, such as subsidence, can be evaluated.

Application of these conceptual facies models is best accomplished for sequences that are sufficiently exposed to determine facies and erosion-surface geometries over large areas and where diagenesis has not obscured recognition of weathered horizons. Even where these conditions are not met, inferences can sometimes be made on the basis of careful vertical-section analysis alone.

Volcaniclastic-apron sequences contain facies that are often ascribed to alluvial-fan deposition in nonvolcanic settings. Low depositional gradients, lack of abrupt lateral facies and clast-size changes reflecting gradation from channelized to unconfined flow help to distinguish alluvial-plain aprons from alluvial fans (Smith, 1987a, b). Volcaniclastic alluvial fans do exist, and, although not yet as well studied as aprons, their facies geometries appear to be distinct from those of aprons. These differences result from 1) lack of incision following syneruption deposition because of shifting loci of depositional lobes on the fan surface, and 2) lack of significant differences between inter-eruption and syneruption hillslope-sediment-supply systems and hydrology in small, ephemeral, arid watersheds.

In studying volcaniclastic-apron sequences deposited coevally with volcanism, we should recognize the episodicity of eruptions and resulting fluctuations in grain size, composition, and volume of sediment and the processes that transport that sediment as the most important difference from nonvolcanic settings. These fluctuations will occur at too high a frequency ($10^1$ to $10^3$ years) to be ascribed to tectonism, and climatic changes are not likely abruptly to alter the landscapes sufficiently to mimic the extremely large volumes and long runout distances of debris flows and floods that occur in response to volcanism.

## ACKOWLEDGMENTS

Ongoing research on volcaniclastic sedimentation is sponsored by the Donors to the Petroleum Research Fund of the American Chemical Society. Travel to New Zealand was supported by the Caswell Silver Foundation at the University of New Mexico. The manuscript was improved by the helpful review of J. D. L. White.

## REFERENCES

ABITZ, R. J., AND SMITH, G. A., 1989, Stratigraphy and depositional features of the Peralta Tuff, Jemez Mountains, New Mexico: Continental Magmatism Abstracts, New Mexico Bureau of Mines and Mineral Resources Bulletin 131, p. 1.

BULL, W. B., 1972, Recognition of alluvial fan deposits in the stratigraphic record, *in* Rigby, J. B., and Hamblin, W. K., eds., Recognition of Ancient Sedimentary Environments: Society of Economic Paleontologists and Mineralogists Special Publication 16, p. 63–83.

CAMPBELL, I. B., 1973a, Late Pleistocene alluvial pumice deposits in the Wanganui Valley: New Zealand Journal of Geology and Geophysics, v. 16, p. 717–721.

CAMPBELL, I. B., 1973b, Recent aggradation in Whangaehu Valley, central North Island, New Zealand: New Zealand Journal of Geology and Geophysics, v. 16, p. 643–649.

CRANDELL, D. R., 1987, Deposits of pre-1980 pyroclastics flows and lahars from Mount St. Helens volcano, Washington: U.S. Geological Survey Professional Paper 1444, 91 p.

DAVIES, D. K., VESSEL, R. K., MILES, R. C., FOLEY, M. G., AND BONIS, S. B., 1978, Fluvial transport and downstream sediment modification in an active volcanic region, *in* Miall, A., ed., Fluvial Sedimentology: Canadian Society of Petroleum Geologists Memoir 5, p. 61–83.

DILL, H., 1987, Environmental and diagenetic analyses of Lower Permian epiclastic and pyroclastic fan deposits—Their role for coal formation and uranium metallogeny in the Stockheim trough (F.R.G.): Sedimentary Geology, v. 52, p. 1–26.

DISBROW, A. E., AND STOLL, W. C., 1957, Geology of the Cerrillos area, Santa Fe County, New Mexico: New Mexico Bureau of Mines and Mineral Resources Bulletin 48, 73 p.

FISHER, R. V., 1961, A proposed classification of volcaniclastic sediments and rocks: Geological Society of America Bulletin, v. 72, p. 1409–1414.

FISHER, R. V., AND SCHMINCKE, H-U., 1984, Pyroclastic Rocks: Berlin, Springer-Verlag, 472 p.

FRASER, G. S., AND SUTTNER, L. J., 1986, Alluvial fans and fan deltas: A guide to exploration for oil and gas: Boston, International Human Resources Development Corporation, 199 p.

FRITZ, W. J., 1982, Geology of the Lamar River Formation, Northeast Yellowstone National Park: Wyoming Geological Association Guidebook, 33rd Annual Field Conference, p. 73–101.

GESSLER, J., 1971, Aggradation and degradation, *in* Shen, H. W., ed., River Mechanics: Fort Collins, Colorado, v. 1, p. 8-1–8-24.

GILBERT, G. K., 1917, Hydraulic-mining debris in the Sierra Nevada: U.S. Geological Survey Professional Paper 105, 154 p.

GUEST, J. E., DUNCAN, A. M., AND CHESTER, D. K., 1988, Monte Vulture volcano (Basilicata, Italy): An analysis of morphology and volcaniclastic facies: Bulletin of Volcanology, v. 50, p. 244–257.

HACKETT W. R., AND HOUGHTON, B. F., 1989, A facies model for a Quaternary andesitic composite volcano: Ruapehu, New Zealand: Bulletin of Volcanology, v. 51, p. 51–68.

HUME, T. M., SHERWOOD, A. M., AND NELSON, C. S., 1975, Alluvial sedimentology of the upper Pliocene Hinuera Formation, Hamilton basin, New Zealand: Journal of the Royal Society of New Zealand, v. 5, p. 421–462.

JUDSON, S., AND RITTER, D. F., 1964, Rates of regional denudation in the United States: Journal of Geophysical Research, v. 69, p. 3395–3401.

KESEL, R. H., AND LOWE, D. R., 1987, Geomorphology and sedimentology of the Toro Amarillo alluvial fan in a humid tropical environment, Costa Rica: Geografiska Annaler, v. 69A, no. 1, p. 85–99.

KUENZI, W. D., HORST, O. H., AND MCGEHEE, R. V., 1979, Effect of volcanic activity on fluvial-deltaic sedimentation on a modern arc-trench gap, southwestern Guatemala: Geological Society of America Bulletin, v. 90, p. 827–838.

LEEDER, M. R., 1978, A quantitative stratigraphic model for alluvium, with special reference to channel deposit density and interconnectedness, *in* Miall, A., ed., Fluvial Sedimentology: Canadian Society of Petroleum Geologists Memoir 5, p. 587–596.

LEHRE, A., COLLINS, B. D., AND DUNNE, T., 1983, Post-eruption sediment budget for the North Fork Toutle River drainage, June 1980–June 1981: Zeitschrift Geomorphology Neues, v. 46, p. 143–163.

LISLE, T. E., LEHRE, A. K., MARTINSON, H. A., MEYER, D. F., NOLAN, K. M., AND SMITH, R. D., 1982, Stream channel adjustments after the 1980 Mount St. Helens eruptions: Proceedings of the Symposium on Erosion Control in Volcanic Areas, Public Works Institute of Japan Technical Memorandum 1908, p. 33–72.

MCPHERSON, J. G., SHANMUGAM, G., AND MOIOLA, R. J., 1987, Fan-deltas and braid deltas: Varieties of coarse-grained deltas: Geological Society of America Bulletin, v. 99, p. 331–340.

MEADE, R.H., AND PARKER, R. S., 1985, Sediment in rivers of the United States: U.S. Geological Survey Water-Supply Paper 2275, p. 49–60.

MILLIMAN, J. D., AND MEADE, R. H., 1983, World-wide delivery of river sediment to the oceans: Journal of Geology, v. 91, p. 1–22.

MORGAN, P. G., AND GIBSON, W., 1927, The geology of the Egmont subdivision: New Zealand Geological Survey Bulletin 29, 92 p.

NEALL, V. E., 1979, Geologic Map of New Zealand (New Plymouth, Egmont, and Manaia) New Zealand Geological Survey, Sheets P19, P20, P21, scale 1:50,000 plus 35 p. text.

NEMEC, W., AND MUSZYNSKI, A., 1984, Volcaniclastic alluvial aprons in the Tertiary of Sofia District (Bulgaria): Annales Societatis Geologorum Poloniae, v. 52, p. 239–303.

NILSEN, T. H., 1982, Alluvial fan deposits, *in* Scholle, P. A., and Spearling, D., eds., Sandstone Depositional Environments: American Association of Petroleum Geologists Memoir 31, p. 49–86.

OLLIER, C. D., AND BROWN, M. J. F., 1971, Erosion of a young volcano in New Guinea: Zeitschrift fur Geomorphologie, v. 15, p. 12–28.

PALMER, B. A., AND WALTON, A. W., 1990, Accumulation of volcaniclastic aprons in the Mount Dutton Formation (Oligocene-Miocene), Marysvale volcanic field, Utah: Geological Society of America Bulletin, v. 102, p. 734–748.

PARKER, D. F., KRYSTINIK, J. G., AND MCKEE, B. J., 1988, Provenance of the Gueydan Formation, south Texas: Implications for the late Oligocene-early Miocene tectonic evolution of the Trans-Pecos volcanic field: Geology, v. 16, p. 1085–1089.

PRINGLE, P., 1990, Mount St. Helens—A ten-year summary: Washington Geologic Newsletter, v. 18, no. 2, p. 3–10.

RUST, B. R., AND KOSTER, E. H., 1984, Coarse alluvial deposits, *in* Walker, R. G., ed., Facies Models: Geological Association of Canada Reprint Series 1, p. 53–70.

SCHMID, R., 1981, Descriptive nomenclature and classification of pyroclastic deposits and fragments: Recommendations of the IUGS Subcommission on the Systematics of Igneous Rocks: Geology, v. 9, p. 41–43.

SCHOFIELD, J. C., 1965, The Hinuera Formation and associated Quaternary events: New Zealand Journal of Geology and Geophysics, v. 8, p. 772–791.

SEGERSTROM, K., 1950, Erosion studies at Paricutin, State of Michoacan, Mexico: U.S. Geological Survey Bulletin 965-A, 164 p.

SMITH, G. A., 1986, Stratigraphy, sedimentology, and petrology of Neogene rocks in the Deschutes basin, central Oregon: A record of continental margin volcanism and its influence on fluvial sedimentation in an arc-adjacent basin: Corvallis, Oregon State University, Unpubl. Ph.D. Dissertation, 467 p.

SMITH, G. A., 1987a, Sedimentology of volcanism-induced aggradation in fluvial basins: Examples from the Pacific Northwest, U.S.A., *in* Ethridge, F. G., Flores, R. M., and Harvey, M. G., eds., Recent Developments in Fluvial Sedimentology: Society of Economic Paleontologists and Mineralogists Special Publication 39, p. 217–228.

SMITH, G. A., 1987b, The influence of explosive volcanism on fluvial sedimentation: The Deschutes Formation (Neogene) in central Oregon: Journal of Sedimentary Petrology, v. 57, p. 613–629.

SMITH, G. A., 1988, Sedimentology of proximal to distal volcaniclastics dispersed across an active foldbelt: Ellensburg Formation (late Miocene), central Washington: Sedimentology, v. 35, p. 953–977.

SMITH, G. A., 1989, Observations on the record of Cenozoic pyroclastic volcanism in nonmarine sediments, northwest and southwest United States: Continental Magmatism Abstracts, New Mexico Bureau of Mines and Mineral Resources Bulletin 131, p. 247.

SMITH, G. A., CAMPBELL, N. P., DEACON, M. W., AND SHAFIQULLAH, M., 1988, Eruptive style and location of volcanic centers in the Miocene Washington Cascade Range: Reconstruction from the sedimentary record: Geology, v. 16, p. 337–340.

SMITH, G. A., SNEE, L. W., AND TAYLOR, E. M., 1987, Stratigraphic, sedimentologic, and petrologic record of late Miocene subsidence of the central Oregon High Cascades: Geology, v. 15, p. 389–392.

SMITH, G. A., AND VINCENT, K. R., 1987, Rates of sedimentation, subsidence, and volcanism as controls on facies architecture in terrestrial volcaniclastics: Geological Society of America Abstracts with Programs, v. 19, p. 845.

SMITH, G. A., VINCENT, K. R., AND SNEE, L. W., 1989, An isostatic model for basin formation in and adjacent to the central Oregon High Cascade Range, *in* Muffler, L. J. P., Weaver, C. S., and Blackwell, D. D., eds., Geological, Geophysical, and Tectonic Setting of the Cascade Range: U.S. Geological Survey Open-File Report 89-178, p. 411–429.

SMITH, R. C. M., 1989, Landscape response to a major ignimbrite eruption: Post-Taupo eruption sedimentation and landform evolution: Auckland, New Zealand, University of Auckland, Unpubl. M.S. Thesis 138 p.

SWANSON, F. J., COLLINS, B., DUNNE, T., AND WICHERSKI, B. P., 1982, Erosion of tephra from hillslopes near Mt. St. Helens and other volcanoes: Proceedings of the Symposium on Erosion Control in Volcanic Areas, Public Works Research Institute, Japan, Technical Memorandum 1908, p. 183–221.

TURBEVILLE, B. N., WARESBACK, D. B., AND SELF, S., 1989, Lava dome growth and explosive volcanism in the Jemez Mountains, New Mexico: Evidence from the Plio-Pleistocene Puye alluvial fan: Journal of Volcanology and Geothermal Research, v. 36, p. 267–291.

U.S. ARMY CORPS OF ENGINEERS, 1982, Mount St. Helens, Cowlitz and Toutle Rivers sedimentation study, 1980–1982: Portland District, 126 p.

VAN HOUTEN, F. B., 1976, Late Cenozoic volcaniclastic Andean foredeep, Colombia: Geological Society of America Bulletin, v. 87, p. 481–495.

VESSELL, R. K., 1979, Recent and ancient volcaniclastic sedimentation on an active continental margin: Lubbock, Texas Technological University, Unpubl. Ph.D. Dissertation, 127 p.

VESSELL R. K. AND DAVIES, D. K., 1981, Non-marine sedimentation in an active fore arc basin, *in* Ethridge, F. G., and Flores, R. M., eds., Recent and Ancient Nonmarine Depositional Environments: Models for Exploration: Society of Economic Paleontologists and Mineralogist Special Publication 31, p. 31–45.

VICARS, R. G., AND BREYER, J. A., 1981, Sedimentary facies in air-fall pyroclastic debris, Arikaree Group (Miocene), northwest Nebraska, U.S.A.: Journal of Sedimentary Petrology, v. 51, p. 909–921.

WALDRON, H. H., 1967, Debris flow and erosion control problems caused by the ash eruptions of Irazu Volcano, Costa Rica: U.S. Geological Survey Bulletin 1241-I, 35 p.

WALKER, G. P. L., 1984, Downsag calderas, ring faults, caldera sizes, and incremental caldera growth: Journal of Geophysical Research, v. 89, p. 8407–8416.

WALTON, A. W., 1979, Volcanic sediment apron in the Tascotal Formation (Oligocene?), Trans-Pecos Texas: Journal of Sedimentary Petrology, v. 49, p. 303–314.

WALTON, A. W., 1986, Effects of Oligocene volcanism on sedimentation in the Trans-Pecos volcanic field of Texas: Geological Society of America Bulletin, v. 97, p. 1192–1207.

WARESBACK, D. B., AND TURBEVILLE, B. N., 1990, Evolution of a Plio-Pleistocene volcanogenic alluvial fan: The Puye Formation, Jemez Mountains, New Mexico: Geological Society of America Bulletin, v. 102, p. 298–314.

# LANDSCAPE RESPONSE TO A MAJOR IGNIMBRITE ERUPTION, TAUPO VOLCANIC CENTER, NEW ZEALAND

ROBYN C. M. SMITH[1]
*Departments of Geography and Geology, University of Auckland, New Zealand*

ABSTRACT: At the end of the 1.8-ka eruption from Taupo Volcanic Center, New Zealand, the emplacement of a moderately voluminous, unwelded and highly pumiceous ignimbrite destroyed vegetation and drainage networks over $\approx$20,000 km$^2$ around the vent in the Lake Taupo basin. A recent study of derived sediments and landforms in an area on the southeast shore of the lake has revealed the style and magnitude of the landscape response to this cataclysmic event.

In upland areas away from Lake Taupo the response took two forms. Where the ignimbrite overlies impermeable hardrock, distinctive landforms and a wide range of lithofacies indicate that debris flows, hyperconcentrated flows, flood-prone streams, temporary lakes, and braided and meandering rivers were common and generally occurred in the order given. Where the impermeable rock is absent, however, the range of deposits is limited to those from the first three processes. The differing associations suggest that where the hard rock formed a barrier to the downward movement of ground water, the large volumes of runoff moving through the devegetated landscape continued to flow at the land surface, facilitating the establishment of a wide range of depositional environments, until the re-establishment of a full vegetation cover some 15 to 20 years after the eruption reduced runoff by $\approx$30 percent. Where the ground water base was unimpeded in its downward movement through the relatively permeable ignimbrite, however, surface flow decreased substantially within a few years of the eruption.

Near Lake Taupo, valleys were carved into the ignimbrite by rivers and streams whose waters refilled the lake basin, which had been nearly emptied by the eruption, at a rate of $\approx$10 m yr$^{-1}$. Blockage of the only outlet caused the lake to rise 34 m higher than its present level. Transgressive sequences of beach and nearshore sediments with some debris-flow deposits are now exposed on the resulting lake terrace and within those early valleys, some of which were partially re-excavated as the lake-level fell. The general lack of regressive sequences and a greatly reduced drainage density on the caldera-lake terrace and in upland areas away from the lake indicate both a decrease in sediment supply and runoff and a lake-level fall somewhat greater than 10 m yr$^{-1}$.

## INTRODUCTION

The form of the central North Island landscape of New Zealand is dominated by the presence and activity of the many volcanoes comprising the Taupo Volcanic Zone (TVZ), a continental-margin arc related to subduction off the east coast of New Zealand (Fig. 1). Whereas dacitic and rhyolitic domes, andesitic stratovolcanoes and basaltic-scoria cones form numerous positive topographic features, including the elevated volcanic plateau at the southern end of the TVZ, the large rhyolitic-caldera volcanoes tend to produce areas of negative topography. Relatively frequent minor eruptions from these calderas, of <3 km$^3$ DRE (dry rock equivalent) at intervals of 10$^3$ years, blanket the landscape with tephra layers generally <1 m thick and do not cause major disruption to physical processes in the landscape. The more infrequent large eruptions, however, of >30 km$^3$ DRE at intervals of 10$^4$–10$^5$ years, cause widespread changes in both the topography and physical processes. Lakes are a common feature of these volcanoes (Fig. 1). Some occupy the caldera itself; others result from the blockage of drainage networks by ignimbrite or lava flows (Healy, 1975). Major river systems flow from the volcanic plateau to all coasts of the North Island, extending the influence of large eruptions to well outside the area of major tephra deposition.

The Taupo Volcanic Center, the most southerly of the caldera volcanoes, exemplifies most of these characteristics. Its lake, New Zealand's largest at 616 km$^2$ (Fig. 1), is set in the lowest part of a major topographic depression, bounded by northeast-southwest-trending normal faults. The negative topography causes all runoff over nearly 2,700 km$^2$ to drain into the lake and also makes blockage of the only outlet, the valley of the Waikato River (Fig. 1), relatively easy. This river takes a long course to its mouth on the west coast of the North Island and has had an important effect on sedimentation in low-lying areas, especially after major eruptions (Schofield, 1965; Kear and Schofield, 1978).

The 1.8-ka Taupo eruption from this center was the most recent in a series of eruptions that probably began about 50 ka (Wilson and others, 1986). It was also one of the largest, its total erupted products having an estimated *in situ* volume greater than 105 km$^3$, of which 65 km$^3$ are known or inferred from outcrops on land. Fallout deposits from the early part of the eruption sequence comprise approximately half the latter volume; the nonwelded ignimbrite deposited from the culminating pyroclastic flow forms the remainder (Wilson and Walker, 1985).

This paper examines the effects of the Taupo eruption on the landscape. The evidence for these effects is derived from the depositional and erosional landforms created and the sediments deposited by post-eruption processes. As the Taupo eruption occurred relatively recently and was the last major volcanic event to affect the study area, this physical evidence is unmodified and generally complete. Before the evidence is presented, however, it is necessary to know something of the inherited landscape, of the eruption itself, and of the initial environment created by that eruption, as each controlled some aspects of the post-eruption response.

The study area is on the southeastern shore of modern Lake Taupo, approximately 5 to 30 km from the inferred vent located within Lake Taupo. It covers $\approx$380 km$^2$, between lake level at 357 m and the vicinity of the 700-m contour, and between the Tongariro and Hinemaiaia valleys (Fig. 2). The inland boundary approximately coincides with an inferred major northeast-southwest-trending fault that separates the downfaulted Quaternary volcanics of the study

[1]Present address: Department of Earth Sciences, University of Liverpool, Liverpool L69 3BX, United Kingdom

Sedimentation in Volcanic Settings, SEPM Special Publication No. 45

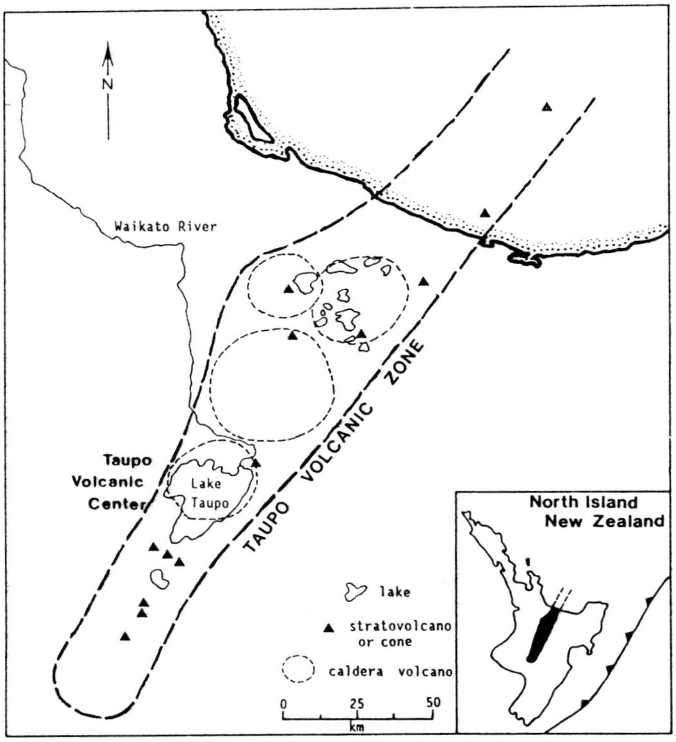


FIG. 1.—Central North Island of New Zealand showing the major volcanic centers and lakes, and the boundary of the Taupo Volcanic Zone.

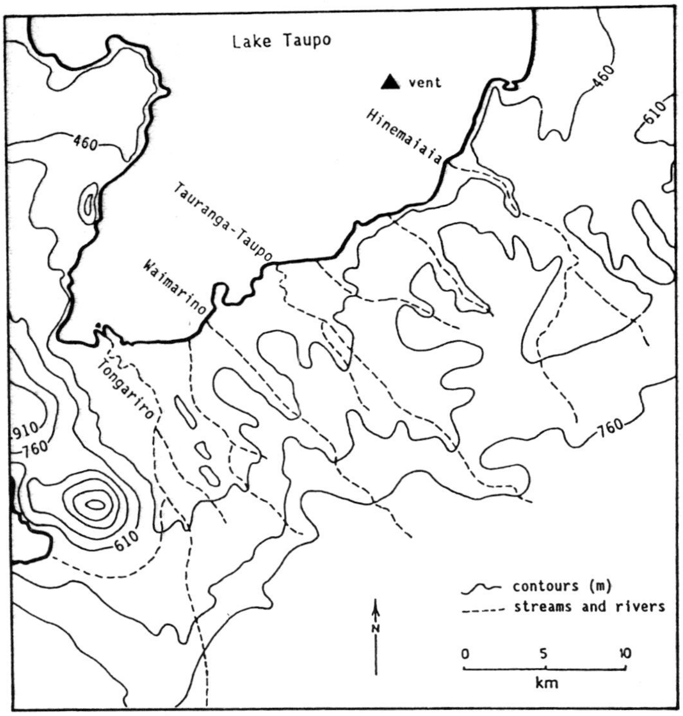


FIG. 2.—Generalized topography and drainage network of southeastern Lake Taupo. The study area lies between the lake shore, at 357 m, and ≈700 m above sea level, and between Tongariro River and Hinemaiaia Stream.

area from the Mesozoic greywacke of the Kaimanawa Ranges to the southeast.

## THE INHERITED LANDSCAPE

The landscape system includes all variables that impinge on the form and processes of the land surface. The most important of those inherited from the pre-eruption environment are the topography, geology, climate and vegetation.

### *Topography*

The skeleton of the modern landscape existed prior to the Taupo eruption, and the pre-eruption land surface is well defined by laterally continuous tephra layers deposited from 9.9 to 2.1 ka without major erosion. These define a topography similar to that of the present, which consists of northwest-southeast-trending ridges and small plateaus separating subparallel valleys (Fig. 2); these have generally moderate slopes of 10° to 20°, although they may be steeper in higher areas. The pre-eruption Lake Taupo shoreline, defined by both sediments and landforms, was near its present position.

### *Geology*

The near-surface geology of the inherited landscape was relatively simple, consisting almost entirely of the unindurated products of the 22.7-ka eruption from the Taupo Volcanic Center, overlain by relatively thin tephras from later eruptions of this and other centers (Grindley, 1960). Limited exposures of more indurated rocks occur throughout the study area. Cropping out mainly in streambeds and nearshore cliffs are late Pleistocene rhyolitic lavas and welded ignimbrites (Grindley, 1960); semi-lithified to well-indurated sedimentary rocks (clays, pumiceous pebbly sandstones and greywacke conglomerates) are exposed along parts of the present lake shore. In the middle Tongariro Valley are andesitic lavas and volcaniclastic conglomerates derived from stratovolcanoes to the south; coarse fluvial sediments dominate the lower valley (Grindley, 1960).

### *Climate and Vegetation*

Vegetation growing at the time of the Taupo eruption has been used to infer that the contemporary climate was broadly similar to that of the present. Analysis of carbonized logs contained within the Taupo ignimbrite and of pollen in soils immediately below and above the 1.8-ka tephras indicates that the area was covered by thick forest and scrub identical to that presently growing in undisturbed parts of the region (Birrell and Pullar, 1973; McGlone and Topping, 1977; Froggatt and others, 1981). In addition, the degree of weathering in younger paleosols suggests their formation when temperatures were close to modern values (Birrell and Pullar, 1973).

Mean annual rainfall in the Taupo region increases with increasing altitude, from about 1,250 mm at lake level to 2,400 mm in the Kaimanawa Ranges. Although monthly totals are similar, rainfall variability is much higher in summer and autumn than in winter and spring (Thompson, 1984) because much summer rain is convective, occurring as intense, localized showers. The area's inland position tends to produce a moderate range of temperatures; normalized data show lakeside means ranging from 17.5° in February to 6.5° in July, cooling by 2° to 3° with increasing altitude and distance from the lake (Anon, 1983; Thompson, 1984). Lower elevations of the east Taupo area are among the least windy in New Zealand; mean monthly wind speeds range from ≈6 to 11 km $hr^{-1}$ and calm prevails for 20 to 50 percent of the time. The prevailing winds are westerly to southwesterly (Thompson, 1984).

## THE TAUPO ERUPTION AND ITS PRODUCTS

In the first part of the Taupo eruption, phreatomagmatic activity alternated with Plinian eruptions as lake water gained access to or was excluded from the vent (Wilson and Walker, 1985). Isopachs for the combined fallout deposits show the effectiveness of the prevailing winds in dispersing the eruptive products to the east of Lake Taupo. Whereas more than 5 m of material was deposited in the northernmost section of the study area, most parts received less than 1 m, and all areas south of the Tauranga-Taupo River less than 0.2 m (Wilson and Walker, 1985).

The fallout tephras are overlain by a widespread ignimbrite, indicating that the eruption column collapsed and a pyroclastic flow moved out radially from the vent. All vegetation remaining after the deposition of the fallout tephras was carbonized, and valleys were partially filled with hot, unconsolidated, dominantly pumiceous material (Wilson and Walker, 1985; Wilson, 1985). The bulk of the ignimbrite was deposited from the body of the flow and has two topographically distinct variants. The ignimbrite-veneer deposit (IVD) is generally 0.5 to 1 m thick and drapes ridges and plateaus; the valley-pond ignimbrite (VPI) drained into valleys where it is 5 to 60 m thick in the study area (Walker and others, 1981b; Wilson and Walker, 1982; Wilson, 1985).

Of the three components in the Taupo eruption material, rhyolitic glass, including vesiculated pumice lapilli and bombs and nonvesiculated glass shards, is by far the most important volumetrically. Lithics form only about 5 percent of the total *in situ* volume (Wilson and Walker, 1985), and the total crystal content in the calculated original eruptive mixture for the ignimbrite was 3.5 percent (Wilson, 1985).

Grain size in the glassy material generally ranges from fine sand to boulders for vesicular clasts and from clay to coarse sand for glass shards. Pumice clast density at any one location varies with both grain size and the eruption unit. The variation with grain size is reasonably systematic, from ≈1.1 g $cm^{-3}$ for pumice cobbles to ≈2.3 g $cm^{-3}$ for silt-sized grains; nonvesicular glass shards have a density of 2.4 g $cm^{-3}$ (Walker, 1980; Froggatt, 1982; Wilson, 1985; Tilly, 1987).

Lithic clasts are of a variety of rock types, the most important numerically being varieties of rhyolite (Ewart, 1963; Walker and others, 1981a). Clast size generally falls within the sand-to-pebble range, with fewer cobbles and boulders. Andesine is the most common mineral, especially in the medium- to very coarse-sand-size range; quartz is a minor constituent (Ewart, 1963). Heavy minerals (hypersthene and magnetite) become more important in the fine- to very fine-sand sizes, but are minor components of the total crystal fraction. Densities in lithic and crystal grains range from ≈2.5 to 2.7 g $cm^{-3}$ for quartz, andesine and most lithics, to 3.4 to 3.5 g $cm^{-3}$ for hypersthene and 5.2 g $cm^{-3}$ for magnetite.

Distribution of clast types within the ignimbrite is not uniform. Denser pumice lapilli and bombs are concentrated with lithics and crystals toward the base of the ignimbrite, which shows a well-developed density grading as well as a coarse-tail normal size grading (Wilson, 1985).

## THE POST-ERUPTION ENVIRONMENT

The three most important effects of the Taupo eruption on the surrounding landscape were the total destruction of vegetation, the mantling of ridges and partial filling of valleys with hot, unconsolidated and noncohesive material having a wide range of grain sizes and densities, and the obliteration of hydrologic networks. In addition, most water was expelled from the lake basin and >30 $km^3$ of fine-grained ash was lost to the atmosphere (Wilson and Walker, 1985). Short-term weather patterns were affected by this ash cloud, producing an increased number of convective rainstorms (Froggatt, 1982; Wilson and Walker, 1985; Grant, 1985) and possibly a drop in air temperature (Taylor and others, 1980) for a few years after the eruption, before reversion to the general climatic conditions described above. The thicker VPI units would also have taken a few years to cool. The most important changes affecting landscape processes, however, were those concerning the erodibility and hydrologic characteristics of the new land surface.

In forested areas with a humid temperate climate, such as existed prior to the eruption, the incoming rainfall is partitioned into several different surficial flow paths and storage sites, including evapotranspiration, interception storage, stem flow, leaf drip, throughfall, and depression storage, before infiltration and any overland flow begin, and subsurface water moves to streams. Removal of the forest cover will immediately alter this partitioning so that rainfall has fewer possible storage sites and flow paths before reaching streams. In the absence of the litter layer, the high soil-moisture deficits and the deeper, more permeable soils associated with forests, infiltration is slowed, and both faster runoff and more frequent overland flow are expected, especially as rain beat on the unprotected ignimbrite would tend to compact the surface (Murata and others, 1966; Waldron, 1967; Pierce and others, 1970; Dunne and Leopold, 1978; Pitty, 1979; Chorley and others, 1984).

Elimination of interception and evapotranspiration will also allow more water to leave the catchment; bare soils are unimportant water users compared with vegetated surfaces (Baumgartner, 1967; Hornbeck and others, 1970; Pierce and others, 1970). More specifically, minimal surface evaporation rates in the post-eruption environment would be ensured by the high albedo of the cream-white Taupo ignimbrite, which reflects a large proportion of incoming shortwave radiation (Oke, 1987), by a high inter- and intragranular porosity, and by the lack of turbulence above the surface, due to low roughness elements and light winds. Surface evaporation would probably have accounted for only 10 to 20 percent of incoming precipitation once secondary fumarolic activity ceased, probably within a few months or years of the eruption. Runoff is therefore expected to have been 80 to 90 percent of incoming precipitation (Pierce and others, 1970), in marked contrast with rates in the forested pre-eruption landscape, which must have resembled the modern value of ≈60 percent (Schouten, 1983).

When soils are covered by vegetation and a thick litter layer, raindrop-impact energy is dispersed and few soil particles are detached. Erosion of particles from bare land thus proceeds at rates two or three orders of magnitude higher than those from well-vegetated or well-protected areas (Schiff, 1951; Noble, 1963; Copeland, 1963; Goudie, 1986). In addition, most erosion of bare soil is caused by a small number of high-intensity storms (Schiff, 1951; Piest, 1963), such as those occurring in the Taupo area during the late summer. Besides needing highly erosive rainfall and an easily erodible soil, rapid erosion is promoted by steeper slopes; these produce faster overland-flow velocities and a net downward movement of particles detached by rain splash. Very high erosion rates would therefore be expected when intense rainfall beats on more steeply sloping, shallow, noncohesive and finely textured bare regolith with low-infiltration capacities or high antecedent-moisture levels. Such conditions prevailed on the ridges of the study area where the relatively thin IVD overlay less permeable older tephras and paleosols, and the low density of pumice particles ensured that even large clasts could be eroded.

## LANDSCAPE RESPONSE IN THE UPLANDS

### *Early Response*

The initial response of landscape processes to the new environment in the upland areas was the erosion of the IVD and its deposition on the flat VPI surface by a variety of mass movements and high sediment-concentration flows. Pumice was stripped from the upper parts of many ridges, and the fine network of subparallel rills and gullies formed on the ridge slopes at that time is still visible in clear-felled areas. The more elevated parts of the VPI surface were also eroded as runoff began to concentrate in dendritic networks. Filling many shallow and, in some places, cross-cutting channels on the VPI surface are deposits from debris flows, hyperconcentrated flows, and flood-prone early streams.

*Debris flows.—*

Several types of debris-flow deposits record the variety of clast types and clast-support mechanisms within the various flows (Table 1). The most widespread deposit is type 1, which shows no internal structure or grain fabric. Individual flow units are generally <0.5 m thick. In type 2 deposits, which involve larger volumes of material, debris has been rearranged to form crudely stratified and graded deposits. Both these types of deposit are highly pumiceous.

Type 3 deposits occur in only one location in the study area; here the topography caused the pyroclastic flow to drop a higher than normal proportion of its lithic load. This local accumulation of lithics in a site close to steep slopes has produced a number of thin (<20 cm), coarse-tail reverse-graded units of lithic gravel in a vitric matrix (Fig. 3), each unit representing a discrete flow event. All three types of debris-flow deposit are texturally and compositionally indistinguishable from their local ignimbrite source.

Debris-flow deposits lacking pebbles and cobbles are uncommon in the study area because of the presence of much coarse material in the ignimbrite source. It is probable that the few gravel-free deposits seen represent material that had been reworked at least once before being remobilized as a mudflow.

*Hyperconcentrated flood flows (HFF).—*

Deposits similar in outcrop to those described by Smith (1986) as the product of hyperconcentrated flood flows were found in many valleys, where they overlie the VPI surface adjacent to present streams or occupy gullies cut into the VPI at the upper ends of the valleys, generally near steep ridges. These deposits are characterized by vague horizontal stratification, lenses of coarser material, isolated outsized clasts and generally poor-to-moderate sorting. Four types of deposit were identified: gravelly, sandy, graded-stratified, and vitually gravel-free muddy-sandy deposits (Table 1). The last type may result from the selective sorting of more poorly sorted flows as finer material remained in suspension and was deposited farther downstream (Pier-

FIG. 3.—Type 3 debris-flow deposits: coarse-tail reverse-graded lithic gravel in a vitric matrix.

son and Scott, 1985). The graded-stratified deposits are both distribution size and density normally graded from a lithic- and crystal-rich basal-gravel lag with prominent rip-up clasts of the underlying ignimbrite (Fig. 4). The dominance of lithic and crystal grains in the coarser deposits indicates that erosion had reached lower levels of the IVD.

*Early streams and rivers.*—

Deposits from early streams and rivers usually overlie or are interstratified with debris-flow and HFF deposits, from which they are distinguished by the presence of cross-bedding. Their textural immaturity, characterized by poor-to-moderate sorting and >5 percent mud content in most samples, distinguishes these deposits from later fluvial deposits, as does the dominance of transition and upper-flow-regime structures (Table 1 and Fig. 5). Deposition prior to major valley incision is indicated by the location of early fluvial deposits on the VPI surface or in small channels on that surface. The flows that deposited these sediments were probably fed by both overland flow and returning subsurface water from nearby ridges. Sheetfloods, which require high-intensity rainfall, low surface permeability and a lack of channelized drainage (Hogg, 1982), would have been common in the post-eruption environment, promoting channel instability and shallow flows.

The dominance of transitional and upper-flow-regime structures in these deposits indicates that frequent flooding, fed by rapid and high-volume runoff, continued after the flows became channelized. Abrasion of pumice pebbles in

FIG. 5.—Early fluvial deposits showing low-angle cross-bedding and scour-fill structures. Hammer is 25 cm long.

FIG. 4.—Basal section of a graded-stratified hyperconcentrated-flood-flow deposit. Note the scoured contact with the underlying ignimbrite and rip-up clast (outlined). Scale bar in centimeters.

a fully turbulent, dilute would facilitate their saturation by removing the narrow vesicle entrances (Whitham and Sparks, 1986) and make them more liable to deposition. That length of transport in these deposits was insufficient to abrade the pumice clasts strongly is indicated however, by the several days or weeks taken by dry pumice pebbles from these deposits to sink in water. This period resembles that required by pebbles from the ignimbrite and from debris-flow deposits; pebbles from later fluvial deposits normally sink within a few hours. The presence of low-density charcoal in several beds reinforces this indication of short transport lengths, which were probably due to channel instability. In their unstable bedforms and channel patterns, and rapid response to storm runoff, these early streams and rivers probably resembled those initially developed on the debris-avalanche deposit after the 1980 Mount St. Helens eruption (Meyer and others, 1982).

## *Later Response*

With the cooling of the ignimbrite and the concomitant lowering of the 100°C isotherm, the water table would begin to fall and any channelized runoff would be able to incise the highly erodible VPI. Where the fall of the groundwater base was unhindered by impermeable rock underlying the VPI, this incision would have been rapid and continuous and has resulted in simple U-shaped valleys with few or no sediments. In valleys where an impermeable hardrock base kept the water table high, however, the large volumes of water moving through the landscape kept sedimentary activity at the VPI surface and promoted the establishment of three additional depositional environments: temporary lakes, braided rivers, and, after additional incision, meandering rivers.

*Temporary lakes.—*

Although all lakes are temporary on geologic time scales, those described in this section were very short-lived compared with the age of Lake Taupo, which may be $10^6$ years (Timperley, 1983a). Small ponds formed in the many depressions on the VPI surface for relatively short periods, as indicated by the limited area and thickness of sediments, generally $<4$ m. In the larger valleys, however, lakes of substantial size (up to 11 $km^2$) and persistence (deposits up to 10 m thick) were formed. In the middle Hinemaiaia valley, the volume of lake sediments deposited is estimated as $\approx 2 \times 10^7$ $m^3$. This represents a sedimentation rate of at least $1 \times 10^6$ $m^3$ $yr^{-1}$, 100 times greater than the rate in a modern reservoir lake occupying the same position (Smith and others, in prep.).

In the smaller valleys, lake deposits overlie eroded ignimbrite, but in some parts of the Hinemaiaia valley, they drape deposits from debris flows or early streams. A thinning upward in beds of the larger lakes, from a maximum of 1.5 m to a few centimeters, in total outcrop thicknesses of 5 to 10 m, probably represents shallowing due to aggradation accompanied by a decrease in sediment supply.

The lacustrine deposits are generally planar and horizontally bedded, although some beds are discontinuous over 20 to 30 m. Most sediments are muddy sands dominated by pumice grains and glass shards, with scattered charcoal fragments and low gravel contents (Table 1). Granule modes in the latter and a scarcity of pebbles $>2$ cm in diameter indicate that clasts larger than medium-pebble size did not readily sink. A normally graded 60- to 70-cm-thick bed of clast-supported pumice pebbles and cobbles overlying lake beds in the Hinemaiaia valley probably represents a floating raft that grounded when the lake waters drained. Stream-influenced, lake-margin beds are dominantly lithic- and crystal-rich gravelly muddy sands, with only a moderate pumiceous content, reflecting their higher energy environment.

*Braided rivers.—*

All conditions promoting the development of braided rivers, i.e., large and highly variable discharges, steep gradients, lack of bank vegetation, and much coarse sediment (Richards, 1982), were present in the post-eruption landscape, but sediments interpreted as having been deposited by a braided river occur in only three stream valleys. Dense streambank vegetation and few roadcuts in the other valleys with impermeable bases may explain the limited outcrops. The most extensive outcrops, 2.5 to 7 m thick, are found in the middle Hinemaiaia valley, where in some places they overlie eroded remnants of lake beds (Fig. 6).

Large-scale, mainly trough cross-bedded sediments in grouped sets characterize the braided-river sediments (Table 1). Individual sets in the lower sections, which are associated with HFF deposits, are up to 2 m high with dips of 12° to 20° on cross-beds and up to 30° on bounding surfaces. An upward decrease in cross-bed size, reflecting a reduction in bedform size, probably results from a shallowing of the flow, due to rapid aggradation and/or decreasing runoff. Interstratified with the cross-beds and eroding some upper surfaces are sediments showing plane lamination characteristic of the transition to upper-flow regime (Fig. 6); scour-fill structures are common in the upper sections.

FIG. 6.—Large-scale cross-beds and plane lamination in braided-river deposits. These larger, basal beds overlie eroded, horizontally bedded lake sediments (behind spade). Spade is 1 m long. The full outcrop is sketched in Figure 11.

Subequal proportions of pumice and crystal grains tend to dominate braided-river sediments (Table 1); lithics are common only in beds closer to ridges, where basal parts of the ignimbrite were more accessible to erosion. Very small proportions of glass shards and mud in these sediments indicate that the finer grains were removed in suspension.

*Meandering rivers.—*

In all valleys where impermeable hardrock underlies the VPI, there is a single, discontinuous, paired or unpaired river terrace, 200 to 400 m in total width and approximately halfway between the VPI surface and the present streamflow level. These are mainly erosional terraces, being composed of ignimbrite with a thin veneer of sediments (Fig. 7), and are located immediately upstream of hardrock outcrops forming rapids or waterfalls up to 30 m high.

The formation of these terraces is attributed to a temporary halt in valley incision, caused by the inability of the still voluminous groundwater to penetrate the impermeable hardrock base. The height of the terrace above the hardrock barrier thus represents the thickness of the water-saturated VPI into which the streams could no longer incise. Any tributary valleys would also have to cease downcutting and become graded to this level (Fig. 7). As their gradient de-

FIG. 7.—Meandering river terrace with dry tributary valley (A) graded to the terrace and incised into temporary lake deposits. Terrace sediments overlying light-colored VPI (B) are 30 to 50 cm thick.

creased, the streams would begin aggrading and meandering in order to maintain equilibrium among the hydraulic variables (Richards, 1982). Cutting laterally instead of downward, the streams then deposited a thin veneer of channel and floodplain sediments over the ignimbrite. The thickness of sediments deposited (generally $<1$ m) indicates that this later period of aggradation did not last very long, and renewed incision followed the reduction in runoff that accompanied the re-establishment of vegetation (see later discussion).

Bedding in the deposits on the meandering-river terraces is seldom sharply defined; contacts tend to be gradational and a crude stratification by texture and/or clast composition is more common. Single fining-upward sequences of structureless or ripple cross-bedded lithic- and crystal-rich gravelly channel sands to pumiceous and vitric floodplain muds with pumice gravel are common (Fig. 8). Channel deposits generally show a reduction in the scale of sedimentary structures or bedding as the depth of water over point bars decreased. The three types of grading seen, coarse-tail normal grading of lithics and reverse of pumice clasts, and density-normal grading of lithic- and crystal-rich to pumiceous sediments (Table 1), result from flow deceleration and the earlier deposition of the denser clasts.

## LANDSCAPE RESPONSE NEAR THE CALDERA LAKE

Blockage of the outlet valley by ignimbrite caused the large volumes of water refilling the Lake Taupo basin to continue to do so long after the lake level reached the equilibrium level determined by resistant formations a few kilometers downstream from the outlet. A prominent lake terrace was formed, its inland margin generally reaching 34 m above the present lake level.

FIG. 8.—Fining-upward sequence in meandering-river sediments, from channel deposits of structureless gravelly sands and ripple cross-bedded slightly gravelly sands, all lithic- and crystal-rich, to vitric floodplain muddy sands. Scale bar in centimeters.

The length of time required to fill the basin to +34 m was $\approx$13 to 15 years. This calculation uses published data on present basin volume (White and Downes, 1977) and on precipitation inputs and evaporation from the lake surface (Timperley, 1983b); the additional volume to +34 m was then calculated from average shore gradients, and land-surface evaporation was estimated above as 10 to 20 percent of incoming precipitation. This time period implies a rise in lake level of $\approx$10 m $yr^{-1}$, and transgression in low-gradient areas of $\approx$1 km $y^{-1}$, rates with important implications for the stability of depositional environments.

A lengthy stillstand at +34 m is unlikely. As soon as the ignimbrite in the area of the outlet was overtopped, discharges somewhat greater than the modern value of $\approx$130 $m^3$ $sec^{-1}$ would begin to incise a new channel rapidly in the highly erodible material. Geomorphologic features in the study area (see later discussion) indicate fairly rapid drainage for the first 18 to 20 m, i.e., $\approx$75 percent of the volume between +34 m and the present level, followed by a possible slowing in the rate of fall. It is likely, given modern discharge values, that the whole period of drainage to present levels was no more than 4 to 5 years.

Landscape processes near the caldera lake responded to both rising and falling water levels, and that response is again reconstructed from both landforms and sedimentary deposits.

### *Response to Rising Lake Levels*

*The +34-m lake terrace.*—

The most obvious and ubiquitous result of rising water levels in Lake Taupo was the formation of the prominent lake terrace remarked upon by early workers (von Hochstetter, 1864; Cussen, 1887). Most of the terrace is of mainly erosional origin with minimal deposition. These sections formed where the rising lake waters planed the Taupo ignimbrite to form a gentle slope of ≈1.3° from the present lake shore to +34 m. The highest limit of these sections is generally marked by a prominent, 1- to 2-m-high scarp; this separates the smooth terrace surface, where depressions and early excavated small gullies were filled with lake sediments, from the more irregular topography inland.

Some of the few depositional sections of the lake terrace are in valleys carved into the ignimbrite by those early rivers whose discharge refilled the lake basin. In and near these valleys, sedimentation kept pace with the rise in lake level, so that a 5-m sediment thickness at the top of the terrace must have been deposited in less than a year, given the probable lack of a prolonged stillstand (Smith and others, in prep.). Other depositional sections are at the base of pre-eruption hardrock cliffs that resisted erosion by the rising lake waters. If these sections were sediment starved because of distance from river valleys, the top of the terrace would be some meters below the inferred maximum height of the lake, indicating that sedimentation did not keep pace with the rise in lake level (Smith and others, in prep.).

Most sediments on the caldera-lake terrace were deposited in a nearshore environment; outcrops of these sediments are up to 4 m thick. Beach deposits are much thinner, generally <1 m in total outcrop thickness, reflecting the narrowness of a lake shoreline. Transgressive sequences occur in most thicker outcrops, especially in exposures normal to the lakeshore. These sequences are both compositionally and texturally graded, with decreasing-upward density and mean-grain size of components (Fig. 9).

Beach deposits are centimeter-bedded gravelly sands and sandy gravels, dominated by lithic and crystal grains (Table 1), with rare pumice cobbles and boulders (Fig. 9). In contrast to the very well- to well-sorted modern sands on the eastern Lake Taupo shore, these post-eruption sediments are only moderately-to-moderately-well sorted, due to the rapid lake transgression. Units are generally structureless, although scour-fill structures and solitary, large-scale cross-beds with an offshore orientation occur near stream valleys (Table 1). These probably represent deposition in small stream-mouth deltas.

Nearshore sediments are dominated by pumice clasts and glass shards, reflecting their less energetic environment. Beds generally <20 cm thick show one of five combinations of structure, texture and composition (Table 1), reflecting the variety of both processes and locations in the depositional environment. Type 1 deposits, matrix-supported pumice

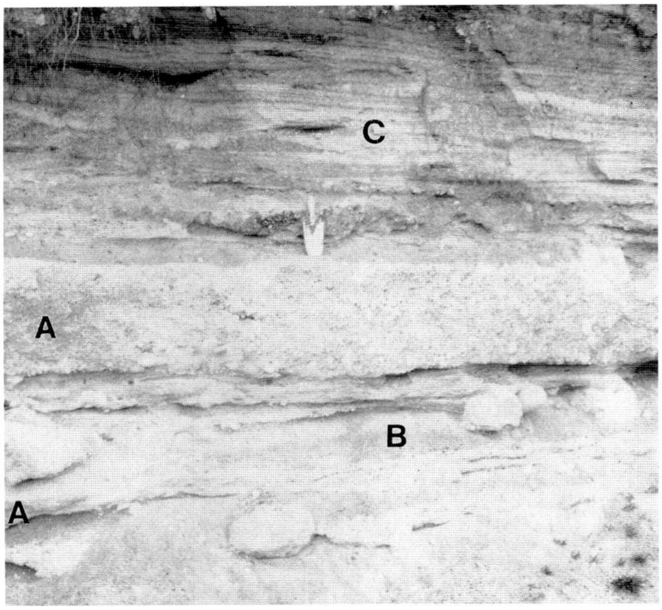


FIG. 9.—Transgressive sequence near the top of the +34-m caldera-lake terrace. Several debris-flow units (A) are interstratified with lithic- and crystal-rich beach sands (B) and large pumice cobbles and boulders. These are overlain by thinly bedded and laminated nearshore deposits of alternating pumice + glass and lithic + crystal units (C). Trowel is 25 cm long.

gravel in muddy sands, probably represent the rapid deposition of flood debris. Reworking and sorting of this material by wave oscillation has produced type 2 (well-sorted, clast-supported pumice gravels) and 3 (ripple cross-bedded muddy sands) deposits. Where sediments had a variety of grain densities, reworking has produced type 4 deposits, alternating pumice + glass and lithic + crystal laminae (Fig. 9); some laminae are also ripple cross-bedded. Type 5 sediments are poorly sorted pumice gravels with large-scale cross-bedding that crop out only near streams and probably formed as currents of cooler stream water flowed below warmer shelf waters. As in the sediments of temporary lakes, clasts larger than ≈2 cm are generally absent from type 1–4 nearshore deposits, indicating that the larger pumice clasts probably had formed floating rafts on the lake rather than having been deposited near the shoreline.

Debris-flow deposits are interbedded with both beach and nearshore sediments on the +34-m terrace (Fig. 9). Eight such units in one 3-m outcrop suggest deposition within a period of 3 to 4 months, given the estimated rate of lake-level rise. Lake-terrace debris-flow deposits form tabular or channel-fill bodies, and units normally thin upward, reflecting a decreasing sediment supply. The most common unit is the structureless type 1, but a few deposits are coarse tail reversely graded. Like the debris-flow deposits found in the uplands, these are dominated by pumice grains and glass shards and have gravel/sand/mud proportions similar to those of the ignimbrite. Interaction of the muddy matrix with lake water, however, has caused the formation of several features not found in such deposits in the uplands. These include numerous, well-defined air-bubble vesicles, a thin

TABLE 1.—SUMMARY OF SEDIMENT CHARACTERISTICS

| Depositional Process or Environment | Position in Time Period[1] | Thickness Range (m) | General Texture[2] | General Composition[3] | Cross-bedding | Grading | Other Structures |
|---|---|---|---|---|---|---|---|
| Debris flows: Type 1 | throughout | 0.3–2 | gmS | PG | — | — | — |
| Type 2 | early | 1–2 | gmS | PG | — | coarse-tail reverse<br>reverse normal | concentration of larger clasts in bands |
| Type 3 | early | 0.1 | gmS | PGL | — | coarse-tail reverse | — |
| Type 4 | later | 1–4 | G | P | — | — | imbrication to 20° |
| Mud flows: | early | 1–2 | mS | G(P) | — | — | — |
| HFF: gravelly | throughout | 1–1.5 | sG | LC(P) | — | coarse-tail reverse (lithics)<br>density-normal (lithic-pumice) | vague stratification<br>isolated boulders |
| sandy | throughout | <1 | (g)S | LC(P) | — | — | vague stratification<br>isolated boulders |
| muddy-sandy | early | 1–1.5 | (g)mS | PG(LC) | — | coarse-tail reverse | vague stratification |
| graded stratified | early | 2–3 | gmS | LCP | — | density and size normal | — |
| Early streams and rivers | early | 0.5–3 | gmS-(g)S | P or LCP | low-high angle, large-scale, planar tabular; low-angle ?antidune | coarse-tail normal, reverse, and reverse to normal | plane lamination<br>scour fill |
| Temporary lakes | moderately early | 1–10 | gmS | PG(LC) | large-scale, low-angle, solitary planar tabular (deltas) | normal (floated pumices)<br>coarse-tail reverse and reverse to normal | clast-supported pumice lenses |
| Braided rivers | moderately early | 1.5–7 | gmS | P(LC) | high-angle, large-scale trough; low-angle ?antidune | coarse-tail reverse | plane lamination |
| Meandering rivers (point bars) | moderately early | 0.2–2 | gS-(g)S | LC(HM) | high-angle, large-scale planar tabular; ripple | coarse-tail normal (lithics)<br>coarse-tail reverse (pumices)<br>density normal (lithic-pumice) | scour fill, gravel lenses, plane lamination |
| 34-m lake terrace: | | | | | | | |
| delta | later | 8 | (g)S | HM(L) | high-angle, large-scale trough; ripple | — | plane lamination |
| beach | later | 0.2–1 | gS-sG | LC(P) | high-angle, large-scale, trough + planar tabular | — | scour fill<br>gravel lenses |
| nearshore: Type 1 | later | 0.2–0.4 | gmS | PGC | — | coarse-tail reverse (pumices) | — |
| Type 2 | later | 0.2–0.3 | G | P | low-high angle, large-scale trough + planar tabular | — | — |
| Type 3 | later | 0.05–0.2 | mS-M | PG | ripple | — | — |
| Type 4 | later | 0.5–1.5 | mS-(g)mS | PL | ripple | — | laminated |
| Type 5 | later | 0.1–1 | G | P | high angle, large-scale planar tabular | reverse, normal and reverse to normal | — |

[1]within the 20-year post-eruption period
[2]m/M = muddy/mud; s/S = sandy/sand; g/G = gravelly/gravel; ( ) = slightly; therefore, (g)mS = slightly gravelly muddy sand
[3]P = pumice clasts; G = glass shards; L = lithics; C = crystals; (HM) = hypersthene and magnetite; ( ) = minor component (<25%)

layer of beach sediment adhering to the base of a unit (Fig. 10), and discrete lithic- and crystal-rich pockets that may be water-escape structures.

*River and stream response.—*

As the lake level rose, rivers were forced to flow at successively higher levels and formed river terraces at positions well above their present beds. Leopold and Bull (1979) have stated that rivers respond to a raised base level by aggrading at a reduced gradient over a wider area. Almost all such aggradational sediments deposited in the study area, however, were later removed by their rivers as the lake level fell, and the best examples of river-mouth response to a rising lake are found only at the highest level.

In the Tauranga-Taupo valley, relatively thin (generally <1 m) meandering-river deposits of lithic- and crystal-rich sand and gravel with a few boulders unconformably overlie eroded Taupo ignimbrite and pre-eruption tephras and paleosols on a terrace remnant ≈20 m above the present river. This terrace extends for ≈1.5 km toward the lake and, at its downstream end, is at the same height as the highest lake deposits. In the lower Tongariro valley, well-rounded river cobbles and boulders were deposited by the river over a wide, relatively flat area 15 to 20 m above the present river. These deposits thicken toward the lake and overlie both Taupo ignimbrite and HFF deposits. Sediments deposited in a delta relating to the highest lake levels are also located in this area. These deposits are very similar in composition and texture to those of the modern Tongariro River delta, being slightly gravelly medium and fine sands dominated by hypersthene and magnetite (Table 1). An 8-m stratigraphic thickness of deltaic sediments overlies 2 to 3 m of pumiceous and vitric silty sand containing pumice gravel lenses and andesite cobbles. The whole thickness probably represents aggradation of the type described by Leopold and Bull (1979), as the river mouth retreated inland over earlier immature river sediments.

Smaller streams also responded to the rising base level and lowered gradients by meandering and thus widening

FIG. 10.—Subaqueous debris-flow deposit on the +34-m terrace, with well-defined vesicles (black spots) and gravelly beach sands adhering to the base.

their incised lower valleys. An abrupt width change in these valleys occurs at approximately the same distance (2.5 km) from the lake shore, marking the inland limit of the effect of base-level rise. The presence of resistant hard rock immediately upstream from the highest river-terrace remnants in the Tauranga-Taupo, Tongariro and Hinemaiaia Rivers probably prevented the effects of rising base level from being propagated farther inland than a few kilometers.

### *Response to Falling Lake Levels*

*Lake terrace.*—

A steady fall in lake level would leave a smooth gradient on the exposed +34-m terrace, whereas stillstands would carve minor subterraces onto the main terrace surface. Only one reasonably continuous subterrace was developed in the study area; this is located in the Tongariro and Waimarino valleys and on the south side of the Tauranga-Taupo valley. From a base at +15 to +16 m above the present lake, it rises sharply to +18 m, or +21 m at the southern end of the Tongariro valley segment, where the additional height is due to the later deposition of sediments. On the northern side of the Tauranga-Taupo valley, a slope slightly steeper (≈2.2°) than the general terrace gradient rises from +5 m to +12.5 m, suggesting that wave action during a relatively slow drop in water level eroded the ignimbite substrate. These features suggest a relatively rapid fall in lake level to ≈+18 m, followed by a slowing in the rate of fall.

Although there are no regressive beach sediments overlying nearshore sediments on the +34-m terrace, reinforcing the suggestion of an initially rapid fall in lake level, beach ridges of pumice gravel with an amplitude of <20 cm and wavelength of 60 to 100 cm are found on both sides of the Tauranga-Taupo River. On the south side of the valley, these ridges extend from the top of the +18-m subterrace to the base of the +30-m scarp; on the northern side they cover the slightly steeper slope between +5 m and +12.5 m. The differing location of the beach ridges may be due simply to the impingement of a floating pumice-gravel raft at a particular point on the shoreline at one water-level height. Erosion of the ignimbrite during water-level rise, producing the planed surface and +30 to +34-m scarp described above, and this moderate amount of deposition during lake-level fall generally conform to Bruun's Rule, originally formulated for changes in sea level (Bruun, 1962; Bruun and Schwartz, 1985).

More extensive beach-ridge plains developed below +5 m (Muir, 1979). In one area the coastline prograded by up to 750 m and formed a tombolo to an offshore rhyolitic dome. These later ridges have a much greater amplitude range than those higher on the terrace (Muir, 1979), and their coarse, clast-supported, pumiceous sediments also contrast strongly with the finer sediments on the +34-m terrace. This suggests that only when the lake level stabilized could substantial volumes of the floating pumice gravel, blown onshore by the prevailing winds, build up along the shoreline.

*River and stream response.*—

Rivers and streams responded to the fall in base level in one of two ways, and the course followed was probably dependent on discharge (as estimated from catchment area or from present measured values). Small streams that meandered over the easily erodible Taupo ignimbrite responded by meander incision, a typical response to falling base levels (Schumm, 1977). The higher discharge Tauranga-Taupo River responded by discontinuous incision of its high-level river terrace to form several shallow (≈1–2 m) subterraces, their benches concave toward the present river. These subterraces, from 3 to 10 m above the present river, are covered by river gravels in a pebbly-sand matrix and strongly resemble those terraces described by Born and Ritter (1970) as having developed in response to a relatively continuous fall in lake level and short periods of unusually high river discharge. The Hinemaiaia Stream, with a somewhat lower modern discharge than that of the Tauranga-Taupo River, although higher than the small streams, has only one subterrace in its more confined lower valley, at ≈13 m above the lake (6 m above the present river).

Other sediments associated with falling lake levels and renewed valley incision are those of a fourth type of debris flow. Whereas the first three types are located on the VPI surface, this last type crops out only within stream valleys, well below that surface but still some meters above present streams. These sediments have the smallest proportion of fines of all debris-flow types, being dominated by large, often clast-supported and imbricated pumice pebbles and cobbles (Table 1). In addition, the sparse muddy matrix of several deposits has a distinctive reddish coloration. These deposits may have resulted from the saturation flowage of coarser material that accumulated at the base of gully walls. Removal of much fine material and weathering of the remainder probably took place while the debris was accumulating.

Other debris-flow deposits from this later period are of the structureless type 1. Their location, overlying eroded lake beds within valleys re-excavated in the +34-m terrace, indicates that debris flows were still being generated some 15 to 16 years after the eruption, when the lake shore had retreated more than halfway to its present position.

## CONTRACTION OF THE DRAINAGE NETWORK

Most of the valleys carved into the ignimbrite immediately after the eruption are now dry. In the study area, this contraction of the drainage network resulted from a reduction in the number of both tributaries to streams and rivers, which still drain the area, and streams rising in the ignimbrite and draining directly to the caldera lake.

Geomorphic evidence in the vicinity of the lower Tauranga-Taupo valley suggests that the reduction in drainage density began at or before the highest stand of Lake Taupo. North of the Tauranga-Taupo River are many small, dry valleys that end at the +34-m scarp. Other, larger, dry valleys cut the +34-m scarp but end before the lake shore. Thus, drainage density probably stabilized near its modern low value as the lake approached its present level, less than 20 years after the eruption.

Drainage-network contraction is the result of decreased surface runoff. This may have several causes, the most important being long-term climatic changes, i.e., a decrease in total rainfall or storminess, or an increase in air temperature, and land surface changes caused by increased infiltration rates associated with soil formation, or by vegetation regrowth. In the post-eruption environment, all but the last cause are unlikely. Soil formation in the area is very slow due to low weathering rates in the pumiceous material (Gibbs, 1980); modern soils on the Taupo ignimbrite are still relatively immature. The increased storminess and possible decreased air temperature that immediately followed the eruption would have been of insufficient duration to affect the network some 10 to 20 years after the eruption.

Vegetation regrowth would have increased substrate infiltration rates and reduced total runoff through evapotranspiration, while heightened roughness elements and a reduction in surface albedo increased both surface evaporation rates and evapotranspiration (Fitzharris, 1974). In addition, vegetation reduces erosion by physically binding the substrate and slowing the velocity of any remaining overland flow (Schiff, 1951; Graf, 1979). Incision of the ignimbrite would have been unable to keep pace with the accelerated drop in the water table, and most valleys became dry. Of the drainage systems rising within the Taupo ignimbrite, only those with impermeable hardrock bases have drainage densities approaching their post-eruption value. Even in these basins, several of the smaller first-order tributaries dried up as the second- or third-order streams began the second period of valley incision, leaving dry valleys graded to the meandering-river terraces (Fig. 7). In the areas with a more permeable substrate, the larger catchments (>10 $km^2$) are drained either by a first-order stream, or by a second-order stream with one or two tributaries at most; catchments smaller than ≈10 $km^2$ lack any modern surface drainage.

## SUMMARY

### *Sequence of Events in the Uplands*

Initial erosional events were rilling and gullying with much of the material transported by debris and hyperconcentrated flows. In many parts of the landscape, especially plateaus, ridges, and valleys with no impermeable base, deposits from these flows dominate, with only minor outcrops of early fluvial deposits.

In valleys underlain by impermeable hard rock, however, there was a wider variety of processes. In the middle Hinemaiaia valley, for example, initial erosion of the VPI and earlier tephras was followed by deposition from debris flows and flood-prone streams (Fig. 11). A lake then filled much of the valley, and the early deposits were covered by large volumes of sediment; these in turn were partly eroded by a braided river (Fig. 11). Most of the braided river and temporary lake deposits were then removed by lateral cut-

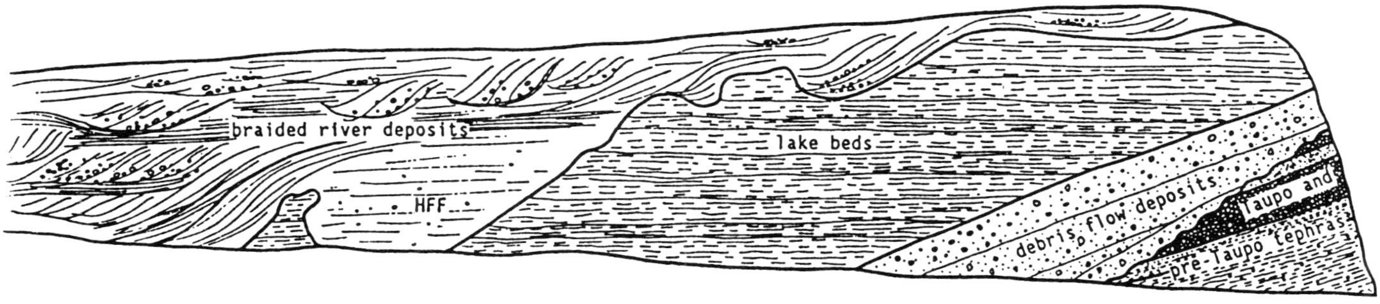


FIG. 11.—Lithofacies associations in the middle Hinemaiaia valley, drawn from field sketches and photographs. Outcrop is 10 m high and 100 m long.

ting as a meandering river flowed at a level approximately halfway between the top of the lake beds and the present stream. The latter is now eroding the terrace as it meanders within a wide floodplain. This sequence of events, the fullest in the study area, is summarized in Figure 12. Deposits in similarly large valleys also record the prior existence of several sedimentary environments, whereas smaller valleys, with a probable lower stream discharge, experienced a more restricted number of environments, omitting either the temporary lake or braided river.

It may be seen, therefore, that erosional processes, and those depositional processes not requiring large volumes of surface runoff, operated everywhere. Only in certain valleys were sediments recording a wider range of depositional processes and environments deposited. These locations were dictated by the geology of the inherited landscape, which ensured that runoff could not seep away rapidly underground, but was available for the continued erosion, transport and deposition of large quantities of sediment.

### *Sequence of Events Near Lake Taupo*

Lake deposits in the lower Hinemaiaia valley, overlying river gravels rich in lithics from the Taupo eruption, typify the drowning of the lower reaches of all valleys by a rising lake that forced rivers to aggrade at successively higher levels. Frequent debris flows moved from steep unvegetated or poorly vegetated slopes onto the beach and nearshore areas of the lake.

The fall in base level was not continuous, as shown by minor subterraces imposed on the slope of the lake terrace, and river incision was similarly discontinuous. Debris-flow activity continued as the lake retreated and a continuing supply of much sediment to the foreshore from both streams and the lake itself is indicated by thick deposits of river gravels near stream mouths and extensive beach-ridge systems both above and near the present lake level. Transgressive-regressive sequences are less common than either type alone and crop out only in stream valleys. In the lower Hinemaiaia valley, the lake beds that overlie early river gravels are themselves overlain by thin meandering-river deposits on the +13-m terrace, and in the northern part of the study area, a transgressive sequence of beach and near-

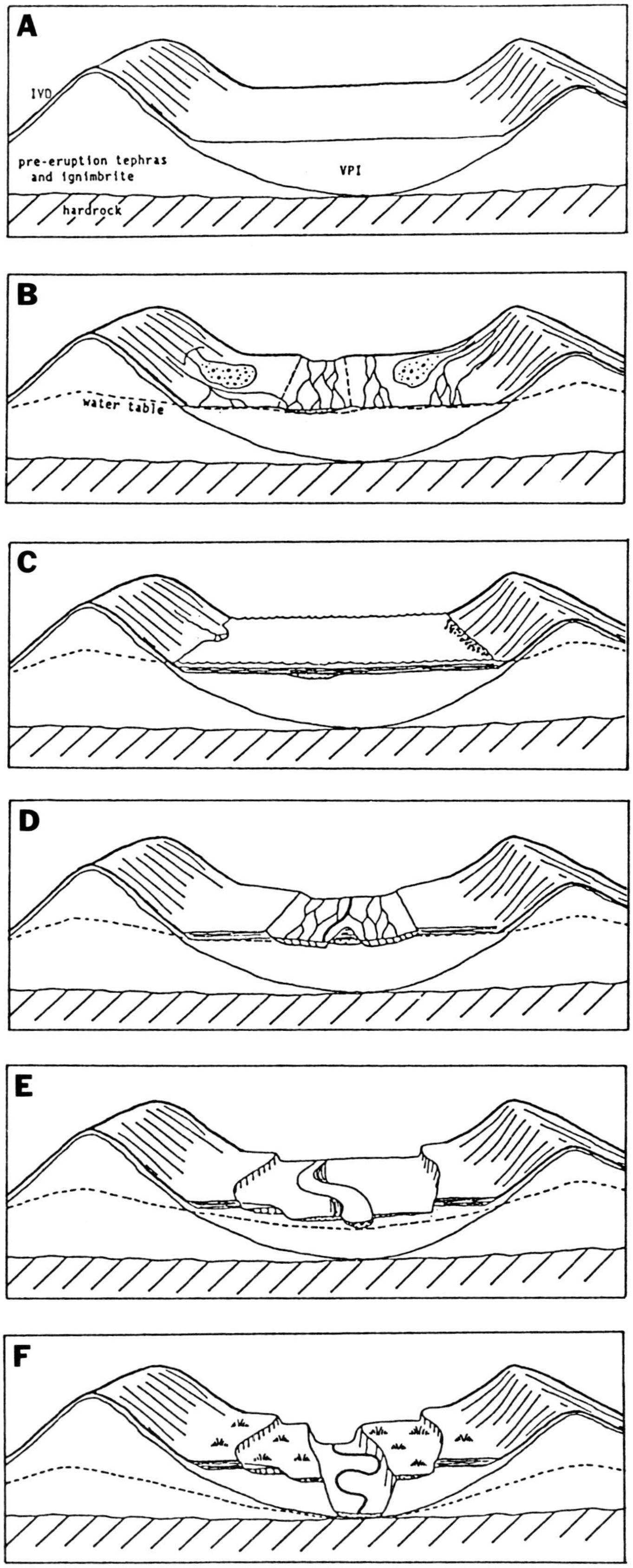


FIG. 12.—Sequence of events in the middle Hinemaiaia valley, reconstructed from cross-cutting relationships between successive deposits. (A) At the end of the eruption, ignimbrite covered the ridges and filled the valley to depths of 5 to 25 m, overlying an impermeable welded ignimbrite, pre-eruption soils and tephras, and Taupo fallout deposits. (B) Tephras from the Taupo and earlier eruptions were eroded and redeposited by debris flows and flood-prone streams in unstable channels. (C) An 11-km$^2$ lake formed as water backed up behind a downvalley constriction, and extensive, horizontally-bedded sediments were deposited. (D) After the lake drained, a braided river partially eroded the lake beds and deposited sediments up to 7 m thick. These include units apparently deposited by hyperconcentrated-flood flows. (E) After further incision, the water-table position stabilized above the impermeable welded ignimbrite, forcing the river to meander over its reduced gradient. A wide floodplain was cut into the lake beds and VPI, and a thin layer of sediments was deposited. (F) With the regrowth of vegetation, groundwater recharge was reduced and the water table dropped. The river quickly incised to the hardrock base over which the present stream meanders

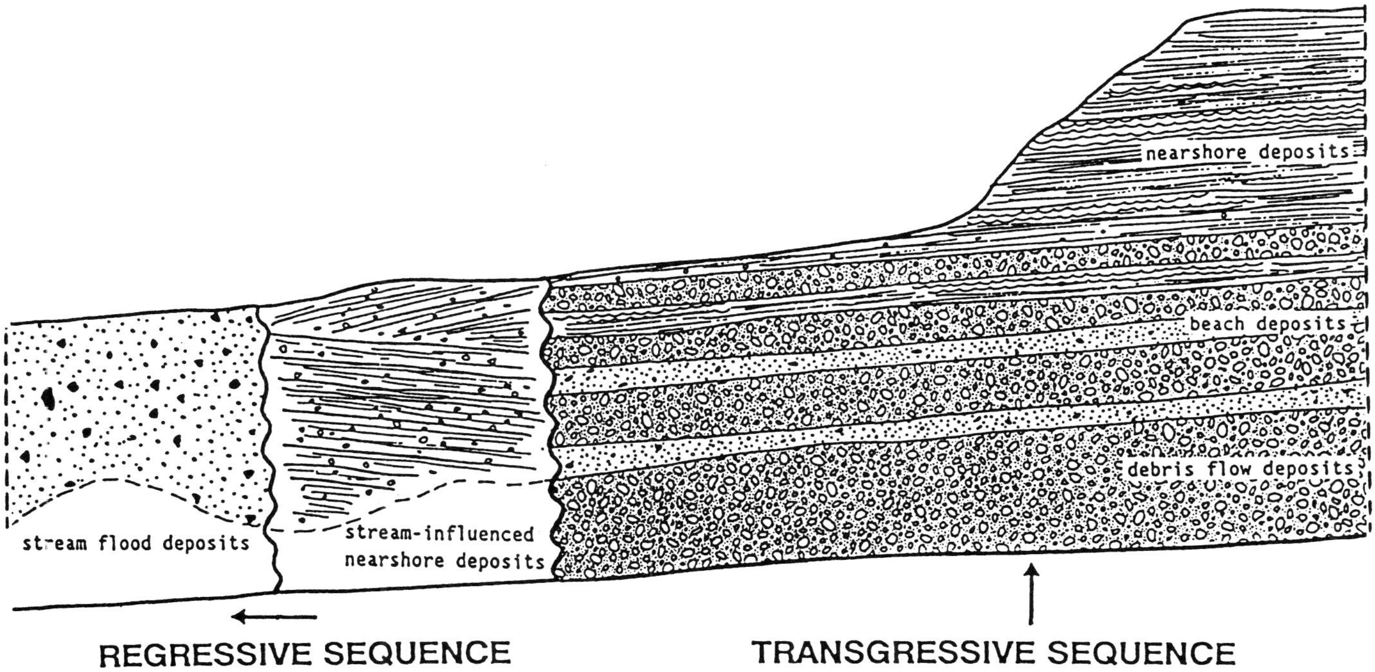


FIG. 13.—Outcrop in the +34-m caldera-lake terrace showing conformable transgressive and unconformable regressive sections (see text). The outcrop is 7 m high and 60 m long, and is aligned approximately parallel to the lakeshore. Stream flow was toward the viewer; the modern stream is to the left (north) of the outcrop. Drawn from field sketches and photographs.

shore deposits interbedded with debris-flow units was partially eroded by the interaction of a nearby stream with lowered lake levels and replaced by pumiceous type 5 nearshore sediments with large-scale, low-angle, planar-tabular crossbeds (Fig. 13). Some of these deposits were then removed by stream floods that deposited massive-bedded sandy lithic gravels (Fig. 13).

### *Integration of Events Throughout the Study Area*

The only link between events in the uplands and those near the lake basin would appear to be the incision of valleys. Leopold and Bull (1979) have shown, however, from studies on both very short (<50 years) and relatively short ($10^4$ years) time scales, that the effects of base-level changes are felt only at the downstream end of a river, and that all other reaches are adjusted solely to local conditions. These findings are especially applicable to those rivers in the study area that have numerous local base-level controls in the form of impermeable hardrock outcrops. Given the relatively short period of time for lake filling, however, it is unlikely that the effects of base-level change would have been propagated throughout those stream valleys lacking such local base-level control.

If, however, renewed incision by meandering rivers in the uplands had the same underlying cause as the drying up of smaller streams and tributaries, i.e., the renewal of the vegetation cover, then both these events occurred about the time Lake Taupo began its retreat from the +34-m level. Thus, it seems that all events in the uplands preceding the abandonment of the meandering-river terraces, i.e., early streamflow, debris flows and hyperconcentrated-flood flows, the filling and emptying of temporary lakes, and the braided rivers, occurred within a period of less than ≈15 years.

## LANDSCAPE RESPONSE OUTSIDE THE STUDY AREA

### *Near the Volcanic Center*

Deposits on the +34-m lake terrace and on fluvial terraces and floodplains along the southwestern and northeastern shores of Lake Taupo are very similar to those in the study area, and dry valleys are ubiquitous in the region covered by thick deposits of Taupo ignimbrite. A large temporary lake, with highly pumiceous sediments similar in texture and structure to those of temporary lakes in the study area, formed in a basin several kilometers north of Lake Taupo, which is also drained by the Waikato River. Narrow terraces suggest that the lake did not last very long (Pain and Pullar, 1975); its blockage was probably breached after the rising Lake Taupo overtopped the confining ignimbrite and the Waikato River was re-established.

### *Downstream Effects*

The destruction of vegetation and deposition of ignimbrite in the headwaters of the many large rivers rising on the elevated central volcanic plateau caused rapid erosion and transport of sediment, and extended the effects of the eruption to areas outside the limits of ignimbrite deposition. Thick deposits of Taupo pumice alluvium are found in the lower valleys of most rivers draining the plateau and there is evidence of major flooding, with tree stumps in growth position covered by pumiceous flood debris in areas to the east, northeast and southwest of the plateau (Fleming, 1953; Grant-Taylor and Rafter, 1963, 1971). River terraces up to 5 m above the present floodplains of rivers to the northeast and west of the vent were built by the aggradation of pumice debris (Grange, 1927; Pain and Pullar, 1968). Similar aggradation of the Waikato River bed in a lowland area some 200 km downriver from Lake Taupo blocked many tributaries, and much pumiceous material was deposited in the resulting temporary lakes (Kear and Schofield, 1978).

At the same time as this major transferral of Taupo eruption products was occurring, large quantities of local material in the central and eastern North Island were redeposited to form the Taupo Alluvium; this contemporary erosion is attributed to rainstorms caused by the eruption (Grant, 1985). The alluviation of both pumiceous and local material that followed the Taupo eruption is considered the most voluminous of all such episodes in New Zealand during the Holocene (Grant, 1985).

## CONCLUSION

Post-eruption sedimentation and geomorphic activity in the southeast Taupo area was both rapid and varied, and has resulted in a number of distinctive lithofacies and landforms. The scale of the response of physical processes must be ascribed collectively to the three factors listed in the introduction. In the post-eruption environment, accelerated and high-volume runoff provided the erosional and transporting medium for the large quantities of highly erodible material deposited during the eruption. The latter had also destroyed the vegetation of the inherited landscape, thus increasing the proportion of rainfall converted to runoff. The geology of the inherited landscape, with its contrasting lithologies, ensured that those large volumes would, in some valleys at least, be available for continued sedimentary and geomorphic activity.

In contrast to both the scale and rate of the post-eruption response, modern, inter-eruption sedimentary activity in the area is slow and is restricted to a small number of depositional environments (Smith and others, in prep.) The short-lived acceleration of physical processes in the post-eruption landscape probably accounts for a greater proportion of sedimentary deposits in the geologic record than those deposited in the long periods between major eruptions.

## ACKNOWLEDGMENTS

Funding by the New Zealand Department of Scientific and Industrial Research, in the form of a Postgraduate Study Award, is gratefully acknowledged, as is supervision by Dr. Peter F. Ballance, Department of Geology, and Professor Paul W. Williams, Department of Geography, both at the University of Auckland, New Zealand. The manuscript was improved after reviews by Geoff Orton and David Buesch, and editing by Gary Smith.

## REFERENCES

ANON, 1983, Temperature Normals 1951–1980: New Zealand Meteorological Service Miscellaneous Publication 183, 25 p.

BAUMGARTNER, A., 1967, Energetic bases for differential vaporization from forest and agricultural land, *in* Sopper, W. E., and Lull, H. W., eds., Forest Hydrology: Pergamon Press, Oxford, p. 381–389.

BIRRELL, K. S., AND PULLAR, W. A., 1973, Weathering of paleosols in Holocene and Late Pleistocene tephras in central North Island, New Zealand: New Zealand Journal of Geology and Geophysics, v. 16, p. 687–702.

BORN, S. M., AND RITTER D. F., 1970, Modern terrace development, Nevada, and its geologic implications: Geological Society of America Bulletin, v. 81, p. 1233–1241.

BRUUN, P., 1962, Sea level rise as a cause of shore erosion: Proceedings of the American Society of Civil Engineers, Journal of the Waterways and Harbors Division, v. 88, p. 117–130.

BRUUN P., AND SCHWARTZ, M. R., 1985, Analytical predictions of beach profile change in response to a sea level rise: Zeitschrift fur Geomorphologie, v. 57, p. 33–50.

CHORLEY, R. J., SCHUUM, S. A., AND SUGDEN, D. E., 1984, Geomorphology: Methuen, London, 605 p.

COPELAND, O. L., 1963, Land use and ecological factors in relation to sediment yields: Proceedings of the Federal Inter-Agency Sedimentation Conference, U.S. Department of Agriculture Miscellaneous Publication 970, p. 72–84.

CUSSEN, LIONEL, 1887, Notes on the physiography and geology of the King Country: Transactions and Proceedings of the New Zealand Institute, v. 20 (new series 3), p. 316–332.

DUNNE, T., AND LEOPOLD, L. B., 1978, Water in Environmental Planning: Freeman, San Francisco, 818 p.

EWART A., 1963, Petrology and petrogenesis of the Quaternary pumice ash in the Taupo area, New Zealand: Journal of Petrology, v. 4, p. 392–431.

FITZHARRIS, B. B., 1974, Land-use change and the water balance—An example of an evapotranspiration simulation model: Journal of Hydrology (New Zealand), v. 13, p. 98–114.

FLEMING, C. A., 1953, The Geology of the Wanganui Subdivision: New Zealand Geological Survey Bulletin 52, 362 p.

FROGGATT, P. C., 1982, A study of some aspects of the volcanic history of the Lake Taupo area, North Island, New Zealand: Wellington, New Zealand, Victoria University of Wellington, Unpubl. Ph.D. Dissertation, 349 p.

FROGGATT, P. C., WILSON, C. J. N., AND WALKER, G. P. L., 1981, Orientation of logs in the Taupo ignimbrite as an indicator of flow direction and vent position: Geology, v. 9, p. 109–111.

GIBBS, H. S., 1980, New Zealand Soils: An Introduction: Oxford University Press, Wellington, 115 p.

GOUDIE, ANDREW, 1986, The Human Impact on the Natural Environment: Blackwell, Oxford, 316 p.

GRAF, W. L., 1979, The development of montane arroyos and gullies: Earth Surface Processes, v. 4, p. 1–14.

GRANGE, L. I., 1927, The Geology of the Tongaporutu-Ohura Subdivision, Taranaki Division: New Zealand Geological Survey Bulletin 31, 63 p.

GRANT, P. J., 1985, Major periods of erosion and alluvial sedimentation in New Zealand during the Late Holocene: Journal of the Royal Society of New Zealand, v. 15, p. 67–121.

GRANT-TAYLOR, T. L., AND RAFTER, T. A., 1963, New Zealand natural radiocarbon measurements I–IV: Radiocarbon, v. 5, p. 118–162.

GRANT-TAYLOR, T. L., AND RAFTER, T. A., 1971, New Zealand natural radiocarbon measurements - 6: New Zealand Journal of Geology and Geophysics, v. 14, p. 364–402.

GRINDLEY, G. W., 1960, Taupo: New Zealand Geological Map Series 1:250,000: New Zealand Geological Survey, Sheet 8, Wellington.

HEALY, JAMES, 1975, Volcanic lakes, *in* Jolly, V. H., and Brown, J. M. A., eds., New Zealand Lakes: Auckland University Press, Auckland, p. 70–83.

HOCHSTETTER F. VON, 1864, Geology of New Zealand: Fleming, C. A., trans. and ed., 1959, Government Printer, Wellington, 320 p.

HOGG, S. E., 1982, Sheetfloods, sheetwash, sheetflow, or . . .?: Earth-Science Reviews, v. 18, p. 59–76.

HORNBECK, J. W., PIERCE, R. S., AND FEDERER, C. A., 1970, Streamflow changes after forest clearing in New England: Water Resources Research, v. 6, p. 1124–1132.

KEAR, D., AND SCHOFIELD, J. C., 1978, Geology of the Ngaruawahia Subdivision: New Zealand Geological Survey Bulletin 88, 167 p.

LEOPOLD, L. B., AND BULL, W. B., 1979, Base level, aggradation and grade: Proceedings of the American Philological Society, v. 123, p. 168–202.

MCGLONE, M. S., AND TOPPING, W. W., 1977, Aranuian (post-glacial) pollen diagrams from the Tongariro region, North Island, New Zealand: New Zealand Journal of Botany, v. 15, p. 749–760.

MEYER, D. F., JANDA, R. J., AND MARTINSON, H. A., 1982, Initial and subsequent drainage basin and channel response to the 1980 eruptions of Mt. St. Helens: EOS, Transactions, American Geophysical Union, v. 63, p. 174.

MUIR, M., 1979, The shore geomorphology of Lake Taupo: Auckland, New Zealand, University of Auckland, Unpubl. M.A. Thesis, 75 p.

MURATA, K. J., DONDOLI, D., AND SAENZ, R. 1966, The 1963–65 eruption of Irazu volcano, Costa Rica (the period of March 1963 to October 1964): Bulletin of Volcanology, v. 29, p. 765–793.

NOBLE, E. L., 1963, Sediment reduction through watershed rehabilitation: Proceedings of the Federal Inter-Agency Sedimentation Conference, U.S. Department of Agriculture Miscellaneous Publication 970, p. 114–123.

OKE, T. R., 1987, Boundary Layer Climates, 2nd edition: Methuen, London, 327 p.

PAIN, C. F., AND PULLAR, W. A., 1968, Chronology of fans and terraces in the Galatea Basin: Earth Science Journal, v. 2, p. 1–14.

PAIN, C. F., AND PULLAR, W. A., 1975, Chronology of 'paleosurfaces' and present land surfaces in the Reporoa Basin, North Island, New Zealand: New Zealand Journal of Science, v. 18, p. 313–322.

PIERCE, R. S., HORNBECK, J. W., LIKENS, G. E., AND BORMAN, F. H., 1970, Effect of elimination of vegetation on stream water quantity and quality: International Association of Hydrological Sciences Publication 96, p. 311–328.

PIERSON, T. C., AND SCOTT, K. M., 1985, Downstream dilution of a lahar: Transition from debris flow to hyperconcentrated streamflow: Water Resources Research, v. 21, p. 1511–1524.

PIEST, R. F., 1963, The role of the large storm as a sediment contributor: Proceedings of the Federal Inter-Agency Sedimentation Conference, U.S. Department of Agriculture Miscellaneous Publication 970, p. 98–108.

PITTY, A. F., 1979, Geography and Soil Properties: Methuen, London, 287 p.

RICHARDS, K., 1982, Rivers: Form and Process in Alluvial Channels: Methuen, London, 358 p.

SCHIFF, L., 1951, Surface detention, rate of runoff, land use, and erosional relationships of small watersheds: Transactions, American Geophysical Union, v. 32, p. 57–65.

SCHOFIELD, J. C., 1965, The Hinuera Formation and associated Quaternary events: New Zealand Journal of Geology and Geophysics, v. 13, p. 767–824.

SCHOUTEN, C. S., 1983, Budget of water and its constituents for Lake Taupo (New Zealand), *in* Dissolved Loads of Rivers and Surface Water: Quantity/Quality Relationships: International Association of Hydrological Sciences Publication 141, p. 277–297.

SCHUMM, S. A., 1977, The Fluvial System: Wiley, New York, 338 p.

SMITH, G. A., 1986, Coarse-grained nonmarine volcaniclastic sediment: Terminology and depositional process: Geological Society of America Bulletin, v. 97, p. 1–10.

TAYLOR, B. L., TVZI, G-C., AND SCHNEIDER, S. H., 1980, Volcanic eruptions and long-term temperature records: An empirical search for cause and effect: Quarterly Journal of the Royal Meteorological Society, v. 106, p. 175–199.

THOMPSON, C. S., 1984, The Weather and Climate of the Tongariro Region: New Zealand Meteorological Service Miscellaneous Publication 115(14), 35 p.

TILLY, C. R., 1987, The sedimentology of the Taupo Pumice Alluvium Formation occurring in the lower region of the Hamilton Basin: Hamilton, New Zealand, University of Waikato, Unpubl. M.S. Thesis, 404 p.

TIMPERLEY, M. H., 1983a, Geological history, *in* Forsyth, D. J., and Howard-Williams, C., eds., Lake Taupo: Department of Scientific and Industrial Research, Wellington, p. 5–15.

TIMPERLEY, M. H., 1983b, Climate and hydrology, *in* Forsyth, D. J., and Howard-Williams, C., eds. Lake Taupo: Department of Scientific and Industrial Research, Wellington, p. 45–54.

WALDRON, H. H., 1967, Debris Flow and Erosion Control Problems caused by the Ash Eruptions of Irazu Volcano, Costa Rica: U.S. Geological Survey Bulletin 1241-I.

WALKER, G. P. L., 1980, The Taupo pumice: Product of the most powerful known (ultraplinian) eruption?: Journal of Volcanology and Geothermal Research, v. 8, p. 69–94.

WALKER, G. P. L., SELF, S., AND FROGGATT, P. C., 1981a, The ground layer of the Taupo ignimbrite: A striking example of sedimentation from a pyroclastic flow: Journal of Volcanology and Geothermal Research, v. 10, p. 1–11.

WALKER, G. P. L., WILSON, C. J. N., AND FROGGATT, P. C., 1981b, An ignimbrite veneer deposit: The trail-marker of a pyroclastic flow: Journal of Volcanology and Geothermal Research, v. 9, p. 409–421.

WHITE, E., AND DOWNES, M. T., 1977, Preliminary assessment of nutrient loads on Lake Taupo, New Zealand: New Zealand Journal of Marine and Freshwater Research, v. 11, p. 341–356.

WHITHAM, A. G., AND SPARKS, R. S. J., 1986, Pumice: Bulletin of Volcanology, v. 48, p. 204–224.

WILSON, C. J. N., 1985, The Taupo eruption, New Zealand. II. The Taupo Ignimbrite: Philosophical Transactions of the Royal Society of London, A, v. 314, p. 229–310.

WILSON, C. J. N., AND WALKER, G. P. L., 1982, Ignimbrite depositional facies: The anatomy of a pyroclastic flow: Journal of the Geological Society, London, v. 139, p. 581–592.

WILSON, C. J. N., AND WALKER, G. P. L., 1985, The Taupo eruption, New Zealand. I. General aspects: Philosophical Transactions of the Royal Society of London, A, v. 314, p. 199–228.

WILSON, C. J. N., HOUGHTON, B. F., AND LLOYD, E. F., 1986, Volcanic history and evolution of the Maroa-Taupo area, Central North Island, *in* Smith, I.E.M., ed., Late Cenozoic Volcanism in New Zealand: Royal Society of New Zealand Bulletin 23, p. 194–223.

# CHANGES IN DEPOSITIONAL ENVIRONMENTS RESULTING FROM EMPLACEMENT OF A LARGE-VOLUME IGNIMBRITE

DAVID C. BUESCH

*Department of Geological Sciences, University of California, Santa Barbara, California 93106*

ABSTRACT: Emplacement of the widespread 18.5-Ma Peach Springs Tuff (PST) ignimbrite over a region of varied topography in southeastern California, western Arizona, and southern Nevada, caused adjustments in local drainage and depositional systems. Deposition of the post-PST sediment occurred shortly after, and probably in response to, the emplacement of the PST, where the sediment was deposited on nonwelded tuff or co-ignimbrite fallout tuff, and rills or channels were not eroded into the tuff. Only 12 of 32 exposures show a lack of erosion. The remaining exposures record a period of erosion with post-PST sediment deposited on partially to densely welded or vapor-phase-crystallized PST. At all 32 locations, pre- and post-PST sedimentary sections within 20 m stratigraphically of the PST are interpreted as representing alluvial-fan, braided trunk stream, and lacustrine-to-playa depositional environments. Pre- to post-PST sedimentary rocks do not change in a 1- to 2-km-wide, 500-m-relief valley; thus, local relief was more influential in controlling sedimentation than change in stream gradients caused by deposition of 30 m of PST. In and along the margins of 5- to 25-km-wide valleys, finer grained lithofacies were deposited in post-PST compared to pre-PST rocks. Locally, changes of nearly 90° occur in the flow direction from pre-PST alluvial-fan deposits to post-PST braided-trunk-stream deposits. At 21 of 32 locations, a fining- and thinning-upward sequence occurs between the pre- and post-PST section. The pre- to post-PST change in sedimentation is not simply a result of the time needed for alluvial fans to prograde over the tuff, but is a function of lowered stream gradients caused by deposition of 60 to 200 m of PST in broad valleys. Welding and vapor-phase crystallization in the PST formed erosionally resistant rock that slowed incision and promoted lateral cutting of streams into nonwelded tuff.

## INTRODUCTION

Explosive volcanic eruptions produce fallout, surge, or pyroclastic-flow deposits that can have a significant impact on sedimentary processes both in the area of the initial deposit and at great distances from the volcano. Several studies of historical eruptions have provided insight into the sedimentary processes associated with volcanic eruptions. Collins and Dunne (1986) demonstrated that at Mount St. Helens the greatest erosion and redistribution of the May 18, 1980, tephra occurred within a few years of the eruption, and most of the reworked tephra was deposited in local alluvial fans and terraces (Smith and Swanson, 1987). Eruptions in Guatemala have caused dramatic changes that lasted 20 to 30 years in the river systems downstream from the volcanoes (Kuenzi and others, 1979).

It is more difficult to reconstruct the sedimentologic response of a system to a particular eruption in older deposits. The major influence on the sedimentation after the moderate-size eruption of the 1.8-ka Taupo ignimbrite may have lasted 20 years (R. Smith, this volume). Waresback and Turbeville (1990) interpret the spatial variations of lithofacies in an alluvial-fan and fan-delta sequence as the results of increased sediment load in response to periodic volcanic eruptions. Smith (1987, 1988, and this volume) discusses the effects of volcanic eruptions on the aggradation of a debris apron by deposition from debris flows and hyperconcentrated flows, and periods between eruptions that are marked by nondeposition and incision as streams attempt to re-establish the pre-eruptive stream grade.

This paper describes the changes in depositional environments that resulted from the emplacement of the widespread 18.5-Ma Peach Springs Tuff (PST) ignimbrite. The PST ignimbrite has a minimum volume of 640 $km^3$ (Buesch, in prep.), is exposed over approximately 32,000 $km^2$, and extends across 360 km from the Colorado Plateau, Arizona, to the central Mojave Desert, California (Fig. 1). The deposition of the PST represents a single eruption (Buesch, in prep.). Alluvial fans formed the most abundant pre-PST depositional environments. Many of the post-PST sediments contain ash as part of the matrix, but most of the sediment was derived from the rocks exposed in the highland areas that were not covered with the PST ignimbrite. This study focuses on the local changes between pre- and post-PST sedimentation and does not attempt to evaluate the long-range transport of PST tephra that is discussed in other volcanic areas (Kuenzi and others, 1979; Smith, 1987, 1988).

## REGIONAL GEOLOGY

### *Peach Springs Tuff*

*Correlation and age.—*

The Peach Springs Tuff was originally described on the Colorado Plateau by Young (1966) and Young and Brennan (1974). Additional descriptions of exposures near and east of Kingman, Arizona, are included in Buesch and Valentine (1986), Valentine and others (1989), and Buesch (1987a, b, 1989). Glazner and others (1986) correlated the Peach Springs Tuff across 360 km of California and Arizona (Fig. 1) on the basis of stratigraphic position, the occurrence of adularescent sanidine as the most abundant phenocryst and megascopic sphene, similarity of K-Ar ages, and regional similarities in mineral chemistry. Gusa and others (1987) demonstrated that the PST has a distinctive heavy-mineral suite compared to other tuffs in the region. Wells and Hillhouse (1989) used the characteristic paleomagnetic declination of the PST reported by Young and Brennan (1974) to confirm the regional correlation. A detailed study of regional mineralogic and geochemical trends in phenocrysts (Buesch, 1989, in prep.) supports the correlation. The Peach Springs Tuff has a sanidine $^{40}Ar/^{39}Ar$ laser-fusion age of 18.5 ± 0.2 Ma (Nielson and others, 1990) and was erupted during a period of large-scale crustal extension in the Colorado River extensional corridor (CREC in Fig. 1; Howard and John, 1987). As much as 50 km of the east-west extent

Sedimentation in Volcanic Settings, SEPM Special Publication No. 45

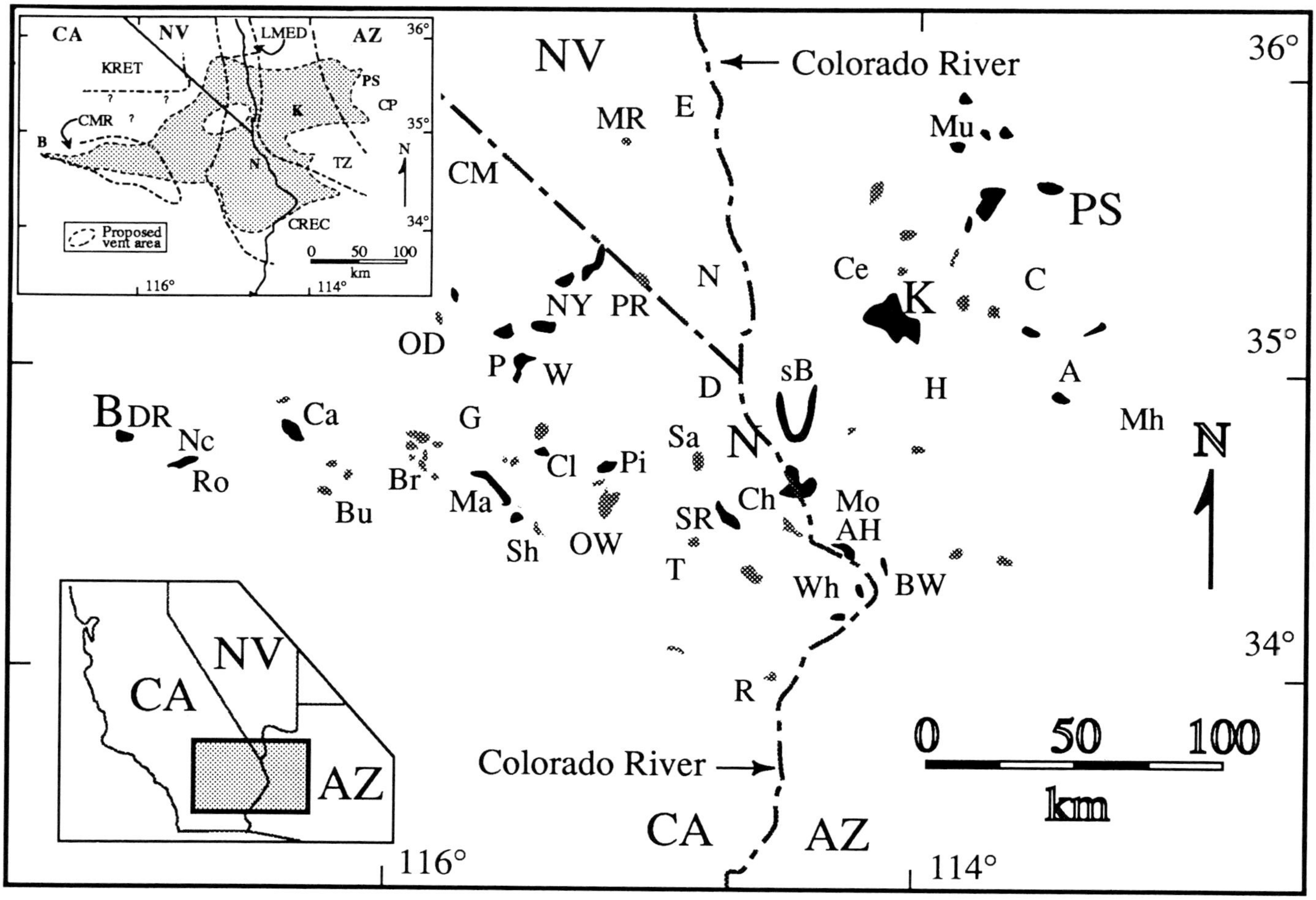


FIG. 1.—Insert: Shaded area is the minimum distribution of the Peach Springs Tuff (modified from Glazner and others, 1986). White ellipse is the region where the proposed PST vent is located (from Buesch, 1989, and in prep.; Hillhouse and Wells, 1989, and pers. commun.). Extensional domains include Central Mojave Rift (CMR; Dokka, 1986), Colorado River extensional corridor (CREC; Howard and John, 1987), Lake Mead extensional domain (LMED; Spencer and Reynolds, 1986; Faulds and others, 1988), and Kingston Range extensional terrane (KRET; Burchfiel and Davis, 1988). Question marks indicate areas of faulted rock, but whether these areas are part of an extensional domain is uncertain. The strike-slip terrane coincides with much of the CMR and is not shown. Other structural provinces: CP = Colorado Plateau, TZ = Transition Zone to the CP. States: CA = California, NV = Nevada, AZ = Arizona. For reference, the cities of Barstow (B), Needles (N), Kingman (K) and Peach Springs (PS) are included. The Colorado River forms the western state line of Arizona. Enlarged Map: Exposures of PST referred to in text (black); all other PST locations (gray). Present mountain ranges: *In California*—Dagget Ridge (DR), Newberry Mtns. (Nc), Rodman Mtns. (Ro), Cady Mtns. (Ca), Bristol Mtns. (Br), Bullion Mtns. (Bu), Old Dad Mountain, (OD), Granite Mtns. (G), Providence Mtns. (P), Marble Mtns. (Ma), Clark Mtn. Range (CM), New York Mtns. (NY), Woods Mtns. (W), Clipper Mtns. (Cl), Ship Mtns. (Sh), Piute Range (PR), Piute Mtns. (Pi), Old Woman Mtns. (OW), Dead Mtns. (D), Sacramento Mtns. (Sa), Turtle Mtns. (T), Chemehuevi Mtns. (Ch), Sawtooth Range (SR), Whipple Mtns. (Wh), Riverside Mtns. (R). *In Nevada*—McCullough Range (MR), Eldorado Mtns. (E), Newberry Mtns. (N). *In Arizona*—southern Black Mtns. (sB), Mohave Mtns. (Mo), Auberry Hills (AH), Bill Williams Mtns. (BW), Cerbat Mtns. (Ce), Hualapai Mtns. (H), Music Mtns. (Mu), Cottonwood Mtns. (C), Aquarius Mtns. (A), Mohon Mtns. (Mh).

of the PST is attributed to extension across the CREC (Howard and John, 1987).

Eruption and deposition of the Peach Springs Tuff occurred over a short time interval, indicated by the presence of only one co-ignimbrite fallout tuff, which locally caps the non- to densely welded tuff, and the occurrence of only one cooling unit (welding and cooling unit terminology after Smith, 1960). At each location, only a single flow unit has been recognized on the basis of the textural subdivisions described in Sparks and others (1973) and Fisher and Schmincke (1984). Thus, emplacement of the Peach Springs Tuff represents a single explosive event that deposited large amounts of pyroclastic material into many drainage-depositional systems of different sizes and shapes.

*Thickness variations.*—

Regionally, the Peach Springs Tuff ignimbrite ranges from 2 to >200 m in maximum thickness (Table 1). The thickest deposits occur in exposures near the center of the regional distribution and fill paleovalleys. Some of the most distal sections (Dagget Ridge, California, and Bill Williams Mountains, Arizona; Fig. 1) are as much as 30 m thick,

TABLE 1.—LOCATION AND THICKNESS OF PST, AND INFERRED DEPOSITIONAL ENVIRONMENTS IN PRE- AND POST-PST SEDIMENTARY ROCKS

| Exposure Location | Layer L2(m)[a] | | Top of PST[b] | Pre-PST Environ[c] | Pre-PST Grain size[d] | Fining[e] Thinning | Post-PST Grain size | Post-PST Environ |
|---|---|---|---|---|---|---|---|---|
| **Central Mojave Rift (CMR)** | | | | | | | | |
| Dagget Ridge, CA | 23 | 0 | NW | UF | G ≥ S | Y | S > F + M | LF > LP |
| S Newberry Mtns, CA | 30V | 0 | NW | UF + AB | G + S | N | G + S | UF + AB |
| Kane Wash | | | | | | | | |
| N Cady Mtns, CA | 12V | 2 | NW ± PW | LF | S ≥ P | Y | S | LF |
| S Cady Mtns, CA | 60 | 5L | PW | LF > UF ≫ LP | S + P > G ≫ F | Y | S > G | LF ≥ UF |
| Ludlow, CA | 6 | | PW | LF | S ≥ P | N | S ≥ P | LF |
| Bullion Mtns, CA | 20 | | PW | LF | S ≥ P ≥ G ≥ F | — | — | ME |
| Central Bristol Mtns, CA | | | | | | | | |
| Bristol Mtn Wash | 35 | 9L | PW | LF | S + G | — | — | ME |
| Siberia Wash | 25 | 11 | PW | LF | S + G | Y | S | LF + ME |
| Orangeblossom Wash | 16 | | PW | LF | S + P ≥ G | — | — | ME |
| Granite Spring, CA | 15V | 3 | PW | LF | S + P | Y | S | LF |
| Old Dad Mountain, CA | 15 | | PW | LF | S + P | N | S + P | LF |
| Ship Mtns, CA | 35 | 6 | PW | LF | S + P | — | — | ME |
| **East Mojave Block (EMB)** | | | | | | | | |
| NW Marble Mtns, CA | 33 | 20L | PW | LF | S + P | — | — | ME |
| S Marble Mtns, CA | 30 | 5L | PW | LF | S ≥ G | — | — | ME |
| Pinto Mtn, CA | 22 | 10L | PW | LF | S ≥ P | Y | C + M + S | ID ≥ LF |
| Providence Mtns, CA | 35 | 2L | NW | UF | G ≥ S | Y | S + F + M > G | LF + AB + LP |
| Barber Cyn | | | | | | | | |
| Clipper Mtns, CA | 30 | 2L | PW | LF | S | — | — | ME |
| New York Mtns, CA | | | | | | | | |
| Barnwell | 30 | 12 | PW | LF | S + P | N | S + P | LF |
| Dove Spring | 22 | 8 | PW | LF | S + P | N | S + P | LF |
| Piute Mtns, CA | 220V | 20 | PW | LF | S + P | N | S + P | LF |
| **Colorado River extensional corridor (CREC)** | | | | | | | | |
| Old Woman Mtns, CA | 20 | 5 | NW ≥ PW | LF ≥ LP | S + P > G ≥ F + M | N | S + P + G | LF |
| Little Piute Mtns, CA | 55 | 6 | PW | UF > LF | G + S | Y | S + G | LF |
| Piute Range, CA | 30 | 3 | PW | LF | S ≥ G | Y | S > C | LF > ID |
| Sacremento Mtns, CA | 85 | | PW | LF | S + P | N | S + P | LF |
| Snaggletooth, CA | 140 | 35 | NW | LF | S + F | Y | F + C | LP + ID |
| Chemehuevi Wash, CA | 50 | 10 | PW | LF | S + P | N | S + P | LF |
| Pyramid Butte, CA | 25 | 12 | PW | LF | S + P | N | S + P | LF |
| Moovalya Lake, CA | 12V | 3E | PW ≥ NW | UF | P + G > S | Y | S + F + M + C | LF + LP + ID |
| Tumarian Hills, AZ | 200V | 20 | PW > NW | LF | S + P > G | Y | S +> G | LF |
| Auberry Hills, AZ | 60V | 10 | PW | LF > UF | S + P + G | Y | S ≥ P ≥ F + M | LF > LP |
| Bill Williams Mtns, AZ | 30V | 4 | NW | LF > UF | S + P > G | Y | S > P > G | LF > UF |
| Parker Dam, CA | 6 | 0E | PW > NW | LF | S + P > G | N | S + P > G | LF |
| **Lake Mead extensional terrane (LMET)** | | | | | | | | |
| McCulloch Mtns, NV | 2 | 0 | NW | UF | G + S | Y | S | LF |
| **Tranzition Zone (TZ)** | | | | | | | | |
| SW Black Mtns, AZ | 100 | 20 | PW | LF | S + P | N | S + P | LF |
| SE Black Mtns, AZ | 100V | 4L | NW > PW | LF | S + P | Y | S ≫ G | LF |
| Flat Top, AZ | 16 | 10 | PW | LF | S + P | — | — | ME |
| Kingman, AZ | 85 | 11L | PW > NW | LF ≫ UF | S + P > G | Y | S + P ≫ G | LF |
| **Colorado Plateau (CP)** | | | | | | | | |
| Hackberry, AZ | 24V | 4 | PW | LF | S + P | Y | S | LF |
| Milkweed Canyon, AZ | 12 | 0 | PW | LF | S + P > G | — | — | ME |
| Valentine, AZ | 24V | 6 | PW | LF | S + P > G | Y | S + P | LF |
| Peacock Mtns, Az | 15 | 8 | PW | LF | S + P | — | — | ME |
| Penitentiary Mtn, AZ | 22V | 5 | PW | LF | S > P | — | — | ME |
| Trout Creek, AZ | 22V | 0 | PW | LF | S + P > G | — | B | |
| Willows Ranch, AZ | 12V | 3 | PW | LF | S + P > G | Y | S + P | LF |
| Peach Spring Cyn, AZ | 16V | 8 | PW | LF | S > G | Y | S | LF + ME |

[a]Maximum and minimum thickness. Data from east of Kingman, Arizona, are from Valentine (per. commun.) and unpublished mapping by Buesch. All other data from Buesch (unpubl. mapping). Paleovalley fill (V), onlapping of topographic high (L), post-PST erosion (E).

[b]Zone of welding at the top of the PST. NW = nonwelded or co-ignimbrite fallout, PW = partially welded.

[c]UF = upper fan, LF = lower fan, AB = axial braided stream or regional trunk braided stream, LP = lacustrine/playa, ID = isolated depression, ME = mesa forming exposure, NE = not exposed.

[d]Grain size of the most abundant beds. G = gravel, P = pebbly sandstone, S = sandstone, F = fine to medium sandstone, M = mudstone, C = carbonate, B = basalt flow.

[e]Y = section has a fining- and thinning-upward sequence, N = no fining- and thinning-upward sequence. - = no information, mesa forming exposures or overlain by basalt.

which signifies that the distribution is a conservative minimum extent. At many exposures, the tuff pinches out from the local maximum to zero thickness (Table 1). Based on depositional and welding facies in the PST, most of these local thickness variations are interpreted as reflecting preexisting topography, with the tuff filling paleovalleys and onlapping paleomountain flanks. In several sections near Parker Dam, California, the lateral thinning is a result of post-depositional erosion of the PST. These pre-existing topographic and post-tuff erosional controls on the thickness of the PST ignimbrite are indicated in Table 1.

Where the PST filled a valley, the top of the tuff gently slopes toward the valley center and most of the slope resulted from compaction during welding, but minor onlapping also occurs. The pyroclastic flow that emplaced the PST must have been highly energetic in order to flow more than 170 km and surmount mountains more than 1 km in relief (Buesch, in prep). This indicates that nonwelded tuff was probably deposited on many of the topographically high areas where the ignimbrite was not preserved.

*Welding facies.—*

Regional and local variations occur in the zones of welding in the PST. The thickest zones of densely welded tuff are near the Colorado River area. The densely welded tuff is thin or completely absent in the distal exposures, a regional trend. Most PST has undergone devitrification and vapor-phase crystallization. Vitric zones always occur in the lower parts of the PST, but the upper vitric zone has been removed by erosion at many locations.

Local variations in the welding facies are superimposed on the regional trend and include "open valley," "edge," and "asymmetric" welding facies (Buesch and Valentine, 1986, 1989, and in prep.). These welding facies have been applied to many exposures throughout the PST and aid in reconstruction of paleovalleys and onlap relationships on the flanks of paleomountains.

*Vapor-phase lithification.—*

Lithification of non- to partially welded tuff resulting from vapor-phase crystallization is well developed in the upper parts of the Peach Springs Tuff and causes the formation of two cliff-forming horizons in the single cooling unit (Buesch and Valentine, 1986, and in prep.). The contact from devitrified and lithified tuff to slightly oxidized but vitric tuff can be very sharp (Fig. 2). Vapor-phase lithification occurred in the upper non- to partially welded zones, but was most dramatic in the nonwelded tuff.

A vapor-phase-lithified cap can form very near the top of the tuff and have an irregular, hummocky topographic surface. In the southeastern Black Mountains, Arizona, the vapor-phase-lithified cap occurs 2 to 10 m below the top of the PST, and the top of the cap has about 8 m of relief. Exposures near Kingman, Arizona, have a relief of about 5 m on the top of the vapor-phase-lithified tuff. Several of the mounds near Kingman are slightly orange or pink in color along irregular fracture surfaces and are interpreted as fossil fumaroles. The irregular geometry and alteration along fractures is comparable to the fossil fumarole mounds in the Bishop Tuff, California (Sheridan, 1970).

*Location of Vent.—*

A vent area for the Peach Springs Tuff has not been recognized and may have been obscured by faulting associated with Neogene extensional tectonics, volcanism or sedimentation. The proposed eruptive center for the PST (Fig. 1) is inferred from flow directions derived from field relationships (Buesch, in prep.), magnetic anisotropy studies (Hillhouse and Wells, 1989), and geochemical variations in phenocrysts (Buesch, 1989, and in prep.). Because of the limited distribution of exposures near the proposed eruptive center, confidence in this location is ±25 km and the northeast elongation of the ellipse may reflect tectonic strain across the CREC. The eruptive center either flanked the western margin of, or was within, the CREC.

FIG. 2.—Closeup of contact between vitric-nonwelded tuff (VNW) and vapor-phase-lithified tuff (VPL), in the southeastern Black Mountains. Tape = 50 cm.

## *Neogene Tectonics*

Tectonic activity during the last 25 Ma in southeastern California, southern Nevada, and southwestern Arizona has formed four highly extended tectonic regions, normal faults in the Transition Zone to the Colorado Plateau, and a strike-slip region in the Mojave Desert of California. At least five of the tectonic provinces were overlapped by the Peach Springs Tuff, and because the known distribution of the PST is a minimum, the tuff may extend across all of these domains.

*Extensional terranes.*—

The PST occurs in three, and possibly four, extensional terranes. Figure 1 illustrates the location of the four extensional terranes. The Central Mojave Rift (CMR; Dokka, 1986) experienced the greatest extension between 22 and 20 Ma. The CREC (Howard and John, 1987) extension was most active between 21.5 and 18 Ma (Nielson and Beratan, 1990). East of the CREC, the present Colorado Plateau and the associated Transition Zone have developed since the eruption of the PST (Young and Brennan, 1974). The Lake Mead extensional domain (LMED) was active 15 to 10 Ma (Smith and others, 1987; Weber and Smith, 1987). The terrane underlain by the Kingston Range detachment, herein referred to as the Kingston Range extensional terrane (KRET), has been active from 12 Ma until present (Burchfiel and Davis, 1988). The KRET does not have a well-defined southern boundary. Thus, PST exposures near Granite Springs and Old Dad Mountain, California, may be in the KRET, or in a relatively nonextended block between the KRET, CREC, and LMED, which is referred to here as the East Mojave Block (EMB).

*Strike-slip terrane.*—

Right-lateral strike-slip faulting has been active in the Mojave Desert of California (Dibblee, 1961). Geographically, the strike-slip terrane coincides with much of the CMR, but Dokka (1983, 1989) has argued that most of this strike-slip faulting is post-extension (<13 Ma).

## CRITERIA USED TO INFER CONTINUOUS SEDIMENTATION IN PRE- TO POST-PST SEDIMENTARY ROCKS

It is difficult to know how long a bed was exposed before burial, but the longer the sediment remained at the geomorphic surface, the more features developed that provide an estimate of the duration of exposure. Processes, such as formation of soil horizons, desert varnish, desert pavement, and lithification of tuff by carbonate and zeolite mineralization, may take hundreds to thousands of years to be recognizable in the stratigraphic record, especially in arid regions. Vapor-phase lithification of a tuff, rill erosion, and revegetation begin very shortly after deposition and peak activity continues for years to tens of years. Depositional and welding facies in the tuff can be critical in evaluating continuity of sedimentation from pre- to post-tuff rocks.

*Criteria between pre-tuff sediments and the tuff.*—

Continuity in sedimentation between the pre-tuff rocks and the PST is inferred where there is a lack of evidence for a hiatus in deposition. Three types of evidence that indicate an extended time of exposure of the pre-tuff rocks are 1) the development of pedogenic features (Bull, 1977; Birkeland, 1984; Retallack, 1986), 2) horizons of pebbles and cobbles that may reflect desert-pavement surfaces (Hooke, 1967; McFadden and others, 1987), and 3) rock varnish on the upper side of clasts that projected above the ground surface (Perry and Adams, 1978; Retallack, 1986; Dorn and Dickinson, 1989). Possible desert-payment surfaces and rocks with rinds of varnish do occur sporadically throughout the pre-PST stratigraphic section, but have not been observed on the surface directly beneath the PST.

Detailed pedological studies have not been done on the pre-PST rocks, but several sections have features that are suggestive of incipient soil development in the pre-tuff sediments. The most common feature in the pre-tuff sediment is an increase in red or orange color and a slight decrease in the preservation of textures in individual beds. These features are more strongly developed in the upper part of beds and grade downward to the parent rock. Locally, the color and textural changes cut across several beds and therefore are not primary sedimentary features. The increase in red color and disruption of primary sedimentary textures are consistent with the early stages of formation of an argillic B horizon (Buol and others, 1980; Birkeland, 1984). Carbonate nodules and rhizoliths occur at some locations, but are not common. These possible pedogenic features occur in varying degrees in the pre-PST stratigraphic section but are typically poorly developed. In several locations the PST was deposited on a surface that may have undergone incipient soil development. Even at these locations, there are usually some sediments that do not show any sign of soil development and are interpreted as being part of a geomorphic surface that was active just prior to deposition of the PST.

*Criteria within the tuff.*—

Post-tuff sediments that are deposited in response to the emplacement of a tuff should be deposited on nonwelded tuff, before the tuff can be eroded away or lithified. Many sections contain post-PST sediments deposited on lithified PST. Therefore, it is important to determine the type, timing, and cause of lithification of the nonwelded tuff. Nonwelded tuff can be lithified by a number of processes, which include: 1) cementation by calcite that is transported into the tuff by surface runoff or ground water, 2) cementation by zeolites and clays that can form essentially *in situ* from alteration of the vitric ash component (Fisher and Schmincke, 1984), and 3) vapor-phase crystallization. Calcite, clay, and zeolite formation is most likely to occur thousands to millions of years after deposition of the tuff (WoldeGabriel and Goff, 1989; WoldeGabriel, pers. commun.), whereas vapor-phase crystallization takes place very shortly after deposition. Sediments deposited on tuff that has been lithified by vapor-phase crystallization are interpreted to represent a hiatus in deposition because erosion must have removed the nonwelded vitric tuff. Approximation of the nonwelded texture and lack of vapor-phase crystallization can be made in the field, but petrography is very important for rocks that have broad gradational contacts between non- and partially welded tuff and for determining the lithifying component.

*Contact between nonwelded tuff and post-tuff sediments.*—

In many areas with high rainfall, revegetation of tuff can occur rapidly (see R. Smith, this volume), but in arid regions the recovery may be much slower. The lack of plant matter on top of the tuff may not be indicative of rapid burial. No fossil vegetation has been observed along the contact between the PST and post-PST sediments.

Collins and Dunne (1986) demonstrated that at Mount St. Helens the greatest erotion of the May 18, 1980, tephra by rills and channels occurred within a few years of the eruption. Rills and channels developed on the May 18, 1980, tephra are narrow and steep sided, and some of the rills were up to 1.5 m deep only 10 months after the eruption (Collins and Dunne, 1986). With time, individual rills became wider and the sidewall slope decreased.

In the PST, establishing lateral continuity of the co-ignimbrite fallout tuff or nonwelded tuff is one of the main criteria that indicates rapid burial by post-PST sediment, and therefore the inference of continuous sedimentation. Talus and colluvium obscure the PST/post-PST sediment contact in many locations. In the southern Newberry Mountains, California, the co-ignimbrite-fallout-tuff and post-tuff-sediment contact can be traced almost continuously for about 1 km along strike. In the Providence Mountains, California, the nonwelded-PST and post-PST-sediment contact can be traced intermittently over several kilometers. Lack of erosion into the tuff is established by the planar contact between the nonwelded and the post-PST sediment. In locations where no steep-sided rills are observed, continuous sedimentation is inferred.

## LITHOFACIES AND DEPOSITIONAL ENVIRONMENTS

Studies of modern alluvial fans (Hooke, 1967; Bull; 1977; Hubert and Filipov, 1989) and ancient alluvial fans (Bull, 1972; Gloppen and Steel, 1981; Mack and Rasmussen, 1984) show that rapid downfan decrease in grain size and slope of the fan surface and an increase in nonchannelized deposits are characteristic of aluvial-fan deposits. Comparison between fans illustrates that each is unique with respect to grain size and depositional processes. Differences in morphology, grain size, and types of deposits can occur on two fans that are only 1.5 km apart (Hooke, 1967).

In the southern California region during the late Oligocene to Pliocene, alluvial fans varied from debris-flow- and sheetflood-dominated deposition (Kerr, 1984), to sheetflood-dominated deposition (Ballance, 1984), to braided-stream-dominated deposition (Brady, 1984; Kerr, 1984; Link, 1984). Some alluvial fans changed from debris-flow-dominated deposition, to sheetflood-, hyperconcentrated-flood-flow- and braided-stream-dominated deposition (Hendrix and Ingersoll, 1987). The lower parts of most of these alluvial fans contain braided-stream deposits. These fan-related braided-stream deposits are differentiated from braided-stream deposits in an axial drainage system by differences in the grain size and paleoflow directions (Kerr, 1984; Link, 1984; Hendrix and Ingersoll, 1987). Locally, fan deltas (Brady, 1984; Link, 1984; Hendrix and Ingersoll, 1987), and braid deltas (Link, 1984; Hendrix and Ingersoll, 1987) developed in some of the basins. These examples illustrate some of the complexities associated with interpreting depositional environments of alluvial fans that formed in a similar climate, but differed in 1) the rocks exposed in the drainage basins, 2) the sizes of drainage basins, and 3) tectonically controlled sedimentation.

This study focuses on the stratigraphic interval within 20 m above and below the PST. Although there are extensive areas of early Miocene rocks exposed in the study area, there are only limited exposures (1 to 100 km$^2$) of rocks that were deposited very close in time to the PST, and these locations are typically 5 to 30 km apart. Thus, only relatively small parts of the depositional system are represented at each location. The emphasis in this paper is on the vertical changes between the pre- and post-PST sections with respect to the relative proportion of 1) gravel- versus sand-dominant deposits, 2) debris-flow- and sheetflow- versus channelized-braided-stream deposits, and 3) steep-sided, meter-deep versus sloping-sided, decimeter-deep channels.

The lithofacies terminology used in this study is modified from Miall (1978, 1985) and is summarized in Table 2. Architectural elements (Miall, 1985; Table 3) are used to infer the upper-fan, lower-fan, axial braided-stream, and lacustrine depositional environments that are represented in the pre- and post-PST sedimentary rocks (Table 4). Flow direction indicators include imbrication, clast long-axis orientation, cross-bedding in sandstone (Sp and Sr) and gravel (Gp) deposits, and the geometry of the channel walls. Ten to 40 flow directions were measured within a limited area, usually less than about 400 m$^2$. Where a direction of flow is defined by imbrication or cross-bedding, the nonunique indicators, such as long-axis orientation and the geometry of channel walls, are represented in the same quadrant on the rose diagrams.

*Upper-fan alluvial depositional environment.—*

Upper-fan alluvial deposits consist of channel-form (CH) and nonchannelized (SG and LS) deposits of massive to faintly stratified sandstone (Sm), interbedded with massive gravel and breccia beds (Bms, Gms, and Gm) that form ≥30 percent of the section (Table 4). Channels, which are up to 2 m deep and typically have near-vertical channel walls, are filled with interbeds of coarse-grained deposits (Bms, Gms, Gm, and minor amounts of Gp, and Gt) and sandstone (Sm, Sh, and minor amounts of Sp, St, and Ss). Locally, channel-form clast-supported conglomerates are capped by cross-bedded (St) or ripple-marked (Sr) sandstone. Some matrix-supported beds (Bms and Gms) and massive to faintly stratified sandstone beds (Sm) fill an entire channel, and overbank deposits form tabular beds of these lithofacies. Locally, matrix-supported deposits form flat-bottomed deposits that have a relief of 0.3 to 1 m above the substrate, forming constructional debris-flow lobes that are similar in shape and texture to the type I sediment flows of Lawson (1982). Textures, structures and interpretation of the lithofacies that comprise the upper-fan depositional environment (Table 4) are summarized in Table 2. Imbrication is poorly to moderately developed in about 30 percent of the coarse-grained sandstone (Sm and Sh) and gravel (Bms, Gms, and Gm) deposits. Each lithofacies has planar, nonerosive basal contacts, but eroded basal contacts with scour depths ≤1 dm are common, especially in the clast-supported deposits. Locally, possible pedogenic features (Table 2) indicate areas on the fan that may have remained inactive for hundreds to thousands of years.

Flow-direction indicators from the upper-fan deposits tend to vary less than 30° from the average direction, but where exposures are laterally extensive, a radial pattern is formed.

TABLE 2.—LITHOFACIES IN PRE- AND POST-PST SEDIMENTARY ROCKS

| Lithofacies Code[a] | Grain Size | Size Graded | Imbrication | | |
|---|---|---|---|---|---|
| | | | Stratification | Clast Orientation | Interpretation |
| Bms | Breccia, matrix-supported fine-grained sand to pebble matrix | Massive±reverse-graded base | Massive to poorly developed | None to moderately developed | Debris-flow deposits |
| Gms | Gravel, matrix-supported fine-grained sand to pebble matrix | Massive±reverse-graded base | Massive to poorly developed | None to moderately developed | Debris-flow deposits |
| Gm | Gravel, clast-supported, sand and pebble matrix, some beds are capped by Sr | Reverse-to-normal | Poor to moderately developed | None to moderately developed | Longitudinal bars, lag deposits, sieve deposits |
| Gt | Gravel, clast-supported, sand to pebble matrix | Massive, normal or reverse | Trough cross-beds | | Minor channel fills |
| Gp | Gravel, clast-supported, sand to pebble matrix | Massive, normal, or reverse | Planar cross-beds | | Linguiodal bars or deltaic growths from older bar |
| Sm | Sand, fine to v. coarse, commonly pebbly (also in conglomerate) | Massive, minor normal | None to horizontal, no cross-beds | Poor-imbrication pebbles, long-axis elongation | Hyperconcentrated-[b] flood-flow deposits |
| Sh | Sand, fine to v. coarse, commonly pebbly | None or normal | Horizontally laminated | None to poorly developed in coarse grains | Sheetflood, bar top, planar-bed flow (lower- and upper-flow regime) |
| Sp | Sand, medium to v. coarse, may be pebbly | None or normal | Planar cross-beds | None | Linguoidal, transverse bars, sand waves (lower-flow regime) |
| St | Sand, medium to v. coarse, may be pebbly | None or normal | Trough cross-beds | None | Dunes (lower-flow regime) |
| Sr | Sand, v. fine to coarse | Massive to faint reverse or normal | None to faint | Ripple marks of all types | Ripples (lower-flow regime) |
| Ss[c] | Sand, fine to coarse (rare intraclasts) | None to faint | Planar and trough | None | Scour fills |
| Fl | Sand, slit, mud | Massive | Laminated | None | Overbank or wanning flood deposits |
| Fm (Fmc) | Mud, silt (with dessication cracks) | Massive | Massive to laminated | None | Overbank or drape deposits |
| E | Carbonate | None | Massive to laminated | None | |
| T | Volcanic ash, fine to medium | Massive to normal | Massive | None | Fallout tuff, subaerial or subaqueous deposition |
| P | Possible pedogenic features[d]: red coloration, finer grained (more clay?) near top, concretionary carbonate, root clasts | | | | |

[a]Tuffaceous matrix composition is indicated by adding V to the Lithofacies Code. Symbols and interpretations after Miall (1985).
[b]After Smith (1986).
[c]Combines the Ss and Se lithofacies of Miall (1985).
[d]These are field observations and must be laboratory tested to confirm inferrence as to incipient paleosols.

TABLE 3.—ARCHITECTURAL ELEMENTS AND HIERARCHY OF BEDDING CONTACTS IN THE PRE- AND POST-PST SEDIMENTARY ROCKS

| Element | Symbol | Principal Lithofacies Assemblage | Geometry and Relationships |
|---|---|---|---|
| Channels | CH | Any combination | Lens; concave-up erosional base; scale and shape highly variable; internal concave-up secondary-erosion surfaces common |
| Gravel bars and bedforms | GB | Gm, Gp, Gt | Lens, blanket; usually tabular bodies; commonly interbedded with SB |
| Sand bedforms | SB | St, Sp, Sh, Sr, Ss | Lens, sheet, blanket, wedge; occurs as channel fills, minor bars |
| Foreset macroforms | FM | St, Sp, Sh, Sr | Lens resting on flat or channelled base, with convex-up second-order internal erosion surfaces and upper bounding surface |
| Lateral accretion deposits | LA | St, Sp, Sh, Sr, Ss | Wedge, sheet; characterized by internal lateral accretion surfaces |
| Sediment gravity | SG | Bms, Gms, Gm | Lobe, sheet; typically interbedded with GB |
| Laminated sand | LS | Sh; minor St, Sp, Sr | Sheet, blanket |
| Overbank fines | OF | Fm, Fl | Thin to thick blankets; commonly interbedded with SB; may fill abandoned channels |

Hierarchy of bedding contacts

| Order | Contacts bound— |
|---|---|
| first | individual cross-bed sets |
| second | co-sets or genetically related lithofacies |
| third | groups of elements or complexes, well-defined erosion surfaces |

TABLE 4.—CHARACTERISTICS OF THE DEPOSITIONAL ENVIRONMENTS IN PRE- AND POST-PST SEDIMENTARY ROCKS

| Characteristics | Alluvial Fan: Upper | Alluvial Fan: Lower | Axial Braided-Stream | Lacustrine |
|---|---|---|---|---|
| **Geomorphology** | | | | |
| Slope angle[a] | 5–15° | ≥5° | ≥0.5° | horizontal, local delta fronts |
| Channels | | | | |
| Morphology | steep sided; 40–90°, 1 or 2 main channels, others abandoned | swale; ≤40°, many, act as distributaries | variable shapes highly variable | subdued, few across small deltas |
| Depth | meters deep | ≤5 dm deep | ≤5 dm deep | decimeters |
| Flow directions[b] | systematic, ±30° radial from fanhead | moderately systematic, ± 60° overlapping fan-lobes | moderately systematic, ±60° high angles to fan | systematic, ±30° (near-braid delta) |
| **Deposits** | | | | |
| Lithofacies | | | | |
| Abundant | Bms, Gms, Gm, Sm | Sm, Sh, Sp, Sr | Sh, Sp, St | Fl, Fm, E, Sh, Sr |
| Minor | Gp, Gt, Sp, St, Ss, P | Gms, Gm, Ss, St, P | Sr, Fl, Fm, Fmc | Sp, T |
| Rare | | Bms | P | |
| Geometry | | | | |
| Thickness | Bms, Gms, Gm ≤ 2m Sm, Sp, St ≤ 2 dm | Gms, Gm, Bms ≤ 3 dm Sm, Sh, Sp ≤ 2 dm St ≤ 1 dm | Sh, Sp, St ≤ 2 dm Fl, Fm, Fmc ≤ 1 cm co-sets ≤ 2 dm | E, Sp ≤ 3 dm Fl, Fm, Fmc, T ≤ 1 cm co-sets ≤ 2 dm |
| Shape | mostly channel-form, some tabular (overbank), ±constructional lobes (debris-flow deposits) | sheetflood and hyperconcentrated-flood-flow deposits are tabular, braided-stream deposits are channel-form | mostly channel-form, minor tablular (overbank) | tabular, ±constructional delta deposits |
| Architectural elements | | | | |
| Abundant | CH, SG, GB > SB | LS ≈ CH > SB, SB | CH, SB, FM | OF ≥ SB |
| Minor | LS | FM, GB | GB, OF, LA, LS | FM |

[a]Slope angles for modern alluvial fans and braided rivers summarized in McPherson and others (1987). These values represent the inferred slopes in the lower Miocene depositional environments.
[b]Flow directions are from an area less than 400 $m^2$ and therefore present the local variations.

Radial flow directions are typical of alluvial-fan deposits (Nilsen, 1982), but if sampling is from a limited area of the fan, the variation in directions may be small (Kerr, 1984).

The abundance of coarse-grained clasts (in Bms, Gms, and Gm), angularity of clasts in the breccia, the occurrence of massive to faintly stratified sandstone and pebble conglomerates (Sm), and depth of channels indicate high-energy discharge of ephemeral streams that ranged from high to low sediment/water ratios (Miall, 1981; Lawson, 1982; Smith, 1986). Repeated incision and deposition are indicated by the abundance of channels and can result from active autocyclic repositioning of channels (Hooke, 1967; Schumm, 1981; Lawson, 1982; Hubert and Filipov, 1989).

*Lower-fan alluvial depositional environment.—*

Lower-fan alluvial deposits consist of beds with a sheet, or tabular, geometry (LS) and channel-form (CH) deposits of sandstone (Sm, Sh, Sp, and Sr) interstratified with coarse-grained lithofacies (Gms and Gm) that typically comprise ≤10 percent of the section (Table 4). Matrix-supported breccia (Bms) is rare and forms thin, tabular deposits. Channels are typically less than 0.5 m deep and have sloping, and locally steep-sided, channel walls. Most channels are filled with 1 to 5 beds below overbank deposits, forming tabular beds (LS) that are indicative of unchannelized deposits. Lithofacies Sp, Ss, St and Gm are typically confined to channels, whereas tabular beds of Sh, Sp, and rarely Ss, occur in sheetflood deposits, and Sm is deposited by hyperconcentrated-flood-flow deposits. The relative abundances, thicknesses and geometry of the lithofacies associated with the lower-fan deposits are summarized in Table 4. Possible pedogenic features (Table 2) occur in the lower-fan deposits, but are less frequent than in the upper-fan deposits. This decrease in surfaces that developed incipient soil horizons is attributed to the greater number of sheetflow events and a decrease in the amount of inactive area in the lower fan.

Flow-direction indicators from the lower-fan deposits tend to range between ±60° of the average direction. The increase in the variation of flow directions in the lower-fan compared to the upper-fan deposits is interpreted as overlapping of successive distributary lobes and an increase in the amount of braided-stream channels.

The abundance of sandstone versus conglomerate and shallow channels indicates a lower energy environment than in an upper fan (Miall, 1981). Interbedding of channel-form and sheetflood deposits demonstrates that both braided streams and sheet floods were active at different times on the same fan. The variation in the flow directions and the lack of well-developed pedogenic features suggest that lateral migration of the distributary channels commonly occurred and that parts of the fan surface did not remain inactive for long periods of time.

*Axial braided-stream depositional environment.—*

Axial braided-stream deposits are similar to the lower-fan braided-stream deposits except the axial deposits are slightly finer grained and have mean flow directions at a high angle to the adjacent fan-related deposits (Fig. 3). The axial braided-stream deposits consist of channel-form (CH) sandstone deposits (Sh, Sp, St and minor amounts of Sr) with minor amounts of fine-grained sandstone and mudstone (Fl, Fm, and Fmc) that form overbank and abandoned-channel-fill deposits (OF). Locally, pebble con-

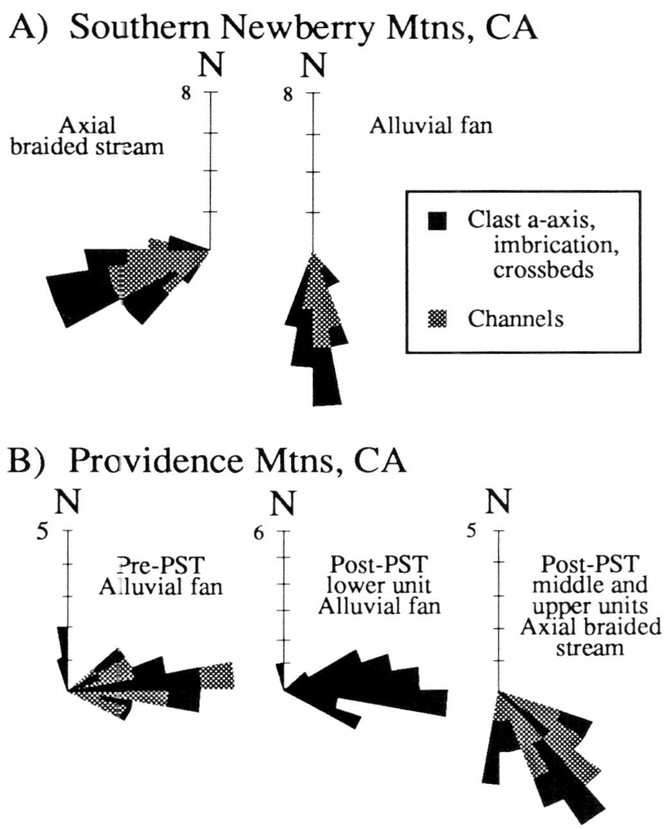


FIG. 3.—Paleoflow directions in the pre- and post-PST sedimentary rocks. All data from the alluvial-fan deposits are from the inner-fan depositional environments and represent the central axis of the fan (see text for discussion). Axial braided-stream deposits occur only in the southern Newberry Mountains and Providence Mountains, California. In all other locations the paleoflow directions trend in the same direction.

glomerates occur as part of the channel-filling sequence and probably represent lag deposits (GB). The minor amount of tabular, laminated sand deposits (LS) may result from the abundance of channels at a given horizon. Channels are ≤5 dm deep, and typically more than five beds fill a channel. The walls of one channel commonly truncate another channel to produce a broad, complex fill geometry that is typical of mobile channels (Friend, 1983). Locally, channels have foreset macroforms (FM). These macroforms, which consist of a sequence of Sh at the base, Sp, and St at the top, form cosets ≤2 dm thick that have second-order bedding contacts. Relative abundances, thicknesses, and bed geometry of the axial braided-stream deposits are summarized in Table 4.

The high percentage of sand, mobile channels, forset macroforms, and the rare occurrence of pedogenic features indicate that stream banks and beds were constantly being reworked by migrating channels that are typical of sandy braided-steam deposits (Miall, 1977, 1985). Axial braided-stream deposits form a very small part of the pre- and post-PST sediments in the southern Newberry Mountains, California. In the Providence Mountains, California, the axial braided-stream deposits occur only in the post-PST section, are laterally equivalent to limestone beds of the lacustrine environment and probably form braid deltas (McPherson and others, 1987), but deposits from shoreline environments are not exposed.

*Lacustrine to playa depositional environments.*—

Lacustrine to playa deposits consist of medium- to fine-grained sandstone and mudstone (Fl, Fm, Sr, and minor amounts of Sh) that are interbedded with limestone (E). Depending on the orientation of the section, individual lake beds extend for 50 m to 4 km. Thickness and relative abundance of lithofacies are summarized in Table 4. Partial Buoma sequences (b to e) occur in about 5 percent of the beds, and a few beds have a full sequence with pebble clasts at the base of the a division. Tuffaceous rocks form a small part of several sections and consist entirely of pyroclasts, or are admixed into epiclastic sediments. Tuff and tuffaceous sandstone beds have normal grading of lithic grains and reverse grading of pumice lapilli. Locally, laminated limestone beds contain algal mats and a few have stromatolitic structures (Beratan, pers. commun.). Some sandstone beds preserve desiccation cracks as casts at the base of the bed.

Deposition in a lake is indicated by the lateral continuity of thin beds, lack of erosional truncations or fluvial scour surfaces, the occurrence of limestone beds, and stromatolitic limestones. The massive-to-normal graded sandstone (Sh) and mudstone (Fl and Fm) were probably deposited from subaqueous suspension sedimentation, possibly as resedimented, dilute mass flow (Gloppen and Steel, 1981; Hendrix and Ingersoll, 1987). Bouma sequences were deposited from turbidity currents, some of which could have formed by the transition of subaerial debris flows after entering the lake (Weirich, 1989). Massive-to-normal graded tuffs were deposited by fallout tephra settling through standing water. Tuffaceous sandstone beds that have normal grading of lithic grains and reverse grading of pumice lapilli were deposited from sediment-gravity flows.

Clastic-sediment starved deposits including limestone, travertine, tufa, or calcareous siltstone are considered a subset of the lacustrine or playa depositional environment. The original aerial extent of these deposits is difficult to determine, but some of the deposits could have been as much as several square kilometers in extent. The lack of clastic grains indicates that they formed by debris-flow deposits (Kerr, 1984) or lava flows far from the basin margin, possibly in the center of the valley, or in an area isolated from the main stream channel. Lack of well-defined cross-laminations, which indicate deposition by traction currents, suggests that the siltstone represents distal eolianites formed by fallout of dust storms. Post-PST clastic-sediment-starved deposits could accumulate in isolated depressions that formed on top of the ignimbrite. On the basis of the relief on top of the vapor-phase-lithified cap in the PST, the isolated depressions could be 10 or more meters in relief.

*Regional summary.*—

Wide-spread distribution of the Peach Springs Tuff allows correlation between deposits in numerous basins that otherwise would be speculative. Figure 4 summarizes the inferred depositional environments for strata that occur within 20 m, and locally within 40 m, of the PST. A fining- and thinning-upward sequence through the pre- and post-Peach Spring Tuff sedimentary rocks occurs in 62 percent of the locations where both sections are preserved (Table 1). This sequence is developed at 10 of 12 locations where the top of the PST is nonwelded. The shift to finer grained rocks occurs where the post-tuff sediment is deposited on nonwelded, or partially welded and vapor-phase-lithified tuff (Figure 4). Several sections show the initial fining-upward sequence between pre- and post-PST sediments and continuation of the fining-upward sequence through the post-PST section (the Providence Mountains, for example). The regional prevalence of the fining- and thinning-upward sequence indicates that a fundamental control on the post-tuff sedimentation was the deposition of the tuff, and locally the control lasted long enough for the nonwelded sediment to be eroded. If the fining-upward sequence were simply a result of the time needed for a fan to prograde over the PST, then the initial pre- to post-PST fining sequence would be followed by a coarsening-upward sequence. Coarsening-upward sequences do not occur; therefore, another process controlled distribution of the post-PST depositional environments.

## MODEL

Studies of Miocene basins in southern California (Brady, 1984; Link, 1984; Hendrix and Ingersoll, 1987; Nielson and Beratan, 1990) indicate that many basins were isolated from adjacent basins throughout much of their history, and the occurence of ephemeral lakes suggests local control of base level. The abundance of lower-fan environments represented in the pre-PST rocks (Fig. 4) indicates that alluvial fans formed the dominant geomorphic surface in early Miocene valleys. Thus, the changes between pre- and post-PST depositional environments result from deposition of a large-volume welded tuff in a semi-arid climate, across varied topography that included high-relief mountain ranges, broad alluviated valleys, and internally drained basins.

*Changes in stream gradient.—*

One of the most fundamental changes that resulted from deposition of the PST was a decrease in the local stream gradient (Fig. 5A). The size of the "local" area varies from tens to hundreds of square kilometers. The influence of the valley-filling PST ignimbrite on stream gradient depended on the area and topography of the basin, whether the basin was internally drained or integrated into a regional drainage

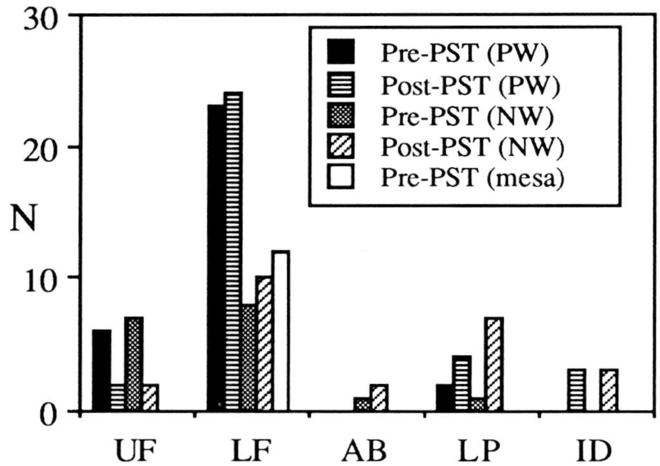


FIG. 4.—Frequency (N) of pre- versus post-PST depositional environments. Data presented in this histogram are from individual sections, but where two sections from the same location contain the same stratigraphic sequence, only one section is used. Where a location has both nonwelded and partially welded tuff at the top of the PST, and more than one depositional environment is represented in the pre- or post-PST rocks, each sequence occurs in the histogram. From the 45 PST locations, 71 stratigraphic sequences are represented on the figure. See Table 1 for symbols of depositional environments and zones of welding at the top of the PST. PW includes vapor-phase lithified tuff and Mesa = ME of Table 1.

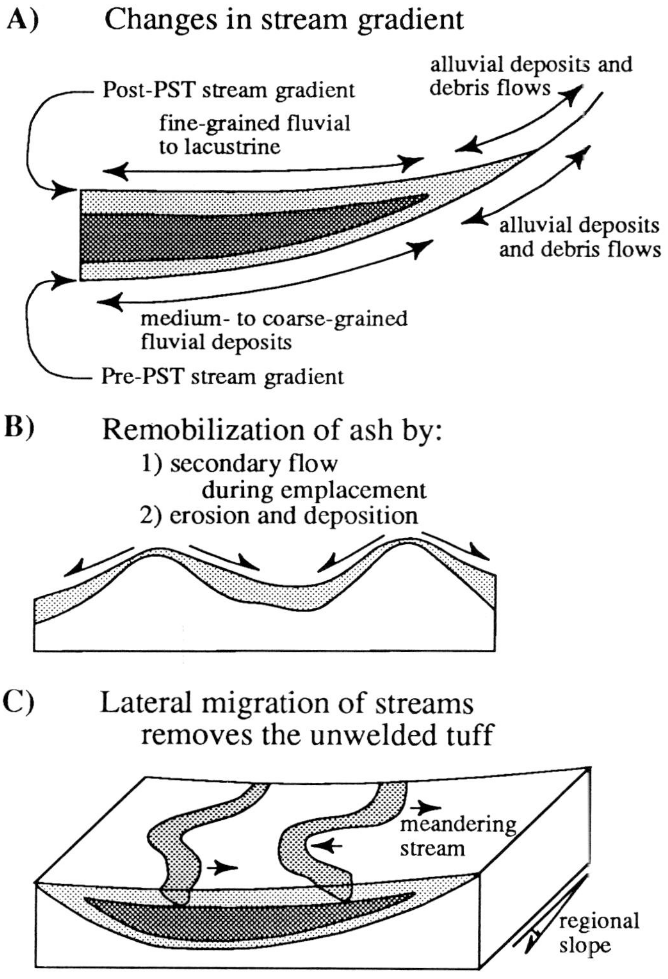


FIG. 5.—(A) Lower stream gradient and the associated shifts in the position of depositional environments result from a thick deposit of tuff in the valley. Formation of a densely welded horizon will prolong this change in gradient. (B) Processes of remobilizing ash from hillslopes. (C) Lateral migration of streams due to formation of a well-developed welding profile. In each diagram, light gray = nonwelded tuff and dark gray = partially to densely welded tuff. See text for discussion

system, the amount of tuff deposited, and the degree of welding of the tuff. Where the PST was deposited in a narrow valley between high-relief mountains, the local stream gradient was relatively unaffected and the type of sedimentation did not change from pre- to post-tuff environments (insert in Fig. 6).

Where the PST was deposited in broad valleys, especially if internally drained, a large decrease in the local stream gradient and base level caused significant pre- to post-tuff changes in the depositional environment. Along the mountain front and valley margin, deposition of the PST buried the pre-tuff stream gradients and formed a new gradient with a lower slope (Fig. 5A). As post-PST sedimentation adjusted to the lowered stream gradient, fans prograded across the top of the tuff. The pre-tuff depositional environment was re-established in only two of the 12 locations, where post-tuff sediments were deposited on nonwelded PST, and in 11 of the 27 locations, where post-tuff sediments were deposited on partially welded PST (Table 1). The pre- to post-PST shift to finer grained deposits is demonstrated by the fining- and thinning-upward sequences at many locations (Table 1, Fig. 4). Away from the valley margin and in the center of broad valleys (Fig. 6), the stream gradient may have been so low that only the finest grained material, if any at all, was delivered to the area. Differential compaction of the the tuff formed isolated depressions in which limestone and carbonaceous siltstone accumulated.

*Remobilization of ash.—*

Ash from a pyroclastic flow can be remobilized during emplacement (Buesch, in prep.), or reworked from the adjacent hillsides (Fig. 5B). Because nonwelded tuff can cover much, if not all, of a mountain, the formation of tuffaceous debris-flow deposits records the rapid stripping of the nonwelded tuff from the adjacent hillsides. Decrease in the tuffaceous component in the first 10 to 15 m of post-PST sediment in many locations indicates that the influx of reworked tuff into sediments decreased quickly after initial deposition on the hillside.

*Preservation of the nonwelded tuff.—*

Preservation of the nonwelded part of an ignimbrite or the co-ignimbrite fallout requires burial by sediment soon after tuff deposition. High-relief paleotopographic settings adjacent to areas with low stream gradients result from pre-tuff topography, or a change in base level due to deposition of the tuff. In these areas, debris flows deposited large quantities of tuffaceous material on the nonwelded tuff. Many debris flows did not erode into the substrate during emplacement; thus, the otherwise easily eroded nonwelded tuff was blanketed by a rapid accumulation of material.

Shallow lakes, even ephemeral lakes, helped preserve the nonwelded tuff by reducing the area susceptible to deflation by wind and depositing a protective layer of sediment that would need to be removed before any of the nonwelded tuff could be eroded. Even if the amount of water in the basin was insufficient to form lakes, shallow groundwater levels could reduce deflation of the nonwelded tuff. Development of shallow lakes and rise of the groundwater table were probably augmented by the formation of a densely welded horizon in the basin-filling tuff that acted as an aquaclude to form a perched water table. Locally, zeolitization of the

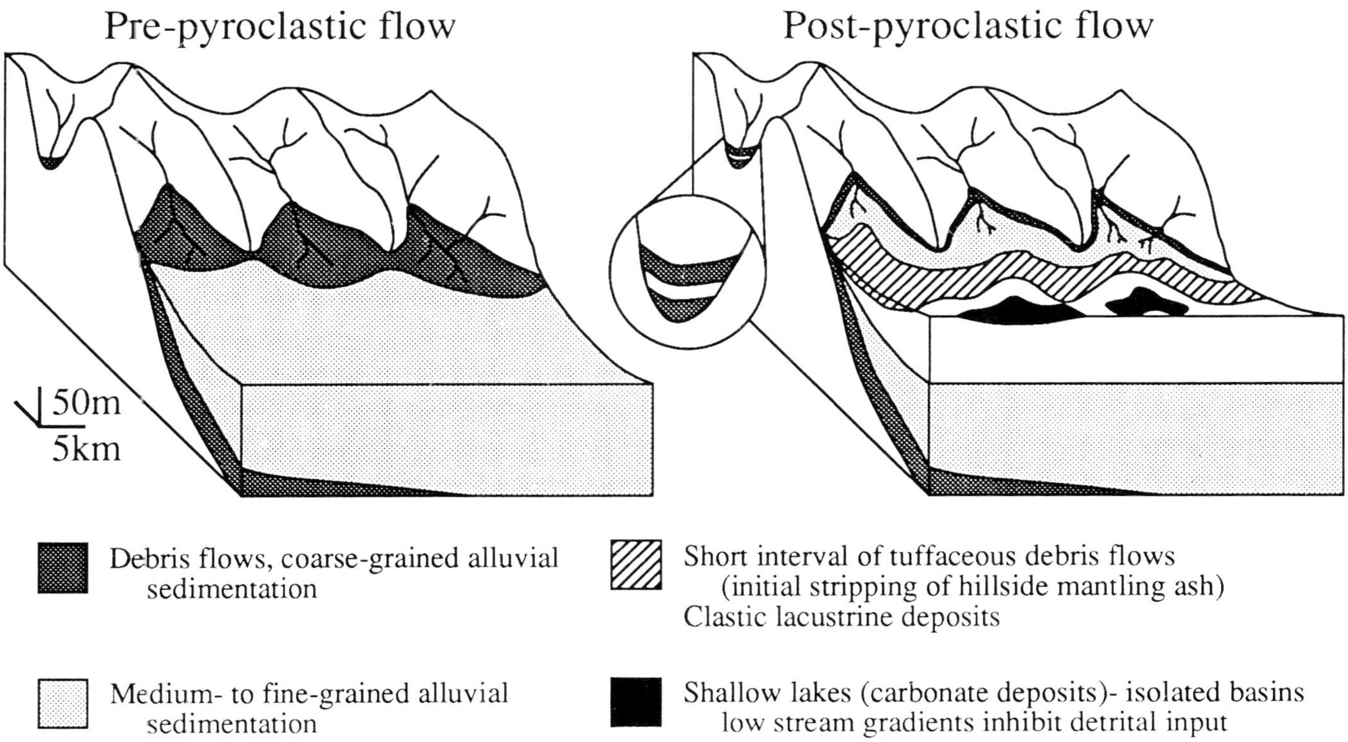


FIG. 6.—Block diagram illustrating pre- and post-PST depositional environments.

nonwelded tuff indicates shallow groundwater alteration, and although the timing of this alteration has not been established, alteration and carbonate deposition in a shallow lake can be coincident (Sheppard and Gude, 1973).

*Erosion of the nonwelded tuff.—*

Deflation could have eroded the nonwelded tuff from any location. The only evidence that suggests deflation from post-PST deposits occurred is the probable aeolianite, laminated, very fine-grained sandstones in sediment-starved depressions.

Water runoff stripped the tuff from hillsides and formed the debris-flow deposits in the alluvial-fan deposits. Where stream-flow formed braided streams in the lower-fan and axial braided-stream environments, channel incision and migration eroded the nonwelded tuff. Lateral migration has been discussed in sandy braided streams (Walker and Cant, 1984). Miall (1985, p. 298) described several "misleading assumptions" about river processes that include "braided streams are always of low sinuosity." The sinuosity of braided-stream deposits exposed at many locations is difficult to demonstrate. Direct evidence of intermediate to high sinuosity and lateral migration of stream channels is lacking; but, the extensive stripping of the nonwelded tuff from the top of the PST suggests that lateral migration, or meandering, of the stream occurred prior to, or concurrently with, deposition of the post-PST sediment.

On the Colorado Plateau, meanders develop where the stream gradient is low and lateral migration of the meanders occurs where the rocks are not resistant to erosion (Harden, 1990). This condition was met in valleys where deposition of the PST caused the stream gradient to lower, the welding profile was well developed, and vapor-phase lithification occurred in the non- to partially welded tuff. Where groundwater moves along the contact between nonwelded tuff, erosionally resistant tuff seeps form along the stream bank. Howard and McLane (1988) showed that sapping is very effective at eroding and undermining cliff faces of soft, cohesionless material. Nonwelded tuff can be approximated as a cohesionless material. Where easily eroded nonwelded tuff was underlain by erosionally resistant, partially to densely welded tuff, or vapor-phase-lithified tuff, lateral migration of streams removed to nonwelded tuff (Fig. 5C).

## DISCUSSION

The large-volume Peach Springs Tuff was deposited in a semi-arid climate across variable topography with many internally drained basins. This combination of variables does not make the PST unique, but it does provide contrasts with other studies of sedimentation in volcanic terranes.

Most sedimentologic studies related to volcanic eruptions involve the sedimentation and reworking of pyroclasts (Kuezni and others, 1979; Collins and Dunne, 1986; Smith, 1987, 1988; Waresback and Turbeville, 1990; R. Smith, this volume). Post-PST sediments are mostly epiclastic grains with admixed ash, which was eroded from the nonwelded tuff deposited on the highlands, and this influx of ash was highest during accumulation of the first 10 to 15 m of the post-PST section.

Deposition of the PST in many isolated basins resulted in the dominance of local processes on post-PST sedimentation that were highly dependent on the basin geometry. This contrasts with regionally integrated river systems that exhibit aggradation and incision cycles for 50 to 100 km in response to volcanic eruptions, and the erosion and resedimentation of the initial deposits (Smith, 1987, and this volume). Regionally integrated drainages are graded to a base level that does not change as a result of deposition of a thick tuff; thus, headward erosion through the tuff would be expected to occur, even in densely welded tuff.

Deposition of the PST in internally drained basins probably caused the base level to rise and significantly changed the local stream gradients in the basins. Leopold and Bull (1979) showed that for a stream gradient of 0.045, upstream aggradation occurred in the horizontal:vertical ratio of 55:1. Stream gradients of 0.017 to 0.087 (1°–5°) are common in the lower parts of alluvial fans (Table 4) and numerous PST deposits are 60 to 100 m thick, with a few sections up to 200 m thick (Table 1). Assuming the top of the tuff forms a flat, horizontal surface, and 100 m of tuff were deposited in a basin, the change in stream gradient and the resulting aggradation of the stream occurs for 5.5 km upstream. The increase in stream gradient in the upper-fan environment limits the upstream aggradation. Momentum of a pyroclastic flow can deposit ignimbrite on slopes at a higher elevation than the top of the ignimbrite in the basin center. This mantling of topography, which locally occurs in the PST, aids the aggradation of the stream. Even a change in stream gradient that causes a shift in the depositional environment of several kilometers can result in the vertical fining- and thinning-upward sequence of the pre- and post-PST sediments.

Vapor-phase lithification has not been discussed very much in the volcanologic literature, but may be an important factor in the erosion of a welded tuff and deposition of subsequent sediments. Because vapor-phase lithification can occur within meters of the top of a tuff, the formation of a well-developed vapor-phase-lithified cap can significantly reduce the amount of tuff that is easily removed by erosion. It is difficult to estimate the volume of nonwelded and nonlithified tuff that may have been eroded from the PST. An example where a reasonable estimate can be made is the Bishop Tuff, California, where 150 $km^3$ of ignimbrite was originally exposed across 1,300 $km^2$ (Sheridan, 1970; Hildreth, 1979). Sheridan (1970) reports fossil fumarole mounds up to 30 m above the surrounding top of the Bishop Tuff, but most mounds average 10 m in relief. Assuming an average thickness of 15 m, 20 $km^3$ of nonwelded, nonlithified tuff occurred at the top of the Bishop Tuff. This volume is less than the volume of the Taupo ignimbrite discussed by R. Smith, (this volume). Even though an eruption produces a large volume of tephra, the formation of welded or vapor-phase lithified tuff may limit the easily eroded, nonwelded, nonlithified tuff to a fraction of the original volume of tephra.

## CONCLUSIONS

Interbasin correlation of sedimentary rocks and depositional environments has always been the goal and challenge of stratigraphers. The widespread distribution of the Peach Springs Tuff provides a rare opportunity to document sedimentary facies in a number of coexisting basins and depositional environments. Regions of high local relief show no change between pre- to post-PST depositional environments, whereas the margins and central parts of broad valleys show a shift to more lower-fan and lacustrine-to-playa depositional environments. A shift to finer grained rocks in post-PST sediments probably resulted from the rising of the local base level as a result of partial basin filling by the PST. Deposition of the PST in internally drained basins caused the local base level to rise and resulted in a lowering of stream gradients. Lower stream gradients promoted the lateral migration of streams and eroded the nonwelded top of the PST.

Formation of a well-developed welding profile, or vapor-phase-lithified cap, influenced the response of the depositional system. Partially to densely welded tuff is more resistant to erosion; therefore, shallow stream gradients were prolonged. Erosion of the nonwelded tuff in valley-filling areas was promoted by lateral migration of streams where erosion into the soft, nonwelded tuff required less stream energy than to cut down into the erosionally resistant tuff. Sapping from springs formed at the contact between nonwelded and erosionally resistant tuff probably promoted the lateral erosion and migration of the stream channel.

Continuous sedimentation in a section is inferred where there is 1) a lack of evidence for a hiatus in deposition between the PST and the pre- and post-tuff sediments, and 2) deposition of post-PST sediment was on nonwelded PST. Deposition on an active geomorphic surface is indicated where pedogenic features, buried desert pavements, and varnished surfaces are lacking. The criteria used to infer continuous post-PST sedimentation without prolonged periods of erosion, or simply no deposition, include: 1) deposition on nonwelded tuff or co-ignimbrite fallout tuff, 2) cementation of nonwelded tuff by calcite or zeolites that probably formed after burial, and 3) lack of rills or channels eroded into the nonwelded tuff. A break in sedimentation, with an unknown period of erosion, is inferred where post-PST sediment overlies vapor-phase-lithified tuff, or rills and channels formed on the nonwelded tuff.

ACKNOWLEDGMENTS

Field and technical support for this work was provided by Jane Nielson of the U.S. Geological Survey, and grants from the Geological Society of America, the UCSB Graduate Division, the UCSB Department of Geological Sciences, and NSF Grant EAR-16532 to R. V. Fisher. I thank all people who showed me their favorate PST locations. Greg Valentine, Jane Nielson, Alan Glazner, Keith Howard, Ray Wells, Jack Hillhouse, and Kathi Beratan provided support and informative discussions. Kathi Beratan reviewed an early manuscript, and Eugene Smith and Gary Smith provided constructive reviews. I am indebted to Greg Valentine for use of data from areas east of Kingman, and for helping formulate ideas on the various facies within the PST. R. V Fisher deserves my greatest appreciation for his support, advice and for asking questions to which I did not have the answers.

REFERENCES

Ballance, P. F., 1984, Sheet-flow-dominated gravel fans of the non-marine middle Cenozoic Simmler Formation, central California: Sedimentary Geology, v. 38, p. 337–359.

Birkeland, P. W., 1984, Soils and Geomorphology: Oxford University Press, New York, 372 p.

Brady, R. H., III, 1984, Neogene stratigraphy of the Avawatz Mountains between the Garlock and Death Vallley fault zones, southern Death Valley, California: Implications as to the late Cenozoic tectonism: Sedimentary Geology, v. 38, p. 127–157.

Buesch, D. C., 1987a, Compositional stratification of the Peach Springs Tuff near Kingman, Arizona: Geological Society of America Abstracts with Programs, v. 19, p. 362.

Buesch, D. C., 1987b, Stratification in layer 2 deposits in the Peach Springs Tuff, Arizona: Response to topography and lateral variations: International Union of Geodesy and Geophysics, XIX General Assembly, Vancouver, Canada, August 9–22, Abstract Volume 2, p. 394.

Buesch, D. C., 1989, Geochemical variations in ignimbrite phenocrysts: Implications for eruption history and vent location: Transactions, EOS, American Geophysical Union, v. 70, p. 1413.

Buesch, D. C., and Valentine, G. A., 1986, Peach Springs Tuff and the volcanic stratigraphy of the southern Cerbat Mountains, Kingman Arizona: Geological Society of America, Cordilleran Section, Field Trip Guide book p. 7–14.

Buesch, D. C., and Valentine, G. A., 1989, Thickness and flow dynamics as factors controlling welding variations in ignimbrites: International Association of Volcanology and Chemistry of the Earth's Interior, General Assembly, Santa Fe, New Mexico, USA, June 25–July 1: New Mexico Bureau of Mines & Mineral Resources Bulletin 131, p. 32.

Bull, W. A., 1972, Recognition of alluvial-fan deposits in the stratigraphic record, *in* Rigby, J. K., and Hamblin, W. K., eds., Recognition of Ancient Sedimentary Environments: Society of Economic Paleontologists and Mineralogists Special Publication 16, p. 63–83.

Bull, W. A., 1977, The alluvial fan environment: Progress in Physical Geography, v. 1, p. 222–270.

Buol, S. W., Hole, F. D., and McCraken, R. J., 1980, Soil genesis and classification: Ames, Iowa, Iowa State University Press, 406 p.

Burchfiel, B. C., and Davis, G. A., 1988, Mesozoic thrust faults and Cenozoic low-angle normal faults, eastern Spring Mountains, Nevada, and Clark Mountains thrust complex, California, *in* Weide, D. L., and Farber, M. L. eds., This Extended Land, Geological Journeys in the Southern Basin and Range: Geological Society of America, Cordilleran Section, Field Trip Guidebook, p. 87–106.

Collins, B. D., and Dunne, Thomas, 1986, Erosion of tephra from the 1980 eruption of Mount St. Helens: Geological Society of America Bulletin, v. 97, p. 896–905.

Dibblee, T. W., 1961, Evidence of strike-slip movement on northwest-trending faults in the Mojave Desert, California: U.S. Geological Survey Professional Paper 424-B, p. B197–B199.

Dokka, R. K., 1983, Displacements on late Cenozoic strike-slip faults of the central Mojave Desert, California: Geology, v. 11, p. 305–308.

Dokka, R. K., 1986, Patterns and modes of early Miocene crustal extension, central Mojave Desert, California, *in* Mayer, L., ed., Extensional Tectonics of the Southwestern United States: A Perspective on Processes and Kinematics: Geological Society of America Special Paper 208, p. 75–96.

Dokka, R. K., 1989, The Mojave extensional belt of southern California: Tectonics, v. 8, no. 2, p. 363–390.

Dorn, R. I., Dickinson, W. R., 1989, First paleoenvironmental interpretation of a pre-Quaternary rock-varnish site, Davidson Canyon, southern Arizona: Geology, v. 17, p. 1029–1031.

Faulds, J. E., Hillmeyer, F. L., and Smith, E. I., 1988, Geometry and kinematics of a Miocene "Accommodation Zone" in the central Black and southern Eldorado Mountains, Arizona and Nevada, *in* Weide, D. L., and Farber, M. L., eds., This Extended Land, Geological Journeys in the Southern Basin and Range: Geological Society of America, Cordilleran Section, Field Trip Guidebook, p. 293–310.

Fisher, R. V., and Schmincke, H.-U., 1984, Pyroclastic Rocks: Berlin, Springer-Verlag, 472 p.

Friend, P. F., 1983, Towards the field classification of alluvial architecture or sequence: International Association of Sedimentologists Special Publication 6, p. 345–354.

GLAZNER, A. F., NIELSON, J. E., HOWARD, K. A., AND MILLER, D. M., 1986, Correlation of the Peach Springs Tuff, a large-volume Miocene ignimbrite sheet in California and Arizona: Geology, v. 14, p. 840–843.

GLOPPEN, T. G., AND STEEL, R. J., 1981, The deposits, internal structure and geometry in six alluvial fan-fan delta bodies (Devonian-Norway)—a study in the significance of bedding sequence in conglomerates, *in* Ethridge, F. G., and Flores, R. M., eds, Recent and Ancient Non-Marine Depositional Environments: Models for Exploration: Society of Economic Paleontologists and Mineralogists Special Publication 31, p. 49–69.

GUSA, S., NIELSON, J. E., AND HOWARD, K. A., 1987, Heavy mineral suites confirm the wide extent of the Peach Springs Tuff in California and Arizona, U.S.A.: Journal of Volcanology and Geothermal Research, v. 33, p. 343–347.

HARDEN, D. R., 1990, Controlling factors in the distribution and development of incised meanders in the central Colorado Plateau: Geological Society of America Bulletin, v. 102, p. 233–242.

HENDRIX, E. D., AND INGERSOLL, R. V., 1987, Tectonics and alluvial sedimentation of the upper Oligocene/lower Miocene Vasquez Formation, Soledad basin, southern California: Geological Society of America Bulletin, v. 98, p. 647–663.

HILDRETH, W., 1979, The Bishop Tuff: Evidence for the origin of compositional zonation in silicic magma chambers, *in* Chapin, C. E., and Elston, W. E., eds., Ash-Flow Tuffs: Geological Society of America Special Paper 180, p. 43–76.

HILLHOUSE, J. W., AND WELLS, R. E., 1989, Magnetic fabric and source area of the lower Miocene Peach Springs Tuff in Calif., Nevada, and Arizona (abst.): EOS, Transactions, American Geophysical Union, v. 70. no. 43, p. 1067.

HOOKE, R. L., 1967, Processes on arid-region alluvial fans: Journal of Geology, v. 75, p. 438–460.

HOWARD, A. L., AND MCLANE, C. F., III, 1988, Erosion of cohesionless sediment by ground water: Water Resources Research, v. 24, p. 1659–1674.

HOWARD, K. A., AND JOHN, B. E., 1987, Crustal extension along a rooted system of imbricate low-angle faults: Colorado River extensional corridor, California and Arizona, *in* Coward, M. P., Dewey, J. F., and Hancock, P. L., eds., Continental Extensional Tectonics: Geological Society of London Special Publication 28, p. 299–311.

HUBERT, J. F., AND FILIPOV, A. J., 1989, Debris-flow deposits in alluvial fans on the west flank of the White Mountains, Owens Valley, California, U.S.A.: Sedimentary Geology, v. 61, p. 177–205.

KERR, D. R., 1984, Early Neogene continental sedimentation in the Vallecito and Fish Creek Mountains, western Salton Trough, California: Sedimentary Geology, v. 38, p. 217–246.

KUENZI, W. D., HORST, O. H., AND MCGEHEE, R. V., 1979, Effect of volcanic activity on fluvial-deltaic sedimentation in a modern arc-trench gap, southwestern Guatemala: Geological Society of America Bulletin, v. 90, p. 827–838.

LAWSON, D. L., 1982, Mobilization, movement and deposition of active subaerial sediment flows, Matanuska Glacier, Alaska: Journal of Geology, v. 90, p. 279–300.

LEOPOLD, L. B., AND BULL, W. B., 1979, Base level, aggradation and grade: Proceedings of the American Philosophical Society, v. 123, p. 168–202.

LINK, M. H., 1984, Fluvial facies of the Miocene Ridge Route Formation, Ridge Basin, California: Sedimentary Geology, v. 38, p. 263–285.

MACK, G. H., AND RASMUSSEN, K. W., 1984, Alluvial-fan sedimentation of the Cutler Formation (Permo-Pennsylvanian) near Gateway, Colorado: Geological Society of America Bulletin, v. 95, p. 109–116.

MCFADDEN, L. D., WELLS, S. G., AND JERCINOVICH, M. J., 1987, Influences of eolian and pedogenic processes on the origin and evolution of desert pavements. Geology, v. 15, p. 504–508.

MCPHERSON, J. G., SHANMUGAM, G., MOIOLA, R. J., 1987, Fan deltas and braid deltas: Varieties of coarse-grained deltas: Geological Society of America Bulletin, v. 99, p. 331–340.

MIALL, A. D., 1977, A review of the braided river depositional environment: Earth Science Reviews, v. 13, p. 1–62.

MIALL, A. D., 1978, Lithofacies types and vertical profile models in braided river deposits: A summary, *in* Miall, A. D., ed., Fluvial Sedimentology: Canadian Society of Petroleum Geologists Memoirs, 5, p. 597–604.

MIALL, A. D., 1981, ed., Sedimentation and tectonics in alluvial basins: Geological Association of Canada Special Paper 23, 272 p.

MIALL, A. D., 1985, Architectural-element analysis: A new method of facies analysis applied to fluvial deposits: Earth-Science Reviews, v. 22, p. 261–308.

NIELSON, J. E., AND BERATAN, K. K., 1990, Tertiary basin development and tectonic implications, Whipple detachment system, Colorado River extensional corridor, California and Arizona: Journal of Geophysical Research, v. 95, p. 599–614.

NIELSEN, J. E., LUX, D. R., DALRYMPLE, G. B., AND GLAZNER, A. F., 1990, Age of Peach Springs Tuff, southeastern California and western Arizona: Journal of Geophysical Research, v. 95, p. 571–580.

NILSON, T. H., 1982, Alluvial fan deposits, *in* Scholle, P. A., and Spearing, D., eds., Sandstone Depositional Environments: American Association of Petroleum Geology, Tulsa, Oklahoma, p. 49–86.

PERRY, R. S., AND ADAMS, J. B., 1978, Desert varnish: Evidence of cyclic deposition of manganese: Nature, v. 278, p. 489–491.

RETALLACK, G. J., 1986, The fossil record of soils, *in* Wright, V. P., ed., Paleosols: Their recognition and interpretation: Oxford, England, Blackwell, p. 1–57.

SCHUMM, S. A., 1981, Evolution and response of the fluvial system, sedimentological implications *in* Ethridge, F. G., and Flores, R. M., eds., Recent and Ancient Nonmarine Depositional Environments: Models for Exploration: Society of Economic Paleontologists and Mineralogists Special Publication 31, p. 19–30.

SHEPPARD, R. A., AND GUDE, A. J., III, 1973, Zeolites and associated minerals in tuffaceous rocks of the Big Sandy Formation, Mohave County, Arizona: U.S. Geological Survey Professional Paper 830, 36 p.

SHERIDAN, M. F., 1970, Fumarolic mounds and ridges of the Bishop Tuff, California: Geological Society of America Bulletin, v. 81, p. 851–868.

SMITH, E. I., ANDERSON, R. E., BOHANNON, R. G., AND AXEN, G., 1987, Miocene extension, volcanism, and sedimentation in the eastern Basin and Range province, southern Nevada, *in* Davis, G. N., and VanderDolder, E. M., eds., Geologic Diversity of Arizona and Its Margins: Excursions to Choice Areas: Arizona Bureau of Geology and Mineral Technology Special Paper 5, p. 383–398.

SMITH, G. A., 1986, Coarse-grained nonmarine volcaniclastic sediment: Terminology and depositional process: Geological Society of America Bulletin, v. 97, p. 1–10.

SMITH, G. A., 1987, Sedimentology of volcanism-induced aggradation in fluvial basins: Examples from the Pacific Northwest, U.S.A., *in* Ethridge, F. G., Flores, R. M., and Harvey, M. A., eds., Recent Developments in Fluvial Sedimentology: Society of Economic Paleontologists and Mineralogists Special Publication 39, p. 217–228.

SMITH, G. A., 1988, Sedimentology of proximal to distal volcaniclastics dispersed across an active foldbelt: Ellensburg Formation (late Miocene), central Washington: Sedimentology, v. 35, p. 953–977.

SMITH, R. D., AND SWANSON, F. J., 1987, Sediment routing in a small drainage basin in the blast zone at Mount St. Helens, Washington, U.S.A.: Geomorphology, v. 1, p. 1–13.

SMITH, R. L., 1960, Zones and zonal variations in welded ash flows: U.S. Geological Survey Professional Paper 354-F, F149–159.

SPARKS, R. S. J., SELF, S., AND WALKER, G. P. L., 1973, Products of ignimbrite eruption: Geology, v. 1, p. 115–118.

SPENCER, J. E., AND REYNOLDS, S. J., 1986, Some aspects of middle Tertiary tectonics of Arizona and southeastern California: Arizona Geological Society Digest, v. 16, p. 102–107.

VALENTINE, G. A., BUESCH, D. C., AND FISHER, R. V., 1989, Basal layered deposits of the Peach Springs Tuff, northwestern Arizona, USA: Bulletin of Volcanology v. 51, p. 395–414.

WALKER, R. G., AND CANT, D. J., 1984, Sandy fluvial systems, *in* Walker, R. G., ed., Facies Models: Geological Association of Canada Publishers, Toronto, Ontario, p. 71–91.

WARESBACK, D. B., AND TURBEVILLE, B. N., 1990, Evolution of a Plio-Pleistocene volcanogenic-alluvial fan: The Puye Formation, Jemez Mountains, New Mexico: Geological Society of America Bulletin, v. 102, p. 298–314.

WEBER, M. E., AND SMITH, E. I., 1987, Structural and geochemical constraints on the reassembly of disrupted mid-Miocene volcanoes in the

Lake Mead-Eldorado Valley area of southern Nevada: Geology, v. 15, p. 553–556.

WEIRICH, F. H., 1989, The generation of turbidity currents by subaerial debris flows, California: Geological Society of America Bulletin, v. 101, p. 278–291.

WELLS, R. E., AND HILLHOUSE, J. W., 1989, Paleomagnetism and tectonic rotation of the lower Miocene Peach Springs Tuff: Colorado Plateau, Arizona, to Barstow, California: Geological Society of America Bulletin, v. 101, p. 846–863.

WOLDEGABRIEL, G., AND GOFF, F., 1989, Temporal relations of volcanism and hydrothermal systems in two areas of the Jemez volcanic field, New Mexico: Geology, v. 17, p. 986–989.

YOUNG, R. A., 1966, Cenozoic geology along the edge of the Colorado Plateau in northwestern Arizona: St. Louis, Washington University, unpubl. Ph.D. Dissertation, 167 p.

YOUNG, R. A., AND BRENNAN, W. J., 1974, The Peach Springs Tuff—Its bearing on the structural evolution of the Colorado Plateau and development of Cenozoic drainage in Mohave County, Arizona: Geological Society of America Bulletin, v. 89, p. 1745–1750.

# THE DEPOSITIONAL RECORD OF SMALL, MONOGENETIC VOLCANOES WITHIN TERRESTRIAL BASINS

JAMES D. L. WHITE

*Department of Geological Sciences, University of California, Santa Barbara, California 93106*

ABSTRACT: Repeated small eruptions of basaltic to intermediate magma produce monogenetic volcanic fields, which are common worldwide. The variety of volcanic landforms produced in such fields is controlled by whether or not water is present at or near the surface at the time of eruption. Basinal settings are typically "wetter," and hydrovolcanic eruptions produce tuff cones, maars and tuff rings instead of the scoria cone fields developed in dry, upland settings. Hydrovolcanic fields produce less lava and relatively more clastic debris than scoria cone fields and often contain large craters excavated by maar eruptions. Tephra eroded from volcanoes within hydrovolcanic fields tends to be reworked and redistributed within the basin in sheetform deposits and as crater-filling sequences, whereas the tephra eroded from scoria cone fields is largely carried from the fields in deeply incised streams skirting the fields' lava flows. The stratigraphic records of both groups of monogenetic fields differ from that of large, long-lived polygenetic volcanoes: the record of monogenetic volcanism is chiefly the result of long periods of inter-eruption erosion and sedimentation, whereas that of larger volcanoes is typically dominated by rapidly deposited units produced in immediate response to large eruptions.

## INTRODUCTION

When small eruptions of basaltic- to intermediate-composition magma take place, the types of volcanic edifice produced depend in part on the pre-eruptive character of the area into which eruptions occur. Eruptions involving ground water or surficial water are common in lowland, basinal settings, and the resulting hydrovolcanic eruptions produce tuff cones, tuff rings and maar volcanoes. Areas lacking available ground water or surface water are typically found in well-drained, upland settings, and eruptions into such areas produce scoria cone fields. The sedimentary characteristics of these two types of fields contrast strongly; the following pages address both the nature of this contrast and the reasons for it.

## MONOGENETIC VOLCANISM

Groups of small co-genetic volcanoes erupt over limited periods of time, sometimes within sedimentary basins. Such "monogenetic" volcanism is commonly associated with basaltic- to basaltic-andesite magmas, which erupt to produce lava flows, scoria cones, and hydrovolcanic features (Wood, 1979). A typical monogenetic volcanic field includes all of these landforms, and the relative proportions of each are controlled by a combination of magmatic properties and the environment into which the volcanoes erupt.

### *Controls on Types of Eruption*

Basaltic magmas are known for producing quiet eruptions dominated by lava flows (Basaltic Volcanism Study Project, 1981). Scoria cones are volumetrically minor associates of many such eruptions, generally formed during early stages of eruption when excess gas escapes the erupting magma. For quiet basaltic eruptions, the relative proportion of scoria versus fluid lava is determined by the magma's gas content and the rate of eruption; lava flows form from gas-poor magma at all rates of eruption, but only very high rates of eruption produce lava flows from gas-rich magma (Head and Wilson, 1989). Within monogenetic volcanic fields, scoria cones and lava flows are formed in dry areas.

In contrast, magma erupting in wetter parts of a monogenetic volcanic field will encounter and mix with water to produce violent "hydrovolcanic" eruptions (Heiken, 1971; Wohletz and Sheridan, 1983; Fisher and Schmincke, 1984; Cas and Wright, 1987), which build tuff cones, tuff rings and maars along streams and in lowlands (Hamilton and Myers, 1963; Heiken, 1971; Lorenz, 1973, 1986; Nairn, 1979; Kienle and others, 1980; (Fig. 1). "Tuff cones" are high-standing, cone-shaped volcanoes that are genetically related to tuff rings (Fisher and Schmincke, 1984), although often composed of somewhat coarser grained, more strongly palagonitized tephra (Wohletz and Sheridan, 1983). The factors determining whether a tuff ring or maar forms during a hydrovolcanic eruption are not well established but probably reflect the interplay between magma vesiculation, eruptive flux rate, and water source and supply rate during the eruption (Heiken, 1971; Lorenz, 1986; Table 1).

In sum, the availability of near-surface water at the site of eruption is the most important determinant of the explosivity of a given basaltic monogenetic eruption, and thus of the morphology of the resulting volcanic edifice. When little or no water interacts with the erupting magma, scoria cones and lava flows result. Tuff rings, tuff cones and maars form under wetter conditions, when magma interacts with water to produce explosive eruptions (Kienle and others, 1980; Wohletz and Sheridan, 1983; Kokelaar, 1986). Variations in magma composition, particularly gas content, also affect the types and proportions of the volcanoes (including lava flows) produced, but these effects are minor in comparison to those of water.

### *Eruption Types and Volcanic Landforms*

The volume of tephra composing individual monogenetic volcanoes ranges from 0.0001 km$^3$ (0.1 $\times$ 10$^6$m$^3$) to approximately 0.15 km$^3$ (100 $\times$ 10$^6$m$^3$; Fisher and Schmincke, 1984; Wood, 1980a; Tables 2, 3). The eruption of Paricutin spanned nine years, but most monogenetic volcanic eruptions, whether affected by hydrovolcanic processes or not, are short-lived, with activity commonly lasting from a few

---

Present Address: Centre for Marine Geology, Dalhousie University, Halifax, NS B3K 5L3 CANADA

Sedimentation in Volcanic Settings, SEPM Special Publication No. 45

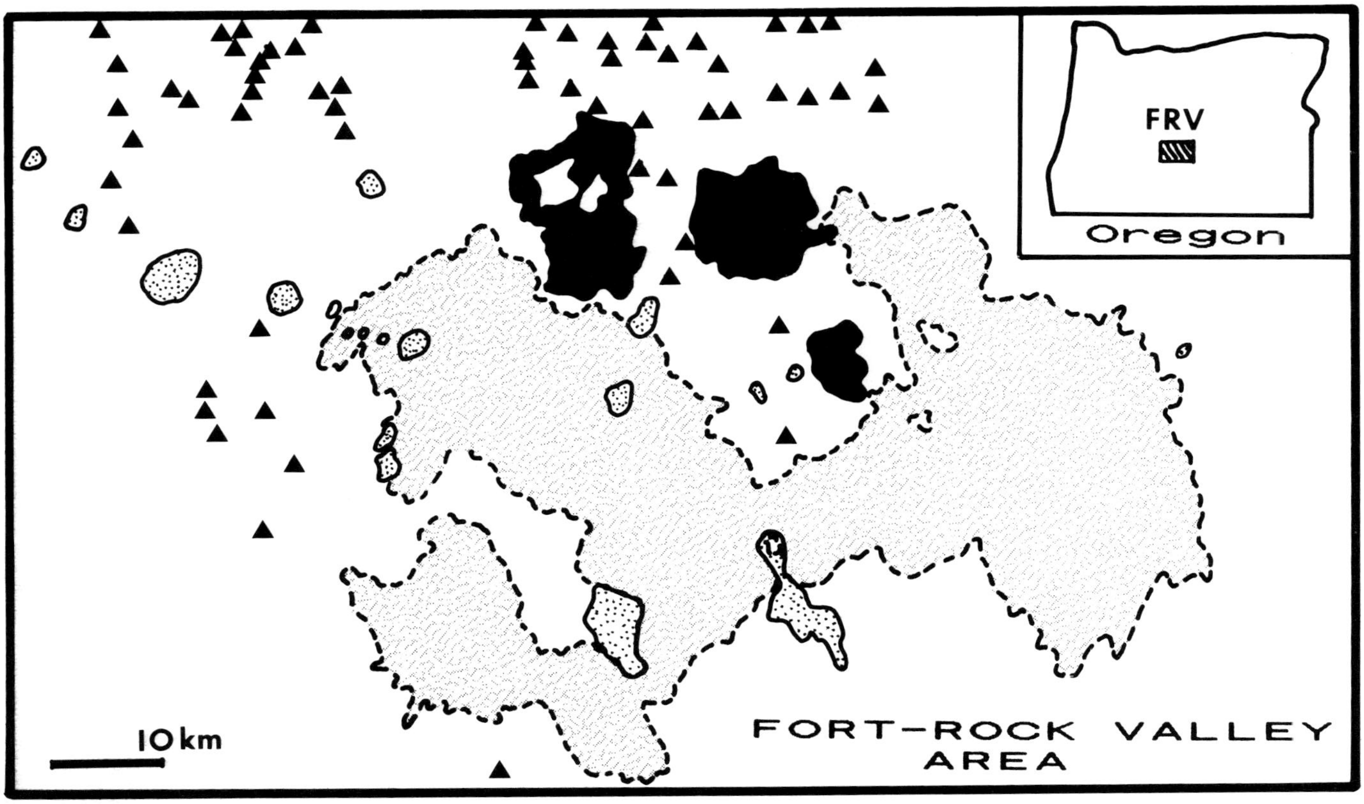


FIG. 1.—Maars and tuff rings (stipple-filled shapes) formed in basinal lowlands of a monogenetic volcanic field in central Oregon, as well as along an ancient stream course feeding into the basin from the northwest. Eruptions in surrounding parts of the field formed scoria cones (triangles) and lava flows (solid black). Such within-field variations reflect the influence of water on eruption style and are common in volcanic fields erupted into basinal settings. Diagram after Heiken (1971).

days to several months (Wood, 1979, 1980a; Fisher and Schmincke, 1984; Cas and Wright, 1987). Volumes of magma erupted in tuff-ring and maar eruptions are believed to be similar to those of scoria cone eruptions (.01 and 0.06 $km^3$ versus .02 $km^3$; Basaltic Volcanism Study Project, 1981) because the eruptions are " . . . merely cinder cone eruptions modified (drastically) by the magma's interaction with water" (Wood, 1979, p. 2821). The increased violence of hydrovolcanic eruptions, however, causes so much of the tephra erupted in tuff-ring or maar eruptions to be carried away as fine-grained, strongly fragmented ash that direct comparisons between maar/tuff-ring volumes and those of tuff and scoria cones are misleading (Fisher and Schmincke, 1984, p. 260; Wohletz and McQueen, 1984).

*Scoria cones and lava flows.*—

In dry, well-drained settings, monogenetic volcanoes most commonly erupt a combination of pyroclasts and fluid lava, producing lava flows, steep-sided scoria cones, and widespread layers of fallout scoria. Scoria cones are piles of coarse-grained, vesicular pyroclastic debris that form during relatively slow eruption of slightly gas-charged magma. They contain a central facies composed largely of dense lava and welded spatter (Head and Wilson, 1989; Houghton and Schmincke, 1989) and may also include finer grained beds produced by hydrovolcanic episodes (Houghton and Hackett, 1984; Houghton and Schmincke, 1986).

Fallout sheets comprise the finer grained ejecta carried away from the scoria cone by wind during the eruptions. Although rarely documented in studies of ancient scoria cones or volcanic fields, the volume of fallout deposits formed by scoria cone eruptions can far exceed that of the cone itself. Foshag and Gonzalez (1956) estimated that fallout deposits from Paricutin accounted for almost three quarters of the total tephra erupted, and Fries (1953) estimated that the volume ratio between fallout ash and ash in the cone ranged between 7.4:1 and 9.2:1 (fallout 88 percent–90 percent of total ash). There is no indication of phreatomagmatic activity at Paricutin, and increased fragmentation due to hydrovolcanism could still further increase the proportion of ash deposited in fallout sheets for mixed Strombolian-hydrovolcanic eruptions (see later discussion).

Lava flows and associated rubble are solid or very coarse-grained and form low-lying, sheetform units that are preserved essentially *in situ*. In contrast, high-standing cones of scoria are eroded, and their components reworked, after eruptions cease. The debris removed from their slopes and associated fallout sheets is either removed from the field or redeposited in low points within it. Scoria is glassy, fine- to coarse-grained and variably vesicular (Heiken, 1978), and

TABLE 1.—SUMMARY OF SEVERAL EXAMPLES OF MONOGENETIC ERUPTIONS THAT WERE AFFECTED BY THE PRESENCE OF NEAR-SURFACE WATER

| Volcano | Type | Water Source | Explosivity | Features |
|---|---|---|---|---|
| Ukinrek (Self and others, 1980) | maars | alluvium | high | thin surge beds |
| Ubehebe (Crowe and Fisher, 1973) | maars | alluvium | high | thin surge beds |
| Pematang Bata (Stehn, 1934) | maars | alluvium | high | unknown |
| Menan Buttes (Hamilton and Myers, 1963) | tuff cones | alluvium | medium-high | palagonitic tuff, thick and thin surge beds |
| Eifel (Lorenz, 1984) | maars | fractures under stream valleys | high | diatremes, maar rims |
| Christmas and Fort Rock Lake (Heiken, 1971, 1972a) | maars<br>scoria cones | lake, ground water<br>no water | medium-high<br>low | subaqueous mudflows, surges<br>scoria, spatter beds |
| Split Butte (Womer and Greeley, 1980) | maar | shallow aquifer | high<br>(late effusive) | thin surges, lava<br>lake formed in crater |
| Rinahue Valley (Müller and Weyl, 1957) | maars | valley alluvium | high | thin surges, deep crater |
| Rothenberg (Houghton and Schmincke, 1986, 1989) | scoria cone | shallow aquifer | low-high | mixed scoria, spatter and phreatomagmatic tuff beds |
| Barcena (Richards, 1959) | ash cone* | marine aquifer | low-medium | high cone, little scoria, weak surges, dry-deposited tephra |
| VOLCANOS ERUPTED INTO STANDING WATER | | | | |
| Volcano | Type | Surface $H_2O$ Depth | Explosivity | Features |
| Marion and Prince Edward Islands, Indian Ocean (Verwoerd and Chevallier, 1987) | tuff cones | 5–50m(?); littoral and wave-cut platform | moderate to high | "mudflows," bomb-sags, thick, coarse-grained palagonite beds, bases of cones are not exposed |
| Anatolia (Keller, 1974) | maars | about 15 m | low (underwater<br>high (emergent) | massive sideromelane beds at base, subaerial surges |
| Rotomahana (Nairn, 1979) | maars | shallow, partly hydrothermal | high | widespread surges, ash "rain," deep crater |
| Christmas and Fort Rock Lake (Heiken, 1971, 1972a) | maars, tuff cones | 0–20 m; groundwater effects? | low (underwater)<br>med-high (emergent) | massive sideromelane beds at base, much variation w/$H_2O$ |
| Pavant Butte (Wohletz and Sheridan, 1983, Gilbert, 1890, Fisher, unpubl. data) | tuff cone | 50 m | low (underwater)<br>med-high (emergent) | coarse sideromelane beds at base, palagonitic surge-deposit cone at top |
| Black Point (Custer, 1973) | tuff cone | 50–100 m | low (underwater)<br>med-high (emergent) | palagonite surge beds at top, at base coarse sideromelane |

petrographic identification of scoria grains allows interpretation of the types of volcanoes originally present from the reworked deposits of ancient fields (cf. Heiken, 1972b, 1974).

*Tuff rings and maars.*—

Most monogenetic volcanic fields also contain some tuff-ring and/or maar volcanoes, which form in wet, poorly drained areas. Tuff rings are small volcanoes with a very wide central crater and a low, broad cone (or "rim"). Maar volcanoes are similar, except that the base of the central crater lies below the level of the original ground surface (Lorenz, 1973). If limited water is available throughout an eruption, a maar volcano is produced and a deep crater remains open to the surrounding area. In other cases, only the first phase of eruption is hydrovolcanic, with subsequent eruptions filling the initial crater with scoria or lava. In many lowland fields (Westeifel, Lorenz, 1986; Hopi Buttes, White, 1990), scoria cones are built above small craters formed by early hydrovolcanic stages of the eruptions (Fig. 2; Lorenz, 1973, 1986).

Turbulent pyroclastic density currents (base and/or pyroclastic surges) accompany tuff-ring and maar eruptions, and construct the bulk of the low rims of ash typically surrounding these volcanoes. The tephra forming the low rims of tuff rings and maars is much finer grained, and tends to be better consolidated, than that of scoria cones (Moore, 1967; Sheridan and Wohletz, 1983). Tuff rings produce abundant very fine-grained ash together with fine- to medium-grained blocky pyroclasts (Wohletz, 1986). Maar ejecta is much like tuff-ring ejecta (Wohletz, 1983), but it also contains abundant non-juvenile debris torn from the pre-eruptive substrate (Lorenz, 1975; Heiken and Wohletz, 1986).

Large, commonly deep, central craters are the characteristic morphological feature of maar volcanoes (Ollier, 1967). The craters are surrounded by low rims of ejecta, but the proportion of ejecta carried completely away from

TABLE 2.—SUMMARY OF TYPES AND VOLUMES OF TEPHRA PRODUCED BY SOME RECENT ERUPTIONS, ALONG WITH EROSION RATES AND STYLE (WHERE AVAILABLE)

| Volcano | Tephra | Volume | Erosion |
|---|---|---|---|
| Mount St. Helens, 1980<br>Stratovolcano eruption<br>(Collins and Dunne, 1986) | 22% silt/clay<br>63% sand<br>15% gravel+ | 0.07 $km^3$ (mantle)<br>(+0.19 blast) | First 2 yrs: 20K tons/$km^2$/yr (14mm/yr)<br>First 10 yrs: 4.2K tons/$km^2$/yr (3mm/yr)<br>Rainfall: 150–300 cm/yr, gentle |
| Volcan Fuego, 1974<br>Stratovolcano<br>(Davies and others, 1978; Vessell and Davies, 1981) | Fallout, avalanche and lahar deposits (cf. St. Helens) | 0.42 $km^3$ (mantle)<br>(+0.18 $km^3$ in avalanches) | Initial, approx. 20K tons/yr/$km^2$ (6,000K tons/yr, est. 300 $km^2$ watershed)<br>Rainfall: 300–600 cm/yr, torrential |
| Volcan Barcena<br>Ash Cone*<br>(Richards, 1959, 1965) | 7% silt<br>48% sand<br>45% gravel | 0.27 $km^3$ | Sea cliff: 0.7 tons/month<br>Terrestrial: 10-m-deep arroyos eroded in months after eruption<br>Rainfall: >20 cm/yr |
| Vulcan<br>Ash Cone*<br>(Ollier and Brown, 1971) | Pumiceous, vesicular ash,<br>Size distribution like that at Barcena | | First 5 yrs: 30K tons/$km^2$/yr<br>Succeeding years: minimal<br>Rainfall: 200 cm/yr, torrential |
| Paricutin<br>Scoria cone<br>(Segerstrom, 1950, 1966 Foshag and Gonzalez, 1956) | Lapilli (gravel) and larger (>2cm) in cone<br>Surrounding ash sheet is sandy to silty vesicular tephra | Cone: 0.125 $km^3$<br>Ash Sheet: 0.525 $km^3$ | Cone: No significant erosion in first 20 years<br>Medial ash sheet (20 cm–165 cm : from 20–100% removed by "sheet erosion" in first 3 yrs; gullies to >10 m cut into ash during single storms<br>Overall rate of removal undetermined<br>Rainfall: 150–200 cm/yr, mostly in intense summer thunderstorms |
| Cima volcanic field<br>Scoria cones and lava<br>(Dohrenwend and others, 1986) | coarse vesicular ash to cobble-size bombs | Cones average approximately 0.008 $km^3$ | Cones: erosion insignificant in 15,000 yrs<br>Lava flows: flow morphology little modified in first several thousand years<br>Rainfall: 10–15 cm/yr |
| Ukinrek maars<br>Maar craters<br>(Self and others, 1980) | very fine ash to coarse lapilli, and accidental boulders | Maar rim contains ca. 0.001 $km^3$ juvenile ejecta (.0005 DRE) (but most beds >75% accidental lithics) | Unknown |
| Nilahue<br>Maar crater<br>(Müller and Veyl, 1957) | very fine ash to coarse lapilli, and accidental boulders | Maar rim contains ca. 0.024 $km^3$ juvenile** ejecta | Rim partly eroded by river during first years after eruption. See text. |

*Barcena's eruption was to an unknown degree phreatomagmatic—the cone is finer-grained than most scoria cones
**I interpret the "black pumice" and "red pumice" of Müller and Veyl (1957) as juvenile material.

TABLE 3.—SUMMARY OF PROPORTIONAL RELATIONSHIPS AMONG LAVA, SCORIA CONES, CRATERS AND FALLOUT LAYERS FOR SEVERAL VOLCANOES

| Volcano | Edifice | Sheet | Lava |
|---|---|---|---|
| Paricutin<br>(Segerstrom, 1950)<br>(Foshag and Gonzalez, 1956) | Scoria cone<br>vol: 0.125 $km^3$<br>volume: 0.525 $km^3$ | Ash sheet (12 m–1 mm)<br>area: 60,000 $km^2$<br>volume: est. 0.57 $km^3$ | Paricutin flows area: est 6 $km^2$ |
| Sierra Chichinautzin<br>(Martin-del Pozzo, 1982) | Scoria cones (vol:)<br>Xitle: 0.025 $km^3$<br>Jumentos: 0.034 $km^3$<br>Pelado: 0.073 $km^3$<br>Huilotito: 0.029 $km^3$<br>Ololica: 0.046 $km^3$ | No data | Flow area/volumes<br>75 $km^2$/3.2 $km^3$<br>3.0 $km^2$/0.15 $km^3$<br>63.5 $km^2$/1.9 $km^3$<br>1.5 $km^2$/0.12 $km^3$<br>??/0.6 $km^3$ |
| Barcena<br>(Richards, 1959) | Tephra cone<br>vol: 0.20 $km^3$ | Mostly in ocean surrounding island | Lava delta and crater dome<br>area: 0.014 $km^2$ |
| Menan Buttes<br>(Hamilton and Myers, 1963) | Two tuff cones<br>combined vol:<br>est. 0.06 $km^3$ | No data | No lava flows |
| Ukinrek<br>(Kienle and others, 1980) | Maar craters<br>vol: 0.00425 $km^3$<br>Maar rims<br>vol: 0.0057 $km^3$ DRE<br>est. 0.0092 $km^3$* | Ash sheet<br>area: several thousand $km^2$<br>vol: not estimated<br>vol: 0.0009 $km^3$ | Stubby flow in crater<br>area: ca. 1.1 $km^2$ |
| Nilahue<br>(Müller and Veyl, 1957) | Maar crater<br>vol: 0.18 $km^3$<br>Maar rim (to 2.5 km)<br>vol: 0.030 $km^3$ | Ash sheet (1.3 m–1 mm)<br>area: a few ten-thousands $km^2$<br>vol: 0.226 $km^3$ | No lava erupted |
| | | | vol: 0.028 $km^3$ |

*For the Ukinrek maars, the authors presented volume data in DRE (dense rock equivalent) form. The ratio 2.9:1.8 (grams/$cm^3$) was assumed here to back calculate from the given "volume DRE" to an equivalent volume of porous, vesicular tephra.

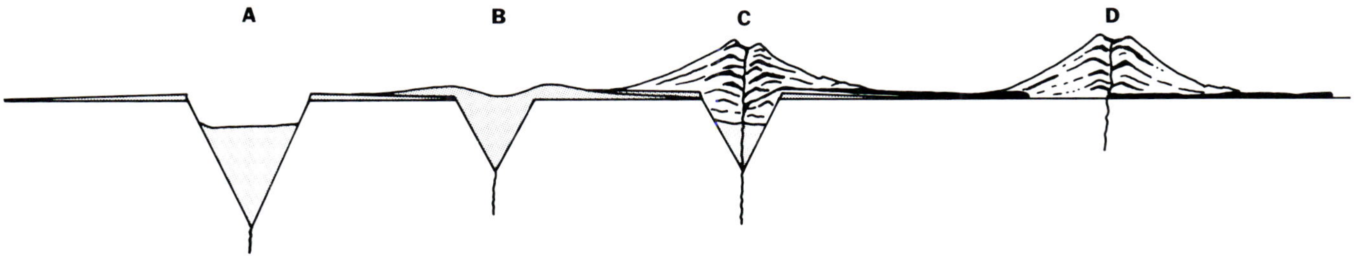


FIG. 2.—Schematic illustration of the morphologic range of monogenetic volcanic landforms discussed in text. From left to right are: A) maar volcano; B) tuff ring [tuff cone similar but with higher rim]; C) scoria cone with initial hydrovolcanic activity, and D) scoria cone. Scoria cones are almost invariably associated with lava flows, which are commonly several times more voluminous than the cones. Note that although the volcanic roots of A, B and C each lie below the syneruptive surface, only the maar (A) forms topographic depression following eruption and is thus the only one that can subsequently capture sediment entering the field.

the eruption site is very large. Müller and Veyl (1957) estimate that just over 90 percent of the eruptive products of Nilahue maar were deposited beyond the maar rim, including measurable fallout 1,500 km downwind. Overall, maar volcanoes are predominantly negative topographic features, in contrast to the significant constructional topography of scoria cones and tuff cones (Table 3; Fig. 3).

*Tuff cones.*—

When erupting magma encounters abundant water, hydrovolcanic eruptions may be only moderately explosive and produce tuff cones rather than tuff rings or maars (Wohletz and Sheridan, 1983). Tuff cones consist of steeply dipping beds of coarse-grained, weakly bedded, generally palagonitized tuff (Wohletz and Sheridan, 1983). Deposition is inferred to occur from steam-laden pyroclastic density currents (surges), and abundant water is present within the tephra upon deposition, as attested to by such phenomena as vesiculated tuff (Lorenz, 1974a; Leys, 1982, 1983), vertical bed accretion (Waters and Fisher, 1971; Heiken, 1971), and syndepositional "mudflows" (Lorenz, 1974b; Verwoerd and Chevallier, 1987).

Tuff cones can form in water-rich subaerial environments (Hamilton and Myers, 1963; Lorenz and others, 1970), but many tuff cones comprise the emergent portions of volcanoes initiated in subaqueous settings, both marine (Thorarinsson, 1967; Lorenz, 1974b; Leys, 1983; Kokelaar, 1986; Verwoerd and Chevallier, 1987) and lacustrine (Barberi and Varet, 1970; Heiken, 1971; Wohletz and Sheridan, 1983). The subaqueous parts of such emergent volcanoes are distinct from typical tuff cones described above and consist of thick-bedded, often shallow-dipping, coarse-grained, glassy, unpalagonitized tephra (Christensen and Gilbert, 1964; Waters and Fisher, 1970; Heiken, 1971; Custer, 1973).

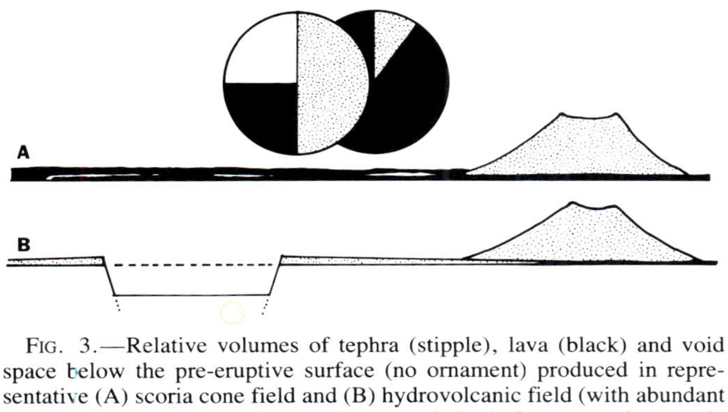


FIG. 3.—Relative volumes of tephra (stipple), lava (black) and void space below the pre-eruptive surface (no ornament) produced in representative (A) scoria cone field and (B) hydrovolcanic field (with abundant maars). The proportions of these three morphologic features varies greatly among hydrovolcanic fields, depending upon the type and degree of eruptive interaction between magma and water.

### *Aggregate, Field-Wide Characteristics*

Groups of small monogenetic volcanoes occur in both upland and lowland settings. Upland areas are well drained, with deep water tables and generally a significant regional drainage gradient. Volcanic fields formed in such areas are dominated by scoria cones and overlapping lava flows (Fig. 4). Lowland areas, in contrast, are areas with shallow water tables and generally poor drainage, often within enclosed basins. Magma erupted into such settings commonly interacts with the available water, and hydrovolcanic tuff cones, tuff rings and maar volcanoes can be the predominant volcanic landforms (Fig. 5).

*Scoria cone fields.*—

Scoria cones are the most visible result of eruptions in well-drained upland settings, but in terms of volume they are merely ". . . beacons, announcing that 5 to 100 times as much lava has probably been extruded at that vent, and that significant quantities of pyroclastic material were spread beyond the cone" (Wood, 1980a, p. 403). The volume of lava flows in a group of monogenetic basalt to basaltic-andesite volcanoes in central Mexico is 10 times that of the associated scoria cones (Hasenaka and Carmichael, 1985). Compilations of scoria cone, fallout layer, and lava volumes from several fields indicate that 1:10 is a reasonable average ratio between fragmental (cones and fallout), and non-fragmental (lava) eruptive products (Table 3). There is a trend for larger scoria cones to have a relatively smaller proportion of total erupted material contained in the associated lava flows (Wood, 1980a; Hasenaka and Carmichael, 1985).

Scoria cone fields occur both in the flank regions of large coeval central-vent volcanoes ("volcano cone fields") and as separate "platform fields" (Settle, 1979). Platform scoria

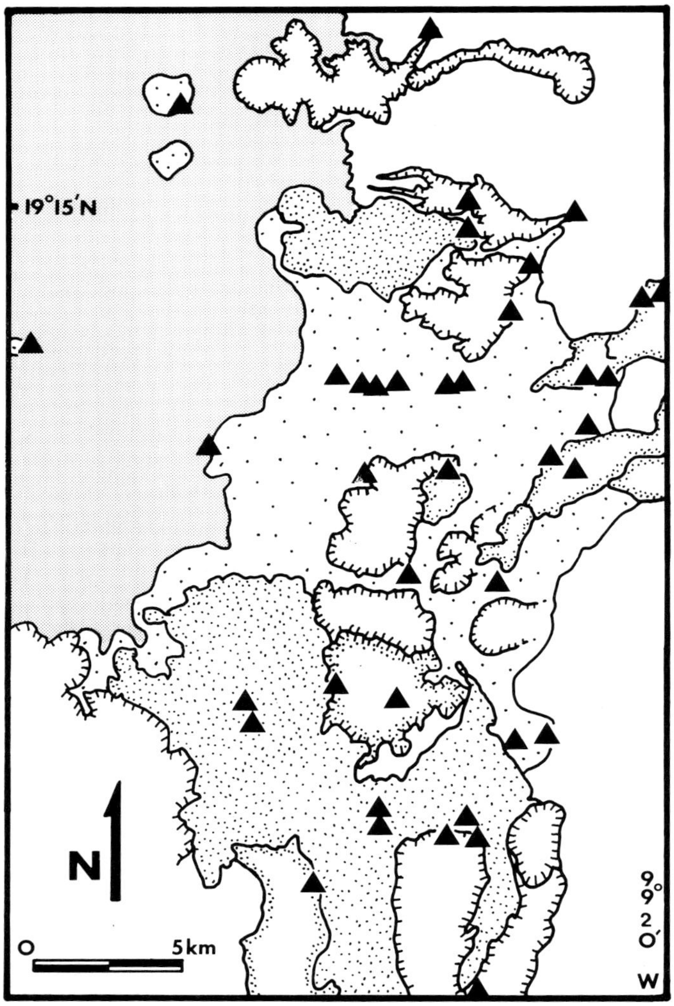


FIG. 4.—Simplified map of a small Quaternary scoria cone field in the uplands of central Mexico, illustrating overlapping association of lava flows associated with some 40 scoria cones. Six ages of lava flow are distinguished, from oldest to youngest: 1 = unpatterned (Tertiary); 2 = entire flows lightly stippled; 3 = flow boundaries hacured; 4 = flow boundaries stippled; 5 = flow boundary stippled & hacured; 6 = whole flows stippled. The shaded area is recent alluvium. Drawn after Bloomfield (1975).

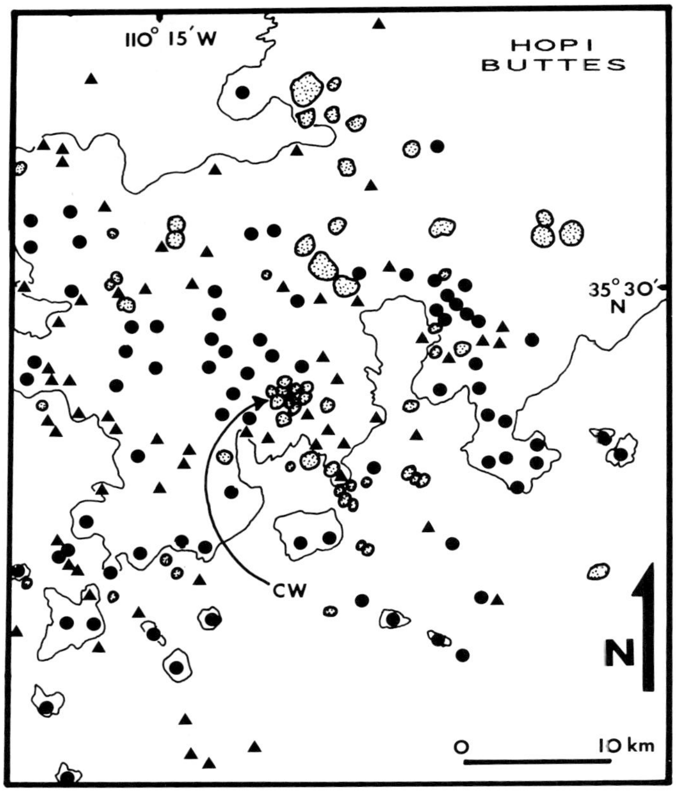


FIG. 5.—Simplified map of Hopi Buttes volcanic field, illustrating abundance of Mio-Pliocene hydrovolcanic features. Triangles represent eroded scoria cone and lava flow vents, solid circles are identifiable hydrovolcanic vents covered by lava flows, and stipple-filled shapes are exposed crater deposits and eroded hydrovolcanic vents. The limits of the northeast-trending outcrop belt of Mio-Pliocene basinal sedimentary rocks are outlined. Drawn after Hack (1942).

cone fields are not associated with contemporaneous stratocones, and their stratigraphic record can thus be appropriately compared with that of independently developed lowland fields.

*Hydrovolcanic fields.*—

Hydrovolcanic eruptions are characteristic of lowland areas, and the volcanic landforms produced and their relative proportions depend upon the permanence and distribution of water at and/or immediately beneath the ground surface during the course of eruptions (Wohletz and Sheridan, 1983). Heiken (1971) describes a hydrovolcanic field developed in Pleistocene time in central Oregon. Eruptions that occurred within and along drainages entering the lake produced maars, tuff rings, and tuff cones, whereas those beyond the basin formed cinder cones and lava flows. The ratio of tephra to lava within the basin is very large: the only lava present occurs as small lava lakes inside tuff rings. Beyond the lake basin, the eruptions formed a typical upland association of scoria cones and lava flows, except where eruptions occurred along major streams. Over the field as a whole, the ratio of tephra to lava was controlled by the availability of near-surface water.

Basins occupied by lakes, particularly large, deep lakes, provide abundant water, and small eruptions commonly produce tuff cones in such settings. Because most lakes are shallower than the volatile fragmentation depth for even basaltic magmas (Fisher and Schmincke, 1984), sublacustrine eruptions produce abundant hydrovolcanic tephra and are generally associated with only small volumes of lava, typically in the form of pillow lavas of various types. Pavant Butte, Utah, exhibits a representative sequence of deposits produced by sublacustrine eruption (Waters and Fisher, 1970; Wohletz and Sheridan, 1983). The lower levels of the Pleistocene edifice consist of loosely consolidated glassy tephra that was terraced and reworked by wave activity during eruption. Built above the platform-like lower level is a breached palagonitic tuff cone. The combination of subaerially produced palagonitic tephra with reworked glassy tephra of subaqueous origin is typical of lacustrine settings and can form complex assemblages that may be associated with pillow lavas, hyaloclastite deltas, and crater-filling lava ponds (e.g , Godchaux and others, 1989).

In lowland areas having less abundant or less accessible water near the surface, hydrovolcanic fields are characterized by the common occurrence of *negative relief* maar volcanoes. Maars are not major sources of sediment within a volcanic field but can play an importanat role in determining the final destination of tephra reworked from other volcanoes in the area (White, 1989). Because of the predominant influence of initial topography on the preservation of small volcanic landforms, maar and low tuff-ring volcanoes are readily preserved in even modestly aggradational settings (White, 1990).

## SEDIMENTATION IN MONOGENETIC VOLCANIC FIELDS

### *Syneruption Sequences*

Sedimentary sequences associated with active volcanoes can be considered to consist of two elements (Smith, this volume): 1) syneruption sequences, which result from primary volcanic deposition and immediate post-eruptive reworking, and 2) inter-eruption sequences, which record tephra redistribution, plant recolonization, pedogenesis and stream regrading during the periods between eruptions (Smith, 1987, 1988a). Both the absolute length and relative duration of syneruption and inter-eruption periods can vary greatly, depending largely upon a) the frequency of eruptions within a field, b) the climate, and c) the nature of erupted lava and tephra (Table 2). The inter-eruption sedimentary regime can be considered to continue beyond a field's last eruptions until the time unlithified tephra is no longer available for sedimentary redistribution.

*Scoria cone fields.—*

Variations in the timing and rates of syneruption erosion are illustrated by comparison of the rapid gullying of Vulcan in New Guinea with the slow initial erosion of Volcan Paricutin Net denudation of 56 cm, accomplished largely by gully erosion, was recorded three decades after the cinder cone Vulcan in New Guinea stopped erupting, and oral reports indicate that virtually all of this erosion took place in only the first five years after eruption ceased (Ollier and Brown, 1971). This stage of rapid erosion ceased before plants recolonized the cone and may have been brought to a close by the development of a widespread, stable network of rills and gullies (Collins and Dunne, 1986). In contrast, Segerstrom (1966) reported neither erosion nor gullying of the Paricutin cone 13 years after eruption ceased. One major influence on the degree of syneruption erosion is climate, particularly rainfall amounts and precipitation rates (Wood, 1980b). Although the area around Paricutin generally receives about 200 cm of rainfall annually, similar to the 220 cm average at Vulcan, New Guinea, the latter commonly receives torrential downpours in which 25 cm of rain fall in a single day (Ollier and Brown, 1971). Segerstrom (1950) estimated that it would be one to two centuries before gully erosion becomes important on Paricutin, as was the case at nearby Volcan Jorullo. In arid regions syneruption erosion is negligible, with minor gullies developing only after thousands of years (Dohrenwend and others, 1986; Scott and Trask, 1971).

Tephra grain size is also strongly correlated with erosion rates, largely because the presence of fine-grained ash greatly decreases surface permeability (Segerstrom, 1950; Waldron, 1967; Collins and Dunne, 1986). This effect is apparent from the contrast between erosion rates for Paricutin itself and for the finer grained fallout sheets surrounding the volcano (Table 2). A decade after eruption ceased, the coarse-grained, highly permeable tephra of Paricutin's cone remained virtually uneroded, but sheetwash alone removed 10 to 50 cm of fallout tephra from sloping surfaces in the Paricutin area within <3 years, and qualitative observations indicate that much more was removed by gully erosion, creep, mudflows, and caving along deeply incised streams (Segerstrom, 1950, 1966). Rainfall on freshly deposited fine-grained tephra typically causes a rigid surface crust to develop (Segerstrom, 1950), but increased runoff from the impermeable surface helps produce flash floods that undermine the crust and cause rapid headward extension of rills and gullies (Waldron, 1967). Richards (1965) noted that the development of such a surface crust effectively stopped eolian redistribution of tephra at Barcena, but the erosive power of runoff from this surface is suggested by the development of 9-m gully headwalls in ash following one or more "heavy showers" over a 2-week period.

The eruption of one scoria cone within a field can temporarily increase erosion rates of surrounding cones by more than an order of magnitude (Wood, 1980b). Volcan Paricutin's eruption blanketed surrounding, older cones with fine-grained tephra and also killed the plants growing on their slopes (Segerstrom, 1950). Fallout ash on the slopes of Cerro de Cutzato was rapidly gullied, and the gullies continued to erode into the pre-eruption slope of the volcano after cutting through the ash mantle, although the gully floors are also the first place plants recolonize (Eggler, 1963). As pointed out by Wood (1980b), fallout mantling of neighboring, inactive volcanoes is likely a common process in most monogenetic volcanic fields.

One of the unique aspects of volcanic deposits is that they can be lithified upon deposition, as are lava flows and some spatter beds of scoria cones. True scoria cones consist largely of loose pyroclasts (Macdonald, 1972), but there is a continuous gradation between scoria cones and small shield volcanoes (Basaltic Volcanism Study Project, 1981), which are highly resistant to erosion. Most scoria cones have a central core consisting of dense lava and welded spatter (Houghton and Schmincke, 1989) that is much more resistant to erosion than the rest of the volcano (Dohwenrend and others, 1986). Thin layers of welded spatter intercalated with loose scoria of the outer cone may inhibit short-term erosion to some degree but could also have the converse effect of enhancing the rate of initial erosion by facilitating runoff from otherwise too-permeable cone slopes.

Much of the ash initially stripped following scoria cone eruptions in humid regions accumulates temporarily in drainage ways blocked by lava flows issued during the eruption. Segerstrom (1950) illustrates a stream 2 km from Paricutin in which more than 10 m of tephra had been redeposited 3 years after the eruption began. This early-redeposited tephra is commonly re-eroded as streams bypass the margins of the lava dam, but subsequent lava flows may cover and thus preserve thick channel-filling sequences. Stream profiles at this stage are highly stepped. Tephra is repeatedly eroded and redeposited as sediment dams fail and lava dams are bypassed or overtopped.

Arroyos within the Paricutin field expose pre-Paricutin strata that consist largely of stacked layers of flat-bedded scoria (Segerstrom, 1950), which probably represent both primary fallout and syneruption flood deposits associated with eruptions of nearby volcanoes. Erosional contacts, weathered horizons and weakly developed paleosols record depositional hiatuses that probably represent inter-eruption periods within the field.

In the stream courses draining the Paricutin area, Segerstrom (1950) described debris flows, hyperconcentrated-flood flows (p. 67–68), and upstream-migrating antidunes (p. 96), all of which have been well documented in syneruption deposits of sand-size tephra associated with stratovolcano eruptions (Smith, 1986). The channelized deposits of syneruption sequences, however, are confined to stream courses between and beyond the field's lava flows, and thus are generally eroded when sediment-supply rates decrease and the streams regrade.

Most tephra eroded from scoria fallout sheets (and cones, in sufficiently rainy settings) during syneruptive periods is eventually carried to through-flowing drainages and removed from upland scoria cone fields. In humid settings not drained by through-flowing streams, hydrovolcanic fields develop in place of scoria cone fields. Arid areas are typified by uneroded cones and only minor syneruptive redistribution of fallout tephra, although eolian processes may redistribute ash locally soon after eruptions cease (Richards, 1965).

*Hydrovolcanic fields.—*

Very little information is available concerning the immediate (1 to 100 years) aftermath of most terrestrial hydrovolcanic eruptions. Wave reworking can be important during and after sublacustrine eruptions (see later discussion), but it seems that, in general, little significant tephra redistribution occurs for some centuries following maar and tuff-ring eruptions (cf. Keller, 1974; Fisher and Waters, 1970). The reasons for this minimal syneruption response are: 1) small volumes of tephra are initially deposited by the eruptions, 2) the initial constructional relief of tuff rings and maars is small, and 3) eruptions occur in low-relief lowlands where background erosion rates are small. The lattermost factor is very much the most important, as illustrated by the rapid syneruption destruction of maar volcanoes that erupt into stream valleys, some of which have been leveled (rim removed, crater filled) in only a few decades (Müller and Veyl, 1957; Illies, 1959). Hydrovolcanic eruptive sequences also tend to contain many fine-grained, compact beds that are resistant to eolian redistribution, but which probably mimic the erosion-enhancing effects of the rain-formed crusts discussed earlier.

Sublacustrine eruptions, like submarine ones, deposit piles of easily eroded tephra both below and above the water surface. At the level of emergence, lacustrine volcanoes are typically marked by wave-cut benches (terraces), and/or steep cliffs, and/or reworked littoral deposits (Gilbert, 1890, p. 325–329; Oviatt and Nash, 1989). Depending upon the history of the lake's water level, large parts of the volcano can be eroded (Heiken, 1972a; Wohletz and Sheridan, 1983), or the edifice may simply be notched by wave-cut terraces (Custer, 1973). The fragility of newly formed emergent volcanoes is well illustrated by the histories of various marine volcanoes (Hoffmeister and others, 1929; Thorarinsson, 1967; Johnson and Tuni, 1987), but most lakes have far lower levels of wave energy (Sly, 1978), and the degree of syneruption reworking is much lower than for these marine examples. Moreover, tephra eroded from lacustrine volcanoes is unlikely to be transported very far and typically is resedimented as "mudflows" or downcurrent sediment fans, or bars in the immediate vicinity of the volcano (cf. Heiken, 1972a; Allison, 1979). Pliocene lacustrine volcanism in southwestern Idaho produced sequences of basaltic tuff beds up to 200 m thick, but the deposits thin rapidly away from areas near known source vents (Jenks and Bonnichsen, 1989).

Subaerial parts of tuff cones are deposited from steam-laden density currents and consist of poorly sorted deposits with abundant fine-grained ash (cf. Waters and Fisher, 1971; Wohletz and Sheridan, 1983). The whole of a tuff cone is compact and impermeable, with little loose tephra, and although early formed surface crusts are ineffective in preventing erosion because they are undermined by erosion of underlying loose tephra, the impermeable compact body of a tuff cone probably resists early syneruptive erosion. This early lithification, however, is much less effective in inhibiting erosion than is the high permeability of scoria cone tephra (cf. Gilbert, 1890, p. 327)

*Comparison with other syneruption sedimentation.—*

Syneruption erosion and redeposition is relatively unimportant in determining the form of depositional sequences resulting from almost all varieties of monogenetic volcanism (possibly excepting some floodplain deposits). In contrast, very extensive syneruption deposits dominate the stratigraphic record of stratovolcanoes (Davies and others, 1978; Vessell and Davies, 1981; Smith, 1987; Palmer and Walton, 1990) and silicic centers (Turbeville and others, 1989; Waresback and Turbeville, 1990; Buesch, this volume). The differences in initial sedimentary response can be attributed largely to the greater volume of tephra produced by stratocone and silicic eruptions, as compared to those of monogenetic volcanoes (Hasenaka and Carmichael, 1985). Tuffaceous-apron deposits around silicic centers in Texas were apparently fed by repeated small eruptions, with the result that depositional processes were closely similar to those typical of nonvolcanic sequences (Walton 1979, 1986), indicating that aggradation must have been facilitated by subsidence of the receiving area (Smith, this volume).

Syneruption aggradation and stream braiding occurred only in response to eruptions exceeding a threshold value of 0.06

$km^3$, of which approximately one-third was contained in locally deposited block and ash flows, at Volcan Fuego in Guatemala (Vessell and Davies, 1981). This volume is greater than the average volume of tephra contained in many entire scoria cones (0.0001–0.15 $km^3$), although the disparity is somewhat less pronounced if one takes into consideration the surrounding fallout deposits, which are often several times the volumes of the cones (Tables 2, 3). Stratovolcanoes also have much greater initial relief and are commonly subject to topographically induced heavy rainfall (uplands of Volcan Fuego receive as much as 500 cm annually; Kuenzi and others, 1979). A final difference concerns the far greater abundance of more easily eroded fine-grained ash in silicic-center and stratovolcano eruptions, as compared with the generally coarse-grained ejecta produced by scoria cone eruptions (Segerstrom, 1950; Collins and Dunne, 1986).

Further accentuating the importance of syneruption sedimentation around large volcanoes is the enhanced preservation potential of tephra redeposited during syneruption episodes. Eruptions that exceed threshold values induce rapid and widespread aggradation of areas surrounding large volcanoes after major eruptions (Kuenzi and others, 1979; Smith, 1988b), whereas revegetation and stream regrading in prolonged inter-eruption periods causes stream incision, rather than additional net sediment deposition (see later discussion).

### *Inter-eruption Sedimentation*

Although syneruption erosion and redeposition can produce impressive short-term degradation and local aggradation, it is during the more prolonged inter-eruption and initial post-eruption periods that the form of a monogenetic field's rock record is determined. The importance of high permeability in inhibiting erosion during syneruption periods was discussed previously, and decreased surface permeability may be the most important single factor in initiating inter-eruption erosion as well.

*Scoria cone fields.—*

When a scoria cone eruption has ceased, ash produced by the eruption is initially eroded and redistributed by syneruption processes dependent upon the presence of unconsolidated surficial tephra. The duration of this syneruptive stage depends largely on the local climate, in part because climate strongly influences soil development (Wood, 1980b). Soil plays two important roles during inter-eruption periods: 1) it decreases surface permeability, and 2) it facilitates plant growth. In arid climates eolian silt and fine sand accumulate at the surface of scoria cones and on lava flows (Dohwenrend and others, 1987), and the soil ". . . progressively changes the near-surface hydrology (infiltration) which causes regionally stable pervious landscapes to evolve into regionally unstable impervious landscapes. . . ." (Wells and others, 1985, p. 1529). Silt accumulation on scoria cone surfaces facilitates piping and thus gully development, but in the Mojave Desert it requires some 15,000 years for the process to initiate widespread gullying (Dohrenwend and others, 1986). In this environment 400 thousand years may pass before the scoria cones are eroded to their agglomeritic core (Dohrenwend and others, 1986), a time exceeding the usual duration of volcanism within monogenetic fields (about 100 thousand years; Hasenaka and Carmichael, 1985). Wood (1980b) similarly attributes the onset of scoria cone erosion to soil development, but notes that in humid settings weathering and plant growth alter the scoria itself to clayey soils that are susceptible to gullying and mass wasting.

Although climate is very important in determining the rate and style of syneruption erosion and sedimentation, the end result of inter-eruption processes for all climates is similar: pyroclastic debris is eroded and most is transported beyond the scoria cone field. Ash deposits are preserved beneath lava flows and in local areas where lava flows have covered valley-filling deposits, or ponded streams and forced aggradation. The resulting deposit is characterized by steep-sided channels that often skirt lava flows, patchily distributed tephra preserved between lava flows, and the lava flows and volcano necks. Despite the abundance of ash produced by scoria eruptions and the impressive visual presence of scoria cones in upland fields, the rock record of such fields consists almost entirely of overlapping lava flows (Fig. 6).

*Hydrovolcanic fields.—*

Inter-eruption erosion in hydrovolcanic fields is controlled by initial topography, background erosion rates, and erodibility of the erupted tephra. The Hopi Buttes area in northeastern Arizona preserves the record of scores of maar volcanoes that erupted in Mio-Pliocene time. Small lava flows cap the mesas of the Hopi Buttes area, and the lower levels of many eroded scoria cones dot the field. The deposits of this exhumed volcanic field illustrate the different sedimentary signatures of lava flows, scoria cones, tuff rings, maars and lava flows (White, 1990).

During prolonged inter-eruption periods at Hopi Buttes, fallout sheets from scoria cone eruptions were redistributed. The cones were eventually eroded and their remains redeposited in the generally aggradational, low-relief lowland setting of the field (White, 1990). Tephra at Hopi Buttes accumulated in a broad basin-floor playa as thin, overlapping sheets of fallout, eolian and sheetflood-redeposited ash (Fig. 7). Alternating wet and dry periods in the basin produced interlayered playa-lake, eolian and ephemeral-stream deposits, the latter two composed almost entirely of scoria eroded from the cones and fallout sheets within the basin. Open maar craters, the lowest features in the basin, collected much of the tephra eroded from the scoria deposits, as illustrated by the predominance of scoriaceous ash in crater-filling successions (White, 1989).

Redistribution of tephra from scoria cone eruptions across the floor of the aggrading basin floor and into negative-relief maar craters during inter-eruption periods is characteristic of lowland hydrovolcanic fields like Hopi Buttes and can be documented by petrographic methods. At Hopi Buttes, many crater-filling deposits are composed of coarse-grained juvenile fragments that are variably vesicular, often tachylitic (opaque glass), and crystal-rich (Fig. 8). The volcanic clasts in primary deposits of maar volcano rims, in contrast, are fine- or very fine-grained, sparsely or non-vesicular, and often of clear glass. The maar-rim tephra

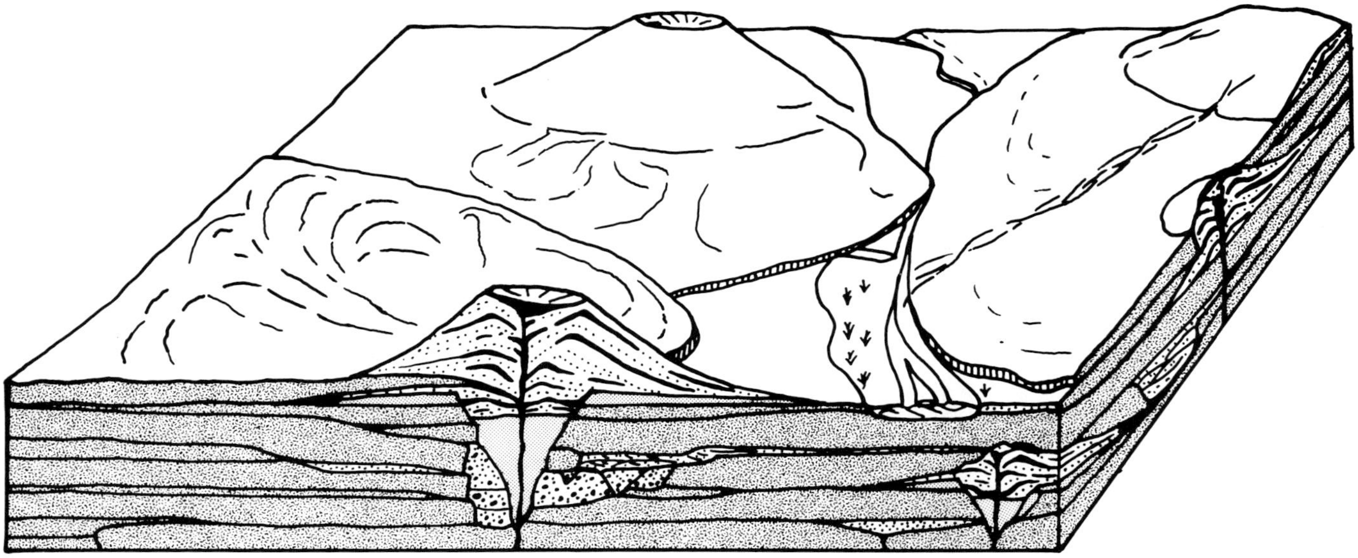

FIG. 6.—Schematic illustration of surface features and stratigraphy of a scoria cone field. Most of preserved sequence is overlapping lava flows (dark stipple) with intercalated sedimentary veneers, channel-filling deposits, and largely eroded scoria cones. Initial scoria cone eruptions are locally hydrovolcanic, producing tuff-ring beds and tuff-filled vents (light stipple). Streams drain lava flows only after soil development and rarely incise the flows deeply. Antecedent drainage is rearranged by flows, but streams skirting flows may form strongly incised channels in which ash washed from the cones and fallout sheets is deposited.

was not the source of tephra now filling the maar craters (White, 1989). High concentrations of crystals, the low percentage of delicate, highly vesicular clasts, and minor rounding of the grains distinguish reworked scoria from fresh scoria fallout and cone-forming deposits.

A closely spaced group of volcanic vents near the center of the Hopi Buttes field illustrates the development of complex stratigraphic relationships within some hydrovolcanic fields (Fig. 9). The close association of eroded scoria cones and maar volcanoes is common, and tephra from the cones

FIG. 7.—Lava-capped mesa in Hopi Buttes: cliff face is approximately 15 m high. The bedded sequence consists of intercalated base-surge deposits from a nearby maar, planar-bedded fallout tuff, sheets of tuff reworked from scoria cones and fallout layers, and playa-lake claystone deposited during periods when the playa was inundated. This sequence is representative of much of the Hopi Buttes volcanic field deposit, and similar sections are exposed almost continuously over hundreds of square kilometers.

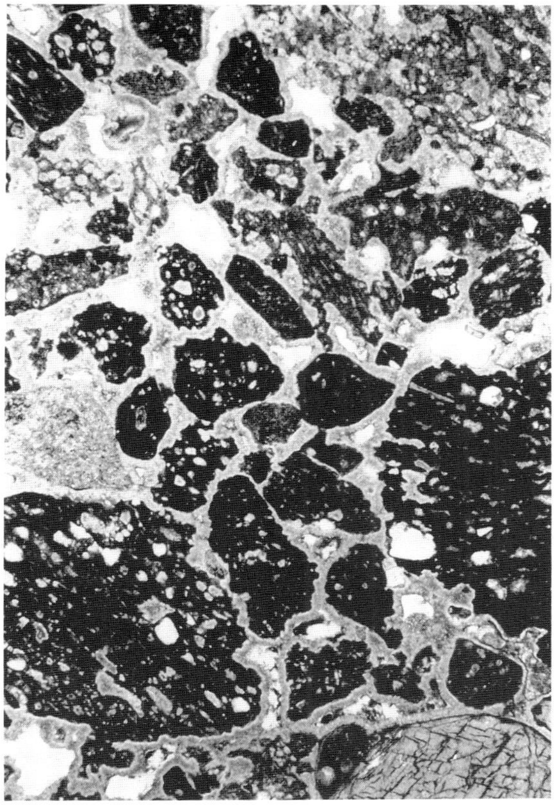

FIG. 8.—Thin section of sand-size tuff filling a crater in the Hopi Buttes volcanic field. The clasts in this eolian deposit are mostly vesicular grains eroded from scoria cones and fallout sheets in the vicinity of the crater and subsequently redeposited in the crater. Tephra from the rim of this maar is unlikely to have been the source for this tuff, because 1) the rim deposits contain mostly fine-grained nonvesicular ash, and 2) this sample is from eolian beds that top a complex assemblage of crater-filling strata deposited over a long period of time (White, 1989).

was deposited into the maars both during eruptions and long afterward. Base surges emanating from maar vent #3 can be followed to the rim of maar vent #7 (Fig. 9A), and in some vents complex deposits of overlapping base-surge deposits indicate that older maar craters collected base-surge ejecta erupted from neighboring maars. The lacustrine, playa and braided-stream sequences that fill most of vent #7 are composed largely of reworked tephra (Fig. 9B) but contain no primary tephra units, implying that deposition postdated eruption of all surrounding vents. In contrast, vent #6 contains thick tongues of primary scoria deposits erupted from vent #4, which are intercalated with lacustrine limestone near the central parts of the vent. An outcrop belt north-northeast of vent #7 exposes base-surge deposits overlying fluvial deposits at a level several dekameters below the pre-eruptive basin floor (Fig. 9C). This sequence indicates that part of the area around vent #7 collapsed well before the eruption of vent #3: the collapsed area was entered by streams that fed into a crater lake in #7, and the base surges exiting vent #3 later followed the same path into vent #7. Such vent-margin collapse is a common feature of maar volcanoes, reflecting instability in the maar-crater walls both during and after eruption (cf. Lorenz, 1986).

Because maar craters are effective sediment traps, the general form of a maar field's sedimentary record is a sort of inverted sequence in which scoria redistributed from fallout sheets and removed from high-standing cones has been redeposited in craters, while the low-relief maar-rim ejecta is largely preserved beneath lava flows and aggrading basinal deposits at the eruptive surface (Fig. 10).

Remains of a Miocene hydrovolcanic field have been documented in east-central Oregon (Cummings and Growney, 1988) and indicate a sequential evolution from tuff cones to maar volcanoes as near-surface water became less available. The preserved sequence consists of three separate tephra units interbedded with fluvial and fossiliferous lacustrine sedimentary rocks, subaerial lava flows, and pillow-palagonite deposits. Tuff cone deposits of the lower unit are recognized by their glassy tephra and poor-to-absent bedding, whereas the maar deposits are well bedded and include base-surge bedding structures. The maar crater also cuts into underlying units and contains some basalt flows (Cummings and Growney, 1988).

Stratigraphic studies of fields erupted into deep lakes have yet to be completed, but the basic elements of lacustrine tephra deposits are beginning to emerge from current work on Pliocene lacustrine tephra in southwestern Idaho. Local accumulations of glassy to palagonitic ash are produced by subaqueous to emergent volcanoes during highstands of long-lived lakes. As the lake evolves over the duration of volcanism within the field, the highstand deposits become interbedded with wiedspread beds of finer grained ash produced by maar and tuff-ring eruptions that occur during intermittent lowstands (Jenks and Bonnichsen, 1989; Godchaux and others, 1989). The result is a multistory deposit with maar volcanoes, tuff rings, lava flows and pillow deltas that closely reflects the variations in lake level over time (Fig. 11).

*Comparison with other inter-eruption sedimentation.—*

In both scoria cone and non-lacustrine hydrovolcanic fields, inter-eruption tephra remobilization is more important than syneruption sedimentation in determining the nature of a field's stratigraphic record. In scoria cone fields, syneruption aggradation is largely confined to stream valleys among the predominant lava flows, and inter-eruption stream regrading removes most of the syneruption tephra accumulations. In hydrovolcanic fields, syneruption erosion and resedimentation are commonly unimportant, but inter-eruption sedimentation may be associated with substantial, though very localized, net sediment deposition within maar craters and general degradation of high-standing elements. Lacustrine fields may suffer significant syneruption reworking, depending upon the life span and energy levels of the lake.

Major volcanic edifices are high-standing features and almost invariably experience significant erosion during inter-eruption periods (Francis, 1983). As discussed in some detail by G. Smith (this volume), inter-eruption periods may be marked by aggradation of epiclastic sediment or by widespread stream incision and paleosol development, de-

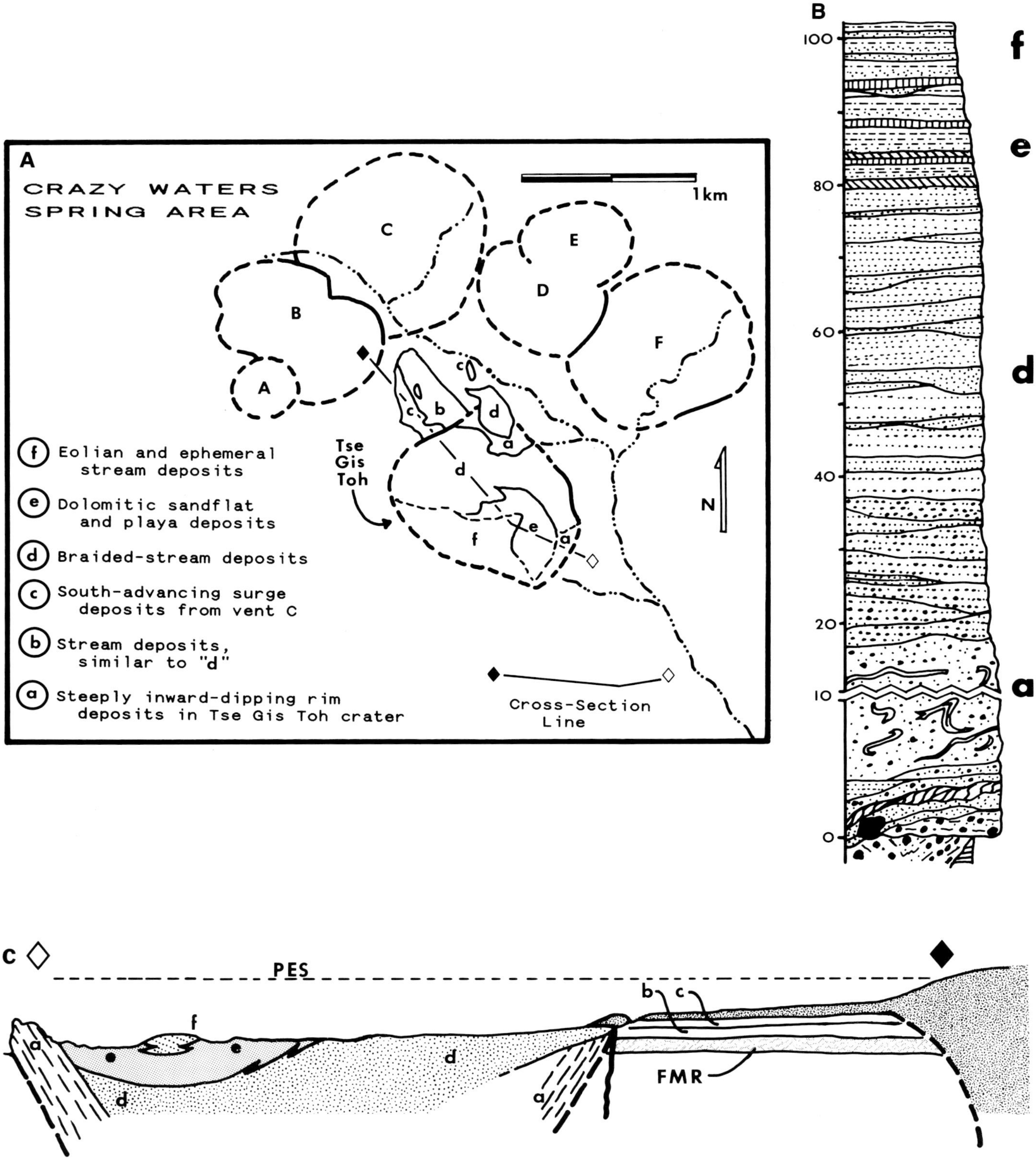


FIG. 9.—Overlapping maar and scoria cone deposits from central Hopi Buttes (see Fig. 5 for location). (A) Note the large size and close spacing of the vents in this area. Vents 'A,' 'B,' 'D' and 'E' are the remnants of scoria cones, with 'B' and 'D' having hydrovolcanic bases. TGT and vents 'C' and 'F' are maars containing thick crater-filling sequences, partly fed by erosion of the nearby scoria cones (see text). (B) Strata within Tse Gis Toh maar show a typical progression from steeply dipping deltaic-rim deposits to an onlapping sequence of alluvial and eolian facies. (C) Cross section shows effects of syneruptive collapse in this area and illustrates general geometry of crater-filling deposits. "PES" indicates level of pre-eruptive surface, and "FMR" is the foundered maar rim of TGT vent. Facies 'c' consists of base surge deposits erupted from vent 'C.'

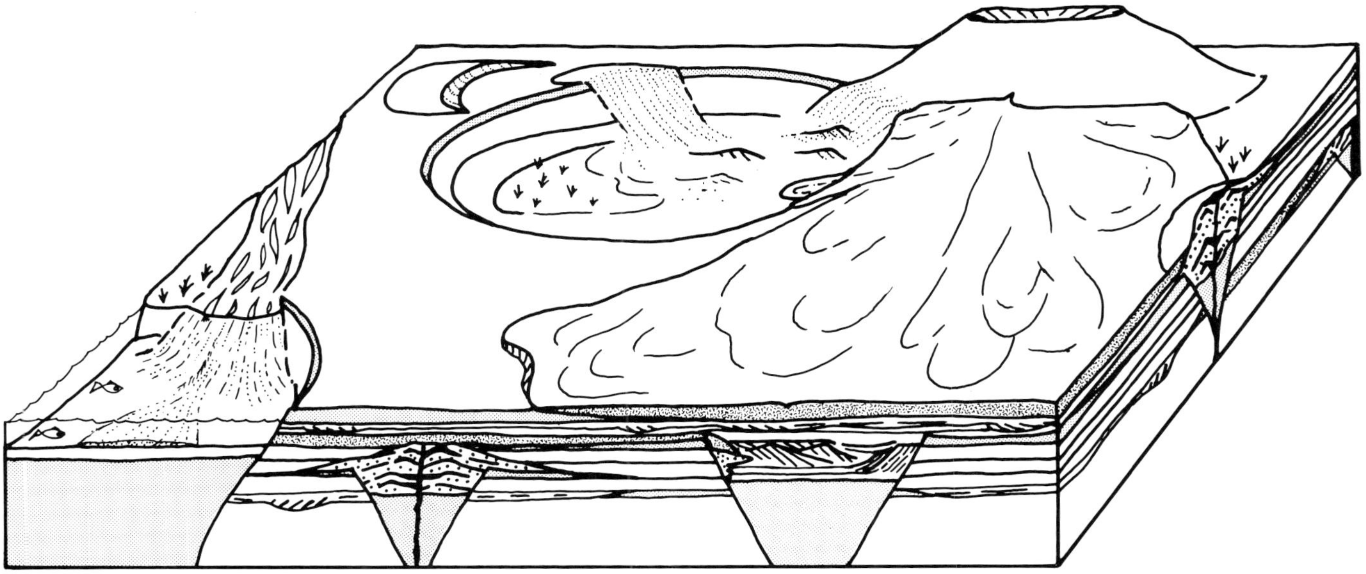

FIG. 10.—Schematic illustration of surface features and stratigraphy of a subaerial hydrovolcanic field. Two maars are shown, one with a delta building into a water-filled crater and the second showing eolian sand accumulating as passing dunes enter the crater. Scoria beds and a lava flow have entered one edge of the second maar from an adjacent scoria cone. The stratigraphy is dominated by widespread sheets of ash and deep crater-filling deposits, but ongoing "background" sedimentation commonly continues to aggrade the floors of lowland basins during inter-eruption periods. Most of the tephra and lava flows erupted into hydrovolcanic fields are preserved, but high-standing elements like scoria cones are eroded and their debris redistributed to fill craters and other low-lying areas.

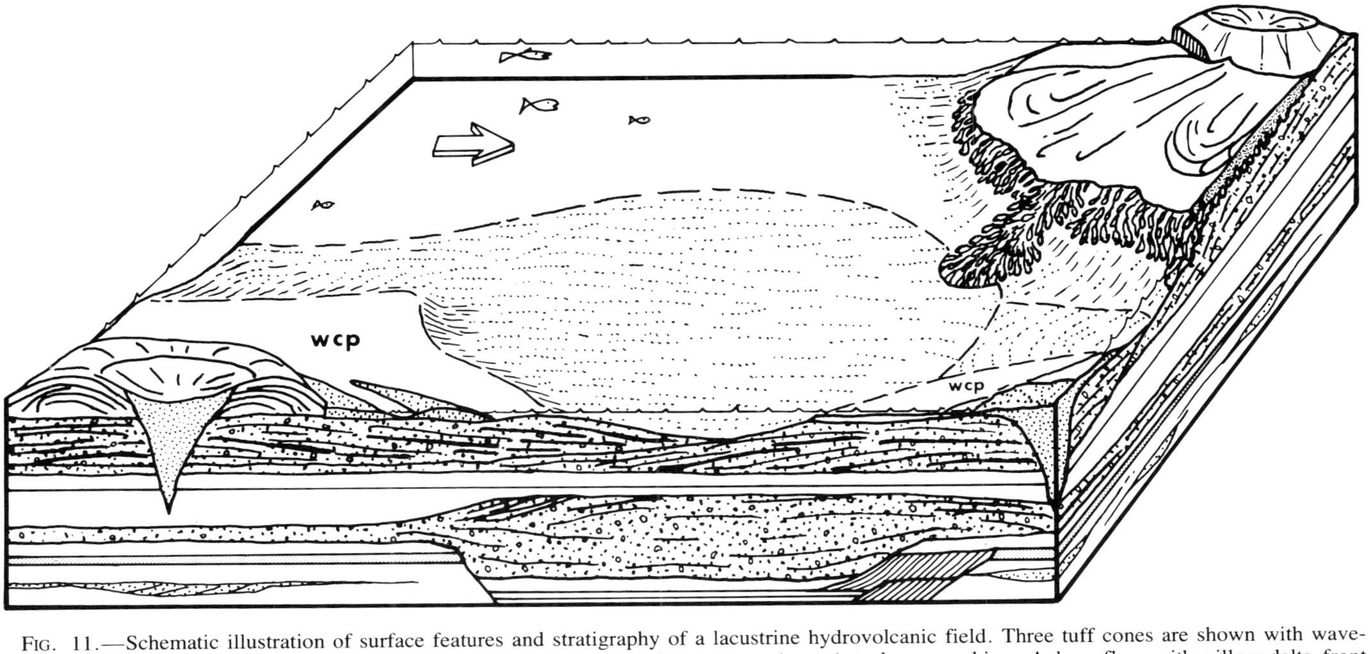


FIG. 11.—Schematic illustration of surface features and stratigraphy of a lacustrine hydrovolcanic field. Three tuff cones are shown with wave-cut platforms (wcp), beach ridges (on nearer cone's platform), and downcurrent (arrow) tephra reworking. A lava flow with pillow-delta front overrides the platform of the second cone, and the third cone has been largely eroded at the waterline. Shallow-dipping, coarse-grained tephra composes the subaqueous parts of the cones (gravel pattern), which are topped by more steeply dipping beds of the emergent cone. Unbedded tephra fills the cones (stipple), and nonvolcanic sediment accumulates in the background (unornamented). Early eruptions of the sequence illustrated here occurred under subaerial conditions, producing a maar crater that was then partly filled by deltaic deposits and finally overlapped by the base of a second volcano, which erupted later under subaqueous conditions.

pending upon the characteristics of depositional sites associated with major volcanoes.

PREFERENTIAL PRESERVATION OF HYDROVOLCANIC SEQUENCES

Despite the abundance of modern scoria cones and scoria cone fields (Wood, 1979), I know of no studies of the sedimentary sequence produced by such a field. The reasons for this lack are probably: 1) the link between scoria cone fields and upland settings, 2) the predominance of lava over scoria deposited in upland fields, and 3) the very small proportion of erupted scoria that is ultimately preserved within an upland field. Thus, even where large-scale subsidence allows upland fields to be buried, pyroclastic debris constitutes only a few percent of the preserved sequence, forming small pockets of tephra along flow edges and thin veneers between flow layers (Fig. 6).

Descriptions of the remnants of hydrovolcanic fields are, unlike those of scoria cone fields, comparatively common in the literature (Shoemaker and others, 1962; Francis, 1968; Leys, 1982; Cummings and Growney, 1988; White, 1989, 1990). Two factors particularly enhance the preservation potential of hydrovolcanic sequences: 1) the link between hydrovolcanism and aggrading lowland settings, and 2) the hydrovolcanic excavation of deep craters into the pre-eruption substrate. The latter factor is particularly important for maar fields, as illustrated by the extensive preservation of crater-filling deposits in more deeply eroded parts of the Hopi Buttes field (Hack, 1942). Francis (1970) described numerous bedded crater-filling deposits from a lowland hydrovolcanic field in which little tephra was preserved at the surface, and no surficial deposits remain associated with the crater-filling lacustrine kimberlite investigated by Smith (1986). Crater-filling sequences are the most likely element of hydrovolcanic field successions to be preserved because the floors of maar craters initially lie at levels well below the eruptive surface and may further subside to hundreds of meters below that surface (Francis, 1970; Hearn, 1968).

SUMMARY AND CONCLUSIONS

Monogenetic volcanism produces scoria cone fields and hydrovolcanic fields, the former commonly in well-drained upland sites, the latter in water-rich lowlands and lakes.

The rock record of monogenetic volcanism is strongly shaped by inter-eruption redistribution of tephra. Syneruption aggradation is limited or nonexistent, largely because most individual monogenetic volcanoes do not produce sufficient ejecta to rearrange existing drainage patterns or to overload major streams for significant periods of time. Each eruption fails to surpass the geomorphic threshold necessary to induce significant aggradation. In contrast, studies of tephra resedimentation associated with the eruptions of stratovolcanoes have distinguished syneruption and inter-eruption deposits (e.g., Vessell and Davies, 1981; Smith, this volume). Syneruption sequences form in direct response to volcanic eruptions, and reflect extensive aggradation adjacent to the volcanic edifice. They dominate the rock record of stratovolcanism, with inter-eruption periods marked largely by stream incision and minor sediment reworking.

In the simplest case of an isolated stratocone, unaffected by sector collapse or other catastrophic events, the volcano's cone and its surrounding volcaniclastic ring plain preserve complementary aspects of the eruptive history, with the plain composed of the pyroclastic debris washed from the cone, which is itself built up largely by accumulation of relatively immobile lava flows (Hackett and Houghton, 1989). Pyroclastic debris is only locally preserved within the cone because it is easily eroded unless buried by lava flows or captured in declivities (Wood, 1978). Similar preservation biases affect the records of monogenetic volcanism. In scoria cone fields, most scoria is rapidly removed from the field unless buried by co-genetic lava flows, or captured in lava-dammed drainages: the fields can be likened to a stratocone edifice. Lowland hydrovolcanic fields are more akin to the ring plains associated with stratocones. They are commonly aggradational and retain the hydro/pyroclastic debris produced during eruptions. Maar craters form down-gradient accumulation sites for debris, as do ring plains adjacent to stratovolcanoes.

## *Future Research*

Sedimentary sequences of reworked tephra are commonly produced within hydrovolcanic fields and provide excellent records of both eruptive activity and the contemporaneous surface environment during the fields' histories. Deposits associated with long-lived lakes are especially promising research topics, as the growth and erosion of lacustrine volcanoes preserve a sensitive record of lake-level variations throughout the field's eruptive history. Little information on contemporaneous surficial processes within scoria cone fields has been identified from studies of such fields to date, and it seems that the deposits of scoria cone fields may preserve only a very limited record of the syneruptive environment. Nevertheless, their potential is poorly known, and sedimentologic and stratigraphic studies of such fields are a promising area of future research. Multiple soil horizons described from the flanks of some "monogenetic" scoria cones indicate that they erupted repeatedly over long periods of time (Wells and others, 1990), a behavior that might also be documented by careful analysis of sedimentary deposits derived from the cone. We should learn more about the interplay between lava flows and the antecedent drainage. In some fields, lakes impounded by lava flows may collect thick sedimentary sequences and/or provide surficial water to fuel subsequent hydrovolcanic eruptions. Where is the ash, washed from these fields, deposited, and in what form?

ACKNOWLEDGMENTS

This paper was written during the tenure of an I. W. Killam postdoctoral fellowship at Dalhousie University. My work at Hopi Buttes was supported by NSF grant #EAR 88-16281 to R. V. Fisher, whose encouragement throughout my tenure at UCSB is greatly appreciated.

Review comments by Grant Heiken helped improve the manuscript substantially.

REFERENCES

ALLISON, I. S., 1979, Pluvial Fort Rock Lake, Lake County, Oregon: Oregon Department of Geology and Mineral Industries Special Paper, v. 7, 72 p.

BARBERI, F. AND VARET, J., 1970, The Erta Ale volcanic range (Danakil Depression, Northern Afar, Ethiopia): Bulletin of Volcanology, v. 34, p. 848–917.

BASALTIC VOLCANISM STUDY PROJECT, 1981, Basaltic Volcanism on the Terrestrial Planets: Pergamon Press Inc., New York, 1286 p.

BLOOMFIELD, K., 1975, A late-Quaternary volcanic field in central Mexico: Geologische Rundschau, v. 64, p. 476–497.

CAS, R. A. F., AND WRIGHT, J. V., 1987, Volcanic Successions: Modern and Ancient: Allen and Unwin, London, 528 p.

CHRISTENSEN, M. N., AND GILBERT, C. M., 1964, Basaltic cone suggests constructional origin of some guyots: Science, v. 143, p. 240–242.

COLLINS, B. D., AND DUNNE, T., 1986, Erosion of tephra from the 1980 eruption of Mount St. Helens: Geological Society of America Bulletin, v. 97, p. 896–905.

CROWE, B. M., AND FISHER, R. V., 1973, Sedimentary structures in base-surge deposits with special reference to cross-bedding, Ubehebe Craters, Death Valley, California: Geological Society of America Bulletin, v. 84, p. 663–682.

CUMMINGS, M. L., AND GROWNEY, L. P., 1988, Basalt hydrovolcanic deposits in the Dry Creek arm area of the Owyhee Reservoir, Malheur County, Oregon: Stratigraphic relations: Oregon Geology, v. 50, p. 75–82.

CUSTER, S. G., 1973, Stratigraphy and sedimentation of Black Point Volcano, Mono Basin, California: University of California, Berkeley, Unpubl. m.s. thesis, 114 p.

DAVIES, D. K., VESSELL, R. K., MILES, R. C., FOLEY, M. G., AND BONIS, S. B., 1978, Fluvial transport and downstream sediment modifications in an active volcanic region, *in* Miall, A. D., ed., Fluvial Sedimentology: Canadian Society of Petroleum Geologists Memoir, v. 5, p. 61–84.

DOHRENWEND, J. C., ABRAHAMS, A. D., AND TURRIN, B. D., 1987, Drainage development on basaltic lava flows, Cima volcanic field, southeast California, and Lunar Crater volcanic field, south-central Nevada: Geological Society of America Bulletin, v. 99, p. 405–413.

DOHRENWEND, J. C., WELLS, S. G., AND TURRIN, B. D., 1986, Degradation of Quaternary cinder cones in the Cima volcanic field, Mojave Desert, California: Geological Scoeity of America Bulletin, v. 97, p. 421–427.

EGGLER, W. A., 1963, Plant life of Paricutin Volcano, Mexico, eight years after activity ceased: American Midland Naturalist, v. 29, p. 38–68.

FISHER, R. V., AND SCHMINCKE, H.-U., 1984, Pyroclastic Rocks: Springer-Verlag, New York, 472 p.

FISHER, R. V., AND WATERS, A. C., 1970, Base surge bedforms in maar volcanoes: American Journal of Science, v. 268, p. 157–180.

FOSHAG, W. F., AND GONZALEZ-R, J., 1956, Birth and development of Paricutin volcano, Mexico: U.S. Geological Survey Bulletin, v. 965-D. p. 355–489.

FRANCIS, E. H., 1968, Effect of sedimentation on volcanic processes, including neck-sill relationships, in the British Carboniferous *in* Proceedings of the 23rd International Geological Congress, Prague, v. 2, p. 163–174.

FRANCIS, E. H., 1970, Bedding in Scottish (Fifeshire) tuff-pipes and its relevance to maars and calderas: Bulletin Volcanologique, v. 34, p. 697–712.

FRANCIS, E. H., 1983, Magma and sediment—II. Problems of interpreting palaeovolcanics buried in the stratigraphic column: Journal of the Geological Society of London, v. 140, p. 165–183.

FRIES, C. JR., 1953, Volumes and weights of pyroclastic material, lava and water erupted by Paricutin volcano, Michoacan, Mexico: EOS, Transactions, American Geophysical Union, v. 34, p. 603–616.

GILBERT, G. K., 1890, Lake Bonneville: U.S. Geological Survey Monograph, v. 1, 438 p.

GODCHAUX, M. M., BONNICHSEN, B., AND JENKS, M. D., 1989, The volcano in the lake in the plain above the hotline: The Sinker Butte story: New Mexico Bureau of Mines and Mineral Resources Bulletin, v. 131, p. 109.

HACK, J. T., 1942, Sedimentation and volcanism in the Hopi Buttes, Arizona: Geological Society of America Bulletin, v. 53, p. 335–372.

HACKETT, W. R., AND HOUGHTON, B. F., 1989, A facies model for a Quaternary andesitic composite volcano: Ruapehu, New Zealand: Bulletin of Volcanology, v. 51, p. 51–68.

HAMILTON, W. B., AND MYERS, W. B., 1963, Menan Buttes, cones of glassy basalt tuff in the Snake River Plain, Idaho: U.S. Geological Survey Professional Paper, v. 450, p. E114–E118.

HASENAKA, TOSHIAKI, AND CARMICHAEL, I.S.E., 1985, The cinder cones of Michoacan—Guanajuato, central Mexico: Their age volume and distribution, and magma discharge rate: Journal of Volcanology and Geothermal Research, v. 25, p. 105–124.

HEAD, J. W., III, AND WILSON, L., 1989, Basaltic pyroclastic eruptions: Influence of gas-release patterns and volume fluxes on fountain structure, and the formation of cinder cones, spatter cones, rootless flows, lava ponds and lava flows: Journal of Volcanology and Geothermal Research, v. 37, p. 261–271.

HEARN, B. C., JR., 1968, Diatremes with kimberlitic affinities in north-central Montana: Science, v. 159, p. 622–625.

HEIKEN, G. H., 1971, Tuff rings: Examples from the Fort Rock-Christmas Lake Valley Basin, south-central Oregon: Journal of Geophysical Research, v. 76, p. 5615–5626.

HEIKEN, G. H., 1972a, Tuff Rings: Examples from the Fort Rock, Christmas Lake Valley Basin, south-central Oregon: University of California, Santa Barbara, Unpublished Ph.D. Dissertation, 328 p.

HEIKEN, G. H., 1972b, Morphology and petrography of volcanic ashes: Geological Society of America Bulletin, v. 83, p. 1961–1988.

HEIKEN, G. H., 1974, An Atlas of Volcanic Ash: Smithsonian Contributions to the Earth Sciences, v. 12, 101 p.

HEIKEN, G. H., 1978, Characteristics of tephra from Cinder Cone, Lassen Volcanic National Park, California: Bulletin of Volcanology, v. 41–2, p. 1–12.

HEIKEN, G. H., AND WOHLETZ, K. H., 1986, Volcanic Ash: University of California Press, Berkeley, 246 p.

HOFFMEISTER, J. E., LADD, H. S., AND ALLING, H. L., 1929, Falcon Island: American Journal of Science, v. 18, p. 461–471.

HOUGHTON, B. F., AND HACKETT, W. R., 1984, Strombolian and phreatomagmatic deposits of Ohakune Craters, Ruapehu, New Zealand: A complex interaction between external water and rising basaltic magma: Journal of Volcanology and Geothermal Research, v. 21, p. 207–231.

HOUGHTON, B. F., AND SCHMINCKE, H.-U., 1986, Mixed deposits of simultaneous Strombolian and phreatomagmatic volcanism: Rothenberg Volcano, East Eifel volcanic field: Journal of Volcanology and Geothermal Research, v. 30, p. 117–130.

HOUGHTON, B. F., AND SCHMINCKE, H.-U., 1989, Rothenberg scoria cone, East Eifel: A complex Strombolian and phreatomagmatic volcano: Bulletin of Volcanology, v. 52, p. 28–48.

ILLIES, H., 1959, Die Entstehungsgeschichte eines Maares in Sudchile: Geologische Rundschau, v. 48, p. 232–247.

JENKS, M. D., AND BONNICHSEN, B., 1989, Subaqueous basalt eruptions into Pliocene Lake Idaho, *in* Chamberlain, V. E., Breckenridge, R. M., and Bonnichsen, B., eds., Guidebook to the Geology of Northern and Western Idaho and Surrounding Area: Idaho Geological Survey Bulletin, v. 28, p. 17–34.

JOHNSON, R. W., AND TUNI, D., 1987, Kavachi, an active forearc volcano in the western Solomon Islands: Reported eruptions between 1950 and 1982, *in* Taylor, B., and Exon, N. F., eds., Marine Geology, Geophysics, and Geochemistry of the Woodlark Basin—Solomon Islands: Circum-Pacific Council for Energy and Mineral Resources Earth Science Series, v. 7, p. 89–112.

KELLER, J., 1974, Quaternary maar volcanism near Karapinar in central Anatolia: Bulletin Volcanologique, v. 36, p. 378–396.

KIENLE, J., KYLE, P. R., SELF, S., MOTYKA, R. J., AND LORENZ, V., 1980, Ukinrek Maars, Alaska, 1 April (1977) eruption sequence, petrology and tectonic setting: Journal of Volcanology and Geothermal Research, v. 7, p. 11–37.

KOKELAAR, B. P., 1986, Magma-water interactions in subaqueous and emergent basaltic volcanism: Bulletin of Volcanology, v. 48, p. 275–289.

KUENZI, W. D., HORST, O. H., AND MCGEHEE, R. V., 1979, Effect of volcanic activity on fluvial-deltaic sedimentation in a modern arc-trench gap, southwestern Guatemala: Geological Society of America Bulletin, v. 90, p. 827–838.

LEAT, P. T., AND THOMPSON, R. N., 1988, Miocene hydrovolcanism in

NW Colorado, USA, fueled by explosive mixing of basic magma and wet unconsolidated sediment: Bulletin of Volcanology, v. 50, p. 229–243.

LEYS, C. A., 1982, Volcanic and sedimentary processes in phreatomagmatic volcanoes, University of Leeds, Department of Earth Sciences, Unpublished Ph.D. Dissertation, 402 p.

LEYS, C. A., 1983, Volcanic and sedimentary processes during formation of the Saefell tuff-ring, Iceland: Transactions of the Royal Society of Edinburgh: Earth Sciences, v. 74, p. 15–22.

LORENZ, V., 1973, On the formation of maars: Bulletin of Volcanology, v. 37, p. 183–204.

LORENZ, V., 1974a, Vesiculated tuffs and related features: Sedimentology, v. 21, p. 273–291.

LORENZ, V., 1974b, Studies of the Surtsey tephra deposits: Surtsey Research Progress Reports, v. 7, p. 72–79.

LORENZ, V., 1975, Formation of phreatomagmatic maar-diatreme volcanoes and its relevance to kimberlite diatremes: Physics and Chemistry of the Earth, v. 9, p. 17–27.

LORENZ, V., 1984, Explosive volcanism of the West Eifel volcanic field, Germany, *in* Kornprobst, J., ed., Kimberlites 1: Kimberlites and Related Rocks: Elsevier, Amsterdam, p. 299–307.

LORENZ, V., MCBIRNEY, A. R., AND WILLIAMS, H., 1970, An investigation of volcanic depressions, Part III, Maars, tuff rings, tuff cones, and diatremes: Houston NASA Progress Report NGR 38003012, 198 p.

MACDONALD, G. A., 1972, Volcanoes: Prentice-Hall, Inc., Englewood Cliffs, New Jersey, 510 p.

MARTIN DEL POZZO, A. L., 1982, Monogenetic vulcanism in Sierra Chichinautzin, Mexico: Bulletin Volcanolologique, v. 45-1, p. 9–24.

MOORE, J. G., 1967, Base surges in recent volcanic eruption: Bulletin Volcanologique, v. 30, p. 337–363.

MÜLLER, G. AND VEYL, G., 1957, The birth of Nilahue, a new maar type volcano at Rininahue, Chile: 20th International Geological Congress, Mexico, Section I, pt. 2, Cenozoic Volcanism, p. 375–396.

NAIRN, I. A., 1979, Rotomahana-Waimangu eruption, 1886. base-surge and basalt magma: New Zealand Journal of Geology and Geophysics, v. 22, p. 363–378.

OLLIER, C. D., 1967, Maars, their characteristics, varieties and definition: Bulletin of Volcanology, v. 31, p. 45–73.

OLLIER, C. D., AND BROWN, M. J. F., 1971, Erosion of a young volcano in New Guinea: Zeitschrift fur Geomorphologie, v. 15, p. 12–28.

OVIATT, C. G., AND NASH, W. P., 1989, Late Pleistocene basaltic ash and volcanic eruptions in the Bonneville basin, Utah: Geological Society of America Bulletin, v. 101, p. 292–303.

PALMER, B. A., AND WALTON, A. W., 1990, Accumulation of volcaniclastic aprons in the Mount Dutton Formation (Oligocene-Miocene), Marysvale volcanic field, Utah: Geological Society of America Bulletin, v. 102, p. 734–748.

RICHARDS, A. F., 1959, Geology of the Islas Revillagigedo, Mexico, 1. Birth and development of Volcan Barcena, Isla San Benedicto: Bulletin of Volcanology, v. 22, p. 73–123.

RICHARDS, A. F., 1965, Geology of the Islas Revillagigedo, 3. Effects of erosion on Isla San Benedicto 1952–1961 following the birth of Volcan Barcena: Bulletin Volcanologique, v. 28, p. 381–403.

SCOTT, D. H., AND TRASK, N. J., 1971, Geology of the Lunar Crater Volcanic field, Nye County, Nevada: U.S. Geological Survey Professional Paper, v. 599-I, 22 p.

SEGERSTROM, K., 1950, Erosion studies at Paricutin volcano, state of Michoacan, Mexico: U.S. Geological Survey Bulletin, v. 956-A, 164 p.

SEGERSTROM, K., 1966, Paricutin, 1965—Aftermath of eruption: U.S. Geological Survey Bulletin, v. 550-C, p. 93–101.

SELF, S., KIENLE, J., AND HUOT, J.-P., 1980, Ukinrek maars, Alaska, II. Deposits and formation of the (1977) craters: Journal of Volcanology and Geothermal Research, v. 7, p. 39–65.

SETTLE, M., 1979, The structure and emplacement of cinder cone fields: American Journal of Science, v. 279, p. 1089–1107.

SHERIDAN, M. F., AND WOHLETZ, K. H., 1983, Hydrovolcanism: Basic considerations and review: Journal of Volcanology and Geothermal Research, v. 17, p. 1–29.

SHOEMAKER, E. M., ROACH, C. H., AND BYERS, F. M., JR., 1962, Diatremes and uranium deposits in the Hopi Buttes, Arizona, *in* Petrologic Studies: A Volume to Honor A. F. Buddington: Geological Society of America, Boulder, Colorado, p. 327–356.

SLY, P. G., 1978, Sedimentary processes in lakes, *in* Lerman, A., ed., Lakes: Chemistry, Geology, Physics: Springer-Verlag, New York, p. 65–68.

SMITH, G. A., 1986, Coarse-grained volcaniclastic sediment: Terminology and depositional process: Geological Society of America Bulletin, v. 97, p. 1–10.

SMITH, G. A., 1987, Sedimentology of volcanism-induced aggradation in fluvial basins: Examples from the Pacific Northwest, USA, *in* Ethridge, F. G., Flores R. M., and Harvey, M. A., eds., Recent Developments in Fluvial Sedimentology: Society of Economic Paleontologists and Mineralogists Special Publication 39, p. 217–228.

SMITH, G. A., 1988a, Neogene synvolcanic and syntectonic sedimentation in central Washington: Geological Society of America Bulletin, v. 100, p. 1479–1492.

SMITH, G. A., 1988b, Sedimentology of proximal to distal volcaniclastics dispersed across an active foldbelt: Ellensburg Formation (late Miocene), central Washington: Sedimentology, v. 35, p. 953–977.

STEHN, C. E., 1934, Die semivulkanische Explosion des Pematang Bata, 1933: Natuurkundig Tijdschrift voor Nederlandsch-Indie, v. 94, p. 46–69.

THORARINSSON, S., 1967, Surtsey. The new island in the North Atlantic The Viking Press, New York, 47 p.

TURBEVILLE, B. N., WARESBACK, D. B., AND SELF, S., 1989, Lava-dome growth and explosive volcanism in the Jemes Mountains, New Mexico: Evidence from the Plio-Pleistocene Puye alluvial fan: Journal of Volcanology and Geothermal Research, v. 36, p. 267–291.

VERWOERD, W. J., AND CHEVALLIER, L., 1987, Contrasting types of Surtseyan tuff cones on Marion and Prince Edward Islands, southwest Indian Ocean: Bulletin of Volcanology, v. 49, p. 399–414.

VESSELL, R. D., AND DAVIES, D. K., 1981, Nonmarine sedimentation in an active fore arc basin, *in* Ethridge, F. G., and Flores, R. M., eds., Nonmarine Depositional Environments: Models for Exploration: Society of Economic Paleontologists and Mineralogists Special Publication 31, p. 34–45.

VESSELL, R. D., AND DAVIES, D. K., 1986, On the growth of maars and diatremes and its relevance to the formation of tuff rings: Bulletin of Volcanology, v. 48, p. 265–274.

WALDRON, H. W., 1967, Debris flow and erosion control problems caused by the ash eruptions of Irazu Volcano, Costa Rica: U.S. Geological Survey Bulletin, v. 1241-I, 37 p.

WALTON, A. W., 1979, Volcanic sediment apron in the Tascotal Formation (Oligocene?), Trans-Pecos, Texas: Journal of Sedimentary Petrology, v. 49, p. 303–314.

WALTON, A. W., 1986, Effect of Oligocene volcanism on sedimentation in the Trans-Pecos volcanic field of Texas: Geological Society of America Bulletin, v. 97, p. 1192–1207.

WARESBACK, D. B., AND TURBEVILLE, B. N., 1990, Evolution of a Plio-Pleistocene volcanogenic-alluvial fan: The Puye Formation, Jemez Mountains, New Mexico: Geological Society of America Bulletin, v. 102, p. 298–314.

WATERS, A. C., AND FISHER, R. V., 1970, Maar volcanoes, *in* Proceedings of the 2nd Columbia River Basalt Symposium, 1969, Cheney, Washington: Eastern Washington State College Press, p. 157–170.

WATERS, A. C., AND FISHER, R. V., 1971, Base surges and their deposits: Capelinhos and Taal volcanoes: Journal of Geophysical Research, v. 76, p. 5596–5614.

WELLS, S. G., DOHRENWEND, J. C., MCFADDEN, L. D., TURRIN, B. D., AND MAHRER, K. D., 1985, Late Cenozoic landscape evolution on lava flow surfaces of the Cima volcanic field, Mojave Desert, California: Geological Society of America Bulletin, v. 96, 1518–1529.

WELLS, S. G., MCFADDEN, L. D., RENAULT, C. E., AND CROWE, B. M., 1990, Geomorphic assessment of late Quaternary volcanism in the Yucca Mountain area, southern Nevada: Implications for the proposed high-level radioactive waste repository: Geology, v. 18, p. 549–553.

WHITE, J. D. L., 1989, Basic elements of maar-crater deposits in the Hopi Buttes volcanic field, northeastern Arizona, U.S.A.: Journal of Geology, v. 97, p. 117–125.

WHITE, J. D. L., 1990, Depositional architecture of a maar-pitted playa, Hopi Buttes volcanic field, northeastern Arizona, U.S.A.: Sedimentary Geology, v. 67, p. 55–84.

WOHLETZ, K. H., 1983, Mechanisms of hydrovolcanic pyroclast formation. Grain-size, scanning electron microscopy, and experimental studies: Journal of Volcanology and Geothermal Research, v. 17, p. 31–63.

WOHLETZ, K. H., 1986, Explosive magma-water interactions: Thermodynamics, explosion mechanisms, and field studies: Bulletin of Volcanology, v. 48, p. 245–264.

WOHLETZ, K. H., AND MCQUEEN, R. G., 1984, Experimental studies of hydromagmatic volcanism, *in* Explosive Volcanism; Inception, Evolution, and Hazards: National Academy Press, Washington, D.C., p. 158–169.

WOHLETZ, K. H., AND SHERIDAN, M. F. 1983, Hydrovolcanic explosions II. Evolution of basaltic tuff rings and tuff cones: American Journal of Science, v. 283, p. 384–413.

WOMER, M. B., GREELY, R., AND KING, J. S., 1980, The geology of Split Butte—A maar of the south-central Snake River Plain, Idaho: Bulletin of Volcanology, v. 43, p. 453–471.

WOOD, C. A., 1978, Morphometric evolution of composite volcanoes: Geophysical Research Letters, v. 5, p. 437–439.

WOOD, C. A., 1979, Monogenetic volcanoes of the terrestrial planets, *in* Proceedings of the Lunar and Planetary Science Conference, v. 10, p. 2815–2840.

WOOD, C. A., 1980a, Morphometric evolution of cinder cones: Journal of Volcanology and Geothermal Research, v. 7, p. 387–413.

WOOD, C. A., 1980b, Morphometric analysis of cinder cone degradation: Journal of Volcanology and Geothermal Research, v. 8, p. 137–160.

# PART IV
# VOLCANICLASTIC SEDIMENTATION IN OCEANIC SETTINGS

Volcaniclastic sedimentation in marine settings is known mostly from studies of ancient geologic settings, and more recently from ocean bottom drilling and exploration. Part IV discusses Early Proterozoic rocks in Canada, Miocene rocks in New Zealand, and ocean bottom exploration in the Izu-Bonin Arc.

# VOLUMINOUS SHALLOW-WATER TO EMERGENT PHREATOMAGMATIC BASALTIC VOLCANICLASTIC ROCKS, PROTEROZOIC (~1886 MA) AMISK LAKE COMPOSITE VOLCANO, FLIN FLON GREENSTONE BELT, CANADA

L. D. AYRES
*Department of Geological Sciences, University of Manitoba, Winnipeg, Manitoba, Canada, R3T 2N2*

N. A. VAN WAGONER
*Department of Geology, Acadia University, Wolfville, Nova Scotia, Canada, B0P 1X0*

W. S. FERREIRA
*W. S. Ferreira Ltd., 91 Lakeshore Road, Winnipeg, Manitoba, Canada, R3T 4A7*

ABSTRACT: A 7-km-thick basaltic sequence composed of 50 percent flows and 50 percent intercalated volcaniclastic rocks records emergence above sea level of a submarine composite volcano (Amisk Lake Volcano), partial resubmergence, and eventual re-emergence. Both emergence and resubmergence are marked by crudely bedded surf-zone or shore-facies deposits up to 300 m thick, comprising pillowed flows and palagonitized tuff, lapilli tuff, and tuff breccia produced by a combination of quenching as subaerial flows entered the sea and by littoral explosions. Subaerial units form the upper 900 m and up to 1 km of the lower part of the sequence. They are dominantly surge-and-fall tuff and fine lapilli tuff composed of non-amygdaloidal vitric granules and lesser amygdaloidal, in part scoriaceous, vitric ash, lithic basalt, and clinopyroxene and plagioclase crystals; accretionary and armored lapilli and tuff vesicles are common. Subaqueous units contain up to 40 percent intercalated fall and reworked tuff and fine lapilli tuff composed of particles similar to those in subaerial units; some deposits are turbidites. Both subaerial and subaqueous tuff and lapilli tuff were produced largely by long-lived or repeated subaerial phreatomagmatic explosive eruptions generated by rise of basaltic magma through the water-saturated coastal plain of a rapidly subsiding volcanic island; the degree of water-magma interaction was variable. The volume of explosively produced ash is orders of magnitude larger than that in modern volcanoes and reflects the rapid subsidence and consequent low-gradient volcano slopes.

## INTRODUCTION

Volcaniclastic rocks are a minor component of most basaltic volcanoes, but they can be produced by a wide variety of subaerial and subaqueous processes, including magmatic and phreatomagmatic explosions, lava quenching in quiet or agitated water, autoclastic brecciation of lava flows, and slumping and disintegration of lava flows (e.g., Carlisle, 1963; Lonsdale and Batiza, 1980; Fisher and Schmincke, 1984; Staudigel and Schmincke, 1984; Kokelaar, 1986; Busby-Spera, 1987; Cas and others, 1989; Smith and Batiza, 1989; Dolozi and Ayres, in prep.) In many volcanoes, several processes may operate simultaneously, producing complex deposits. Once fragmented, volcaniclastic rocks are subjected to various types and degrees of reworking and resedimentation.

Because of the numerous possible formative processes, volcaniclastic deposits, where present, may better constrain eruptive and post-eruptive processes, volcano morphology, and water depth than associated lava flows. Of particular importance in paleovolcanology are shallow-water or subaerial processes, the products of which record the emergence or near emergence of a volcano. In many ancient oceanic basaltic volcanoes that emerged above sea level as islands, direct evidence of emergence is lacking because the subaerial edifice was destroyed by erosion and was small relative to the total volcano area. Chance sections through such a volcano in a deformed mountain belt or greenstone belt are thus most likely to record the subaqueous history. Consequently, in many paleovolcanoes, evidence of subaerial activity may be recorded only by resedimented volcaniclastic rocks deposited on the subaqueous volcano slope but produced by shallow-water or emergent processes.

## PALEOVOLCANOLOGY OF PRECAMBRIAN GREENSTONE BELTS

Precambrian greenstone belts have many similarities to volcanic roof pendants in plutonic terranes of Phanerozoic mountain belts (e.g., Busby-Spera, 1986). Both are metamorphosed and deformed, volcano-sedimentary assemblages that have been partly dismembered by intrusion of large granitoid plutons. As such, greenstone belts are just as amenable to paleovolcanologic study as comparably deformed and metamorphosed Phanerozoic terranes.

In typical greenstone belts, there is a compositional and volcanologic cyclicity with tholeiitic basaltic- to komatiitic-flow sequences overlain by calc-alkaline andesitic- to rhyolitic-flow and volcaniclastic sequences (Condie, 1981; Ayres and Thurston, 1985). In large belts, volcanic sequences are 10 to 20 km thick and may include one or more cycles. Preserved volcanic sequences are largely subaqueous, but sedimentologic and volcanologic evidence indicates that many volcanoes evolved into long-lived islands.

In many greenstone belts, deformation has resulted in upright isoclinal folds, the eroded limbs of which are sections through ancient volcanoes and associated sedimentary basins. Some rock units are highly flattened and attenuated, particularly near large granitoid plutons, where metamorphic grade is amphibolite facies, but away from plutons, penetrative deformation in many belts is minor and metamorphic grade is greenschist or subgreenschist facies. Thus, although rock units have been tilted to a subvertical position, primary volcanic and sedimentary structures are well preserved, and original grain shapes are maintained by pseudomorphs (e.g., Dimroth and Lichtblau, 1979). Critical textural and structural evidence that constrains processes is thus available for paleovolcanology studies. In the

Sedimentation in Volcanic Settings, SEPM Special Publication No. 45

Canadian Shield, flat, smooth, glaciated outcrops provide good vertical sections, on a horizontal plane, through tilted rock units.

## FLIN FLON—SNOW LAKE GREENSTONE BELT

This typical greenstone belt is an isoclinally folded, metavolcanic-metasedimentary sequence intruded by ultramafic to granitoid plutons (Fig. 1). It is covered by Paleozoic strata on the south but where exposed, the belt is 230 km long and 50 km wide. On the north, the belt is bordered by the Kisseynew sedimentary gneiss belt, an amphibolite-facies terrane that is partly a facies equivalent of volcanic rocks of the greenstone belt (Bailes, 1980).

Lava flows and associated volcaniclastic rocks are assigned to the Amisk Group, which comprises a lower mafic to intermediate sequence at least 7.6 km thick, capped by a felsic to intermediate sequence at least 3 km thick; the base of the group is not exposed. The felsic to intermediate volcanism produced extensive subaerial pyroclastic deposits that were rapidly eroded to form submarine turbidite fans (Bailes, 1980). The only record of this subaerial volcanism is the adjacent sedimentary basins. The Amisk Group is unconformably overlain by sandstone, conglomerate, and local volcanic rocks of the Missi Group (Syme and others, 1982). A U-Pb zircon age of 1886 ± 1.3 Ma from rhyolitic tuff near Flin Flon is the best date for Amisk Group volcanism (Syme and others, 1987).

Our work has been concentrated at Amisk Lake, the type area for the Amisk Group, at the west end of the greenstone belt (Fig. 1). Other volcanological investigations have been made near Flin Flon (Bailes and Syme, 1989; Dolozi and Ayres, in prep.), but correlation between these two parts of the belt and with areas farther east is hampered by large granitoid plutons that almost sever the belt into segments.

## THE AMISK VOLCANO

At Amisk Lake, the Amisk Group has been folded into upright north- to northwest-trending isoclinal folds and cut by several north- and northwest-trending faults (Fig. 2). Our investigations are concentrated on the limbs of the Winterton Island anticline in the south and central part of the lake, where flat shoreline outcrops on numerous islands provide excellent sections through flows and volcaniclastic units, metamorphic grade is low, foliation is only weakly developed, and there are few complications from plutons. The volcanic units were subjected to an early zeolitic metamorphic event during which pore spaces were filled by analcime and other minerals. The analcime has since been converted to albite (Place, 1982), but the early infilling prevented major flattening during subsequent deformation. Consequently, many delicate textures have been preserved. Metamorphic grade and degree of penetrative deformation increase northward and westward, where granitoid plutons become prevalent.

The west limb of the anticline is a homoclinal sequence more than 8.5 km thick that is interrupted by one major semi-concordant fault. In ascending stratigraphic order this sequence comprises about 7 km of interlayered mafic flows

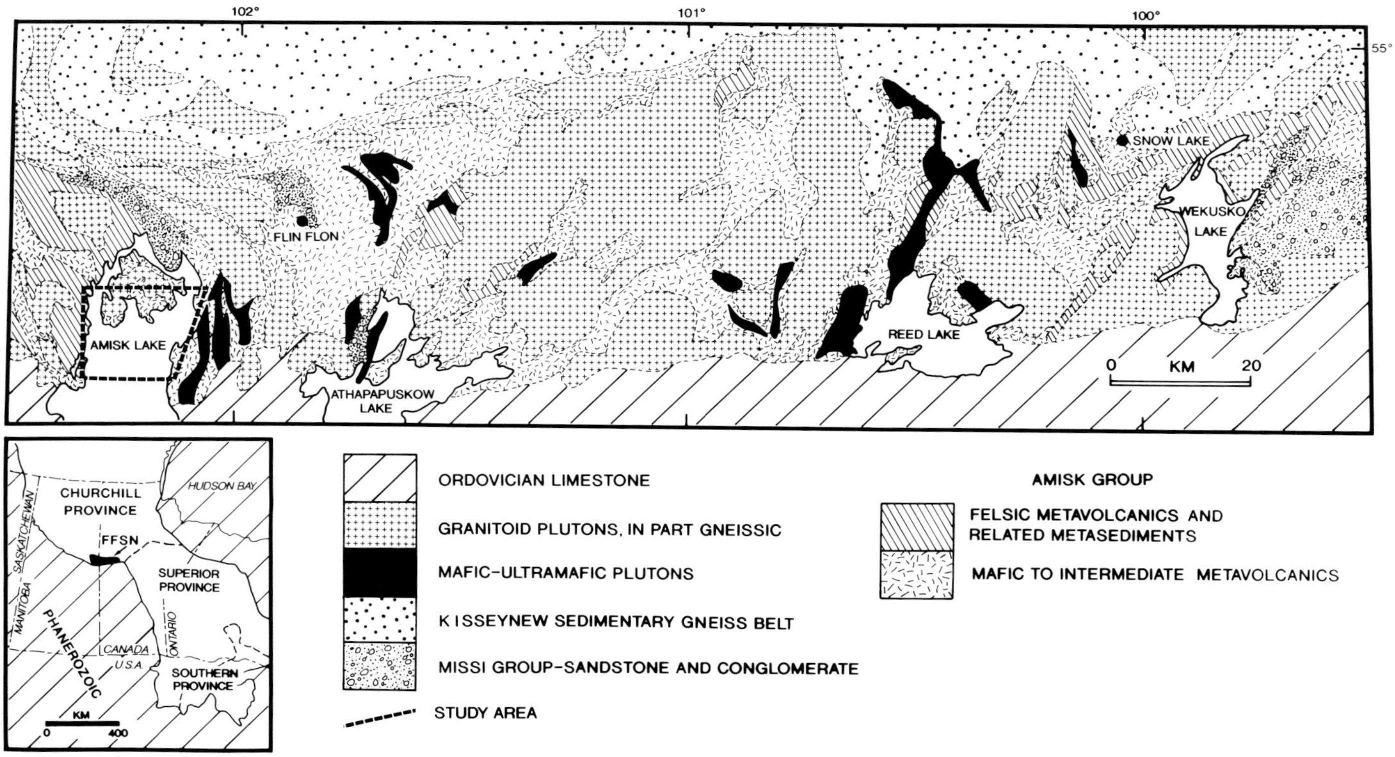


FIG. 1.—Location and simplified geology of the Flin Flon—Snow Lake (FFSN) greenstone belt (modified from Bailes, 1971; Whitaker and Pearson, 1972). Amisk Lake study area is outlined by dashed line.

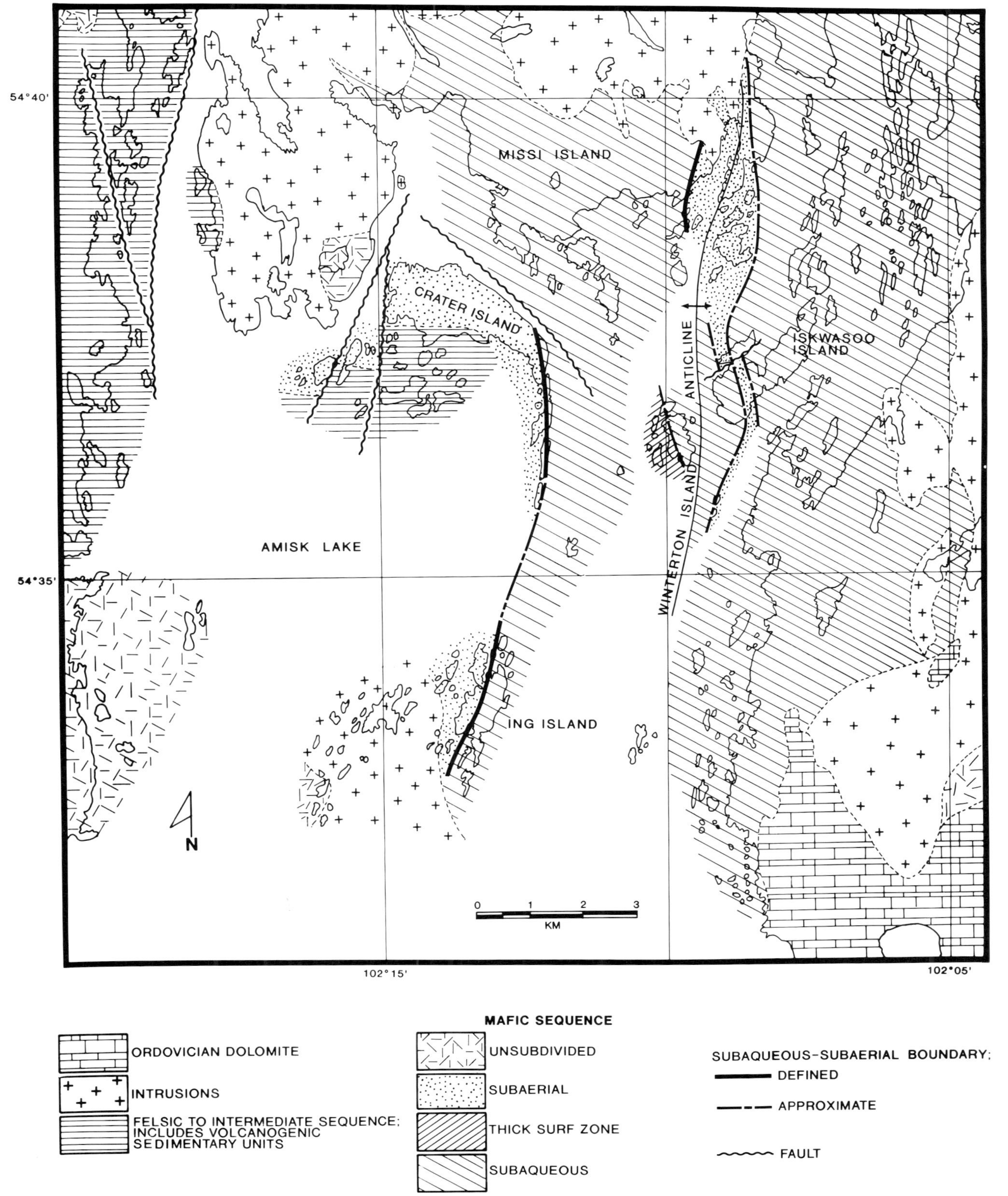


FIG. 2.—Map showing major structures and lithologies, and distribution of eruptive environments, where defined, in the lower mafic part of the Amisk Lake volcano, south-central Amisk Lake. For clarity, gabbro sills and dikes in mafic sequences and many small islands that provide control for lithologic boundaries in the lake have been omitted. Lithologies on the west shore of the lake and intrusions are after Byers and Dahlstrom (1954), Fox (1976), and unpublished mapping by J. Biczok and M. E. Chute.

and volcaniclastic rocks, 0.6 km of intermediate flows and volcaniclastic rocks, and 0.9 km of felsic to intermediate volcaniclastic units and minor flows (Fig. 3). This sequence is separated by a lake and several major north-trending faults from the main part of the upper felsic to intermediate sequence on the west side of the lake (Fig. 2). On the east limb, only the lower 4.4 km of the mafic sequence have been examined. Although not shown on Figures 2 and 3, mafic sequences on both limbs of the fold have been expanded by intrusion of numerous gabbro sills and less common dikes (Byers and Dahlstrom, 1954). The mafic sequence is largely tholeiitic basalt and basaltic andesite (Ferreira, 1984; Gaskarth and Parslow, 1987).

Based on lava-flow characteristics, part of the mafic sequence is subaerial, and the volcano thus differs from most other mafic greenstone-belt volcanoes, which appear to be entirely subaqueous. Subaerial flows lack pillows, a common characteristic of subaqueous flows, but do contain pahoehoe toes, have a high-amygdule content including well-developed basal-pipe amygdules, and have a much thinner chilled selvage than subaqueous flows (Ayres, 1981). Subaerial flows and associated volcaniclastic deposits form sequences up to 1 km thick in the lower and uppermost parts of the mafic sequence (Fig. 3). In the intervening subaqueous interval, both pillowed and sheet flows are common, and most flows have a low- to moderate-amygdule content, indicating eruption in moderately shallow to shallow water. Separating subaerial and subaqueous deposits are distinctive tuff to tuff-breccia units, termed surf-zone deposits (Fig. 3).

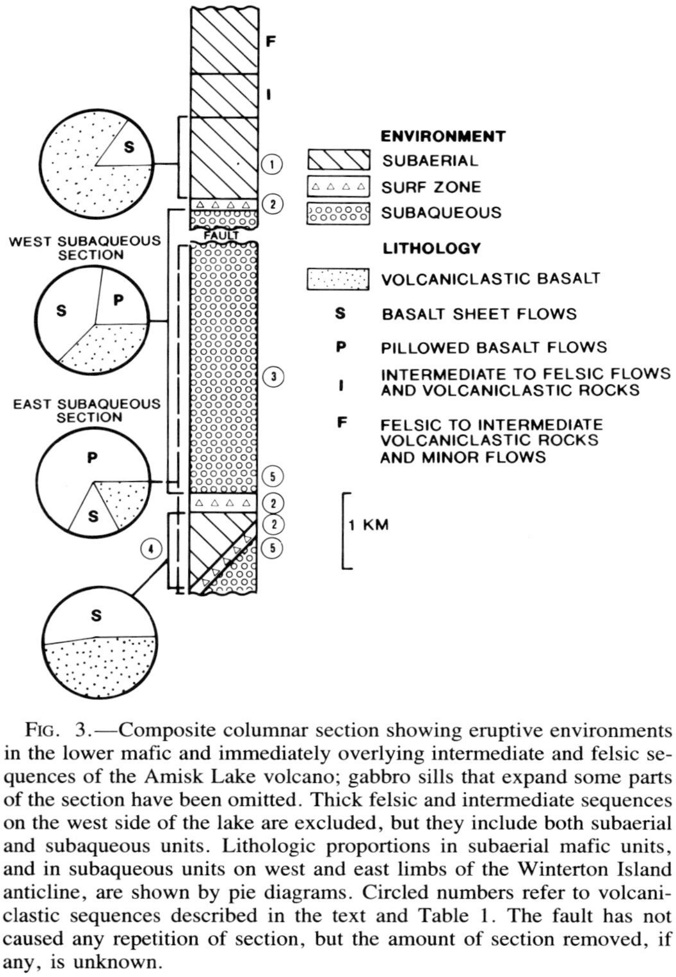


FIG. 3.—Composite columnar section showing eruptive environments in the lower mafic and immediately overlying intermediate and felsic sequences of the Amisk Lake volcano; gabbro sills that expand some parts of the section have been omitted. Thick felsic and intermediate sequences on the west side of the lake are excluded, but they include both subaerial and subaqueous units. Lithologic proportions in subaerial mafic units, and in subaqueous units on west and east limbs of the Winterton Island anticline, are shown by pie diagrams. Circled numbers refer to volcaniclastic sequences described in the text and Table 1. The fault has not caused any repetition of section, but the amount of section removed, if any, is unknown.

DESCRIPTION OF MAFIC VOLCANICLASTIC DEPOSITS

Tuff and lapilli tuff are the dominant components of most subaerial sections, important constituents of subaqueous units, and, along with tuff breccia, the dominant components of surf-zone deposits (Fig. 3; Table 1). Tuff breccia is rare elsewhere except for brecciated tops and bottoms of lava flows (Ferreira, 1984); these are not considered here. In subaqueous units, the abundance of volcaniclastic rocks decreases eastward from about 40 percent on the west limb of the Winterton Island anticline to less than 20 percent on the east limb. On the west limb, volcaniclastic deposits are mainly ungraded bedded to massive units intimately interlayered with flows throughout the section; on the east limb, however, they are dominantly graded, thin-bedded units concentrated in the lower part of the section.

In the following sections, textures and structures of the four main types of volcaniclastic deposits are briefly described and compared to emphasize lateral and vertical similarities and differences and genetic relationships. Each type will be described in more detail in subsequent papers.

*Subaerial Volcaniclastic Deposits*

Volcaniclastic units in both lower and upper subaerial sequences are mostly tuff and fine lapilli tuff with similar component characteristics, but they differ in proportion of intercalated flows and development of bedding (Table 1). In the lower sequence (4 on Fig. 3), bedding is generally poorly developed and flows form about half of the sequence, decreasing in abundance southward. In the upper sequence, bedding is well developed with sharp bed contacts and flows form about 15 percent of the sequence (1 on Fig. 3).

The volcaniclastic rocks are mostly poorly sorted to well sorted fine ash to small lapilli, inferred to be vitric basalt (Table 1; Fig. 4), with lesser holocrystalline lithic basalt (Fig. 5), tuffaceous particles, plagioclase and clinopyroxene fragments, and an early diagenetic cement (Figs. 4, 5, 6, 7). In this and other deposits, the original vitric nature of many grains was inferred from the very fine-grain size of the secondary metamorphic overprint and the scarcity of a recognizable primary crystalline component, but good preservation of tiny crystal particles in fine ash. Vitric grains are angular to subrounded, equant to flattened, aphyric to microporphyritic, and scoriaceous to non-amygdaloidal. Most particles are non-amygdaloidal, and the abundance of amygdaloidal particles is variable from bed to bed; scoriaceous particles (Fig. 6) were observed in only a few beds.

The combined abundance of crystal and lithic particles is generally less than 15 percent in the lower sequence and less than 5 percent in the upper sequence, with crystals commonly more abundant than lithic grains. Local thin to

TABLE 1.—COMPARISON OF BASALTIC VOLCANICLASTIC UNITS IN THE AMISK LAKE VOLCANO

| | Lower Subaerial | Upper Subaerial | Surf-Zone Deposits | Ungraded Subaqueous | Graded Subaqueous |
|---|---|---|---|---|---|
| Number on Figure 3 | 4 | 1 | 2 | 3 | 5 |
| Unit thickness (m) | 1000[1] | 900[1] | 5–300[1] | 0.3–87[2] | 20–350[2] |
| Lateral extent[3] (km) | 9 | 11 | up to 8 | 1 | <1 |
| % Volcaniclastic rocks | 48 | 85 | 57 | trace-40[4] | 40–70[5] |
| Volcaniclastic lithology | tuff, fine lapilli tuff | tuff, fine lapilli tuff, rare tuff breccia | see Table 2 | tuff, fine lapilli tuff | tuff, minor fine lapilli tuff |
| Components[6] | | | | | |
| —total vitric | mostly >85% | >95% | 30–100% | mean ~85% | >98% |
| —vitric, amygdaloidal | rare to common | 0–25% | common | 0–50%; mean ~15% | sparse |
| —hypohyaline juvenile | sparse | sparse | 0–70 | sparse | sparse |
| —holocrystalline lithic (cognate) | to 20%; mostly <5% | <2% | rare | to 25%; mean ~3% | <2% |
| —crystals | to 30%; mostly <10% | <3% | rare | to 30%; mean ~10% | rare |
| Shape of vitric component | mostly angular | angular to subrounded | angular | angular to rounded | angular to rounded |
| Mean component size (mm) | ~1 | ~1 | 5 | ~1 | 0.5 |
| Accretionary structures | common | ubiquitous | rare | rare | minor |
| Bedding development | poor to good | good | poor | poor to good | good |
| Bed thickness | laminated to very thick | laminated to very thick | mostly very thick; locally laminated to thin bedded | laminated to very thick | laminated to thin; locally medium to thick |
| Incidence and type of grading | local to common normal grading; rare reverse grading | ~25%; normal grading, reverse-normal grading, rare reverse grading | rare normal grading | local normal grading; rare reverse grading | ubiquitous normal grading; mostly Bouma ab and upward-fining, laminated layers |
| Other structures | local cross-beds and undulatory beds | local low-angle cross-beds and scours | rare cross-beds | rare cross-beds and scours | load structures, flame structures, rip-ups, ball and pillow structures, and slump folds |

[1]Total thickness of sequence.
[2]Thickness range of intercalations among flows.
[3]Minimum values; this is a function of exposure on scattered islands in the lake.
[4]Within the 3- 4-km-thick sequence between the lower and upper subaerial sequences.
[5]Within 500-m-thick intervals immediately below and immediately above surf-zone deposits that bound the lower subaerial sequence.
[6]Excluding cement, which ranges from 0 to 25%.

medium beds of crystal-rich and lithic-rich tuff occur in the lower sequence (Table 1). Lithic grains include both holocrystalline and hypohyaline varieties, with holocrystalline grains ranging from ophitic to intergranular. Both crystals and lithic grains are texturally and mineralogically identical to lava flows in underlying sequences, the most probable source of these grains. Tuffaceous particles were observed only in the lower sequence and apparently represent reworking of underlying tuff deposits.

Accretionary lapilli and armored lapilli occur in many beds, ranging in abundance from trace to 100 percent (Fig. 7). Accretionary lapilli, some of which are broken, vary in distribution from randomly scattered throughout beds to concentrated in thin layers and lenses (Fig. 8); in places, accretionary lapilli are normally size graded. Rounded to ovoid tuff vesicles up to 1 mm in size were observed in some beds that contain a high abundance of accretionary lapilli (Fig. 7).

Beds range in thickness from 5 mm to 8 m, have sharp to diffuse and poorly defined commonly planar bed contacts, and are generally continuous across the width of shoreline exposures, a distance of 5 to 10 m. Some of the thicker beds have an internal lamination. Most beds are ungraded, but normally graded beds are common, although variable in abundance; reversely graded and reversely to normally graded beds occur locally. Normally graded beds are generally better sorted than ungraded beds, and in places, they have a higher abundance of amygdaloidal particles. Basal scours, low-angle cross-beds, and undulatory-bed contacts were observed in some beds.

### *Surf-Zone Deposits*

Three surf zones have been recognized, two emergent and one intervening submergent zone (Fig. 2; 2 on Fig. 3). The middle and upper zones are best defined and have been traced laterally for 8 km (Fig. 2). Surf-zone deposits are lithologically complex (Fig. 9), and they range in thickness from 5 to 300 m (Table 2). The volcaniclastic component is mostly massive to crudely bedded, clast-supported to matrix-supported tuff breccia, lapilli tuff and minor coarse tuff, some of which contains 10 to 90 percent isolated basaltic pillows or lava lobes and could be termed pillow breccia (Fig. 9); this lithology grades into pillowed flows in which pillows are separated by thin, discontinuous volcaniclastic envelopes. Bedded tuff containing accretionary lapilli, some normally graded beds, and local cross-bedding is a minor component of the lower surf-zone deposits (Table 2).

Typically, the coarser volcaniclastic rocks comprise rounded to angular, non-amygdaloidal to amygdaloidal, hypohyaline, microlitic blocks, sparse bombs, and large lapilli in a matrix of angular vitric coarse ash and small lapilli, rare crystals, and interstitial cement (Figs. 10, 11); maximum block size is 75 cm. Particles in these volcaniclastic

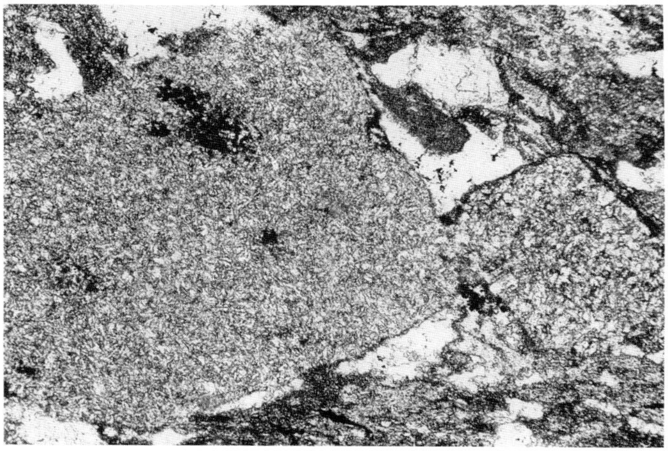

FIG. 4.—Photomicrograph of two angular to subrounded, basaltic vitric-ash grains in coarse tuff of the upper subaerial sequence. The grains have been replaced by very fine-grained actinolite and epidote but primary shapes are well preserved; the thin black rim on the smaller right-hand grain is a coating of very fine ash. A small clinopyroxene fragment pseudomorphed by actinolite occurs in lower left and upper right. The white interstitial material is albite (after analcime) cement. Width of view is 1.7 mm.

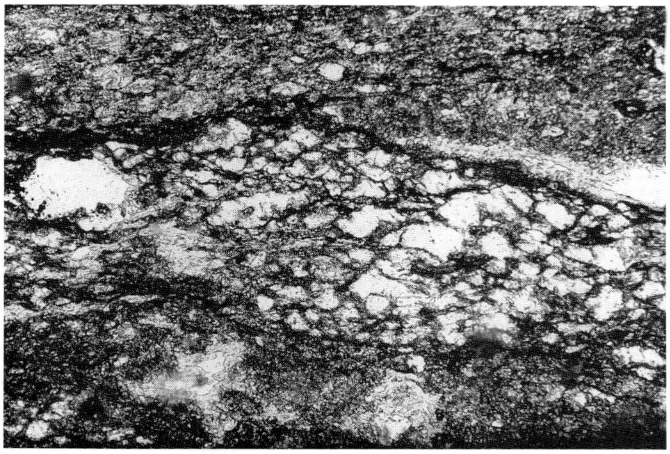

FIG. 6.—Scoriaceous basaltic vitric-ash particle in coarse tuff of the upper subaerial sequence. Vesicles were infilled by analcime that has been converted to albite (white). Rounded shapes of amygdules indicate that infilling occurred prior to major compaction or deformation. Width of view is 1.7 mm.

rocks differ from those in adjacent subaerial sequences in having a much wider size range, more pronounced angularity, more variable crystallinity, and inferred palagonitization (Table 1).

Non-amygdaloidal blocks and large lapilli predominate and commonly have straight to concave, fracture-controlled margins (Fig. 10); many clasts have partial to complete, darker colored 5- to 10-mm-thick rims that appear to be the result of chilling intensified by subsequent alteration. Amygdaloidal blocks and large lapilli contain up to 40 percent amygdules, are more rounded (Fig. 11), and have a more restricted distribution. Many of the larger amygdaloidal clasts have partial to complete chilled rims and are broken and intact bombs. Cored bombs with holocrystalline lithic cores were observed locally.

Matrix ash and small lapilli generally have straight or slightly curved boundaries (Fig. 12), but about 10 percent have vesicle-controlled boundaries aand shard-like shapes (Place, 1982); particles smaller than 0.1 mm are rare. Many particles are zoned (Fig. 12), apparently as a result of palagonitization modified by low-grade metamorphism. The average abundance of interstitial cement is about 25 percent, reflecting the high original porosity of these deposits and the paucity of fine ash.

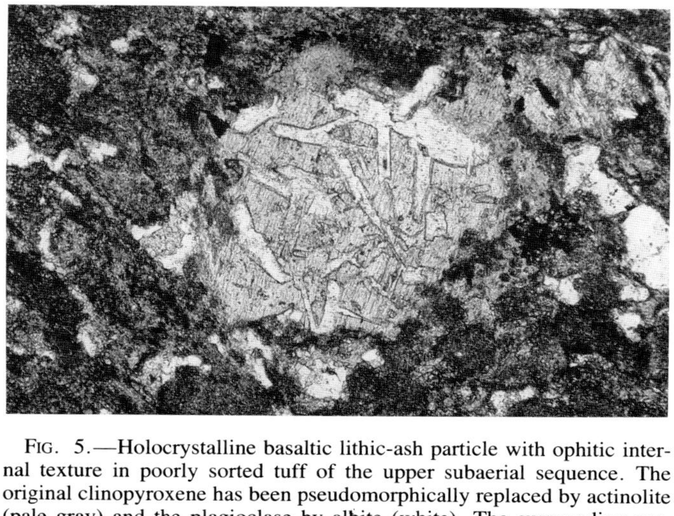

FIG. 5.—Holocrystalline basaltic lithic-ash particle with ophitic internal texture in poorly sorted tuff of the upper subaerial sequence. The original clinopyroxene has been pseudomorphically replaced by actinolite (pale gray) and the plagioclase by albite (white). The surrounding material is fine vitric ash, the boundaries of which are more diffuse because of recrystallization; cement content (white) is minor. Width of field is 1.7 mm.

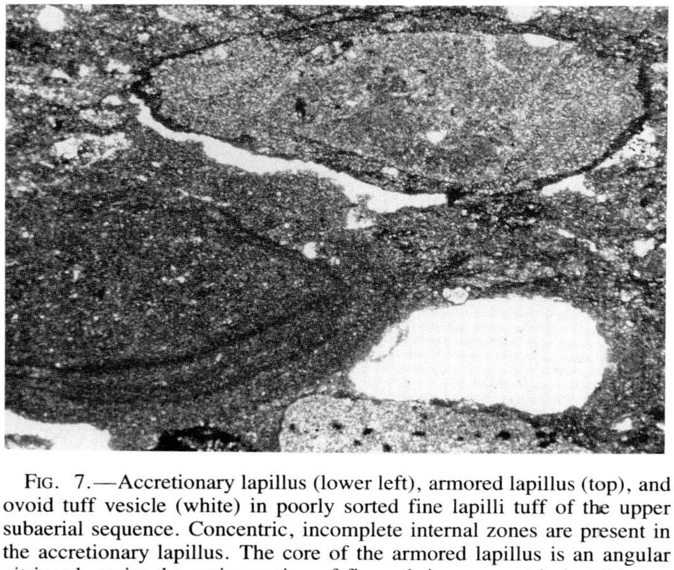

FIG. 7.—Accretionary lapillus (lower left), armored lapillus (top), and ovoid tuff vesicle (white) in poorly sorted fine lapilli tuff of the upper subaerial sequence. Concentric, incomplete internal zones are present in the accretionary lapillus. The core of the armored lapillus is an angular vitric-ash grain; the main coating of fine ash is asymmetric but this has a very thin, symmetric outer rim (black). The tuff vesicle was originally infilled by analcime that has now been converted to albite. Width of field is 4.2 mm.

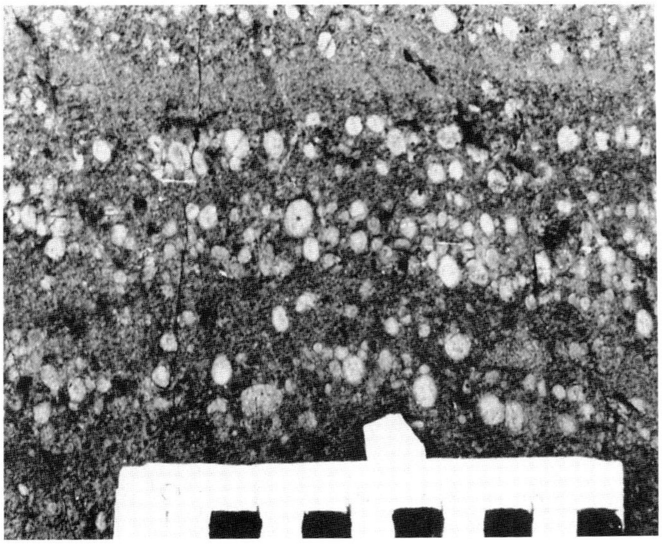

FIG. 8.—Concentrations of accretionary lapilli (white) in fine lapilli tuff, lower subaerial sequence. Black marks at bottom are 1 cm wide.

TABLE 2.—SUMMARY OF LITHOLOGIES AND THICKNESSES OF SURF-ZONE DEPOSITS, AMISK LAKE VOLCANO

| | Lithologic Abundance (%) | | | | |
|---|---|---|---|---|---|
| | Emergent Zones | | | Submergent Zone | |
| | Upper | Lower West[1] | Lower East | West | East |
| Bedded tuff | 0 | 1 | 12 | 3 | 14 |
| Massive tuff and lapilli tuff | 6 | 3 | 11 | 17 | 0 |
| Tuff breccia | 44 | 0 | 0 | 0 | 19 |
| Lapilli tuff and tuff breccia with lava lobes | 23 | 49 | 46 | 34 | 22 |
| Pillowed-lava flows | 27 | 0 | 6 | 41 | 11 |
| Sheet flows, sparsely amygdaloidal | 0 | 15 | 25 | 5 | 18 |
| Sheet flows, moderately to highly amygdaloidal | 0 | 32 | 0 | 0 | 16 |
| Maximum thickness (m) | 40 | 300 | 150 | 190 | 300 |

[1]West and east refer to position relative to axis of Winterton Island anticline.

## *Subaqueous Deposits*

The subaqueous sequence is dominantly sheet and pillowed-lava flows, among which are numerous intercalations of tuff and fine lapilli tuff; the relative proportions of the two flow types and volcaniclastic rocks differ from place to place. Two types of volcaniclastic intercalations can be

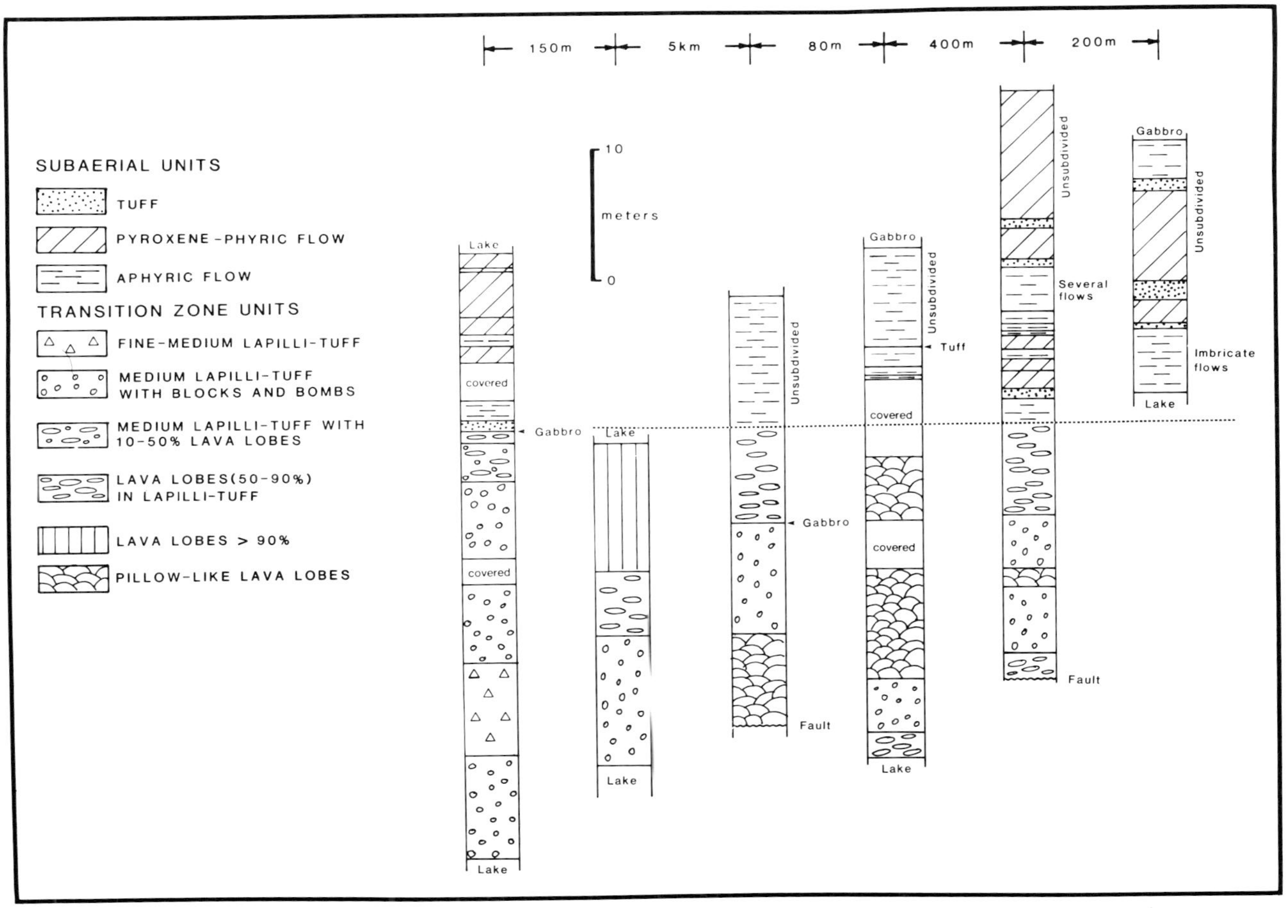


FIG. 9.—Columnar sections through the uppermost surf-zone and overlying subaerial flows. The lack of correlation between sections is a characteristic of the surf zone and probably represents foreset bedding.

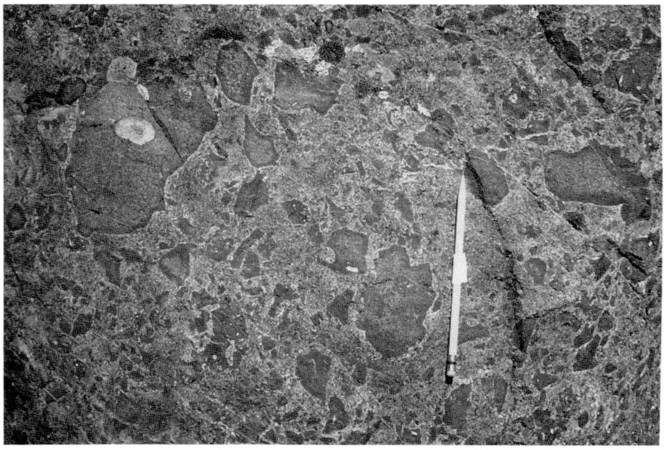

FIG. 10.—Tuff breccia in the uppermost surf zone. Large clasts have fracture-controlled boundaries and partial to complete marginal reaction rims; amygdules are rare. Matrix is tuff and fine lapilli tuff with a high-cement content, as in Figure 12. The white patch on the large clast on left is lichen.

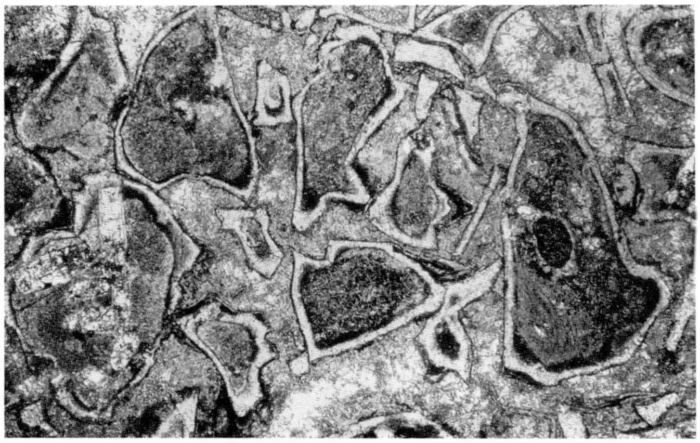

FIG. 12.—Photomicrograph of zoned basaltic vitric granules in matrix of tuff breccia of upper surf zone (from Place, 1982). Angular shape of granules is suggestive of a quench origin. Granules have been replaced by a very fine-grained actinolite, chlorite and epidote; zoning is now defined by an outward decrease in actinolite and increase in epidote and chlorite contents. Material between granules is metamorphosed cement now composed of albite, chlorite, and epidote. Width of field is 5 mm.

defined: 1) ungraded units that occur mostly in the 3- to 4-km-thick sequence between the submergent and upper emergent surf zones on the west limb of the Winterton Island anticline (Fig. 2; 3 on Fig. 3), and 2) graded units that occur mainly in the lower part of the sequence on both limbs of the anticline (Fig. 2; 5 on Fig. 3).

*Ungraded tuff and fine lapilli tuff.—*

On the west limb of the anticline above the submergent surf zone, about half of the subaqueous flows are capped by moderately continuous, massive to bedded, ungraded, tuff and fine lapilli-tuff units; unit thickness ranges from 0.3 to more than 87 m, and averages about 10 m. These volcaniclastic units form about 40 percent of the west limb subaqueous sequence but are rare on the east limb.

Particles are all basaltic and are similar in composition, size, and cementation to those in subaerial units, but there is a higher abundance of crystals, and only rare accretionary lapilli (Table 1). Amygdaloidal particles are common, but amygdule content is generally less than 30 percent; scoriaceous particles are sparse (Fig. 13). Crystals are mostly broken plagioclase with lesser clinopyroxene and possible olivine.

About half of these interflow volcaniclastic units, including some up to 40 m thick, appear to be massive with rare normal grading in the basal few centimeters. Other units are laminated to very thickly bedded, with bed thickness varying among units and along strike within units. In the 3-m width of shoreline exposures, beds are mostly ungraded and planar, with sharp bed contacts. Locally, how-

FIG. 11.—As in Figure 10, but proportion of matrix is lower and many of the clasts are amygdaloidal. Some amygdaloidal clasts have partial chilled rims and appear to be fragments of bombs.

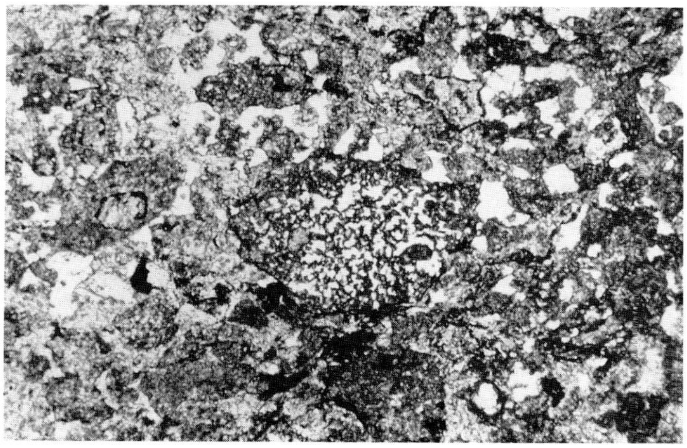

FIG. 13.—Photomicrograph of ungraded tuff from an intercalation between subaqueous flows on the west limb of the Winterton Island anticline. An angular scoria clast in the center is surrounded by smaller angular vitric grains cemented by albite (white; after analcime); vitric grain in left center is microporphyritic. Width of field is 3.5 mm.

ever, beds are lenticular or have diffuse bed contacts. Other structures (Table 1) and soft-sediment deformation features such as slump faults and dismembered beds were observed locally.

*Graded tuff.—*

Laminated to thin-bedded tuff units ranging in thickness from 20 to 350 m form 40 to 70 percent of the subaqueous sequence in the interval extending from 500 m below to 500 m above the lower subaerial sequence (5 on Fig. 3). These units are rare elsewhere. They occur on both limbs of the Winterton Island anticline and form most of the east limb volcaniclastic component. Particles in the tuff are similar to those in subaerial tuff, but they are smaller, crystal abundance is lower, amygdaloidal particles are rare, and accretionary lapilli are a minor component (Table 1). Vitric particles are mostly angular with straight to slightly concave edges. Where present, accretionary lapilli form 20 to 45 percent of local lenses and tabular beds up to 50 cm thick.

Three bed types occur in the tuff: 1) normally graded beds, 2) laminated layers, and 3) sparse internally massive beds (Van Wagoner, 1982). Graded fine to coarse tuff beds have an average thickness of about 1 cm and many beds have characteristics of Bouma ab sequences (Fig. 14); thicker beds comprising Bouma abc and abcd sequences occur locally. Laminated layers comprise graded fine-tuff laminae that range in thickness from 1 to 10 mm and form upward-fining layers up to 10 cm thick. These laminated layers are interbedded with graded and massive beds and are only slightly less abundant than graded beds. Both graded beds and laminated layers contain soft-sediment structures (Table 1; Van Wagoner, 1982), including slump folds, indicating a southeasterly paleoslope.

FIG. 14.—Typical Bouma ab turbidite beds in graded basaltic tuff intercalation between flows in the lower subaqueous unit on the east limb of the Winterton Island anticline (from Van Wagoner, 1982). Scale is in millimeters.

## DISCUSSION

### Genesis of Volcaniclastic Deposits

*Surf-zone deposits.—*

The stratigraphic position between subaerial, highly amygdaloidal flows and subaqueous pillowed flows, in conjunction with the lateral extent and constitution of these units, indicates deposition at sea level. The units have many features in common with flow-foot breccias, produced when subaerial lava entered the ocean (Moore and Fiske, 1969; Jones and Nelson, 1970). These include the wide particle-size range, straight to concave particle margins produced by quenching, variable vesicularity representing different degrees of degassing, variable crystallinity, pillow-like lava lobes, and paucity of crystal and holocrystalline particles. Other features, however, such as the presence of bombs and the irregular vesicle-controlled shapes of some particles, are indicative of concomitant magmatic, possibly littoral, explosions.

The thickness of a surf-zone deposit is an indication of how long a part of the volcano remained near sea level. The variable thicknesses of surf-zone deposits, both within and among deposits, are apparently a function of the interrelationship of processes that acted to build the deposits, such as extrusion and sedimentation rates, and processes that acted to lower the deposits, such as wave erosion, gravity failure, and volcano subsidence.

*Tuff and fine lapilli tuff—fragmentation processes.—*

Unlike most basaltic volcanoes elsewhere, tuff and fine lapilli tuff form about half of the 7-km-thick basaltic sequence in the Amisk Lake volcano. These volcaniclastic rocks occur in both subaerial and subaqueous environments and locally in surf-zone deposits throughout all parts of the sequence, except the upper part of the subaqueous section on the east limb of the Winterton Island anticline. Although the deposits differ widely in bed characteristics, indicative of different modes of deposition (Fig. 3, Table 1), components of tuff and fine lapilli tuff in all parts of the sequence are remarkably similar and were probably produced by a common fragmentation process.

In modern basaltic volcanic islands such as Hawaii, submarine slopes are veneered by volcaniclastic material resulting from downslope movement and reworking of basalt fragments produced by shore or nearshore processes (Moore and Fiske, 1969). A common source of such basalt fragments is flow-foot breccias (Moore and others, 1973), comparable to surf-zone deposits of the present study. In the Amisk Lake volcano, however, surf-zone deposits could not have been a major source of tuff and fine lapilli tuff, although minor contributions cannot be ruled out. This is indicated by 1) the large volume of subaerial tuff and fine lapilli tuff; 2) similar abundances, sizes, and morphologies of vitric, lithic and crystal components in both subaerial and subaqueous deposits; 3) higher vitric, crystal and lithic contents, and smaller particle-size range relative to surf-zone deposits; and 4) smaller particle-size range relative to submarine-slope deposits in Hawaii (Moore and Fiske, 1969). Other submarine processes such as quench granulation or steam explosions (Kokelaar, 1986) were also possible sources for subaqueous tuff and fine lapilli tuff, but could not have produced the subaerial deposits.

The most likely alternative source, particularly for the voluminous on-land deposits, was subaerial explosive eruptions. Such eruptions, possibly augmented by uniquely submarine processes, would have also produced the subaqueous tuff and fine lapilli tuff. The large volume of fine volcaniclastic deposits in the subaqueous sequence indicates that the vent area remained emergent, but upslope of the preserved subaqueous sequence, during the subaqueous phase of this part of the volcano.

The abundant non-amygdaloidal vitric ash, much of which is angular with fracture-controlled boundaries, in conjunction with a variable content of amygdaloidal particles, some of which are scoriaceous, are indicative of eruptions with a varying degree of interaction between vesiculating magma and external water. Non-amygdaloidal particles indicate that the dominant fragmentation process was phreatomagmatic explosions (Wohletz, 1983; Heiken and Wohletz, 1985) of a poorly vesiculated magma. Highly amygdaloidal particles, on the other hand, suggest localized Strombolian-type, magmatic explosions during periods when the role of external water was minimized (Ross, 1986). A high degree of interaction with external water is also supported by abundant accretionary lapilli, armored lapilli and local tuff vesicles in subaerial deposits and local accretionary lapilli in subaqueous deposits (Lorenz, 1974; Sheridan and Wohletz, 1983; Wohletz and Sheridan, 1983; Moore, 1985).

The generally low vesicularity suggests that explosions occurred at some depth below ground surface before magmatic gases were fully able to exsolve. Considering the stratigraphic position of subaerial deposits relative to surf zones (Fig. 3), the explosive eruptions probably occurred on the coastal plain of a volcanic island, possibly from fissure-controlled flank vents (Fig. 15). Underground explosions could have resulted in reworking of particles from underlying volcaniclastic units. Reworking would explain the relatively low abundance of lithic grains in most of the volcaniclastic units compared to many other basaltic phreatomagmatic deposits (e.g., Houghton and Schmincke, 1989).

*Tuff and fine lapilli tuff—deposition.—*

The variable environments and bed characteristics indicate that tuff and fine lapilli tuff were deposited by a number of different processes. In subaerial units, the absence of blocks and large lapilli indicates that tuff and fine lapilli tuff were deposited some distance from the vent or vents. They comprise both distal wet surge-and-fall deposits with some reworked beds, and possibly localized, unwelded pyroclastic-flow deposits in the lower sequence. The wet surge deposits are poorly sorted and ungraded to locally graded, with a high content of accretionary structures and tuff vesicles and localized basal scours, low-angle cross-beds, and undulatory bed planes (Sheridan and Wohletz, 1983; Wohletz and Sheridan, 1983). Fall deposits are normally graded, well-sorted beds with a higher abundance of vesicular particles and lower abundance of accretionary structures. They were probably produced when the magma-to-water ratio was high (Sheridan and Wohletz, 1981).

In preserved subaerial units, reworking appears to be restricted to a few reversely to normally graded, thick to very thick, poorly sorted lapilli tuff beds that contain some accretionary structures. These appear to be debris-flow deposits generated by slumping of more proximal, wet surge-and-fall deposits. Derivation from a more proximal source is indicated by a slightly larger maximum particle size of these beds, whereas the wet source is indicated by ubiquitous accretionary structures that survived mass-flow transport.

Ungraded subaqueous tuff and fine lapilli tuff intercalated with flows throughout most of the subaqueous sequence on the west limb of the Winterton Island anticline probably include fall, some reworked fall, and possibly pyroclastic-flow deposits. Deposition of graded subaqueous tuff in the lower part of the sequence involved reworking and redeposition by turbidity currents. However, the amount of reworking is minor because component ratios and grain angularity are comparable to those in subaerial surge-and-fall deposits (Table 1), and accretionary lapilli survived transport (cf. Boulter, 1987). The close proximity between emergent zones and turbiditic tuff indicates that the tuff was deposited in relatively shallow water, possibly by storm-generated turbidity flows transporting ash originally deposited higher up the volcano by fall or surge (Van Wagoner, 1982).

### *Volcano Morphology and Subsidence*

The Amisk Lake volcano is a composite basaltic volcano that had a shield morphology. The composite nature is indicated by the high abundance of tuff and fine lapilli tuff, which form about half of the 7-km-thick sequence on the west limb of the Winterton Island anticline. Low slopes are indicated by thick accumulations of both subaerial and subaqueous volcaniclastic rocks, particularly relatively shallow-water turbiditic sequences, and paucity of reworking. A number of factors suggests that the examined section represents the eastward-facing upper flank of the volcano (Fig. 15). These include the low abundance of volcaniclastic rocks on the east limb of the anticline, an eastward increase in the ratio of pillowed to sheet flows, easterly paleoslopes indicated by orientation of slump folds, and concentration of upper volcano formations in the west.

The occurrence of both surf-zone and subaerial deposits at several levels of the sequence (Figs. 2, 3, 15) indicates that the volcano subsided more than 4.5 km since it emerged above sea level to form the lower subaerial sequence. This subsidence is a minimum value, because continuing subsidence is indicated by recurrence of subaqueous eruptions in overlying felsic to intermediate sequences. As indicated by the thick subaqueous sequence in the middle of the section, the rate of downward subsidence due to crustal loading during most of the mafic volcanism was greater than the rate of upward volcano construction by eruptions. However, the relative rates of upward construction and downward subsidence were not constant, and variations in these rates resulted in fluctuations in size of the subaerial edifice and shoreline transgression and regression.

Such rapid subsidence is unlike that observed in modern volcanoes constructed on oceanic crust (e.g., Lindwall, 1988; Moore and Thomas, 1988) and may indicate that the Amisk Lake volcano was erupted on a thinner or hotter lithosphere than Recent volcanoes (Van Wagoner and others, 1982). Rapid subsidence would also enable deposition of thick subaqueous turbiditic deposits on the upper volcano slope and accumulation of thick subaerial deposits that underwent only minimal reworking.

Because of the shoreline nature of exposures, the morphology of subaerial deposits is poorly constrained. Typically, subaerial basaltic phreatomagmatic eruptions form tuff cones and rings (Wohletz and Sheridan, 1983). Based on dimensions of Recent tuff cones and rings, the volume, thickness, and lateral extent of each subaerial volcaniclastic sequence are too large to be single tuff cones or rings. Instead the sequences could represent a number of intersecting tuff cones, or they could be blanket deposits unlike those produced by Recent eruptions. A primary depositional slope is indicated by the debris-flow deposits, but the locally intercalated pahoehoe-lava flows and paucity of fluvial reworking indicate that slopes were relatively low.

The preferred model is blanket deposits erupted from flank-fissure vents on the rapidly subsiding coastal plain of a volcanic island. With rapid subsidence, subaerial magma conduits near the shoreline could remain in contact with ground water that was being recharged from the nearby ocean through porous, subaerially erupted but now submerged, volcaniclastic units. Higher up the volcano, beyond the preserved section, lava flows were probably the dominant eruptive mode; lower down, submarine extensions of the fissures may have been the source of preserved pillowed and sheet flows. Although actual vent locations are either covered by the lake or are outside of the preserved section, gabbro intrusions found throughout the mafic sequence may represent magma conduits.

### *Facies Relations*

Four volcaniclastic facies can be recognized in the basaltic part of the Amisk Lake volcano: 1) subaerially erupted tuff and fine lapilli tuff deposited subaerially by distal surge-and-fall processes; 2) subaerially erupted tuff and fine lapilli tuff deposited subaqueously by fall through the water column combined with some downslope reworking; 3) subaerially erupted tuff that was reworked and transported downslope by turbidity currents; and 4) a shore facies of tuff breccia, lapilli tuff, and minor tuff produced by littoral explosions and flow of lava into the sea. The first three facies were produced largely by phreatomagmatic explosive eruptions in the coastal plain of a subsiding volcanic island that was fluctuating in size. Considering the unconsolidated nature of the original ash and lapilli deposits and the fragile nature of many of the particles, the amount of reworking

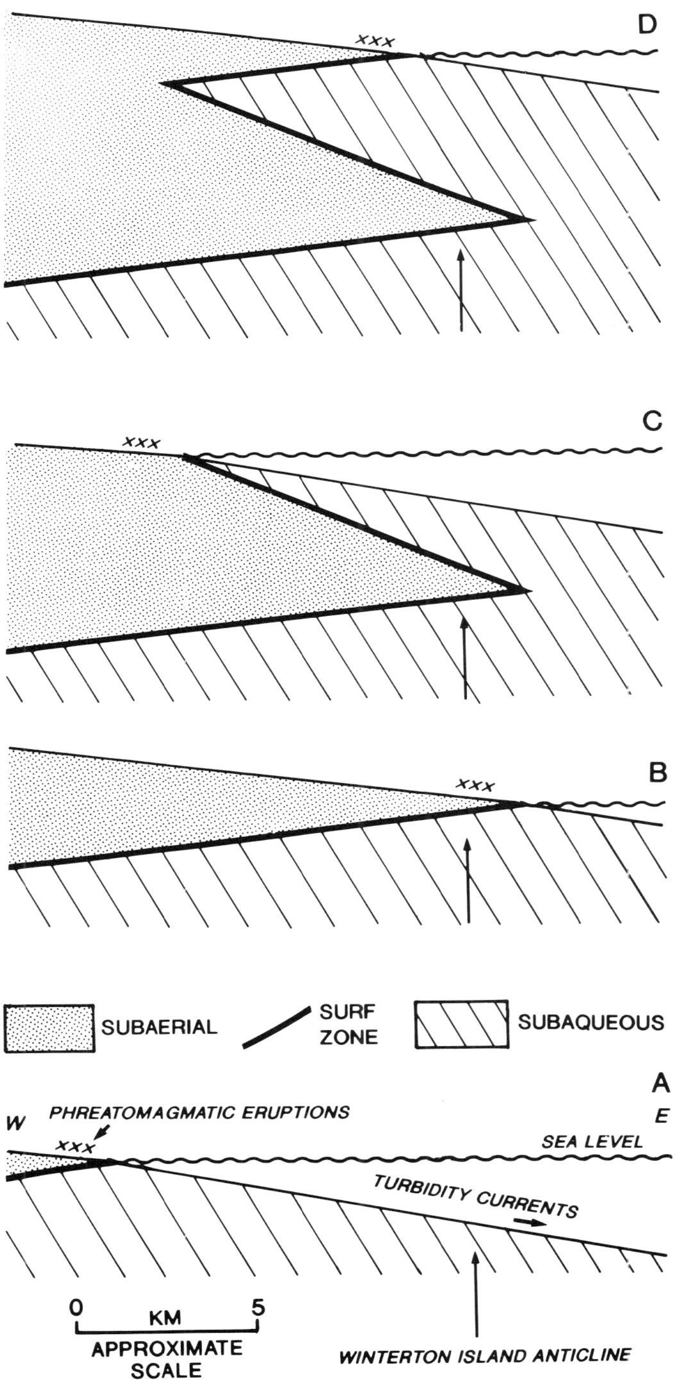


FIG. 15.—Schematic cross sections showing four stages in the inferred evolution of the basaltic part of the Amisk Lake volcano. The preserved sequence directly documents only that part of the volcano near the Winterton Island anticline, but more distant parts can be inferred from stratigraphic variations in the preserved sequence. (A) Early emergence of the volcano. Phreatomagmatic explosive eruptions (x) on the subaerial coastal plain of a volcanic island of low-gradient slope produced both subaerial and subaqueous tuff and lapilli tuff, some of which were reworked by turbidity currents. Vents upslope and downslope from the shore (not shown) had a more effusive history. Only the subaqueous part of this phase is preserved at Amisk Lake; pre-emergence history of the volcano is unknown. (B) Regressive stage. Volcanic island expanded in size because rate of upward construction due to eruptions was greater than rate of downward subsidence. Lower subaerial sequence was produced at this stage. (C) Transgressive stage. Volcanic island decreased in size because rate of downward subsidence was greater than upward construction due to eruptions. Preserved sequence is entirely subaqueous, but subaerial eruptions are indicated by nature of particles in intercalated volcaniclastic units. (D) Second regressive stage. This led to deposition of the upper subaerial sequence.

is surprisingly minor and apparently reflects rapid subsidence of this volcano.

The facies differ in proportion and nature of intercalated lava flows, bed characteristics, and depositional environment and processes. On the volcano flank, there is a downslope transition from subaerial surge-and-fall deposits, through the shore facies, to subaqueous fall-and-reworked deposits. There is also a downslope decrease in volume of volcaniclastic deposits, which are a minor component of the lower volcano slope. The high abundance of intercalated volcaniclastic rocks in parts of the subaqueous sequence is indicative of shallow-water conditions and emergent vents upslope from, and outside of, the preserved sequence. In the Amisk Lake volcano, emergence can also be documented by preserved subaerial units, but in other paleovolcanoes, intercalated volcaniclastic units in subaqueous-flow sequences may be the only evidence of volcano emergence as an island.

## CONCLUSIONS

The Amisk Lake volcano differs from modern basaltic volcanoes in high abundance of volcaniclastic rocks, composite nature but low-gradient slopes, and rapid subsidence. Basaltic tuff and fine lapilli tuff form about half of the upper 7 km of the volcano. These volcaniclastic rocks were erupted largely by phreatomagmatic explosions on the coastal plain of a rapidly subsiding composite basaltic volcano that had a shield morphology with low-slope angles. Some of the tephra was subaerially deposited as blanket-like distal surge-and-fall deposits, but the remainder fell on the upper submarine slopes where it was intercalated with lava flows. Between these two facies is a distinctive surf zone or shore facies composed of coarser volcaniclastic rocks. Downslope reworking of tephra was relatively minor and occurred mostly in the lower part of the sequence.

The basaltic volcaniclastic rocks provide evidence of 1) long-lived emergence of part of the volcano, even when only subaqueous units are preserved, 2) fluctuating size of the volcanic island, 3) the low-slope angle of this composite volcano, 4) rapid and continuous subsidence of the volcano due to crustal loading, 5) variable rates of volcano subsidence and upward construction due to eruptions, and 6) voluminous subaerial phreatomagmatic explosions. Irrespective of deposit morphology or depositional environment, the preserved volume of tephra produced by subaerial phreatomagmatic explosions is orders of magnitude greater than that produced by Recent eruptions of this type. Water-magma interaction must have been a long-lived process that continued throughout eruption of the entire 7-km-thick preserved part of the volcano. Such long-lived interaction may have been the result of rapid subsidence of a volcano erupted on a thinner or hotter lithosphere than Recent volcanoes (Van Wagoner and others, 1982).

## ACKNOWLEDGMENTS

This research was supported by grants from the Natural Sciences and Engineering Research Council, Geological Survey of Canada, and Riocanex, and by logistic support from the Saskatchewan Department of Mineral Resources. Field and laboratory assistance, at various stages of the work, were provided by R. Bezys, J. Fredericks, D. Kobewka, C. Nikols, C. Place, L. Slezak, and S. Van Wagoner. The manuscript has been reviewed by R. V. Fisher and J. G. Moore.

## REFERENCES

AYRES, L. D., 1981, A subaqueous to subaerial transition zone in the early Proterozoic metavolcanic sequence, Amisk Lake, Saskatchewan: Centre for Precambrian Studies, 1981 Annual Report, University of Manitoba, Winnipeg, p. 49–61.

AYRES, L. D., AND THURSTON, P. C., 1985, Archean supracrustal sequences in the Canadian Shield: An overview, *in* Ayres, L. D., Thurston, P. C., Card, K. D., and Weber, W., eds., Evolution of Archean Supracrustal Sequences: Geological Association of Canada Special Paper 28, p. 343–380.

BAILES, A. H., 1971, Preliminary compilation of the geology of the Snow Lake—Flin Flon—Sherridon area: Manitoba Department of Energy and Mines, Geological Paper 1/71, 27 p.

BAILES, A. H., 1980, Origin of early Proterozoic volcaniclastic turbidites, south margin of the Kisseynew sedimentary gneiss belt, File Lake, Manitoba: Precambrian Research, v. 12, p. 197–225.

BAILES, A. H., AND SYME, E. C., 1989, Geology of the Flin Flon—White Lake area: Manitoba Department of Energy and Mines, Geological Report GR87-1, 313 p.

BOULTER, C. A., 1987, Subaqueous deposition of accretionary lapilli: Significance for paleoenvironmental interpretations in Archaean greenstone belts: Precambrian Research, v. 34, p. 231–246.

BUSBY-SPERA, C. J., 1986, Depositional features of rhyolitic and andesitic volcaniclastic rocks of the Mineral King submarine caldera complex, Sierra Nevada, California: Journal of Volcanology and Geothermal Research, v. 27, p. 43–76.

BUSBY-SPERA, C. J., 1987, Lithofacies of deep marine basalts emplaced on a Jurassic backarc apron, Baja California (Mexico): Journal of Geology, v. 95, p. 687–697.

BYERS, A. R., AND DAHLSTROM, C. D. A., 1954, Geology and mineral deposits of the Amisk-Wildnest Lakes area, Saskatchewan: Saskatchewan Department of Mineral Resources, Report 14, 177 p.

CARLISLE, D., 1963, Pillow breccias and their aquagene tuffs, Quadra Island, British Columbia: Journal of Geology, v. 71, p. 48–71.

CAS, R. A. F., LANDIS, C. A., AND FORDYCE, R. E., 1989, A mongenetic, Surtla-type, Surtseyan volcano from the Eocene-Oligocene Waiareka-Deborah volcanics, Otago, New Zealand: A model: Bulletin of Volcanology, v. 51, p. 281–298.

CONDIE, K. C., 1981, Archean Greenstone Belts: Elsevier, Amsterdam, 434 p.

DIMROTH, E., AND LICHTBLAU, A. P., 1979, Metamorphic evolution of Archean hyaloclastites, Noranda area, Quebec, Canada. Part I: Comparison of Archean and Cenozoic sea-floor metamorphism: Canadian Journal of Earth Sciences, v. 16, p. 1315–1340.

FERREIRA, W. S., 1984, A physical comparison of subaerial and subaqueous eruptive environments in the Proterozoic Amisk Group, Saskatchewan, Canada: University of Manitoba, Winnipeg, Unpubl. M.S. Thesis, 199 p.

FISHER, R. V., AND SCHMINCKE, H.-U., 1984, Pyroclastic Rocks: Springer-Verlag, Berlin, 472 p.

FOX, J. S., 1976, Preliminary bedrock geology of the west Amisk Lake area, 63L-9 W1/2 and 63L-16 southwest portion: Saskatchewan Research Council, Geological Division Map.

GASKARTH, J. W., AND PARSLOW, G. R., 1987, Proterozoic volcanism in the Flin Flon greenstone belt, east-central Saskatchewan, Canada, *in* Pharaoh, T. C., Beckinsale, R. D., and Rickard, D., eds., Geochemistry and Mineralization of Proterozoic Volcanic Suites: Geological Society of London Special Publication 33, p. 183–200.

HEIKEN, G., AND WOHLETZ, K., 1985, Volcanic Ash: University of California Press, Berkeley, 246 p.

HOUGHTON, B. F., AND SCHMINCKE, H.-U., 1989, Rothenberg scoria cone, East Eifel: A complex strombolian and phreatomagmatic volcano: Bulletin of Volcanology, v. 52, p. 28–48.

JONES, J. G., AND NELSON, P. H. H., 1970, The flow of basalt lava from air into water—its structural expression and stratigraphic significance: Geological Magazine, v. 107, p. 13–19.

KOKELAAR, P., 1986, Magma-water interactions in subaqueous and emergent basaltic volcanism: Bulletin of Volcanology, v. 48, p. 275–289.

LINDWALL, D. A., 1988, A two-dimensional seismic investigation of crustal structure under the Hawaiian Islands near Oahu and Kauai: Journal of Geophysical Research, v. 93, p. 12107–12122.

LONSDALE, P., AND BATIZA, R., 1980, Hyaloclastite and lava flows on young seamounts examined with a submersible: Geological Society of America Bulletin, v. 91, p. 545–554.

LORENZ, V., 1974, Vesiculated tuffs and associated features: Sedimentology, v. 21, p. 273–291.

MOORE, J. G., 1985, Structure and eruptive mechanisms at Surtsey volcano, Iceland: Geological Magazine, v. 122, p. 649–661.

MOORE, J. G., AND FISKE, R. S., 1969, Volcanic substructure inferred from dredge samples and ocean-bottom photographs, Hawaii: Geological Society of America Bulletin, v. 80, p. 1191–1202.

MOORE, J. G., PHILLIPS, R. L., GRIGG, R. W., PETERSON, D. W., AND SWANSON, D. A., 1973, Flow of lava into the sea, 1969–1971, Kilauea Volcano, Hawaii: Geological Society of America Bulletin, v. 84, p. 537–546.

MOORE, J. G., AND THOMAS, D. M., 1988, Subsidence of Puna, Hawaii inferred from sulfur content of drilled lava flows: Journal of Volcanology and Geothermal Research, v. 35, p. 165–171.

PLACE, C., 1982, A microprobe study of altered basalt glass in an early Proterozoic hyaloclastite, Amisk Lake, Saskatchewan: University of Manitoba, Winnipeg, Unpubl. B.S. Thesis, 55 p.

ROSS, G. M., 1986, Eruptive style and construction of shallow marine mafic tuff cones in the Narakay volcanic complex (Proterozoic, Hornby Bay Group, Northwest Territories, Canada): Journal of Volcanology and Geothermal Research, v. 27, p. 265–297.

SHERIDAN, M. F., AND WOHLETZ, K. H., 1981, Hydrovolcanic explosions: The systematics of water-pyroclast equilibrium: Science, v. 212, p. 1387–1389.

SHERIDAN, M. F., AND WOHLETZ, K. H., 1983, Hydrovolcanism: Basic considerations and review: Journal of Volcanology and Geothermal Research, v. 17, p. 1–29.

SMITH, T., AND BATIZA, R., 1989, New field and laboratory evidence for the origin of hyaloclastite flows on seamount summits: Bulletin of Volcanology, v. 51, p. 96–114.

STAUDIGEL, H., AND SCHMINCKE, H.-U., 1984, The Pliocene seamount series of La Palma/Canary Islands: Journal of Geophysical Research, v. 89, p. 11195–11215.

SYME, E. C., BAILES, A. H., GORDON, T. M., AND HUNT, P. A., 1987, Zircon geochronology in the Flin Flon belt: Age of Amisk volcanism: Manitoba Department of Energy and Mines, Report of Field Activities 1987, p. 105–107.

SYME, E. C., BAILES, A. H., PRICE, D. P., AND ZIEHLKE, D. V., 1982, Flin Flon volcanic belt: Geology and ore deposits at Flin Flon and Snow Lake, Manitoba: Geological Association of Canada, Winnipeg 1982 Field Trip Guidebook 6, 91 p.

VAN WAGONER, N., MOON, W., AYRES, L. D., TANG, R., AND CARSWELL, A., 1982, Preliminary results of rheological modelling of a Proterozoic volcano and Earth: Centre for Precambrian Studies, 1982 Annual Report, University of Manitoba, Winnipeg, p. 128–150.

VAN WAGONER, S., 1982, A probable storm-generated turbidite association—Proterozoic Flin Flon—Snow Lake greenstone belt: Centre for Precambrian Studies, 1982 Annual Report, University of Manitoba, Winnipeg, p. 115–127.

WHITAKER, S. H., AND PEARSON, D. E., 1972, Geological Map of Saskatchewan: Saskatchewan Department of Mineral Resources.

WOHLETZ, K. H., 1983, Mechanisms of hydrovolcanic pyroclast formation: Grain-size, scanning electron microscopy, and experimental studies: Journal of Volcanology and Geothermal Research, v. 17, p. 31–63.

WOHLETZ, K. H., AND SHERIDAN, M. F., 1983, Hydrovolcanic explosions II. Evolution of basaltic tuff rings and tuff cones: American Journal of Science, v. 283, p. 385–413.

# PARNELL GRITS—LARGE SUBAQUEOUS VOLCANICLASTIC GRAVITY FLOWS WITH MULTIPLE PARTICLE-SUPPORT MECHANISMS

PETER F. BALLANCE AND MURRAY R. GREGORY

*Department of Geology, University of Auckland, Auckland, New Zealand*

ABSTRACT: Parnell Grits are volcaniclastic beds 2 to 20 m thick interbedded comformably with bathyal flysch of the Waitemata interarc basin (early Miocene) of northern New Zealand. They comprise lava clasts, crystals, rip-up clasts, shallow-marine fossils, and rare extrabasinal clasts in a matrix of sand and silt. Inverse-to-normal size grading is always present. Other primary structures are generally absent. Rarely, the fine-grained upper levels show horizontal and micro-cross-lamination. Evidence of high post-depositional internal pore pressures consists of intrusions of grit downward into underlying sediments, inward into most rip-up clasts, and upward into upper levels of the Grit and rarely into overlying sediments. Rip-up clasts range up to 90 m long. They comprise more than half of the volume of the deposit in some places, occur at all levels in the flow, and may be folded. Lava clasts up to 20 m long were carried to the center of the basin (>40 km), implying a bulk strength for some moving flows of >1.25 kg/cm$^2$. After deposition, the two largest known lava clasts indented underlying beds, implying a bulk strength after deposition of between 0.5 and 0.78 kg/cm$^2$. The lower strength after deposition is attributed to decay of grain interaction. We interpret Parnell Grits as the deposits of volcaniclastic gravity flows which began as debris avalanches resulting from sector collapses on subaerial slopes of the Waitakere Arc to the west of the Waitemata Basin. The volcanic-debris flows crossed the strand zone, picked up a shallow-marine biota and flowed on to bathyal depths. On entering the sea, there was a change in flow character resulting from the uptake of water. The reduction in matrix strength led to the development of a basal traction carpet of clasts in grain-flow and normal-size grading of dense clasts. Rip-up clasts were acquired by injection of grit into the substrate. There is very little evidence for turbulence in the flows and no evidence of flow transformation. Deposition was by frictional freezing in the traction carpet, cohesive freezing in the bulk of the deposit, and in some beds fluidal settling of the fine elutriated tail. Decay of some clast-support mechanisms after deposition reduced the bulk strength of the deposit. There were probably more than 100 such flows, all with similar properties.

## INTRODUCTION

A moving sediment gravity flow may exhibit more than one mechanism of particle support (Middleton and Hampton, 1976). Flow and particle-support mechanisms may also change during flow (Morgenstern, 1967; Middleton and Hampton 1976; Middleton and Southard, 1978; Weirich, 1988). Fisher (1983, 1984) coined the term "flow transformation" for transitions between laminar and turbulent flow. In this paper we describe distinctive volcaniclastic beds (Parnell Grits) and infer that they were deposited by large subaqueous gravity flows in which matrix strength and particle interaction were the main contributors to particle support. We find little evidence of turbulence or of flow transformation.

The beds were first interpreted as subaqueous gravity flow (mudflow or "submarine lahar") deposits by Kuenen (1950). Fisher (1971) included them in a discussion of coarse-grained high-concentration fluids and their deposits, and subsequently (Fisher, 1983 and 1984) noted that the combination of inverse grading at the base and large rip-up clasts in the body of the deposit may indicate flow transformation (body transformation) from turbulent to laminar flow.

Parnell Grit is a stratigraphic name, first coined by Hochstetter (1864). The deposits are "gritty" in the old sense of having a rough texture arising from the presence of many angular grains of 1 to ~5 mm diameter. However, "grit" as a rock type is not recognized in modern grain-size schemes (e.g., Leeder, 1982; Fisher and Schmincke, 1984). Therefore, whereas it is convenient to refer to the deposits we discuss in this paper as Parnell Grits or Grit beds, it should be recognized that there is considerable variation in lithology both within and between beds, and that Grit is a name and a convenient label.

## GEOLOGICAL SETTING

Beds of Parnell Grit are interbedded in the sandy turbidite facies (flysch) of the early Miocene Waitemata Group of Auckland, New Zealand. The flysch infilled a bathyal basin that lay between parallel northwest-trending volcanic arcs (Ballance, 1974; Hayward, 1979; Ricketts and others, 1989; Fig. 1). Much of the basinal sediment was derived from nonvolcanic rocks located to the north and northeast of the basin. It was transported in a counterclockwise direction, mixed with varying amounts of volcaniclastic sediment, and was finally delivered from input points along the western side of the basin to form a ramp of coalescing submarine fans sloping to the southeast, oblique to the long axis of the basin (Ballance, 1974, fig. 17; Fig. 1). The Parnell Grit volcaniclastic flows, however, appear to have entered the basin directly from the western volcanic chain. They carried coarse volcanic detritus at least 40 km into the basin (Gregory, 1966; Ballance, 1974, fig. 11; Fig. 1). Equivalent stratigraphic levels in the western volcanic chain are mostly below ground, but the petrology of the lava clasts is compatible with a western source (Smith and others, 1989). Two Parnell Grit deposits are preserved at higher stratigraphic levels within the western chain, now exposed at Muriwai.

## DESCRIPTION

Individual sedimentation units of Parnell Grit range from 2 to 20 m thick and generally lie conformably within the enclosing more thinly bedded flysch (Fig. 2). They are always thicker, coarser grained and more richly volcaniclastic than the enclosing sandstones. Maximum dense-clast size varies widely between localities, from <1 cm to 20 m, but

Sedimentation in Volcanic Settings, SEPM Special Publication No. 45

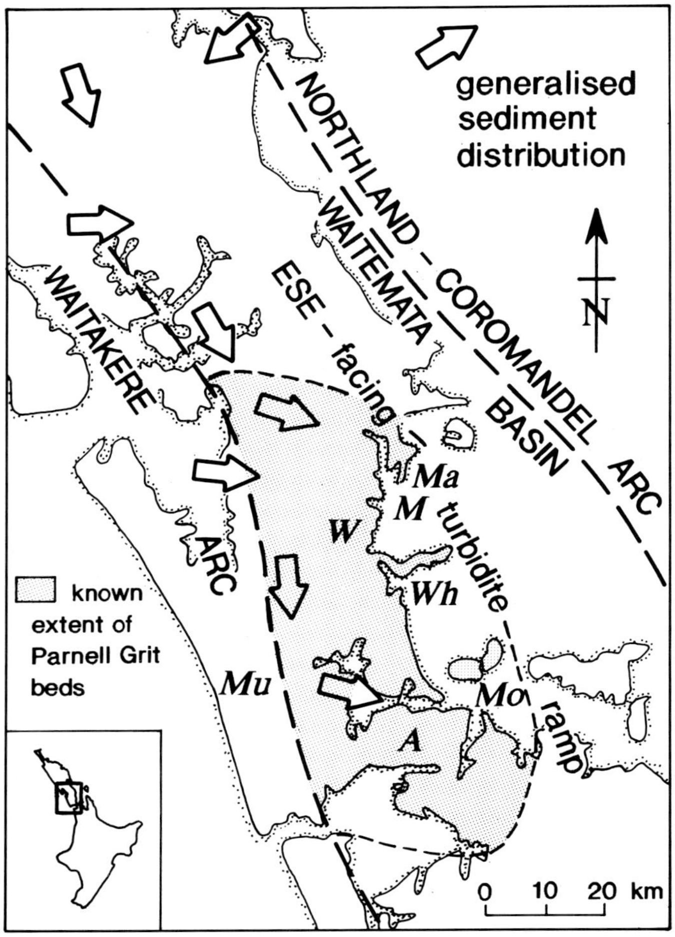


FIG. 1.—Extent of Parnell Grit beds within the early Miocene Waitemata Basin. A = Auckland; M = Motuora Island; Ma = Mahurangi; Mo = Motuihe Island; Mu = Muriwai; W = Waiwera; Wh = Whangaparaoa Peninsula.

FIG. 2.—View of 8 m Parnell Grit bed conformable with enclosing flysch. Waiwera. Person for scale (lower right). Details of this bed are shown in Figures 5a–d.

is generally less than 50 cm (excluding rip-up clasts). There is no correlation between clast size and bed thickness (Fig. 3).

The major clast components in approximate order of importance are 1) porphyritic basalt and andesite (1 cm–20 m), 2) crystal fragments of plagioclase, clinopyroxene, orthopyroxene, olivine and rare brown amphibole (Day, 1986; up to 5 mm), 3) detrital grains of quartz, argillite and chert derived from the basinal flysch (up to 2 mm), 4) rip-up clasts of basinal flysch (up to 90 m), 5) bryozoan colonies and other shallow-marine fossils (Crabb, 1971; up to 30 cm), 6) well-rounded and polished igneous pebbles derived from another Waitemata basin unit, the Albany Conglomerate (Ballance, 1974, fig. 11; up to 10 cm), 7) extrabasinal well-cemented calcarenite (up to 2 m), and 8) zeolite-cemented tuff (up to 1 m).

Basalt clasts range from dense nonvesicular types to highly vesicular scoria and are generally angular to subangular. They have irregular shapes and rough surfaces. Reddened, oxidized clasts are rare. Larger clasts show parallel columnar jointing. Shape measurements of 19 excavated clasts at Waiwera gave an average sphericity index of 0.70, and an oblate/prolate index of +3.00 (method of Dobkins and Folk, 1970). This shape is not typical of beach gravel.

The finest matrix in Parnell Grit deposits has generally been replaced either by calcite cement or by zeolites. The matrix in general is a very poorly sorted muddy sandstone. Sand grains range from very fine- to very coarse-sand size. There is no predominant size. The amount of clay-size material at the time of deposition is unknown because of alteration and replacement, but it is unlikely to have been more than a few percent.

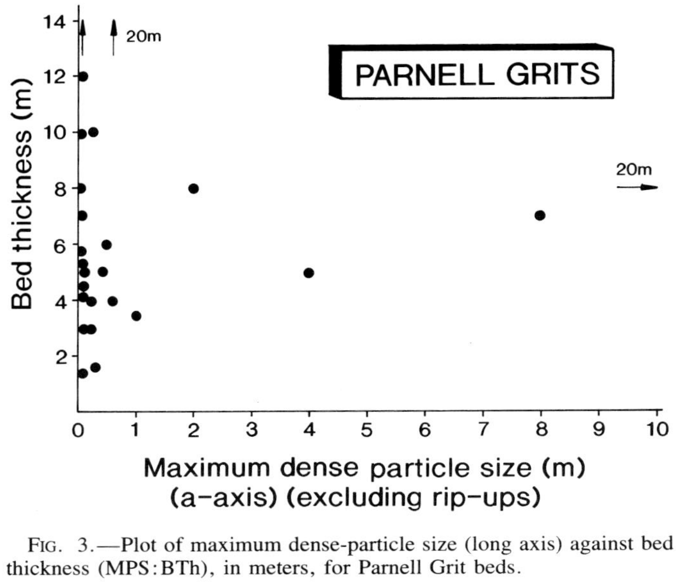


FIG. 3.—Plot of maximum dense-particle size (long axis) against bed thickness (MPS:BTh), in meters, for Parnell Grit beds.

Gravel-size clasts are in clast support in the lower part of some Parnell Grit beds, passing upward into matrix support in the upper levels (Fig. 4). In other instances the larger clasts are in matrix support throughout a bed, but are smaller and more widely spaced in the upper levels (Figs. 4, 5). In still other beds the maximum dense-clast size is only 1 or 2 cm, and rip-up clasts are rare.

Inverse-to-normal size grading, first recorded by Fisher (1971), is an invariable characteristic of the Parnell Grit (Figs. 4, 5). In clast-rich beds it is shown by reverse grading in a clast-supported basal portion, followed upward by a rapidly increasing ratio of matrix to gravel-size clasts and a decreasing average size of the largest clasts; maximum clast size also diminishes, gradually, so that rare 15-cm lava clasts may be present at the top of a bed (Fig. 4d). In some Grit beds mean and maximum clast sizes vary together; in these beds the upper levels are composed of fine sand to silt, without larger clasts.

Other internal sedimentary structures are generally lacking. In the finer grained upper levels, horizontal lamination occurs rarely and micro-cross-lamination is very rare. Clastic dikes, sills, pipes and ill-defined intrusions are common. They comprise injections of Parnell Grit downward into underlying sediment (Fig. 6), inward into all rip-up clasts (Fig. 7), upward from lower to upper levels of a Parnell Grit bed (Fig. 8), and upward into overlying beds (Figs. 4d, 9). Vertical sheet structure, implying post-depositional water movement (Laird, 1970), is common (Gregory, 1966); dish structure, however, is rare.

The largest lava clasts in individual beds are generally <0.5 m, and rarely 1 to 2 m long. Outsize clasts—those several times larger than the surrounding ones—are common in some beds. At two localities east of Army Bay, Whangaparaoa Peninsula, there are particularly large outsize clasts of columnar-jointed lava (Turner and Bartrum, 1929; Gregory, 1966; Figs. 10, 11), one measuring 20 × 12 × 3+ m, one 4 × 2 × 2 m, and another 8 × 5 × 2+ m. All three are associated with matrix-rich Grits, and the two largest sank partially through the base of the bed after deposition depressing underlying strata (Fig. 11B).

The concentration of rip-up clasts varies greatly. In places they comprise more than half of the volume of the deposit. In other beds they are rare. Rip-up clasts reach 90 × 15 × 5+ m on Motuora Island. A wide range of flysch types can be represented in the rip-up clasts at any one locality. They may be folded (Figs. 4C, 12), and they are invariably injected with Grit matrix (Fig. 7). Large rip-up clasts located near the base of a Grit bed may be separated from the planar basal contact by a few centimeters of Grit (Fig. 13). The contact between Grit and large rip-up clasts may be marked by localized clusters or "plugs" of larger lava clasts.

The basal surfaces of Grit beds are normally planar, but in a few localities there is slight erosion into the underlying flysch. Organized scour structures are rare (Fig. 14). The upper surfaces of most beds are also planar, but in one instance mound structures up to 5 m high and 15 m across are present (Fig. 15).

Parnell Grit beds are normally exposed for a few tens or hundreds of meters in two-dimensional cliff exposures (Fig. 2). They generally have parallel tops and bottoms. One such bed can be traced around Motuora Island (Fig. 1), an area of more than 1 km by 2 km, with a uniform thickness exceeding 20 m (Ballance, 1977). A bed on Motuihe Island (Mo on Fig. 1) is also exposed continuously over an area of more than 1 km by 2 km and intermittently over an area of more than 3 km by 2 km. It is 20 m thick. Rarely, a steep marginal termination of a bed is exposed.

### ORIGIN, VOLUME AND FREQUENCY OF PARNELL GRIT FLOWS

We infer that Parnell Grit flows most probably originated on the subaerial slopes of island volcanoes from the following facts: 1) subaerial facies are known from the western arc (Hayward, 1979); 2) no undoubted pillow fragments have ever been seen, except in the upper bed in the western volcanic chain at Muriwai (Fig. 1), which actually laps onto the margins of the deep bathyal, Maori Bay pillow-lava mound (Hayward, 1976); 3) the irregular shape and rough exterior of most lava clasts, together with shape analysis (see earlier discussion), indicate that they have not been exposed to beach processes, whereas 4) the presence of shallow-water organisms, including hermatypic corals and rhodoliths, indicates that they either originated in or passed through the shallow-marine photic zone. 5) Larger lava clasts display regular, parallel, columnar jointing, suggesting an origin from tabular lava flows. Such flows are known in submarine settings, for example on the margins of the two pillow-lava mounds at Muriwai, where they are about 3 m thick, 100 to 150 m long, and have prominent glassy selvedges on both top and bottom surfaces (Hayward, 1976). They are much more common in subaerial settings, however, and since there is no trace of the characteristic submarine glassy selvedge on the Parnell Grit clasts, we think that it is much more likely that the columnar-jointed clasts are subaerial in origin.

Thus, we conclude that Parnell Grit flows most probably originated subaerially and flowed through the nearshore shallow-marine zone without lingering on the beaches. Algae-coated lava clasts clearly stayed in the photic zone for a period of time and were not carried directly from subaerial to bathyal realms. They are very rare, however, and since Parnell Grit flows were clearly very large events, we think it most likely that the bulk of the lava components travelled the 40 or more kilometers from the volcanic islands to the bathyal floor of the Waitemata basin without a stop.

The volume of individual flow events is not known. The two most extensive deposits known (Motuora and Motuihe) comprise 0.04 and 0.06 $km^3$, respectively. The fact that flows travelled 40 km and deposited tabular beds from 2 to 20 m thick implies, however, that they were very large-volume events.

The frequency of Parnell Grit events, on the other hand, was much lower than that of turbidity currents. In most exposures of Waitemata Group flysch, there is no Parnell Grit present. The Takapuna section, for example, records about 180 turbidity-current events without any Parnell Grit events (Ballance, 1964). It is rare for there to be more than one Parnell Grit in an exposed sequence; the maximum number observed is four. At a few localities, two Parnell

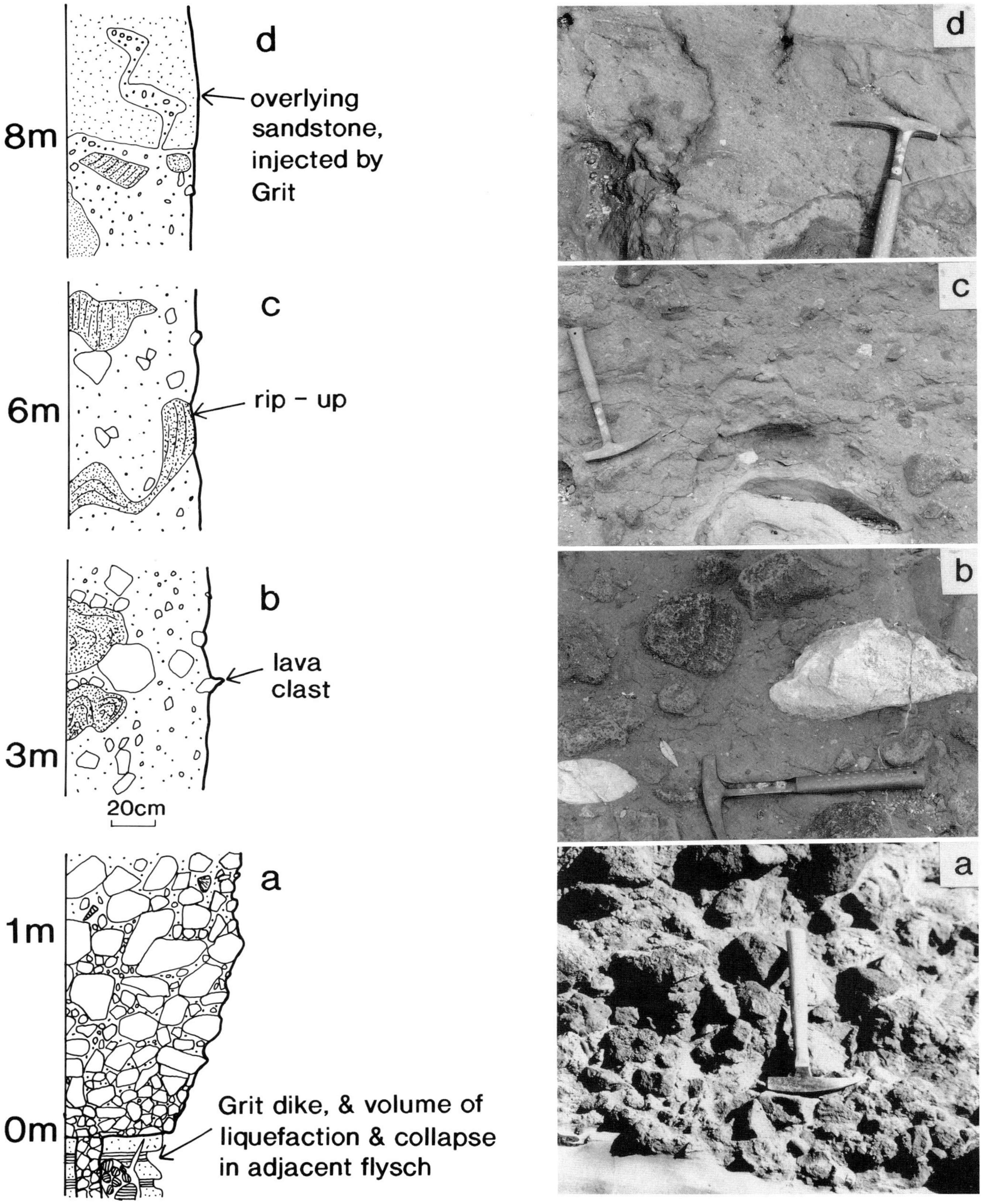
d
8m
overlying
sandstone,
injected by
Grit
c
6m
rip - up
b
lava
clast
3m
20cm
a
1m
0m
Grit dike, & volume of
liquefaction & collapse
in adjacent flysch
d
c
b
a

FIG. 6.—Clastic dike of Grit extending from base of Parnell Grit bed into underlying flysch. Hammer (circled) for scale. Arkles Bay, Whangaparaoa Peninsula.

Grits are in stratigraphic contact without intervening flysch. The total number of Parnell Grit flows that entered the Waitemata Basin from the approximately 130-km length of the western Waitakere Arc (Fig. 1) during the 10 m.y. life of the basin is not known. We estimate that there were certainly more than 100, probably many more.

The large volume and low frequency of Parnell Grit flows suggest that they were not caused by the normal subduction-zone seismicity that triggered lahars, turbidity currents and intrabasinal slump events (Gregory, 1969). Instead, they were probably caused by large-sector collapses and the resulting debris avalanches on emergent stratocones.

Where such debris avalanches come to rest on land, they typically have a mounded upper surface, which reflects the preservation of more-or-less intact blocks of the cone rocks (e.g., Crandell, 1989; Palmer and others, this volume). Parnell Grit deposits are normally flat-topped, and we suggest in later discussion that the flows changed their properties when they entered the sea. Mounds have been identified on debris-avalanche deposits from Mt. Taranaki (formerly Mt. Egmont), New Zealand, over 75 m below present sea level (Gregory, 1974), but they may have been deposited during glacio-eustatic lowstands of −100 m or more. One Parnell Grit bed has mounds (Motuora Island, Fig. 15) that contain many small rip-up clasts; the mounds do not contain cores of undisturbed or little-disturbed cone rocks, and therefore their origin is different from the mounds on subaerial deposits, which are basically stranded blocks.

FIG. 4.—Detail of macroscopic textures through the Parnell Grit at Waiwera, at the levels indicated. See text for discussion.

FIG. 5.—Photographs of the textures depicted in Figure 4. Letter symbols refer to the same levels as in Figure 4. (a) Reverse grading and clast support at base; (b) gravel-size, irregularly shaped dense lava and limestone clasts in matrix support; (c) more widely spaced and smaller dense clasts with a folded rip-up clast; (d) finer grained upper level of typical Parnell Grit.

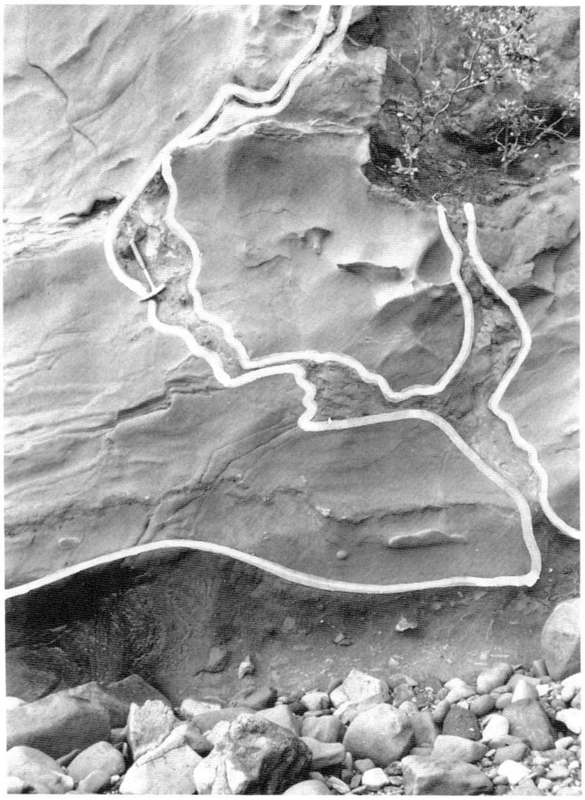

FIG. 7.—Sandstone rip-up clast (center and upper left) in contact with Parnell Grit matrix (bottom and upper right), and injected by dikes of Grit (outlined in ink). Hammer for scale. Motuora Island.

## FLOW AND PARTICLE-SUPPORT MECHANISMS

Parnell Grit flows travelled 40 km or more from subaerial volcanic origins into the center of a bathyal basin. They all attained similar properties and organization. They were able to pick up and transport intrabasinal shell detritus, sand grains and pebbles, as well as erode and carry rip-up clasts of intrabasinal flysch up to 90 m long. Some were able to transport extrabasinal clasts of dense limestone (up to 2 m) and dense lava (up to 20 m) into the center of the basin. They had sufficient residual strength after deposition to maintain steep flow margins, but not enough to prevent extra-large dense clasts from sinking through the base of the deposit.

### *Flow Mechanisms*

Inverse-to-normal size grading in the clast-rich lower part of each deposit implies a concentration of larger clasts and a high degree of particle interaction in the leading/basal regions of the flow. The transport of lava clasts up to 20 m long in the lower portions and up to 15 cm diameter in the matrix-supported upper/rear portions of some flows suggests a high degree of overall strength. Normal-size

FIG. 8.—Calcite-cemented pipe of coarser Grit matrix injected into the finer upper levels of the same bed; 2.5 km east of Army Bay, Whangaparaoa Peninsula.

FIG. 9.—Upward injection of Grit matrix (top left) into overlying sandstone (center and bottom right). Exposed on a loose block, 2.5 km east of Army Bay, Whangaparaoa Peninsula.

grading without associated tractional structures indicates differential settling of the larger clasts during deposition; it implies that the particles were not fully supported by matrix strength, and that there was another factor or factors contributing to particle support during flow.

Fisher (1983, 1984) suggested that rip-up clasts were evidence of turbulence in Parnell Grit flows at an earlier stage of flow, and therefore evidence of flow transformation from turbulent to laminar flow. The Parnell Grit is the only deposit so far named as a possible example of such flow (body) transformation, and it is therefore important to consider the evidence for turbulence carefully. The acquisition of rip-up clasts up to 90 m long occurred before the depositional stage. No incomplete rip-up clasts have been recognized at outcrops. They were probably acquired during the period of accelerating flow or maximum velocity on the basin slopes. The wide variety of rip-up clasts present in the deposit at any one locality indicates that they were gathered over some distance of transport. The sheet-like form of larger rip-up clasts implies that they were lifted bodily into the flow in some way.

Possible mechanisms of rip-up detachment include the following: 1) injection of flow material downward into the substrate during flow, lifting a piece of the channel floor, as documented by Morris and Busby-Spera (1988) for debris flows in a submarine canyon; 2) undercutting and collapse of the side walls of channels; and 3) slope failure along bedding planes, allowing sheets of basinal sediment to slide into channels incised into slopes.

FIG. 10.—Person standing on the largest known lava clast in Parnell Grit beds. The clast measures 20 × 12 × 3+ m. Lava columns dip to the right. The clast has sunk downward into underlying flysch. Base of the Parnell Grit bed is marked by an ink line to the left of the clast. Cf. Figure 11. Location 900 m north northeast of boat ramp, Army Bay, Whangaparaoa Peninsula. Scale 1:50,000, map reference R10/731104.

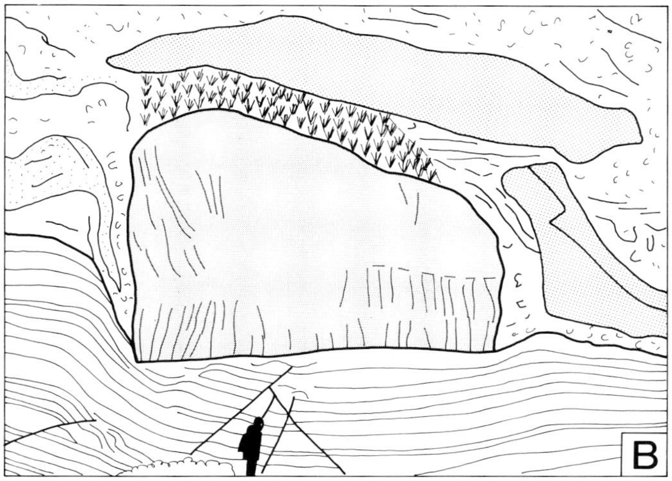


FIG. 11.—(A) Photograph and (B) line drawing of the second largest basalt clast, measuring 8 × 5 × ?m. The clast displays parallel columnar jointing and has sunk 2.5 m into underlying flysch. Location 1.4 km northeast of boat ramp, Army Bay, Whangaparaoa Peninsula. Scale 1:50,000, map reference R10/739104.

Few paleochannels have been seen in the Waitemata basin, perhaps because of limited exposures. A small example, 150 m wide, on the western side of the flysch basin is infilled by a megabreccia of mudstone and pebbly mudstone blocks in a mud matrix (Ballance, 1974). Examples in the Waitakere Arc described by Hayward (1976, 1979) are larger (up to 1,500 m wide) and infilled with displaced shallow-water volcaniclastic sediments. Point sources and hence, feeder channels, are inferred for conglomeratic-fan deposits in the basin (Ballance, 1974). From this independent evidence for channels feeding sediment into the basin, we infer that probably all three of the above mechanisms were active in providing rip-up clasts. This was a conse-

FIG. 12.—Folded rip-up clast of flysch in Parnell Grit, Waiwera.

FIG. 13.—Large rip-up clast located just above base of a Parnell Grit. Base of Grit indicated by lower arrows. Upper arrows mark base of a second Grit bed resting directly on the lower one. Person for scale. Milford.

FIG. 14.—Rare groove casts on the base of a Grit Bed. Army Bay, Whangaparaoa Peninsula.

FIG. 15.—Mound of Parnell Grit, containing mostly rip-up clasts (light colored), rising above the general upper surface of the Grit bed (marked by the rock platform in foreground) and draped by the overlying thick sandstone bed. Back pack and map case for scale. Motuora Island, south side.

quence of the density and momentum of Parnell Grit flows descending the western slope of the basin and occupying channels. The rip-up clasts do not necessarily require turbulence in the flows. The only independent evidence of possible turbulence known to us is a few parallel flute-like protuberances on the base of one bed at one locality (Army Bay, Fig. 14). If these are a result of turbulence, it was present at the moment of deposition and not necessarily throughout the life of the flow. We conclude that there is little evidence for turbulence in Parnell Grit flows, and that the flow transformation suggested by Fisher (1983) is unlikely to have occurred.

On the basis of the classification and nomenclature of Lowe (1979, 1982), there is strong evidence that Parnell Grit flows simultaneously displayed mudflow (or cohesive debris-flow) and grain-flow mechanisms, both of which are laminar in character. In view of the abundant evidence of high pore pressures within Parnell Grit beds after deposition (Figs. 6 to 9), and the evidence that some subaerial debris-flow deposits have elevated pore pressures which dissipate slowly (Pierson, 1981), we contemplated the possibility of a third flow mechanism. Could liquefied or fluidized flow have been operative in Parnell Grit flows? Dikes and water-escape structures are all post-depositional, but there are three depositional features that suggest some addition of water to the moving flows. One is the normal grading of all dense components, implying a cohesive matrix strength lower than that of subaerial flows. The second is the existence of beds up to 20 m thick with a maximum dense-clast size of only 1 or 2 cm, suggesting that these flows may have abandoned larger clasts upslope, as a lag, and hence had a significant reduction in matrix strength. The third is the grading to a fine-grained (muddy, very fine to fine sand) upper portion in some deposits that is free of larger grains (e.g., Fig. 8). The main effect of the additional water was thus to reduce the strength of the cohesive matrix. The flow behavior was still plastic (Bingham) and laminar. In the rare instances where the fine upper sediment is laminated and micro-cross-laminated, the flow may have developed a tail of elutriated fine sediment showing fluidal behavior. Parnell Grit beds lacking the fine upper portion (e.g., Figs. 4, 5) did not develop liquefaction to the same degree but are still graded, implying some reduction in matrix strength.

Thus, the cohesive strength of the matrix in Parnell Grit flows was somewhat reduced by uptake of water, presumably at the head of the flow, implying a degree—perhaps a very small degree—of liquefied-flow in addition to grain-flow and cohesive debris-flow behavior.

The bulk density of the flow allowed the suspension of rip-up clasts of wet sediment (sp. gr. 1.8 to 1.9, McClelland, 1967) at all levels/positions in the flow. It was therefore greater than 1.9. However, it was not as high as some subaerial flows (~2.5, Rodine and Johnson, 1976), because Parnell Grit flows did not suspend large, dense clasts in their upper levels as subaerial flows commonly do (Janda and others, 1981).

### *Particle-Support Mechanisms*

Known particle-support mechanisms (Middleton and Hampton, 1976; Rodine and Johnson, 1976; Hampton, 1975, 1979) can be divided into those that are generated by and during flow (internal fluid turbulence, fluidization by through-moving fluid, particle interaction) and those that are inherent in the sediment and water content of the material (cohesive strength of clay and silt; buoyancy arising from clast content, bulk density, and increased internal-pore pressures caused by the loading effect of large grains on the clay/water matrix).

The first group can be expected to decay directly as the flow comes to rest. As energy inputs from the different mechanisms fall through critical threshhold levels, many gravity flows are subject to cohesive and/or frictional 'freezing' (Lowe, 1982). Of the second group, cohesive strength, clast content and bulk density will be unchanged after deposition, but elevated internal-pore pressures will slowly decay (Pierson, 1981).

The properties of Parnell Grit deposits, and the inferences drawn about flow behavior, suggest that particle support in the moving flows was provided by particle interaction, cohesive matrix strength and buoyancy effects.

One or more of those mechanisms decayed after flow had ceased, allowing two extra-large dense clasts to sink through the base of the deposit. All rip-up clasts were held firmly in the stationary flow, as were dense clasts up to 4 m long. There are no known dense clasts between 4 m and 8 m in length, and thus the critical size/weight at which dense clasts became unsupportable after deposition is not known.

We have tried to use the size/weight relationship of the large dense clasts to estimate the overall suspension competence of the moving flows and the magnitude of the change in competence after deposition. Clearly, particle interaction ceased after deposition, whereas cohesive strength and buoyancy effects would not have changed, apart from the decay of elevated pore pressures, which we have no way of measuring.

A heavy particle exerts a stress on its substrate measurable in $kg/cm^2$. The largest clast that was fully supported after deposition measures approximately $4 \times 2 \times 2$ m (16 $m^3$), weights about 40 tons, has a footprint of 8 $m^2$, and exerts a stress of $\sim$0.5 $kg/cm^2$. Of the two clasts that penetrated downward through the base of the deposit after deposition, the smaller exerts the larger stress because of its smaller footprint, of $\sim$1.25 $kg/cm^2$. The largest clast ($\sim$2,000 tons) exerts a stress of $\sim$0.78 $kg/cm^2$.

Thus, in order to carry the observed clasts into the basin, Parnell Grit flows must have maintained an overall suspension competence equal to an effective bulk strength—combining all the particle-support mechanisms—of more than 1.25 $kg/cm^2$. After deposition, that bulk parameter declined by about one half, to between 0.5 and 0.78 $kg/cm^2$. If we assume that the contribution to particle support made by increased internal pore water pressure—a factor both in the buoyancy equation and in the development of normal grading—was minor, then the decay of particle interaction on deposition caused a drop in effective bulk strength of approximately half. Thus, particle interaction contributed about half of the suspension competence (>1.25 $kg/cm^2$) of moving Parnell Grit flows, the other half having been contributed by cohesive matrix strength plus buoyancy effects.

A very important function of the basal carpet of interacting clasts was to prevent the extra-large dense clasts from touching bottom and grounding.

The support mechanisms for large and extra-large clasts in Parnell Grit flows were therefore different from those discussed by Rodine and Johnson (1976) for subaerial-debris flows, because in the latter, large clasts are suspended at all levels in the moving flows and their support does not decay after deposition.

### *Time Sequence*

Support for large dense clasts must have been present throughout the life of the flows; the fact that large, dense, limestone clasts are restricted to one bed at one locality makes it unlikely that they were transported in stages by successive flows. The wide variety of rip-up clasts was acquired over some distance from the basin sediments west (upslope) of the main belt of coastal exposure (Fig. 1). The carpet of interacting grains was present at the time of deposition and is inferred to have played a major role throughout the duration of subaqueous flow in preventing the large dense clasts from grounding.

## DEPOSITIONAL MECHANISMS

A traction carpet under grain flow is considered to deposit rapidly by frictional freezing, from the base upward (Lowe, 1982). There is only one inversely graded basal layer in Parnell Grits, so there was never the repetitive development of traction carpets observed in some turbidites (Lowe, 1982).

The bulk of each Parnell Grit flow was a laminar cohesive debris flow whose cohesive strength was reduced somewhat by ingestion of fluid. Such flows deposit by cohesive freezing (Lowe, 1982). We envisage 1) frictional freezing of a dilated basal gravel layer in which clast movement was unhindered by the small proportion of cohesive matrix present, 2) cohesive freezing of a middle zone of matrix-rich sediment, and 3) in some, but not all, flows cohesive freezing and compaction of a more water-rich upper zone that lacks large clasts, and 4) in very rare cases, a tail of fine sediment, which may have been elutriated from the body of the flow, was deposited from suspension to form a laminated upper layer, with some traction to form migrating ripples.

Immediately following deposition, two extra-large dense clasts settled through the base of the deposit and impacted underlying sediments. This occurred in response to the decay of basal shear in the traction carpet, which reduced the effective bulk strength of the material from >1.25 $kg/cm^2$ during flow to between 0.5 and 0.78 $kg/cm^2$ after deposition.

### *Post-Depositional Elevation of Porewater Pressure*

As discussed earlier, there is abundant evidence for elevated pore pressures in the bed following deposition. It is of interest to establish whether any of this high pore pressure was present during flow. The only intrusive relationship that could have occurred during flow is that between Parnell Grit material and rip-up clasts. Virtually all rip-up clasts contain intrusions. However, we could not establish any criteria that would distinguish between syn-flow and post-flow intrusion.

Upward intrusions into overlying beds require that the overlying beds were first deposited. Similarly, there is probably a requirement that Parnell Grit beds were confined beneath overlying beds before downward diking could take place. Thus, the post-depositional high pore pressures and intrusions probably developed in confined Parnell Grit beds in response to seismically triggered liquefaction and may be temporally associated with slumping (Gregory, 1969).

## DISCUSSION

The flow and particle-support mechanisms that we infer for subaqueous Parnell Grit flows are very similar to those documented for subaerial-debris flows (e.g., Pierson, 1981, 1985, 1986). The main difference is the lack in Parnell Grits of any outsize dense clasts suspended in upper levels of flows, and of any evidence for the development of rigid plugs. We suggest that these two characteristics are the result of a lowering of cohesive matrix strength upon uptake of water by Parnell Grit flows. This leads to the proposal that the inferred debris avalanches, which initiated Parnell Grit flows and which would have been characterized by numerous rigid plugs and suspended large clasts, underwent considerable change in flow characteristics on entering the sea and taking up additional water (Fig. 16). Plugs dissipated as laminar shear spread through the weakened cohesive matrix, and large dense clasts settled to lower regions of the flows and formed the traction carpet of interacting grains that prevented the extra-large dense clasts from reaching the base of the flow and becoming stranded. Of course, even larger dense clasts may have been stranded upslope in shallow water.

Uptake of water by subaerial-debris flows that enter river channels can result in progressive dilution to the point that the flow characters undergo fundamental change to hyperconcentrated flow (Pierson and Scott, 1985). Similarly, subaerial debris flows that entered San Dimas Reservoir continued initially as debris flows, but then went through a hydraulic jump and a rapid flow transformation from supercritical laminar flow to subcritical turbulent flow (Weirich, 1988). Neither of these changes happened to Parnell Grit flows. It appears that, following the initial change of character as the debris avalanches entered the sea, Parnell Grit flows were stable. Sand, mud, shells and some lava and limestone clasts were taken into the flows from the sea bed. If there was a continuing uptake of water accompanying the ingestion of this material, it did not accumulate in the flows and therefore may have moved upward and out of the top of the flows. This would have introduced some degree of liquefied flow in addition to the grain-flow and cohesive debris-flow behaviors, but there is no independent

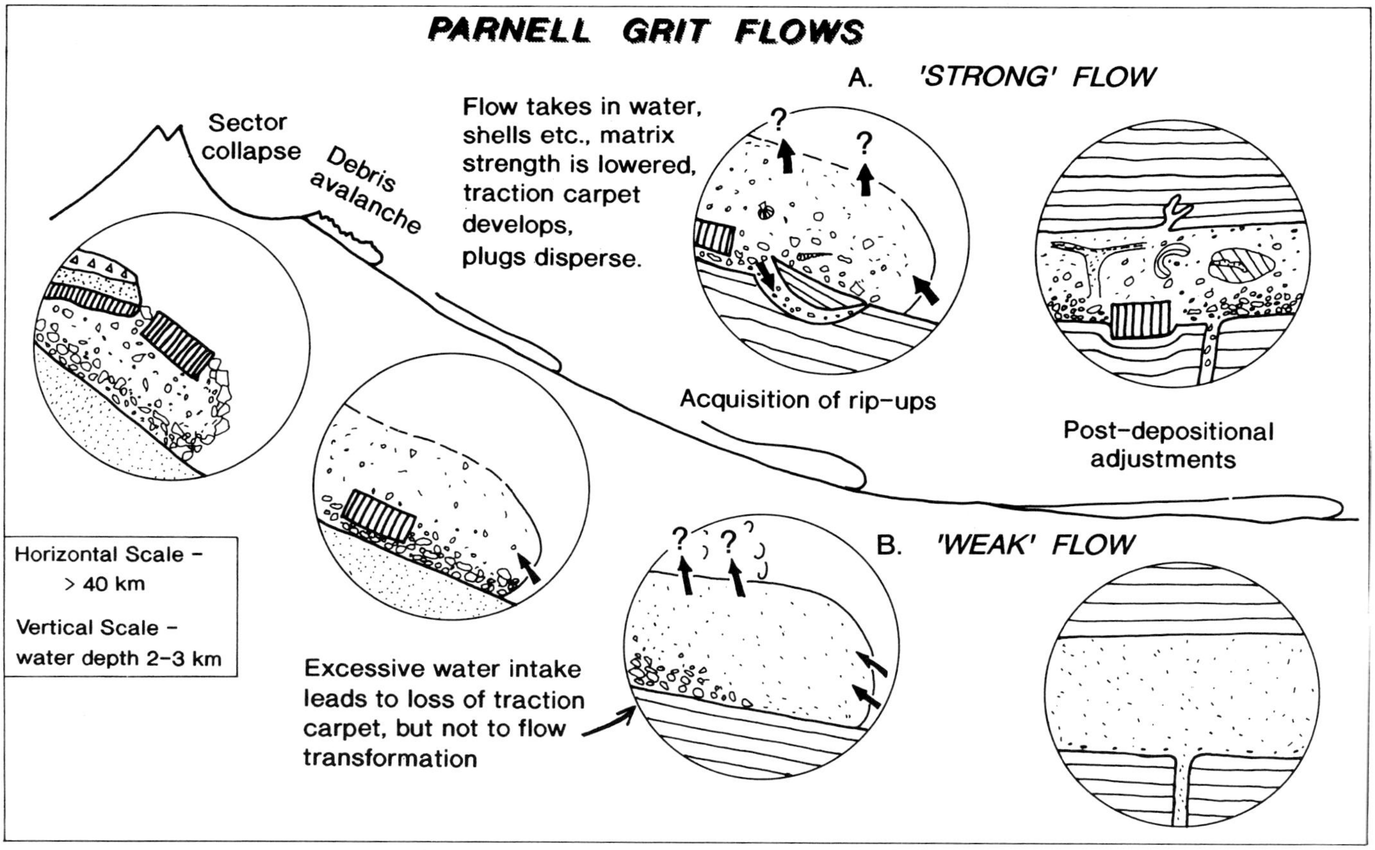


FIG. 16.—Schematic of a Parnell Grit flow.

evidence for its existence and no means of assessing its relative importance. The rate of water uptake by a moving subaqueous-debris flow may be governed by the permeability of the matrix material, and hence by the clay content.

A direct correlation between bed thickness (BTh) and maximum particle size (MPS) has been noted in many mass-flow conglomerates, both subaerial and subaqueous (e.g., Bluck, 1967; Nemec and Steel, 1984; Ballance, 1984). The suspension competence of these flows, as measured by the MPS (Pierson, 1981), is dependent on the thickness of the flow. There may also be strength factors that are independent of the flow thickness, and these are indicated when the MPS/BTh regression line intersects the ordinate above the origin (Nemec and Steel, 1984). They include cohesion of the matrix and grain-to-grain interactions (Walton and Palmer, 1988). The Parnell Grit flows do not show the MPS/BTh correlation; a bed 20 m thick may have MPS (dense clast) of <2 cm (Fig. 3). This fact suggests that Parnell Grit flows differed in some fundamental respect(s) from the more commonly described mass-flow rudites. Our analysis points to water intake and consequent reduction in matrix strength as the cause of the difference. It is possible that Parnell Grit flows lost matrix strength to varying degrees. The flows depositing thick beds with small MPS may have left their larger clasts as a lag deposit upslope as their matrix strength was reduced. The high content of distributed basinal sediment in these deposits supports that suggestion.

## CONCLUSIONS

Parnell Grit flows originated as subaerial-debris avalanches resulting from sector collapses on basaltic/andesitic stratovolcanoes of the early Miocene Waitakere Arc. They moved through the strand zone and entered the Waitemata flysch basin. Ingestion of water and sediment at the head of the still-moving flows changed particle support mechanisms slightly by lowering cohesive strength. Laminar shear then spread to all areas of the flow and destroyed rigid plugs. Dense clasts settled through the flow to form a normally graded layer above a basal traction carpet of dense clasts in grain-flow mode. There may have been an element of raised pore pressure and liquefied laminar flow. Extra-large dense clasts (up to 2,000 tons) were carried more than 40 km into the basin. In order to transport them, the effective bulk strength of the flowing medium, incorporating cohesive strength plus pyramidal buoyancy plus basal-grain flow, was equivalent to $>1.25$ kg/cm$^2$.

Numerous rip-up clasts up to 90 m in length were acquired during passage down the basin slopes. Injection of flow matrix into the substrate and undercutting of channel margins, as a result of density and momentum of flow, were probably the main causes of rip-up acquisition. Rip-up clasts (density 1.8–1.9) were neutrally buoyant in the flows, which therefore had a bulk density of 1.9 to 2.00.

Deposition at bathyal depths took place by frictional freezing in the traction carpet and by cohesive freezing in the body of the flow. The tail of some flows deposited a graded bed of fine sediment without larger clasts and with rare evidence for tractional movement. Parnell Grit deposits lacking larger clasts (MPS of 1 or 2 cm) may have had their strength reduced to a level where larger clasts were abandoned upslope, and a traction carpet was absent. Conical mounds of rip-up clast-rich Grit are preserved on the upper surface of one Grit bed.

After deposition and the decay of the grain-flow contribution to clast support, all rip-up clasts and dense clasts 4 m and smaller (<40 tons) were held in position, implying a residual bulk strength of the Grit matrix greater than 0.5 kg/cm$^2$. However, two larger basalt blocks (one $8 \times 5 \times 2+$ m, one $20 \times 12 \times 3$ m) indented underlying sediments to a depth of 3 or more meters, implying a bulk residual strength less than 0.78 kg/cm$^2$.

At some later time, following burial, pore pressures were raised as a result of seismic shaking, leading to water escape via vertical sheet structures and to injections of Grit—inward into rip-up clasts, downward into underlying sediment, upward from lower to higher levels within Grit beds, and upward from Grit beds into overlying beds.

There were probably more than 100 Parnell Grit flows, all sharing the same general characteristics. The main particle-support factors were cohesive strength, pyramidal buoyancy, and grain flow in the traction carpet. There is evidence that the bulk density and strength were lowered by water uptake when the flows entered the sea, but we can find no evidence to confirm the flow transformation from turbulent to laminar flow suggested by Fisher (1983, 1984).

## ACKNOWLEDGMENTS

We acknowledge the financial support of the Auckland University Research Committee over many years. Donald Lowe and Gary Smith gave valuable reviews. Rosemary Bunker typed the manuscript, and Louise Cotterall drafted the figures.

## REFERENCES

BALLANCE, P. F., 1964, The sedimentology of the Waitemata Group in the Takapuna section, Auckland: New Zealand Journal of Geology and Geophysics, v. 7, p. 466–499.

BALLANCE, P. F., 1974, An inter-arc flysch basin in northern New Zealand: Waitemata Group (Upper Oligocene to Lower Miocene: Journal of Geology, v. 82, p. 439–471.

BALLANCE, P. F., 1977, The geology of Motuora Island, Hauraki Gulf: Tane, v. 23, p. 77–82.

BALLANCE, P. F., 1984, Sheet-flow-dominated gravel fans of the non-marine middle Cenozoic Simmler Formation, central California: Sedimentary Geology, v. 38, p. 337–359.

BLUCK, B. J., 1967, Deposition of some Upper Old Red Sandstone conglomerates in the Clyde area: A study in the significance of bedding: Scottish Journal of Geology, v. 3, p. 139–167.

CRABB, P. L., 1971, Bryozoan zooarial paleoecology of the Waitemata Group: University of Auckland, New Zealand, unpubl. M.S. Thesis, 147 p.

CRANDELL, D. R., 1989, Gigantic debris avalanche of Pleistocene age from ancestral Mt. Shasta Volcano, California, and debris-avalanche hazard zonation: U.S. Geological Survey Bulletin 1861, 32 p.

DAY, R. A., 1986, Miocene volcanic associations, *in* Smith, I. E. M., Day, R. A., and Ashcroft, J., eds., Volcanic Associations of Northland: New Zealand Geological Survey Record, v. 12, p. 10–19.

DOBKINS, J. E., AND FOLK, R. L., 1970, Shape development on Tahiti-Nui: Journal of Sedimentary Petrology, v. 40, p. 1167–1203.

FISHER, R. V., 1971, Features of coarse-grained, high-concentration fluids and their deposits: Journal of Sedimentary Petrology, v. 41, p. 916–927.

FISHER, R. V., 1983, Flow transformations in sediment gravity flows: Geology, v. 11, p. 273–274.

FISHER, R. V., 1984, Submarine volcaniclastic rocks, *in* Kokelaar, B. P., and Howells, M. F., eds., Marginal Basin Geology: Geological Society of London Special Publication 16, p. 5–27.

FISHER, R. V., AND SCHMINCKE, H.-U., 1984, Pyroclastic Rocks: Springer-Verlag, Berlin, 472 p.

GREGORY, M. R., 1966, Rocks of the Waitemata Group, Whangaparaoa Peninsula, Northland: University of Auckland, New Zealand, Unpubl. M.S. Thesis, 233 p.

GREGORY, M. R., 1969, Sedimentary features and penecontemporaneous slumping in the Waitemata Group, Whangaparaoa Peninsula, North Auckland, New Zealand: New Zealand Journal of Geology and Geophysics, v. 12, p. 248–282.

GREGORY, M. R., 1974, Geology, *in* University of Auckland, Maui Development Environmental Study, Report on Phase 1: Shell, B. P. and Todd Oil Services Ltd., New Plymouth, p. 28–44.

HAMPTON, M. A., 1975, Competence of fine-grained debris flows: Journal of Sedimentary Petrology, v. 45, p. 834–844.

HAMPTON, M. A., 1979, Buoyancy in debris flows: Journal of Sedimentary Petrology, v. 49, p. 753–758.

HAYWARD, B. W., 1976, Lower Miocene geology and sedimentary history of the Muriwai—Te Waharoa coastline, north Auckland, New Zealand: New Zealand Journal of Geology and Geophysics, v. 19, p. 639–662.

HAYWARD, B. W., 1979, Eruptive history of the early to mid-Miocene Waitakere Volcanic Arc, and palaeogeography of the Waitemata Basin, northern New Zealand: Journal of Royal Society of New Zealand, v. 9, p. 297–320.

HOCHSTETTER, F., VON, 1964, Geologie von Neu-Seeland. Beitrage zur Geologie der Provinzen Auckland und Nelson: Novara-Expedition, Geologie Thiel 1, v. 1, 274 p. (Transl. C. A. Fleming, 1959, Government Printer, Wellington, New Zealand).

JANDA, R. J., SCOTT, K. M., NOLAN, K. M., AND MARTINSON, H. A., 1981, Lahar movement, effects, and deposits, *in* Lipman, P. W., and Mullineaux, D. R., eds., The 1980 Eruptions of Mount St. Helens, Washington: U.S. Geolological Survey Professional Paper 1250, p. 461–478.

KUENEN, P. H., 1950, Slump structures in the Waitemata Beds around Auckland: Transactions of the Royal Society of New Zealand, v. 78, p. 467–475.

LAIRD, M. G., 1970, Vertical sheet structure—A new indicator of sedimentary fabric: Journal of Sedimentary Petrology, v. 40, p. 428–434.

LEEDER, M. R., 1982, Sedimentology: Process and Product: George Allen and Unwin, London, 344 p.

LOWE, D. R., 1979, Sediment gravity flows: Their classification and some problems of application to natural flows and deposits, *in* Doyle, L. J., and Pilkey, O. H., eds., Geology of Continental Slopes: Society of Economic Paleontogists and Mineralogists Special Publication 27, p. 75–82.

LOWE, D. R., 1982, Sediment gravity flows: II. Depositional models with special reference to the deposits of high-density turbidity currents: Journal of Sedimentary Petrology, v. 52, p. 279–297.

MCCLELLAND, B., 1967, Progress of consolidation in delta front and prodelta clays of the Mississippi River, *in* Richards, A. F., ed., Marine Geotechnique: University of Illinois Press, Urbana p. 22–40.

MIDDLETON, G. V., AND HAMPTON, M. A., 1976, Subaqueous sediment transport and deposition by sediment gravity flows, *in* Stanley, D. J. and Swift, D. J. P., eds. Marine Sediment Transport and Environmental Management: New York, Wiley, p. 197–218.

MIDDLETON, G. V., AND SOUTHARD, J. B., 1978, Mechanics of Sediment Movement: Society of Economic Paleontologists and Mineralogists Short Course No. 3, 242 p.

MORGENSTERN, R. N., 1967, Submarine slumping and the initiation of turbidity currents, *in* Richards, A. F., ed., Marine Geotechnique: University of Illinois Press, Urbana, p. 189–220.

MORRIS, W. R., AND BUSBY-SPERA, C. J., 1988, Sedimentologic evolution of a submarine canyon in a forearc basin, Upper Cretaceous Rosario Formation, San Carlos, Mexico: American Association of Petroleum Geologists Bulletin, v. 72, p. 717–737.

NEMEC, W., AND STEEL, R. J., 1984, Alluvial and coastal conglomerates: Their significant features and some comments on gravelly mass flow deposits, *in* Koster, E. H., and Steel, R. J., eds., Sedimentology of Gravels and Conglomerates: Canadian Society of Petroleum Geologists Memoir 10, p. 1–31.

PIERSON, T. C., 1981, Dominant particle support mechanisms in debris flows at Mt. Thomas, New Zealand, and implications for flow mobility: Sedimentology, v. 28, p. 49–60.

PIERSON, T. C., 1985, Initiation and flow behavior of the 1980 Pine Creek and Muddy River lahars, Mount St. Helens, Washington: Geological Society of America Bulletin, v. 96, p. 1056–1069.

PIERSON, T. C., 1986, Flow behavior of channelized debris flows, Mount St. Helens, Washington, *in* Abrahams, A. D., ed., Hillslope Processes: Allen and Unwin, Boston, p. 269–296.

PIERSON, T. C., AND SCOTT, K. M., 1985, Downstream dilution of a lahar: Transition from debris flow to hyperconcentrated streamflow: Water Resources Research, v. 21, p. 1511–1524.

RICKETTS, B. D., BALLANCE, P. F., HAYWARD, B. W., AND MAYER, W., 1989, Basal Waitemata Group lithofacies: Rapid subsidence in an early Miocene interarc basin, New Zealand: Sedimentology, v. 36, p. 559–580.

RODINE, J. D., AND JOHNSON, A. M., 1976, The ability of debris, heavily freighted with coarse clastic materials, to flow on gentle slopes: Sedimentology, v. 23, p. 213–234.

SMITH, I. E. M., RUDDOCK, R. S., AND DAY, R. A., 1989. Miocene arc-type volcanic/plutonic complexes of the Northland Peninsula, New Zealand, *in* Sporli, K. B., and Kear, D., eds., Geology of Northland: Royal Society of New Zealand Bulletin, 26, p. 205–213.

TURNER, F. J., AND BARTRUM, J. A., 1929, The geology of the Takapuna-Silverdale district, Waitemata County, Auckland: Transactions, New Zealand Institute, v. 59, p. 864–902.

WALTON, A. W., AND PALMER, B. A., 1988, Lahar facies of the Mount Dutton Formation (Oligocene-Miocene) in the Marysvale Volcanic Field, southwestern Utah: Geological Society America Bulletin, v. 100, p. 1078–1091.

WEIRICH, F. H., 1988, Field evidence for hydraulic jumps in subaqueous sediment gravity flows: Nature, v. 332, p. 626–629.

# PLIOCENE-QUATERNARY SUBMARINE PUMICE DEPOSITS IN THE SUMISU RIFT AREA, IZU-BONIN ARC


A. NISHIMURA

*Marine Geology Department, Geological Survey of Japan, 1-1-3 Higashi, Tsukuba, Ibaraki 305, Japan*

K. M. MARSAGLIA

*Department of Geological Sciences, The University of Texas at El Paso, El Paso, Texas 79968*

K. S. RODOLFO

*Department of Geological Sciences, University of Illinois at Chicago, P.O. Box 4348, Chicago, Illinois 60680*

A. COLELLA

*Dipartimento Di Scienze della Terra, Università della Calabria, 87939 Castiglione Cosentino, Scalo (SC), Italy*

R. N. HISCOTT

*Earth Sciences Department, Memorial University, St. John's, Newfoundland, Canada, A1B 3X5*

K. TAZAKI

*Geology Department, Shimane University, Matsue, Shimane Prefecture 690, Japan*

J. B. GILL

*Department of Earth Sciences, University of California, Santa Cruz, California 95064*

T. JANECEK AND J. FIRTH

*Ocean Drilling Program, Texas A & M University, 1000 Discovery Drive, College Station, Texas 77845-9547*

M. ISIMINGER-KELSO

*Department of Geology, Florida State University, Tallahassee, Florida 32306*

Y. HERMAN

*Department of Geology, Washington State University, Pullman, Washington 99164*

R. N. TAYLOR

*Geology Department, University of Southampton, Southhampton, United Kingdom, SO9 5NH*

B. TAYLOR

*Hawaii Institute of Geophysics, University of Hawaii, 2525 Correa Road, Honolulu, Hawaii 96822*

K. FUJIOKA

*Ocean Research Institute, University of Tokyo, 1-15-1 Minamidai Nakano-ku, Tokyo 164, Japan*

AND

LEG 126 SCIENTIFIC PARTY



ABSTRACT: Thick, multiple deposits of Pliocene-Quaternary pumice are the dominant products of central Izu-Bonin Arc volcanism, both prior to and during rifting of the backarc Sumisu Basin. Pliocene pre-extension pulses of magmatism within the arc are recorded by crude coarsening-upward volcaniclastic sequences, and Quaternary synextension volcanism produced thick pumice units separated by intervals of hemipelagic sediment. In particular, the period from about 0.2 Ma to the present may have experienced four major episodes of pumice production and deposition with a periodicity averaging 30,000 years. The pumice was delivered to the basin by a combination of direct deposition of fallout from submarine (and possibly also subaerial) eruptions after settling through the water column, resedimentation from basin slopes by submarine debris flows, and redistribution on the basin floor by bottom currents.


## INTRODUCTION

A principal objective of Leg 126 of the Ocean Drilling Program (ODP) was to establish the timing of initiation and development of the Sumisu Rift (Brown and Taylor, 1988; Murakami, 1988; Nishimura and Murakami, 1988), a backarc basin produced by rifting west of the Izu-Bonin Arc (Figs. 1, 2). Our aim here is to outline the chronology and possible mechanisms of pumice deposition in the basin. Five holes were drilled on the flat floor of the rift basin in water about 2,250 m deep: two at Site 791 near the center of the eastern half graben of the rift, and three above the flank of a basement high at nearby Site 790 (Figs. 1–3). Two other successful holes were drilled atop the eastern flank of the rift basin at Site 788 in water 1,113 m deep. The ages of recovered sediments and rocks were determined biostratigraphically, using nannofossils and planktonic foraminifers, with some assistance from magnetic stratigraphy.

General results of ODP Leg 126, summarized elsewhere (Leg 126 Shipboard Scientific Drilling Party, 1989; ODP Leg 126 Scientific Party, 1989), constrain the beginning of rifting to the interval between 1.1 and 2.7 Ma. The periods both before and after the onset of rifting were characterized by episodic production and accumulation of rhyolitic-pumice lapilli and ash (Figs. 3, 4), with intervening quiescent periods marked by accumulation of carbonate-rich, locally bioturbated hemipelagic mud. This pattern continues to the present.

The pumices range in chemical compositon from dacite to rhyolite (Table 1). There are greater differences in major elements within stratigraphic units than between them. Individual pumice deposits, such as the Pleistocene pumices at Site 788 and Pumice Unit II at Sites 790/1, contain the full range of compositions from dacite to rhyolite. This does not require reworking, however, because a range similar to that between samples 1 and 2 in Table 1 also occurs in pumices from the 1952 eruption of the nearby Myojinsho volcano (Kuno, 1962).

Previous workers (Carey and Sigurdsson, 1984; Karig and Moore, 1975; Klein, 1985) have outlined models for sediment distribution in deep-marine marginal backarc basins, based on drilling results from various Deep Sea Drilling


Sedimentation in Volcanic Settings, SEPM Special Publication No. 45

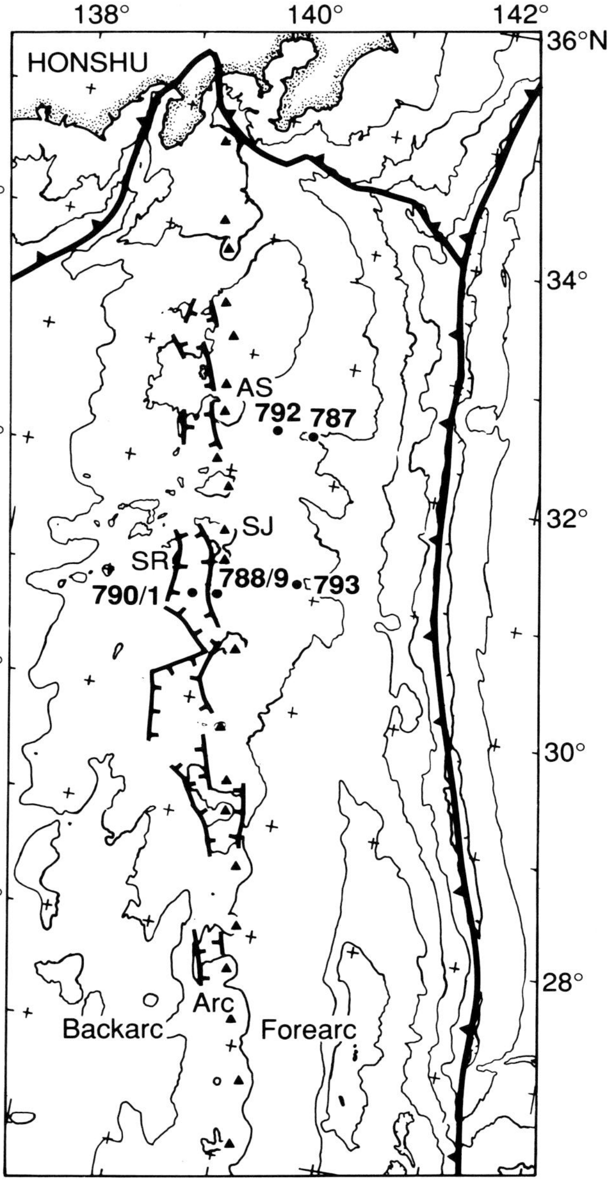


FIG. 1.—Bathymetric map of the Izu-Bonin intraoceanic island arc showing location of Leg 126 sites. SR = Sumisu Rift, SJ = Sumisa Jima, and AS = Aoga Shima.

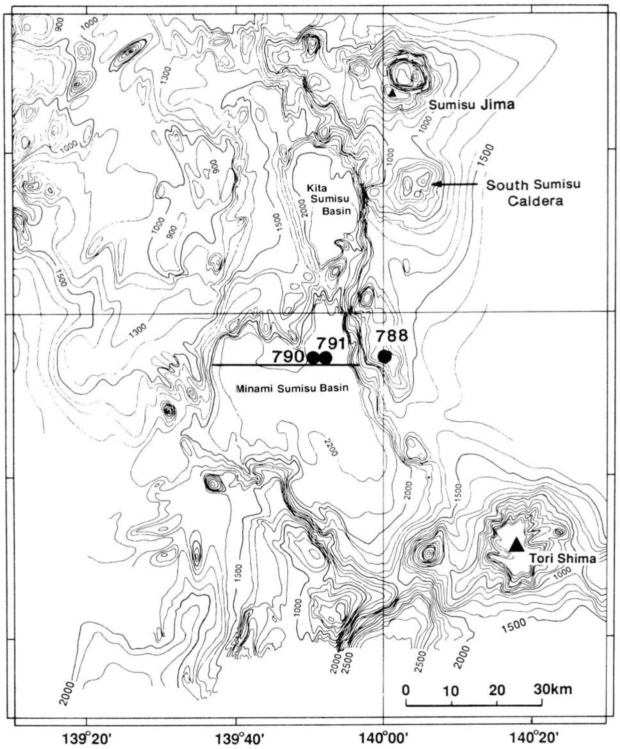


FIG. 2.—Location of ODP Leg 126 Sites 788, 790 and 791 on the basin floor and the eastern uplifted flank of the Sumisu Rift, Izu-Bonin Arc. The rift is subdivided into two basins, the Kita-Sumisu and the Minami-Sumisu Basins, by a ridge of intra-rift volcanoes (Brown and Taylor, 1988). Bathymetry from Murakami (1988). Contour interval is 100 m. Location of the single-channel airgun line of Figure 6 is indicated by an east-west line immediately south of Sites 790 and 791.

Project (DSDP) legs in the western Pacific. Because of penetration problems within the coarse-grained volcaniclastic aprons that flank volcanic arcs, only a few of the DSDP sites used in these models are located along or adjacent to an active-arc axis. For example, on Leg 60 during a transect across the Mariana Arc, one hole was drilled on the volcanic-arc axis (Site 457) and another on the frontal-arc margin of the backarc basin (Site 455). The presence of Pleistocene unconsolidated coarse volcaniclastics led to unstable hole conditions that forced the abandonment of rotary-cored Site 457 after only 52 m of penetration and site 455 after 104 m of penetration (Hussong and others, 1982).

In contrast, thick Pleistocene pumiceous sections were successfully penetrated by piston coring on Leg 126 in the Sumisu rift and on the rift flank (Figs. 1–3; Leg 126 Shipboard Scientific Party, 1989). Thus, the sediments recovered on Leg 126 provide recent analogs for possible ancient equivalent deposits described by Fiske (1986) and Fiske and Cashman (1989) in Miocene-Pliocene units exposed on the Izu Peninsula of Japan by Grechin and others (1981) in the Miocene of the California Continental Borderland, by Busby-Spera (1988) in Middle Jurassic units on Cedros Island, Baja California, and by Niem (1977) in Mississippian units of the Ouachita Mountains of the south-central United States.

## THE FLANK-TOP DEPOSITS: SITE 788

Lower Pliocene (4.2–4.8 Ma) Unit IIB at Site 788 (Fig. 3), which extends from 278.6 to 374 m below the sea floor (mbsf), is predominantly pumiceous, rhyolitic lapillistone with interbeds of coarse and fine vitric tuff and hemipelagic silty claystone. The lapillistone is grayish black and brownish black, contains 20 to 40 percent coarse ash and 60 to 80 percent lapilli with maximum diameters that range from 10 to 35 mm. Maximum bed thicknesses are unknown due to incomplete recovery, but the thickest single interval of

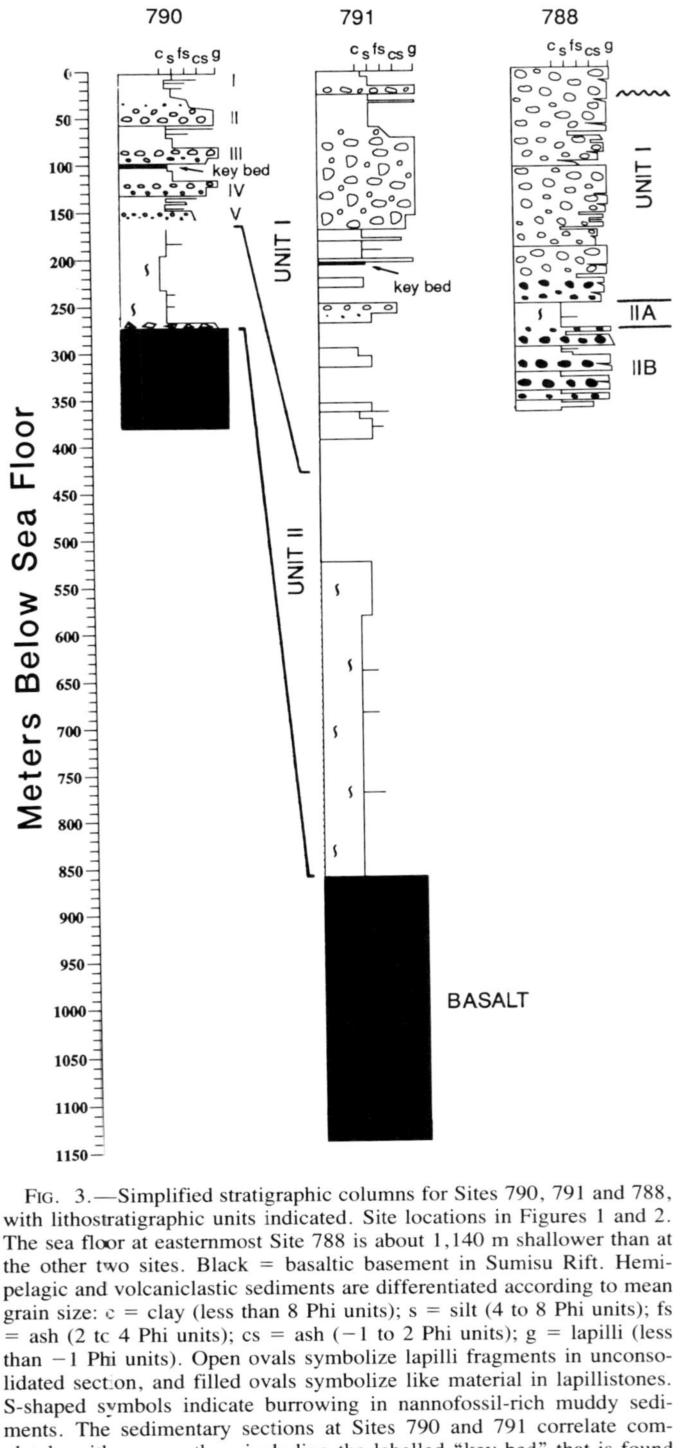


FIG. 3.—Simplified stratigraphic columns for Sites 790, 791 and 788, with lithostratigraphic units indicated. Site locations in Figures 1 and 2. The sea floor at easternmost Site 788 is about 1,140 m shallower than at the other two sites. Black = basaltic basement in Sumisu Rift. Hemipelagic and volcaniclastic sediments are differentiated according to mean grain size: c = clay (less than 8 Phi units); s = silt (4 to 8 Phi units); fs = ash (2 to 4 Phi units); cs = ash (−1 to 2 Phi units); g = lapilli (less than −1 Phi units). Open ovals symbolize lapilli fragments in unconsolidated section, and filled ovals symbolize like material in lapillistones. S-shaped symbols indicate burrowing in nannofossil-rich muddy sediments. The sedimentary sections at Sites 790 and 791 correlate completely with one another, including the labelled "key bed" that is found at both sites. Thick pumice units at Site 790 are numbered downward from I to V, as in Figure 7. The upper 27.5 to 29.7 m at Site 788 is roughly equivalent in age to Unit I at the other sites (late Quaternary); the rest of the Site 788 column is Pliocene and thus older than any of early to late Pleistocene sediments (Unit II) recovered at Sites 790 and 791. The unconformity at Site 788 is shown by a wavy line.

lapillistone and associated lapilli-bearing coarse tuff is about 1.5 m thick. The coarse and fine tuff are grayish black and occur in parallel-laminated beds to 4 to 37 cm thick.

Lower Pliocene (3.9–4.2 Ma) Unit IIA, 249 to 278.6 mbsf, consists primarily of nannofossil-rich claystone that is glauconitic in part with lesser amounts of vitric tuff and hemipelagic silty claystone. The finer grain size of this unit indicates a decrease in the intensity of explosive volcanism or the competence of the sediment-transport process relative to that recorded by Unit IIB. Pumice grains and foraminifer tests throughout Unit II have undergone ductile and/or brittle deformation, indicating compaction by an overlying sedimentary column that must have been originally thicker than it is at present.

Unit I at Site 788, 0 to 249 mbsf, consists primarily (86 percent) of Pliocene and Quaternary ash-bearing pumiceous-lapilli deposits, the lowest 19 m having been pressure compacted to lapillistone. The remaining 14 percent of Unit I is mainly structureless, silicic, coarse vitric ash, much of which occurs in disturbed intervals at the tops and bottoms of cores that may partly represent cavings and flow-in to the borehole. Minor, undisturbed coarse-ash beds within the cores are only 6 to 58 cm thick; two of these grade upward into 4- and 77-cm-thick intervals of fine vitric ash. Apart from its lithification, the lower lapillistone is compositionally and texturally similar to the overlying pumiceous deposits. The unlithified and lithified pumiceous deposits of Unit I contain very little silt and clay (Fig. 4) and therefore very few calcareous microfossils ($CaCO_3$ content mostly less than 0.2 percent).

Most pyroclasts are dacite to rhyolite pumice, typically aphyric, although some contain microlites and plagioclase and quench clinopyroxene. Typical water-saturated specific gravities range from 1.26 to 1.58. Diameters of the largest lapilli, measured as the average of the 10 largest in a core, are 2 to 4 cm. Colors of the pumice fragments are mainly olive gray to light olive gray, although in the uppermost 58 m of the section some of them are oxidized to shades of yellow, orange and brown. In the lower part of this surficial interval (39.2–61 mbsf), a few pumice lapilli show thin (<1 mm) surface coatings of manganese oxide, some in the form of microspherules. The manganiferous interval lies below an unconformity, located between 27.5 and 29.7 mbsf by nannofossil data, which separates the uppermost part of Unit I, latest Quaternary in age (0.275 Ma), from the rest of the Unit, which is upper Pliocene (>2.35 Ma).

Lithic clasts constitute less than 1 percent of the Unit I pumiceous tephra, but are entirely missing only in the interval from 58 to 96 mbsf. Typically, these are angular fragments of scoriaceous andesite with specific gravities of 2.1 to 2.2 and average maximum diameters of only 15 to 37 percent as large as those of associated pumice fragments. Below about 135 m, rare clasts of green-mottled and grayish-black pumiceous breccia are also present.

Crude coarsening-upward cycles of percent lapilli content and maximum fragment size in unconsolidated Unit IA (Fig. 5) are tentatively attributed to four periods during which silicic volcanism built up to climaxes. Each cycle is 30 to 50 m thick; the first three were deposited during the Pliocene from 4.8 Ma to ≥2.35. Pumice deposits of the last

TABLE 1.—CHEMICAL COMPOSITION OF BULK PUMICE ON AN ANHYDROUS BASIS

| Sample | 1 | 2 | 3 |
|---|---|---|---|
| $SiO_2$ | 65.71 | 70.44 | 75.53 |
| $TiO_2$ | 0.84 | 0.60 | 0.37 |
| $Al_2O_3$ | 14.41 | 13.59 | 12.60 |
| $Fe_2O_3$* | 7.17 | 4.54 | 2.62 |
| MgO | 1.50 | 1.02 | 0.43 |
| CaO | 5.24 | 3.84 | 2.26 |
| $Na_2O$ | 4.21 | 5.10 | 5.33 |
| $K_2O$ | 0.76 | 0.73 | 0.82 |
| $P_2O_5$ | 0.15 | 0.13 | 0.05 |

*Total Fe as $Fe_2O_3$
Analyst: J. Gill, UCSC XRF

1. Site 788, hole C, core 19H, section 1, cc; Unit IA; Pliocene; Flank top
2. Site 790, hole B, core 10H, section 1, 0–35 cm; Pumice Unit II; Pleistocene; Rift basin
3. Site 788, hole C, core 3H, section 3, 10–12 cm; Unit IA; Pleistocene; Flank top

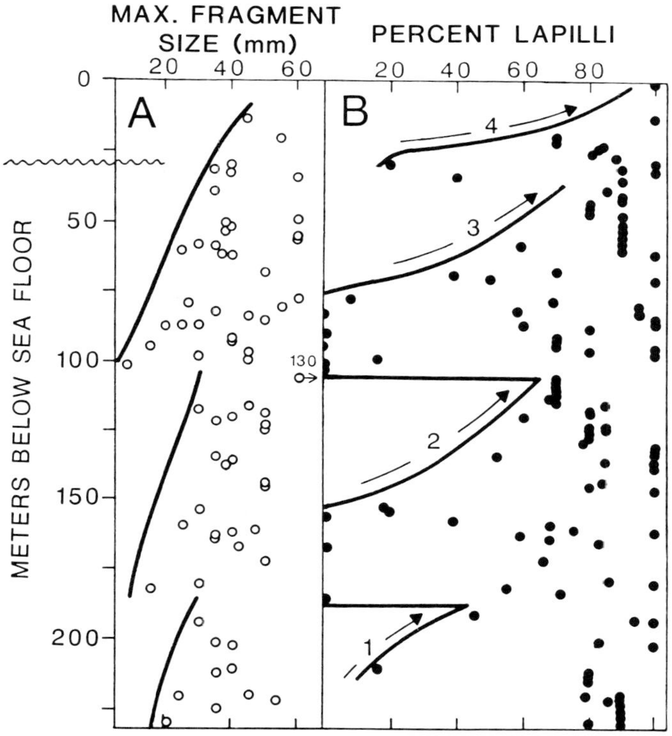


FIG. 5.—Cyclic variation in maximum fragment size of pumice and percent lapilli in Unit IA at Site 788, extracted from shipboard descriptions. Curves to the left of the data points suggest four crude coarsening-upward trends (numbered in B). (A) Circle = maximum fragment diameter per core or section. (B) Percent lapilli per section. Even in parts of the column where lapilli are locally uncommon, some sections consist mainly of lapilli, hence, the trends are only crudely defined. Wavy line represents an unconformity.

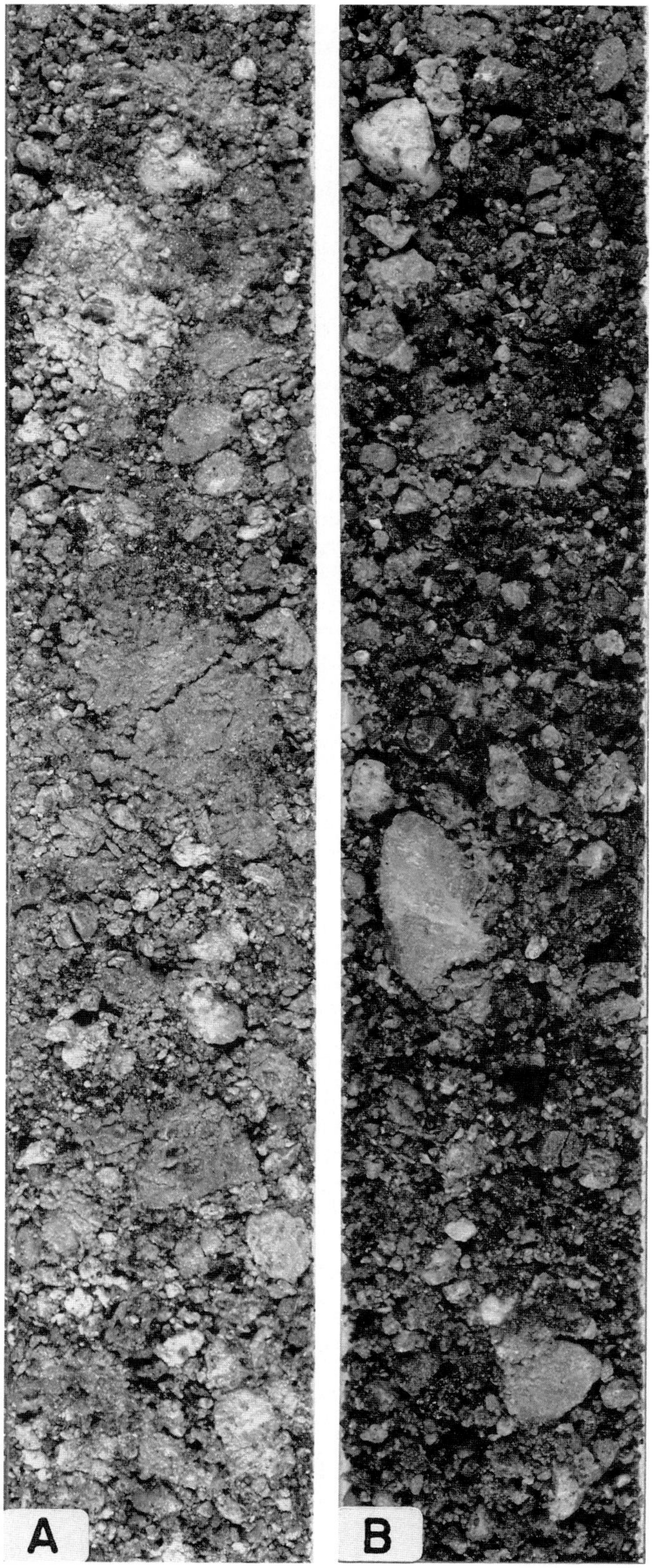


FIG. 4.—Comparison of typical textures of pumiceous deposits from the rift basin and eastern basin flank. Core diameter is 6.5 cm. (A) Structureless, moderately sorted pumice tephra from the upper part of Unit I at Site 791 (sample 126-791A-13H-4, 69-104 cm; 115 mbsf). (B) Structureless, matrix-free pumice tephra from the upper part of Unit I at Site 788 (sample 126-788C,4H-3, 15–50 cm; 48 mbsf).

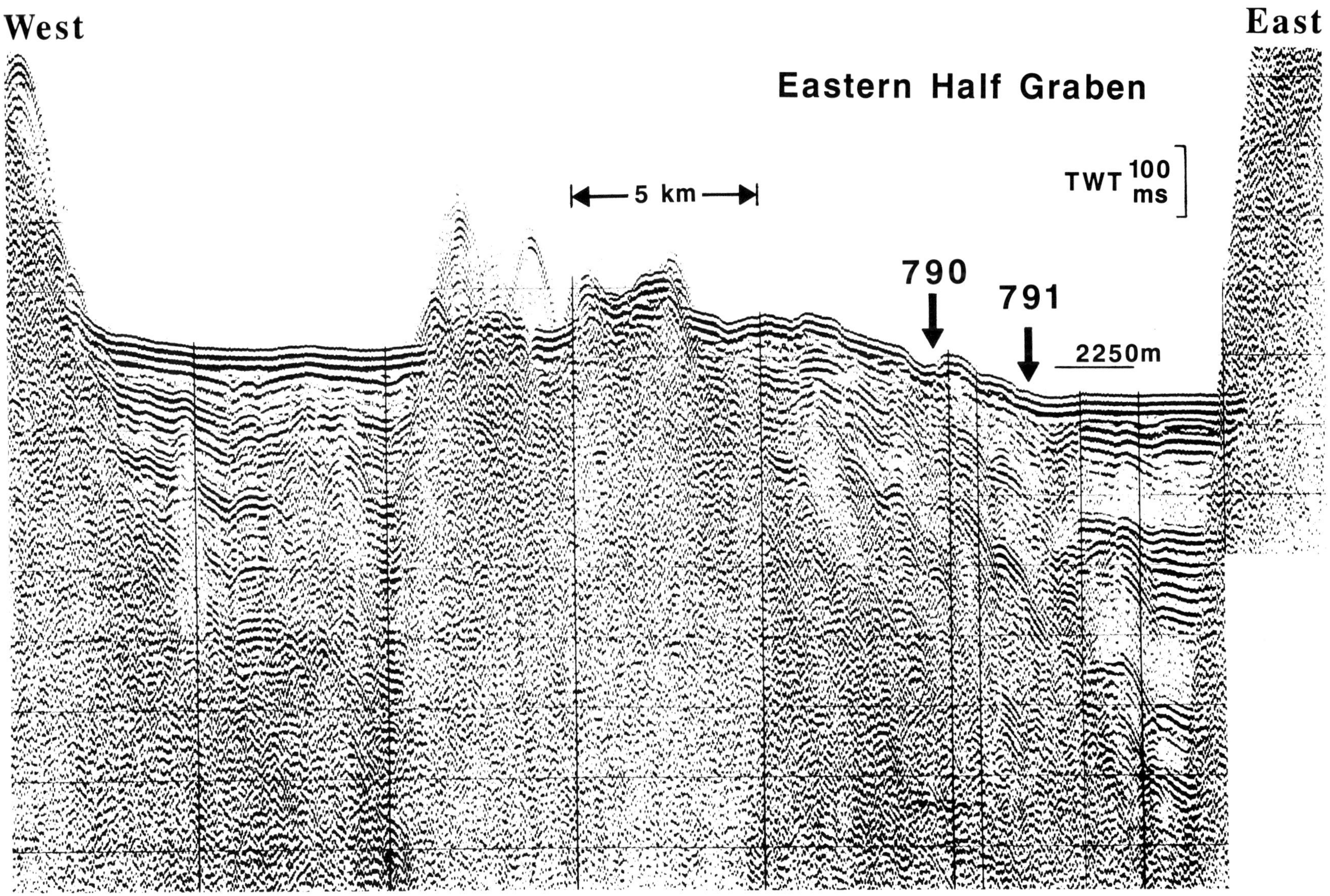


FIG. 6.—Segment of a single-channel airgun profile across the eastern half graben of Sumisu Rift about 2 km south of Sites 790 and 791 (Figs. 1, 2). The seismic data were acquired by the Geological Survey of Japan, R/V Hakurei Maru cruise GH84-2, and are interpreted by Murakami (1988, line 84–56). The wedge-shaped geometry of seismic units is the combined effect of progressive eastward tilting of the basin during rifting and leveling of the basin floor by sedimentation. The vertical scale is in two-way travel time (TWT) of acoustic energy, where 100 ms TWT = 75 m of water. The three uppermost dark traces do not represent sedimentary layering but are artifacts of the airgun configuration used in the survey.

cycle accumulated some time after 0.275 Ma, sufficiently long ago for lapilli to become oxidized and discolored while residing on the sea floor.

## THE RIFT-BASIN PUMICE DEPOSITS: SITES 790 AND 791

Site 791 is only 2.4 km east of Site 790 (Fig. 2); however, accumulation rates of both hemipelagic and volcaniclastic sediments at Site 791 were consistently two to four times as great as at Site 790. This eastward increase in sediment thickness is a general characteristic of the eastern half graben of the backarc basin (Fig. 6). Sites 790 and 791 show two correlative units. The basal Unit II (165–271 mbsf at site 790; 428–834 mbsf at Site 791) is early to late Pleistocene in age (about 0.2–1.1 Ma) and is composed predominantly of burrowed, nannofossil-rich muddy sediments (Fig. 3) punctuated by minor, thin, ash beds. The uncertainty in age of the Unit I/II boundary is about ±20 ka. Unit I (0–165 mbsf at Site 790; 0–428 mbsf at Site 791) is late Quaternary in age (about 0–0.2 Ma) and is composed of thick sequences of pumice lapilli (Fig. 4), vitric ash, and very thin intervals of nannofossil-rich silty clay with interbeds of vitric ash. Recoveries from the three holes at Site 790 provided an essentially complete sequence of Unit I, the lower part of which was not as well sampled at Site 791 (Fig. 3). The thick sequences of pumice at Site 790 are numbered downward from I to V (Fig. 3), with III and IV perhaps constituting a single composite sequence. Each sequence consists of, from the base to the top, pumiceous lapilli, lapilli-bearing ash and fine vitric ash 15 to 30 m thick, separated by 10- to 15-m-thick intervals of hemipelagic clayey silt with 3- to 100-cm-thick interbeds of vitric ash (Fig. 3); at Site 791 the equivalent sequences are two to three times thicker. The pumice deposits at Sites 790 and 791 are more poorly sorted than those on the flank of the rift at Site 788; their matrices contain abundant ash and some nannofossils (Fig. 4). The content of $CaCO_3$, present solely as calcareous microfossils, is 1 to 3 percent, an order of magnitude greater than in flank-top deposits at Site 788.

## DISCUSSION

All but the uppermost 29 m of sedimentary rocks recovered at Site 788 on the rift flank record silicic-pumice eruptions before the onset of backarc rifting in the area, from 4.8–2.73 Ma. The excellent recovery from Unit I at Site 788 yielded an important line of evidence that bears strongly on the origin of the pumiceous deposits of this unit. Some or all of the sparsely recovered Unit II pumiceous deposits may share similar origins. This evidence is the significantly different sizes of the pumice fragments and lithic clasts in the heterolithic intervals. Fiske and Cashman (1989) have analyzed similar Mio-Pliocene deposits in Japan and have theoretically and experimentally simulated the settling of coarser pumiceous fragments and finer lithic clasts. They conclude that these deposits were produced by submarine eruptions. While settling through seawater columns, large low-density pumice grains and smaller and denser lithic clasts settle at the same velocities. This results in a distinct bimodality in grain size, which would be much less pronounced had the pumice fragments and lithic clasts settled through air. In the same manner, the range of sizes of the pumice and lithic lapilli from Unit I of Site 788 on the rift flank also clearly indicate settling through seawater.

The association of similar sizes of pumice fragments and lithic clasts is some indication that the lapilli reached the ocean while freshly erupted and still hot. Experiments by Whitham and Sparks (1986) have shown that hot pumice cubes of comparable size sink almost immediately when dropped into water, because the air in pores is suddenly contracted when cooled, and water is sucked in. If the pumice is still hot, the water drawn into pores is converted to steam that displaces the air; when the pumice cools below 100°C, the steam condenses and more water is drawn in. Cold pumice fragments of similar size, on the other hand, float for considerably longer periods before absorbing enough water to attain bulk densities greater than that of water, so they can sink and thus would not accumulate with lithic clasts of equivalent settling velocity. The interval from 58 to 96 mbsf, in which lithic fragments are missing, may represent pumice from subaerial eruptions that sank after being rafted about for a time. This lithic-free pumice was possibly (but not necessarily) produced by the same subaerial eruption that provided the underlying heterolithic fallout. Another possible origin for these deposits is that they also were fresh fallout, but from eruptions that contributed no lithic component.

The pumice of the basin-flank successions was probably contributed by both subaerial and submarine eruptions. Likely submarine sources include the submarine calderas of Sumisu Jima and Minami Sumisu. In historical times, Torishima Island has experienced subaerial eruptions as well as submarine explosions 1.5 and 9 km offshore (Simkin and others, 1981). Numerous thin ash beds in the upper Pliocene section of Site 792, drilled by Leg 126 in the forearc 150 km to the north, attest to the fact that at least some of the arc eruptions were subaerial, or submarine but in sufficiently shallow water for the eruption columns to break the sea surface and rise high enough in the atmosphere for prevailing westerly winds to transport their fine-ash contents to the forearc and beyond. The shallow submarine eruptions in 1952 that created Myojin-Sho Island, 105 km north of Site 788, produced eruption columns 6 to 8 km high (Niino and others, 1953).

The depletion of fine ash in the flank-site pumice deposits suggests that they were winnowed in some way but does not indicate exclusively subaerial eruptions, because ocean currents may be even more effective than wind as elutriation agents. As an alternative to winnowing by winds or currents, an explanation for the depleted-fines content is that the pumice tephra originated as hot, subaerial pyroclastic flows from which the fines were elutriated by steam as they entered the sea (Sigurdsson and Carey, 1989). Such an origin, however, would not produce the pronounced size differences of pumiceous fragments and lithic clasts.

The possibility that the pumiceous units were remobilized after their initial deposition and travelled as submarine mass movements before arriving at their final depositional site cannot be dismissed. We have seen no evidence for this in the drill cores, but in this tectonically and volcanically active region, earthquakes could easily dislodge fallout deposits that are metastably perched on submarine slopes, generating submarine slides and debris flows. The lack of fines and the bimodal granulometry of the heterolithic deposits are characteristics that would survive such remobilization.

The oxidized-pumice fragments of the uppermost 58 m of the flank accumulation and the manganiferous accumulations in the lower part of this interval are good evidence of one or two extended periods between influxes of lapilli to the flank site. The manganese mineral is stable todorokite, probably hydrogenous and very likely deposited very slowly. If it accumulated at the sea floor, it documents a fairly long time, sometime before 2.73 Ma, during which little or no pumice accumulated. Alternatively, the manganese may have accumulated several meters below the 2.75 to 0.235-Ma hiatus, while that horizon was exposed at the sea floor for an extended period, although the mechanics of this shallow-subsurface manganese deposition are not clear. In any case, the uppermost 30 m of flank deposits appear to have resided at the sea floor long enough for the pumice fragments to become discolored by oxidation. Furthermore, this uppermost interval is heterolithic and thus is probably older than the youngest sediments of the basin sites, which are free of lithic clasts.

The effects of ductile and brittle compaction observed in the Pliocene lapillistones at Site 788 require substantially more overburden than is now present on the flank. The missing column of sediment may have been stripped away by mass wasting during the period marked by the lacuna from ≥2.35 to 0.275 Ma, perhaps coincident with the initial phases of rifting. None of this material is found in the older part of the basin stratigraphy (Unit II of Sites 790 and 791), dated at 0.25 to 1.1 Ma, so that the remobilized material either moved to the east, away from Site 788, or the mass wasting occurred before 1.1 Ma.

Since about 0.2 to 0.25 Ma, silicic-pumice eruptions have been the main source for detritus not only along the eastern flank of the basin but within the basin as well. The main spatial contrasts are in thickness and texture: the pumiceous deposits of the basin are at least an order of magnitude thicker than coeval deposits of the flank and contain a fine-grained matrix including 1 to 3 percent nannofossils. Tectonic tilting of the basin floor is reflected by wedge-shaped units of pumice (Fig. 6).

Many of the slopes surrounding Sumisu Basin have slopes of 20° to 30° (Fig. 1), so that most pumice particles settling onto these areas are likely to be transported downslope toward the floor of the basin, either individually under the influence of bottom currents or via sediment gravity flows. During times of major pumice eruptions at nearby subaerial and submarine volcanoes, the large quantities of pumice settling through the water column onto the basin slopes may feed a more or less continuous cascade of detritus from the slopes onto the basin floor, punctuated by episodic large failures that would generate debris flows. In addition, significant portions of the poorly sorted pumice deposits in the basin may have been emplaced as submarine debris flows, derived directly from pyroclastic or laharic flows issuing from the nearby arc volcanoes. Modern bathymetry (Fig. 2) does not reveal a simple downslope path into the Minami Sumisu Basin for such flows from South Sumisu caldera, unless they were sufficiently thick to sweep across the barrier of basaltic volcanoes (Central Ridge) that separates the Kita and Minami Sumisu Basins. In contrast, the route into the Minami Sumisu Basin from Torishima and associated vents to the southeast is unrestricted. Submarine debris flows, like the Roseau Ash of the Lesser Antilles Arc (Carey and Sigurdsson, 1980), contain nonvolcanic fines, including microfossils, that are incorporated from surficial sediments along the flow path. These fines are quickly buried and cannot be winnowed out of the deposits by bottom currents except at the surface of the deposit.

The bottom-hugging density currents would tend to fill in irregularities on the basin floor and pond in lower parts of the basin to maintain a horizontal basin plain, in spite of continuous eastward tilting of the underlying basin fill (Fig. 6). Once on the basin floor, some low-density pumice fragments from the upper parts of the deposits could be further redistributed by weak bottom currents so as to produce the smooth and flat basin floor.

Incomplete recovery and partial disturbance of the pumice deposits by coring make it impossible to recognize primary features in these facies that can be ascribed to resedimentation. Nevertheless, the striking textural differences between pumice deposits in the basin and those of the same and older age on the flank of the rift (Fig. 4) are compatible with the different modes of deposition suggested.

The thick successions of upper Quaternary pumice in Sumisu Rift indicate episodic input. Volcanism apparently provided pumice five times during the deposition of Unit I at sites 790 and 791. Figure 7 shows our first approximation of the ages of the thick pumice layers of Site 790 (Fig. 7). Our biostratigraphic time controls are the first occurrence of the coccolith *Emiliania huxleii* (0.275 Ma) at the depth of 180 mbsf, and the acme of this species (0.085 Ma) at 130 mbsf. We assume that the thick pumice beds, and the thin ash layers in the hemipelagite intervals, were each deposited within days to months, whereas the strongly burrowed, hemipelagic sediment was deposited slowly at fairly constant rates, of 59 mm/k.y. in the interval between 180 and 130 mbsf, and 150 mm/k.y. for the column above 130 mbsf. This yields ages of 1, 31, 61, 67, and 131 ka for the five thick pumice beds, with only the oldest (Bed V) deposited during the period of slower sedimentation. Alternatively, if the acceleration in hemipelagic sedimentation rate was related to the inception of massive pumic production, we can assign the higher sedimentation rate of 150 mm/k.y. to the hemipelagites between beds V and IV as well, in which case bed V would have an age of only 102 ka. If this second option is exercised, and if the third and fourth events are regarded as close enough in time to represent two phases of a single episode, the period between eruptive episodes is about 30 k.y. Several of the ages correspond to times of relatively high Quaternary sea level (Bloom and others, 1974), so that these eruptions were not triggered by unloading of the flanks of the volcanoes as was suggested for Pantelleria volcano by Wallman and others (1988).

Sumisu Rift is located relatively close to the modern vol-

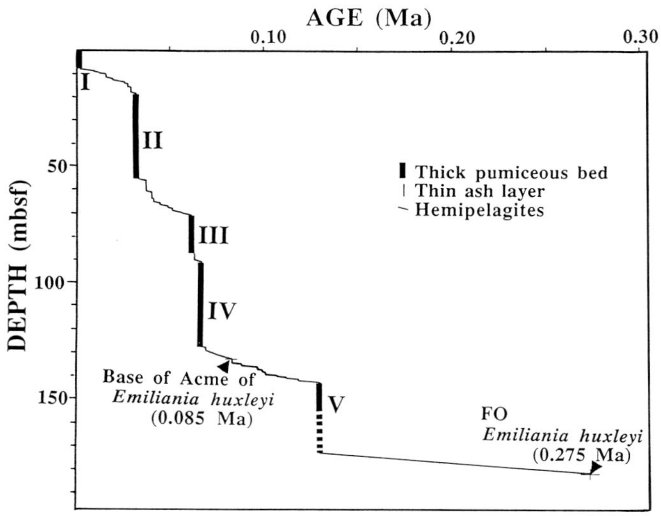


FIG. 7.—Interpretive plot of age versus depth for Unit I sediments recovered at Holes 790B and 790C, illustrating the periodicity of explosive pumice eruptions as recorded by thick pumiceous beds. Assumptions are given in the text. Corresponding depths for site 791 can be obtained from Figure 3.

canic front, and it would be reasonable to expect that this backarc basin and the older adjacent arc contain a record of coarse-grained volcaniclastic sediments and lavas. Lavas younger than 1.1 Ma are not encountered at Sites 788, 790 and 791, however, apparently because late Pliocene and Quaternary volcanism in this part of the Izu-Bonin Arc was restricted to a small number of centers like Sumisu Jima, Torishima, and South Sumisu Caldera (Fig. 2). Before rifting, the arc carapace in the vicinity of Site 788 was constructed by pumice blankets, and within the young rift basin the character of the sedimentary fill was dictated mostly by the intensity of volcanic activity and pumice production of the adjacent arc. During times of volcanic quiescence, mixtures of fine ash and biogenic carbonate dominated the accumulating sedimentary record. During and immediately after major eruptions, thick units of pumice were deposited through a combination of submarine debris flow, direct settling, resedimentation from basin slopes, and bottom-current redistribution.

ACKNOWLEDGMENTS

We thank Richard Fiske, Dan Karig, George DeVries Klein, Mike Ledbetter, Alan Niem and Gary Smith, who reviewed various versions of this manuscript.

REFERENCES

BLOOM, A. L., BROECKER, W. S., CHAPPELL, J. M. A., MATTHEWS, R. K., AND MESOLELLA, K. J., 1974, Quaternary sea level fluctuations on a tectonic coast: New $^{230}Th/^{234}U$ dates from the Huon Peninsula, New Guinea: Quaternary Research, v. 4, p. 185–205.

BROWN, G., AND TAYLOR, B., 1988, Seafloor mapping of the Sumisu Rift, Izu-Ogasawara (Bonin) Island Arc: Bulletin of the Geological Survey of Japan, v. 39, p. 23–28.

BUSBY-SPERA, C. J., 1988, Evolution of a Middle-Jurassic back-arc basin, Cedros Island, Baja California: Evidence from a marine volcaniclastic apron: Geological Society of America Bulletin, v. 100, p. 218–233.

CAREY, S. N., AND SIGURDSSON, H., 1980, The Roseau Ash: Deep-sea tephra deposits from a major eruption on Dominica, Lesser Antilles Arc: Journal of Volcanology and Geothermal Research, v. 7, p. 67–86.

CAREY, S. N., AND SIGURDSSON, H., 1984, A model of volcanogenic sedimentation in marginal basins: Geological Society of London Special Publication 16, p. 37–58.

FISKE, R. S., 1986, Shallow marine volcanic field in the Mio-Pliocene Shirahama Group, Izu-Peninsula, Japan: Geological Society of America Abstracts with Programs, p. 602.

FISKE, R. S., AND CASHMAN, K. V., 1989, Submarine fallout deposit in the Shirahama Group (Mio-Pliocene), Japan: Abstracts Volume, International Association of Volcanology and Chemistry of the Earth's Interior, General Assembly, Santa Fe, New Mexico, June 25–July 1, p. 92.

GRECHIN, V. I., NIEM, A. R., MAHOOD, R. O., ALEXANDROVA, V. A., AND SAKHAROV, B. A., 1981, Neogene tuffs, ashes, and volcanic breccias from offshore California and Baja California: Deep Sea Drilling Project Leg 63: Sedimentation and Diagenesis: Initial Reports of the Deep Sea Drilling Project, v. 63, p. 621–657.

HUSSONG, D. M., UYEDA, S., AND SHIPBOARD SCIENTIFIC PARTY, 1982, Initial Reports of the Deep Sea Drilling Program, v. 60: Washington, D.C., U.S. Government Printing Office, 929 p.

KARIG, D. E., AND MOORE, G. F., 1975, Tectonically controlled sedimentation in marginal basins: Earth and Planetary Science Letters, v. 26, p. 233–238.

KLEIN, G. deV., 1985, The control of depositional depth, tectonic uplift, and volcanism on sedimentation processes in the back-arc basins of the western Pacific Ocean: Journal of Geology, v. 93, p. 1–25.

KUNO, H., 1962, Catalogue of the active volcanoes of the world, part 11. Japan, Taiwan, and Marianas. International Association of Volcanology and Chemistry of the Earth's Interior, 332 p.

LEG 126 SHIPBOARD SCIENTIFIC PARTY, 1989, Arc volcanism and rifting: Nature, v. 342, p. 18–20.

MURAKAMI, F., 1988, Structural framework of the Sumisu Rift, Izu-Ogasawara Arc: Bulletin of the Geological Survey of Japan, v. 39, p. 1–21.

NIEM, A. R., 1977, Mississippian pyroclastic flow and ash-fall deposits in the deep-marine Ouachita flysch basin, Oklahoma and Arkansas: Geological Society of America Bulletin, v. 88, p. 49–61.

NIINO, H., KUMAGORI, T., TSUYA, H., MORIMOTO, R., OSSAKA, G., HAMAGUCHI, H., TATSUMOTO, M., MATSUE, Y., KOMAE, Y., ARIHASA, G., EBINA, K-I., AND TAKAGI, K., 1953, Report of the Submarine Eruption of Myojin-Sho: Tokyo University Fisheries, 52 p.

NISHIMURA, A., AND MURIKAMI, F., 1988, Sedimentation of the Sumisu Rift, Izu-Ogasawara Arc: Bulletin of the Geological Survey of Japan, v. 39, p. 39–61.

ODP LEG 126 SCIENTIFIC PARTY, 1989, ODP Leg 126 drills the Izu-Bonin Arc: Geotimes, v. 34, p. 36–38.

SIGURDSSON, H., AND CAREY, S. N., 1989, Plinian and co-ignimbrite tephra fall from the 1815 eruption of Tambora volcano: Bulletin of Volcanology, v. 51, p. 243–270.

SIMKIN, T., SIEBERT, L., MCCLELLAND, L., BRIDGE, D., NEWHALL, C., AND LATTER, J. H., 1981, Volcanoes of the World: Hutchison Ross, Stroudsburg, Pennsylvania, p. 72.

WALLMANN, P. C., MAHOOD, G. A., AND POLLARD, D. D., 1988, Mechanical models for correlation of ring-fracture eruption at Pantelleria, Strait of Sicily, with glacial sea-level drawdown: Bulletin of Volcanology, v. 50, p. 327–339.

WHITHAM, A. G., AND SPARKS, R. S. J., 1986, Pumice: Bulletin of Volcanology, v. 48, p. 209–233.

# PART V
# PETROLOGY
# OF VOLCANIC SEDIMENTS

The petrology of volcaniclastics reveals the composition of the source-area magmas, the dispersal of sediment, and processes that affect sediment composition during transport and deposition. Petrological characteristics and tectonic implications are presented in Part V using nonmarine Cenozoic volcaniclastic deposits from New Mexico and from synrift and post-rift sediments deposited in the Gulf of California.

# PRE-DIAGENETIC SEDIMENTARY FRACTIONATION OF ANDESITIC DETRITUS IN A SEMI-ARID CLIMATE: AN EXAMPLE FROM THE EOCENE DATIL GROUP, NEW MEXICO

STEVEN M. CATHER[1] AND ROBERT L. FOLK

*Department of Geological Sciences, The University of Texas at Austin, Austin, Texas 78713*

ABSTRACT: The lower Datil Group (40-37 Ma) records the early phase of intermediate-composition volcanism in the northern Mogollon-Datil field of New Mexico. Eruptive products consist almost entirely of high-K andesite and dacite that range in $SiO_2$ content from about 58 to 67 weight percent. Lower Datil conglomerate clasts and correlative(?) lavas are characterized by phenocrystic plagioclase ($An_{20-60}$), amphibole, and titanomagnetite (±biotite, pyroxene) within a groundmass containing alkali feldspar, silica minerals, and plagioclase. Volcaniclastic rocks of the lower Datil Group were deposited by a spectrum of fluvial, debris-flow and lacustrine processes in two semi-arid intermontane basins and differ mineralogically from their volcanic progenitors in the following ways: 1) nonwelded ash is not represented in conglomerate; 2) phenocryst phases are preferentially concentrated in sandstone; 3) groundmass constituents and devitrification products are over-represented in mudstone and detrital matrix in sandstone; 4) opaque reaction rims on free amphibole and biotite grains are poorly developed or not present; 5) amphibole abundance is slightly diminished relative to plagioclase and biotite in sandstone; and 6) potassic rims on plagioclase grains may have been removed by abrasion. Mineralogic fractionation of lower Datil Group volcaniclastic sediments occurred primarily by impact shattering and abrasion during transportation. As a result, provenance determinations for volcaniclastic rocks may be influenced by grain-size effects.

## INTRODUCTION

Late Eocene, dominantly intermediate-composition volcanism in the northern Mogollon-Datil field produced large volumes of volcaniclastic detritus that were deposited in two basins of Laramide ancestry by fluvial, debris-flow, and lacustrine processes (Cather, 1986; Cather and Chapin, 1989). Because of the relatively uniform petrologic characteristics of the andesites, it is possible to evaluate the mineralogic evolution of derivative sediments by comparing the modal characteristics of conglomerates, sandstones, and mudstones. This study considers the combined effects of grain-size diminution and sorting on the mineralogic fractionation of ancient andesitic detritus in a semi-arid environment.

The Tertiary volcanic rocks of the northern Mogollon-Datil field can be broadly divided into two suites: an older (late Eocene) sequence of dominantly intermediate-composition volcanic and volcaniclastic rocks and an Oligocene to Miocene bimodal series composed mostly of silicic-ash-flow tuffs and basalt or basaltic-andesite lava flows (Cather and Chapin, 1989). Throughout the study area (Fig. 1), the older, intermediate-composition rocks are termed the lower Datil Group (Osburn and Chapin, 1983; Cather, 1986) and have yielded isotopic and fission-track ages that range from about 40 to 37 Ma (Cather and Chapin, 1989). Seven measured and sampled sections of the lower Datil Group, ranging from about 200 to 680 m in thickness (Cather, 1986), provide the basis for this report.

### *Depositional Setting*

Paleocurrent indicators and distribution of sedimentary facies indicate that the physiographic framework established during late Laramide crustal shortening persisted throughout early Datil volcanism (Cather, 1986). Lower Datil Group volcaniclastic deposits accumulated in two intermontane basins of Laramide ancestry, the Baca and Carthage-La Joya basins (Fig. 2; Cather and Johnson, 1986). Because of the effects of Neogene rifting, volcanism, and bolson sedimentation, the location of early Datil volcanoes can be determined only approximately. Paleocurrent data and distribution of volcaniclastic facies indicate that eruptive centers were developed along the Laramide Sierra uplift, in what is now the Socorro Basin of the Rio Grande rift. Early Datil volcanoes also were present to the south of the Baca basin, on or near the Laramide Morenci uplift (Cather, 1986).

Lower Datil detritus was transported and deposited by a spectrum of fluvial, debris-flow, and lacustrine processes. Sedimentary sequences representing laterally extensive deposits of volcaniclastic aprons dominate the majority of lower Datil exposures in the study area and consist of varying proportions of braided-stream, hyperconcentrated-flow, and debris-flow deposits. Large-volume debris-flow deposits were emplaced in an actively subsiding portion of the Baca basin, between the Hickman and Puertecito fault zones (Fig. 2; Cather and Chapin, 1989; Harrison, 1980; Broulliard, 1984). A large, closed lake near the eastern end of the Baca basin represented the culmination of lacustrine conditions that began prior to the onset of early Datil volcanism (Cather, 1983).

The climate during the middle and late Eocene in west-central New Mexico was semi-arid, as indicated by the presence in the Baca Formation and lower Datil Group of calcareous paleosols and redbeds, and stratigraphic evidence for closed lacustrine conditions and flashy or ephemeral discharge in rivers (Cather and Johnson, 1984; Cather, 1986; Snyder, 1971). Fossil vertebates indicate that savannas existed in basin areas during Baca time in west-central New Mexico (Lucas and others, 1981; Schiebout and Schrodt, 1981; Lucas, 1983). Although no fossils have been recovered from the lower Datil Group in the study area, vertebrate remains in correlative beds of the Rubio Peak Formation in the Black Range (Fig. 1) indicate the continuance of savannas following the onset of Eocene volcanism (Lucas, 1983, 1986; S. G. Lucas, 1990, pers. commun.).

## PETROLOGY OF THE LOWER DATIL GROUP

Exposures of the lower Datil Group in the study area are dominated by a compositionally uniform sequence of in-

[1]Present address: New Mexico Bureau of Mines and Mineral Resources, Campus Station, Socorro, New Mexico 87801

Sedimentation in Volcanic Settings, SEPM Special Publication No. 45
 ISBN 0-918985-89-7

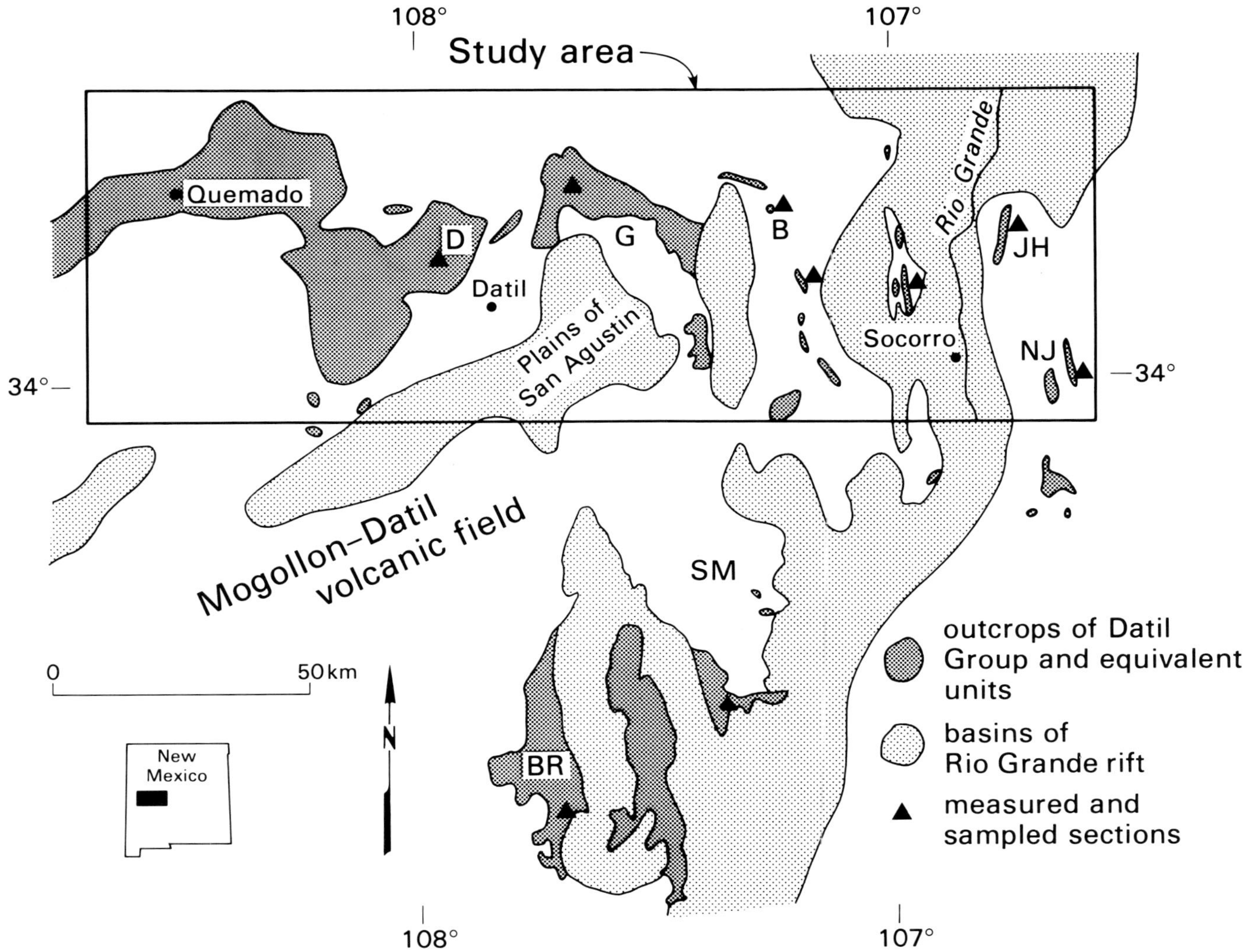


FIG. 1.—Map of west-central New Mexico depicting study area, outcrops of Datil Group, basins of Rio Grande rift, and locations of measured and sampled sections (from Cather, 1986). Selected geomorphic features: JH, Joyita Hills; NJ, northern Jornada del Muerto; B, Bear Mountains; G, Gallinas Mountains; D, Datil Mountains; SM, San Mateo Mountains; BR, Black Range. Modified from New Mexico Geological Society (1982).

termediate-composition volcaniclastic rocks that are characterized by phenocrystic plagioclase, amphibole, and titanomagnetite (±biotite, pyroxene) and groundmass alkali feldspar, silica minerals, and plagioclase. Silica content of chemically analyzed conglomerate clasts (Cather, 1986) ranges from about 58 to 67 weight percent; $K_2O$ averages about 2.9 weight percent. According to the classification of Ewart (1979, 1982), these rocks are dominantly high-K andesite and high-K dacite. Herein, we refer to these rocks simply as andesite. Mafic lithologies are virtually absent in the lower Datil Group.

Extrusive rocks of early Datil age are poorly represented in the northern Mogollon-Datil field because of burial by younger volcanics and rift-related sedimentation. No exposures are known in the study area, but probable lower Datil extrusive rocks in excess of 300 m thick are present in the southern San Mateo Mountains (Fig. 1). These rocks comprise the Whiskey Hill hornblende andesite of the Red Rock Ranch Formation (Farkas, 1969); both units subsequently have been correlated with the Datil Group (*sensu* Osburn and Chapin, 1983) by Foruria (1984).

In this report, we divide the lower Datil Group into three sedimentary lithosomes: conglomerate, sandstone, and mudstone. These subdivisions represent a general progression from proximal to distal paleoenvironments. Because no single locality in the study area contains a complete proximal-to-distal tract, the petrologic characteristics of various lower Datil lithosomes were determined from samples taken from throughout the study area. Samples representative of the spectrum of paleoenvironments in both the Baca and Carthage-La Joya basins were collected and analyzed.

Petrographic characteristics of lower Datil rocks were determined by examination of about 150 thin sections, of which 63 were point counted (average 230 points each; see Appendices A, B). For sandstones, data were tabulated by both

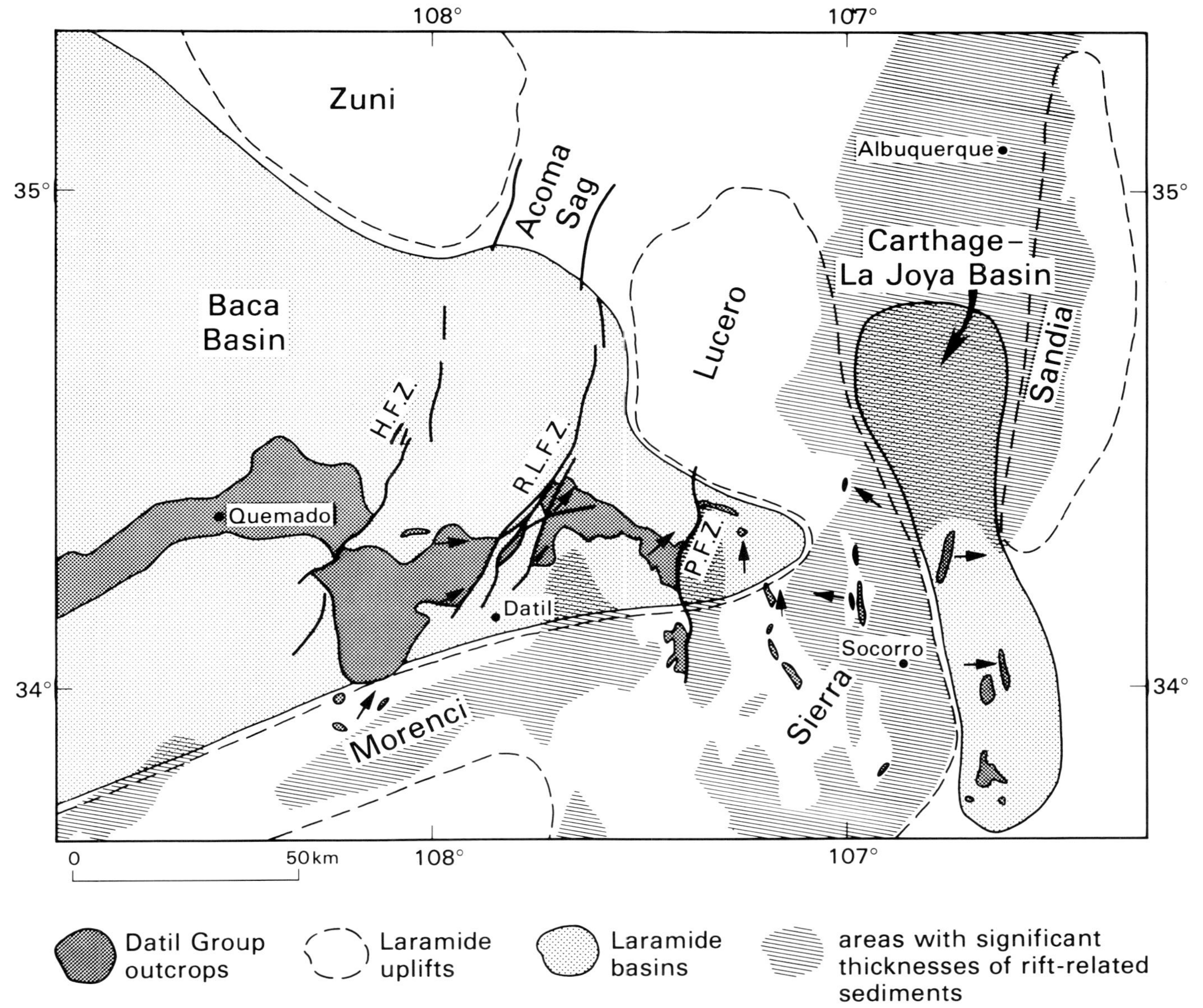


FIG. 2.—Map showing late Laramide (Eocene) basins, uplifts and intrabasin fault zones, and superimposed, younger rift basins. Arrows indicate mean paleocurrent vectors for Datil Group (Cather, 1986). P.F.Z., Puertecito fault zone; R.L.F.Z., Red Lake fault zone; H.F.Z., Hickman fault zone. Modified from Cather and Johnson (1984, 1986) and New Mexico Geological Society (1982).

traditional (QFR) and Gazzi-Dickinson methods (Gazzi, 1966; Dickinson, 1970; see also Ingersoll and others, 1984). Because this study emphasizes the pre-diagenetic evolution of lower Datil detritus, modal data for diagenetic constituents are not reported here. Additionally, an attempt was made to "see through" post-depositional changes. For example, secondary porosity in amphibole was tabulated as amphibole; calcitized plagioclase was counted as plagioclase. The diagenesis of lower Datil volcaniclastic rocks will be described in a forthcoming paper.

Microprobe analyses of feldspar were performed using a three-spectrometer ARL-EMX electron microprobe equipped with a Tracor-Northern TN-2000 energy-dispersive system. Operating conditions were: 15 Kv accelerating voltage; 0.034-0.015 $\mu$A sample current on brass; 2–5 $\mu$m beam diameter; counting time was 30 seconds per point. Primary and secondary standards were minerals and glasses. CaO, $Na_2O$, and $K_2O$ in feldspars were analyzed using wavelength-dispersive spectroscopy; abundances of all other oxides were determined by energy-dispersive spectroscopy. Additional microprobe analyses for amphibole and biotite in lower Datil conglomerate clasts and for feldspar, pyroxene, and zeolite in the upper Datil Group are reported in Cather (1986).

Scanning electron microscopy in conjunction with energy-dispersive spectroscopy (EDS) assisted in characterization of finely crystalline components of the lower Datil Group. In particular, semi-quantitative "point counting" of mudstone and matrix in sandstone using EDS proved a useful technique to distinguish between fine-grained detrital and diagenetic components. The analyses were occasionally ambiguous, such as when the electron beam impinged upon a grain boundary between two minerals. Nonetheless, the technique enables identification and estimation of approximate abundances of fine-grained constituents of the lower Datil Group.

In this report, "mean" or "average" refers to arithmetic mean, and in many cases is accompanied by the value (±) of one standard deviation about the mean. The probabilistic significance of the difference between means is designated by "$p$." For example, the statement "$p = 0.01$" indicates a 1 percent chance that the difference between two means is due simply to random sampling of a homogeneous population. The terms "statistically significant" and "statistically insignificant" represent the cases where $p \leq 0.05$ and $p > 0.05$, respectively.

### *Conglomerate Petrology*

Conglomerate is an important constituent of the lower Datil Group, comprising an estimated 30 percent of the unit. Lower Datil conglomerates were deposited by a spectrum of braided-fluvial, hyperconcentrated-flood, and debris-flow processes. Clasts in debris-flow deposits are typically (but not universally) matrix supported and may exhibit wide ranges of roundness and sphericity. This textural variability presumably results from admixture within the debris flow of angular, untransported detritus from near the eruptive center with rounded clasts entrained from fluvial drainages.

Braided-stream conglomerates are typically clast supported and commonly display imbrication. Such clasts are, on average, better rounded, better sorted, and more spherical than their debris-flow counterparts. According to the roundness scale of Powers (1953), fluvial clasts of the lower Datil Group are generally rounded to well rounded. Estimated average sphericity of fluvial clasts is high (0.7–0.8), which is consistent with a braided-stream depositional environment (R. L. Folk, unpubl. data). Pebbles and cobbles are typically more spherical than associated boulders. The common occurrence of bladed and platy shapes in fluvial boulders, as well as the rarity of diameters greater than about 1 m, suggests that joints and other fractures in parental volcanic rocks may have influenced the development of shape and maximum size of clasts in Datil fluvial systems.

In hand specimen, lower Datil andesite clasts average light gray (N7) when fresh; altered samples commonly exhibit greenish or reddish hues. Most andesites are conspicuously porphyritic; phenocrysts include plagioclase, amphibole, and, less ubiquitously, biotite and clinopyroxene. Andesites are typically holocrystalline, but groundmass devitrification textures were noted in a few thin sections. Trachytic texture is rarely well developed. Vesicles range in abundance from 0 to about 10 percent and are present in about 80 percent of the clasts examined.

The mineralogic and textural attributes of lower Datil andesite clasts are remarkably uniform over a broad region of west-central New Mexico. Comparison of clasts from exposures in the eastern (northern Jornada del Muerto and Joyita Hills) and western (Bear Mountains, Gallinas Mountains, and Datil Mountains) parts of the study area indicates only subtle lithologic differences exist. Figure 3 illustrates nearly complete overlap of relative abundances of major phenocryst phases (plagioclase, amphibole, and biotite) between western and eastern andesite clasts. Eastern andesite clasts, however, contain, on average, slightly less groundmass than their western counterparts (Fig. 4). Limited petrographic data for the Whiskey Hill hornblende andesite in the southern San Mateo Mountains (Farkas, 1969; Cather, 1986) are comparable to those for lower Datil conglomerate clasts (Figs. 3, 4) and support Foruria's (1984) correlation of the units.

*Feldspar phenocrysts (20.5–46.7 percent, mean 37.5 ± 7.2 percent of total rock volume).—*

Virtually all of the phenocrystic feldspar in lower Datil andesite clasts is plagioclase. Although alkali feldspar is an abundant constituent of andesite groundmass, it is rarely seen as a phenocryst phase (it occurs only as rims on plagioclase phenocrysts and, more rarely, as microphenocrysts).

Plagioclase is typically tabular and subhedral to euhedral (Fig. 5). Crystals average about 1.0 mm in length and occur both individually and as glomeroporphyritic masses of two to five crystals. All of the common twin types of plagioclase (albite, pericline, and Carlsbad) were observed; the occurrence of more than one type in an individual crystal is common.

Figure 6a is a plot of microprobe analyses for feldspar phenocrysts in lower Datil andesite clasts. Anorthite content ranges from about $An_{20}$ to $An_{60}$; average feldspar composition is $An_{31}$ $Ab_{64}$ $Or_5$. Although no microprobe data are

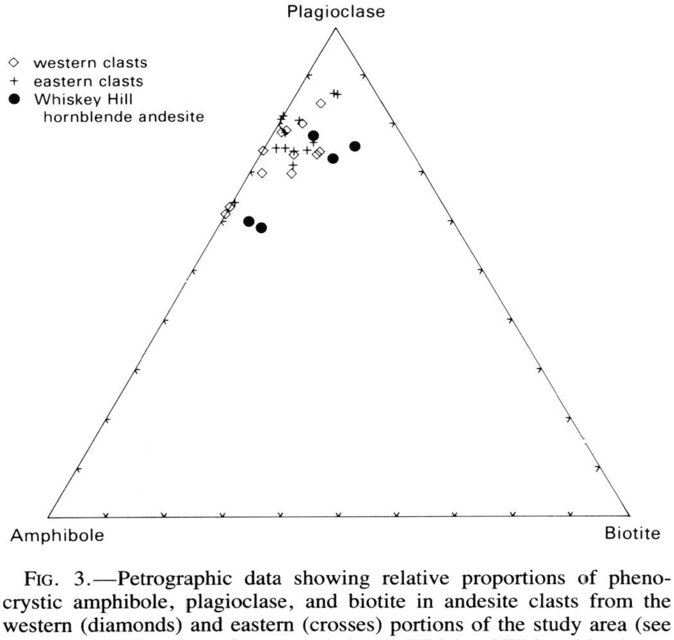


FIG. 3.—Petrographic data showing relative proportions of phenocrystic amphibole, plagioclase, and biotite in andesite clasts from the western (diamonds) and eastern (crosses) portions of the study area (see text). Dots indicate data from correlative(?) Whiskey Hill hornblende andesite (Farkas, 1969; Cather, 1986).

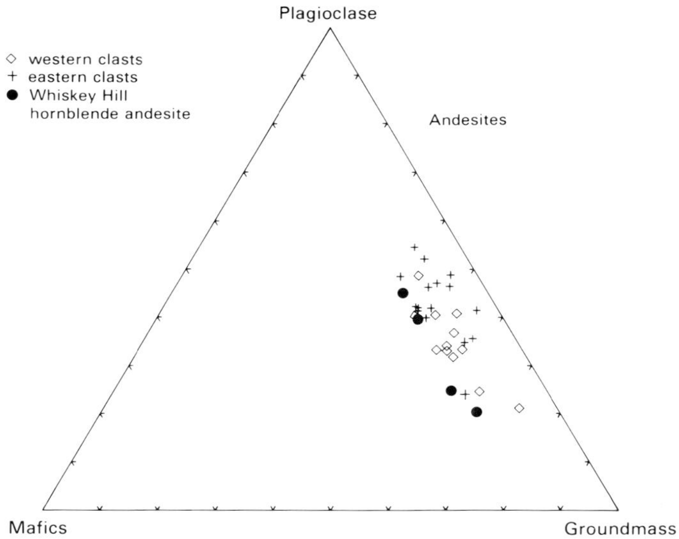


FIG. 4.—Petrographic data showing relative proportions of mafic minerals (amphibole, biotite, opaques, and pyroxene), plagioclase, and groundmass in andesite clasts from the western (diamonds) and eastern (crosses) portions of the study area. Dots indicate data from correlative(?) Whiskey Hill hornblende andesite (Farkas, 1969; Cather, 1986). Sample number 118 of Farkas (1969) is not plotted due to unreasonably low groundmass content (21%).

available for plagioclase phenocrysts in lavas of the correlative(?) Whiskey Hill hornblende andesite, Farkas (1969) reports values between $An_{45}$ and $An_{48}$ on the basis of extinction-angle measurements.

Compositional zonation of plagioclase phenocrysts is nearly ubiquitous in andesite clasts of the Datil Group. Such

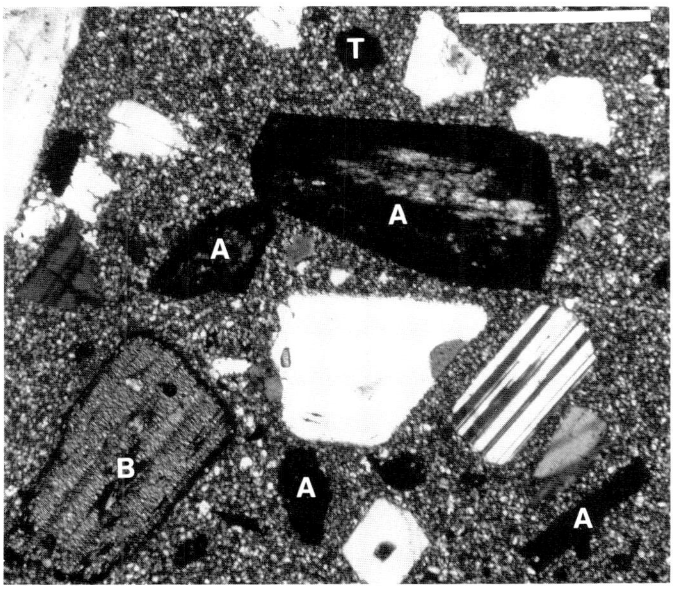


FIG. 5.—Photomicrograph of representative lower Datil andesite clast, showing plagioclase, amphibole (A), biotite (B), and titanomagnetite (T) phenocrysts, and holocrystalline groundmass. Scale bar is 0.5 mm; crossed nicols.

zoning is characteristic of orogenic andesites (Gill, 1981, p. 171–172) and may result from numerous processes including magma mixing, fluctuations in $P_{H_2O}$, thermal convection within the magma body, and diffusion-rate-controlled compositional gradients at the crystal-liquid interface. Microprobe analyses indicate that zonation of plagioclase in Datil andesites is dominantly oscillatory. Comparison of mean anorthite compositions of "core" (actually two-dimensional center) and rim areas of phenocrysts (Fig. 7), however, indicates a statistically significant ($p = 0.02$) overall tendency toward decreased anorthite content in rims, although numerous examples of reversed zoning are present. Similarly, there is a statistically significant ($p = 0.001$) increase in mean orthoclase content of rim areas in feldspar phenocrysts (Fig. 8).

*Amphibole phenocrysts (3.6–19.9 percent, mean 9.5 ± 3.6 percent of total rock volume).—*

Amphibole is the dominant ferromagnesian mineral in andesite clasts of the Datil Group. Amphibole phenocrysts are typically prismatic and euhedral (Fig. 5). Crystals usually occur individually, rarely exhibit twinning, and average about 0.6 mm in diameter. Pleochroism is pronounced; hues of reddish brown predominate, but khaki-green amphiboles are common in a few samples.

Evidence for late magmatic or deuteric alteration of amphibole was noted in about 80 percent of the thin sections examined. Opaque reaction products range from thin coronas on phenocrysts to crystals that have been entirely replaced. The low, positive dP/dT slope of the amphibole stability curve (Gill, 1981, p. 194–196) indicates that amphibole will be resorbed during magma ascent if not rapidly quenched, and probably explains the nearly ubiquitous reaction rims and lack of fine-grained amphibole in lower Datil andesite clasts.

At least 3 weight-percent $H_2O$ is required in the liquid phase for crystallization of amphibole (Green, 1972; Eggler and Burnham, 1973), thus placing an important lower limit on water content of Datil andesite magmas. Although no primary (i.e., unreworked) pyroclastic deposits have been documented in the lower Datil Group, the presence of hydrous minerals (amphibole and biotite) indicates that pyroclastic processes may have played an important role in early Datil eruptions.

*Biotite phenocrysts (0.0–4.5 percent, mean 1.7 ± 1.5 percent of total rock volume).—*

Biotite is present in about two-thirds of the andesite clasts examined. It is typically subhedral and tabular (Fig. 5) and averages about 0.4 mm in length. Crystals most commonly occur individually, but association with other phenocrysts of biotite or amphibole is common.

Pleochroic colors are much like those of associated amphiboles: dominantly reddish brown except in a few samples that contain mostly khaki-green biotites. There is a strong correlation between pleochroic colors of biotite and amphibole within individual samples. Biotite commonly displays opaque reaction rims similar to, but less well developed than, those of associated amphiboles.

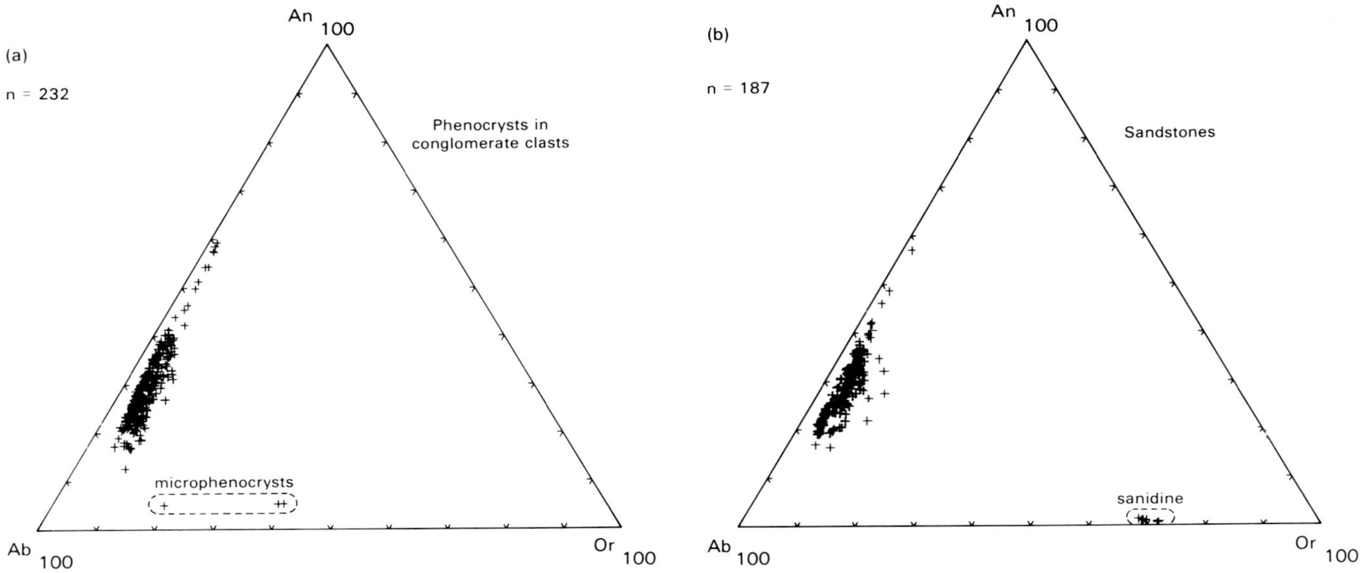


FIG. 6.—Electron-microprobe analyses of (a) feldspar phenocrysts and microphenocrysts from eight lower Datil andesite clasts and (b) sand-size feldspar from six lower Datil sandstones. See text for discussion.

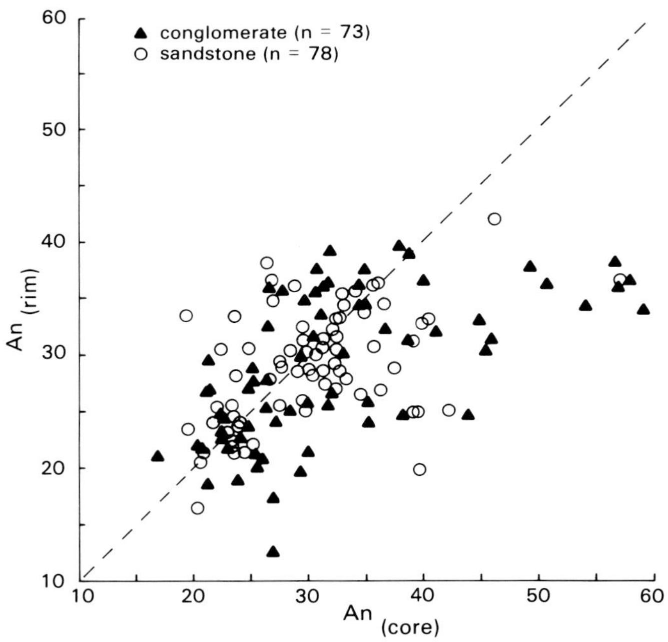


FIG. 7.—Electron-microprobe analyses of anorthite content (mole %) of "cores" and rims of feldspar in lower Datil Group. Phenocrysts in andesite clasts (triangles) based on analyses of 73 core-rim pairs from eight samples; average An content of core = 32.2 ± 9.9, average An content of rim = 28.9±6.4. Sand-size feldspar in sandstones (circles) based on analyses of 78 core-rim pairs from six samples; average An content of cores = 30.2±6.7, average An content of rims = 29.0±4.9. Dashed line indicates equal core and rim values.

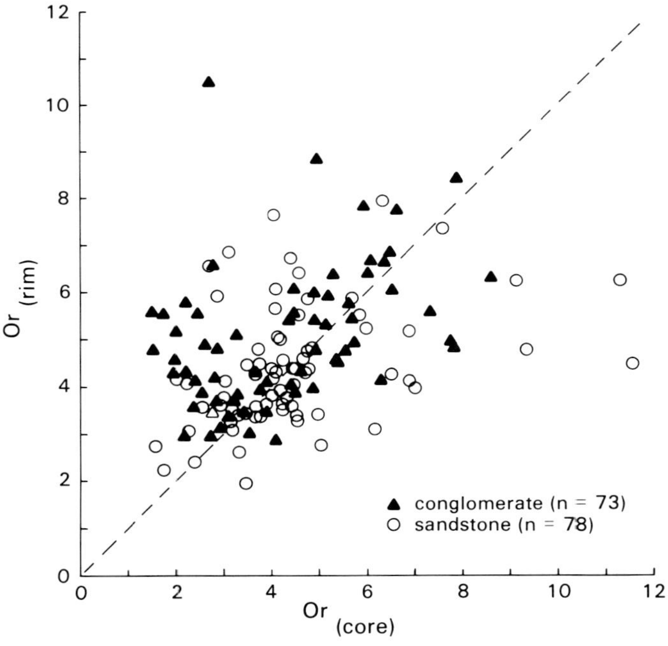


FIG. 8.—Electron-microprobe analyses of orthoclase content (mole %) of "cores" and rims of feldspar in lower Datil Group. Phenocrysts in andesite clasts (triangles), based on analyses of 73 core-rim pairs from eight samples; average Or content of core = 4.2±1.7, average Or content of rim = 4.9±1.5. Sand-size feldspar in sandstones (circles) based on analyses of 78 core-rim pairs from six samples; average Or content of cores = 4.5±1.8, average Or content of rims = 4.4±1.3. Dashed line indicates equal core and rim values.

*Opaque phenocrysts (trace −4.3 percent, mean 1.5 ± 1.0 percent of total rock volume).—*

Phenocrysts and microphenocrysts of opaque minerals are present in virtually all andesite clasts of the lower Datil Group. Phenocrysts are as large as 0.4 mm in diameter and are typically equant and subhedral. Opaque minerals occur both as individual crystals and as clots of crystals and are commonly associated with other non-opaque ferromagnesian phenocrysts.

Only 16 opaque phenocrysts were analyzed with the electron microprobe; all are titanomagnetite (that is, minerals of the magnetite-ulvospinel solid-solution series). The possibility remains that other opaque species exist in the andesite clasts. Examination of polished microprobe sections with reflected light, however, indicates that titanomagnetite comprises the vast majority of opaque minerals in clasts of the lower Datil Group.

*Other phenocrysts (trace).—*

Pyroxenes occur in minor amounts (less than a few percent) in about one-third of the andesite clasts that were examined petrographically. Both clinopyroxene and orthopyroxene are present, but clinopyroxene predominates.

Apatite was noted in a few thin sections, occurring as microphenocrysts and, more rarely, as inclusions in biotite. Crystals of apatite are euhedral and prismatic and typically less than 0.1 mm in length.

Zircon is a trace constituent of some lower Datil clasts and occurs as small, euhedral microphenocrysts. It was rarely observed in thin section but was frequently noted during preparation of mineral separates for K-Ar dating.

*Groundmass (32.7–70.0 percent, mean 47.0 ± 8.1 percent of total rock volume).—*

In plane-polarized light, groundmass in lower Datil andesite clasts is typically colorless to light brown. Groundmass minerals average about 20 μm in diameter, thus rendering their petrographic identification difficult or impossible. The groundmass of most clasts is holocrystalline and felsitic (Fig. 5), although rare examples show spherulitic devitrification features (Cather, 1986). Trachytic texture is typically not present or poorly developed due to the dominant subequant habit of the groundmass minerals.

Semi-quantitative modal analysis of both devitrified and holocrystalline groundmass using EDS in conjunction with the scanning electron microscope indicates the presence and approximate abundances of the following phases: alkali feldspar (60 percent), silica minerals (20 percent), and plagioclase (20 percent). Trace amounts of titanomagnetite(?) and biotite(?) are also present. The mineralogy of groundmass silica in lower Datil andesites is unknown; quartz, cristobalite, or tridymite may be present. Microprobe data for 62 groundmass feldspars are presented in Figure 9.

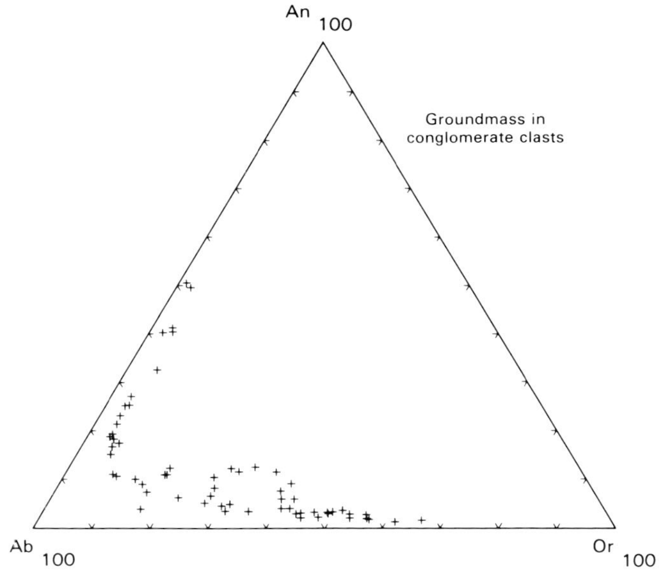


FIG. 9.—Groundmass feldspar compositions of andesite clasts, based on 62 electron-microprobe analyses from four samples.

## Sandstone Petrology

Sandstones comprise about 65 percent of measured sections in the lower Datil Group. Mean sieve-equivalent grain size (Harrell and Eriksson, 1979) for point-counted sections ranges from −0.3 ϕ (very coarse sand) to 4.1 ϕ (coarse silt). Most sandstones, however, range between 0.0 ϕ and 2.0 ϕ, which corresponds approximately to the range of phenocryst sizes observed in lower Datil andesite clasts. Sandstone sorting values range from very poorly sorted to well sorted, and average poorly to moderately sorted. According to Power's (1953) roundness scale, most grains finer than coarse sand are subangular to angular; coarse and very coarse sandstones average subangular to subround.

Because of their essentially quartz-free nature, lower Datil sandstones plot on the feldspar-rock fragment join of traditional QFR diagrams, and F/R ratio is primarily a function of grain size. Lower Datil sandstones plot (not surprisingly) in the undissected magmatic-arc-provenance field of Dickinson and Suczek (1979).

*Feldspar (6.8–70.4 percent, mean 40.4 ± 16.6 percent of framework grains).—*

Plagioclase is the dominant feldspar in lower Datil sandstones; grains of alkali feldspar are present only in trace amounts. Plagioclase exhibits cleavage-controlled shapes and rarely is appreciably rounded (Fig. 10). Twin types are the same as those present in conglomerate clasts.

Figure 6b is a plot of microprobe analyses of feldspars from lower Datil sandstones. Anorthite content ranges from about $An_{20}$ to $An_{60}$, and feldspar composition averages $An_{30}Ab_{66}Or_4$.

Comparison of anorthite contents of "core" and rim areas (Fig. 7) indicates the presence of both normal and reverse zoning. In contrast to that in conglomerate clasts, however, the difference between average anorthite content of core and rim areas of plagioclase grains in lower Datil sandstones is statistically insignificant ($p = 0.20$). Similarly, there is a statistically insignificant difference ($p = 0.50$) between average potassium contents of core versus rim areas in plagioclase grains of lower Datil sandstones (Fig. 8).

FIG. 10.—Photomicrograph of representative lower Datil sandstone, showing over-representation of phenocryst minerals [plagioclase (bright grains), amphibole (A), biotite (B), titanomagnetite (T)]. Note lack of opaque rim on "free" amphibole grain at lower right. Scale bar is 0.5 mm; plane-polarized light.

Alkali feldspar is an important constituent in aphanitic portions of volcanic-rock fragments and in detrital matrix of lower Datil sandstones (see later discussion). Trace amounts of sand-size grains of sanidine, however, also occur in sandstones (Fig. 6b). Because discrete phenocrysts of alkali feldspar were not observed in any of the andesite lavas or conglomerate clasts that were examined petrographically, the provenance of these grains is not known. The common association of alkali feldspar grains with devitrified grains containing shards and rare volcanic quartz (Cather, 1986), however, suggests that all these grains were derived from pyroclastic eruptions that were more siliceous than those typical of early Datil volcanism. It is possible that the anomalous alkali feldspar, quartz, and devitrified grains were reworked from pyroclastic deposits derived from distant sources.

*Amphibole (1.3–13.4 percent, mean 7.1 ± 3.1 percent of framework grains).—*

Detrital amphibole is present in all the lower Datil sandstones that were examined. Grain shapes are largely cleavage controlled; rounding is poorly developed (Fig. 10). Pleochroic colors are similar to those in conglomerate clasts. Amphibole exhibits evidence for relatively rapid grain diminution during transport, as shown by the common occurrence of the fine sand- to silt-size amphibole fragments in lower Datil sandstones. Such fine-grained amphibole is lacking in lower Datil andesite clasts and correlative(?) lavas because of reaction of microphenocrystic and groundmass amphibole during magma ascent and cooling. Opaque reaction rims on "free" amphibole grains (as opposed to those contained within volcanic-rock fragments) are typically thin or not present. This attests to the poor durability of these reaction rims during transport, but also may reflect the over-representation in sandstones of rapidly quenched pyroclastic detritus in which reaction rims are poorly developed (G. A. Smith, 1987, pers. commun.).

*Biotite (0–11.0 percent, mean 1.8 ± 2.3 percent of framework grains).—*

About 90 percent of the sandstones of the lower Datil Group contain biotite. In contrast to amphibole, biotite appears to be relatively resistant to abrasion and does not yield abundant fine-grained fragments. This is probably because of the elasticity and the presence of only one direction of cleavage in biotite, as well as the increased likelihood of suspension transport owing to the hydraulic behavior of biotite flakes. Reaction rims on "free" biotite grains, however, are thin or absent and appear to be quite susceptible to abrasion. Biotite commonly exhibits bent and frayed shapes, resulting from compactional deformation. Pleochroic colors of detrital biotite are similar to those of amphibole. Biotite flakes in the lower Datil Group commonly display partial or complete hexagonal shapes, which are diagnostic of volcanic-derived biotite (Folk, 1974).

*Opaque minerals (0.4–3.5 percent, mean 1.4 ± 0.9 percent of framework grains).—*

Opaque minerals are present in all lower Datil sandstones that were examined. No microprobe data were obtained, but examination of polished sections under reflected light indicates the dominance of titanomagnetite in these rocks. Opaque grains commonly form placer-like layers in laminated sandstones and are typically somewhat smaller than associated non-opaque grains.

*Volcanic rock fragments (18.9–80.9 percent, mean 48.9 ± 21.8 percent of framework grains).—*

All polymineralic grains of volcanic origin, including those consisting solely of aphanitic or vitric groundmass, were assigned to this category. The mineralogy and texture of volcanic-rock fragments (VRFs) are fundamentally the same as those present in Datil andesite clasts and correlative(?) lavas, although vitric and devitrified groundmass is more common in sand grains than in conglomerate clasts. Devitrified grains containing shards were noted in trace amounts in a few samples.

Although abundance of VRFs in the framework fraction of Datil sandstones decreases with diminishing grain size (Fig. 11), the proportion of groundmass constituents within VRFs *increases* with decreasing grain size (Fig. 12) because of the poor representation of phenocrysts within VRFs in fine and very fine sandstones.

*Other minerals (trace).—*

Volcanic quartz is present in trace amounts in a few samples and is commonly associated with grains of alkali feld-

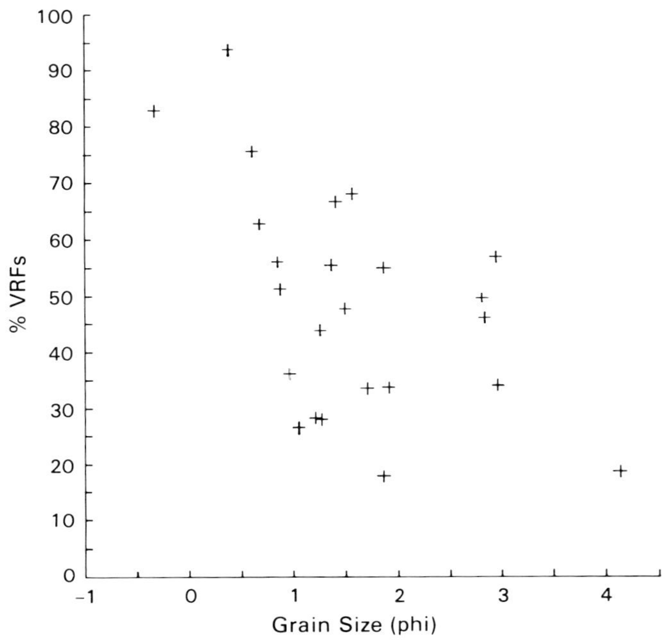


FIG. 11.—Graph showing correlation between mean grain size and percent of VRFs in framework fraction of lower Datil sandstones. Correlation coefficient = −0.47, *n* = 24.

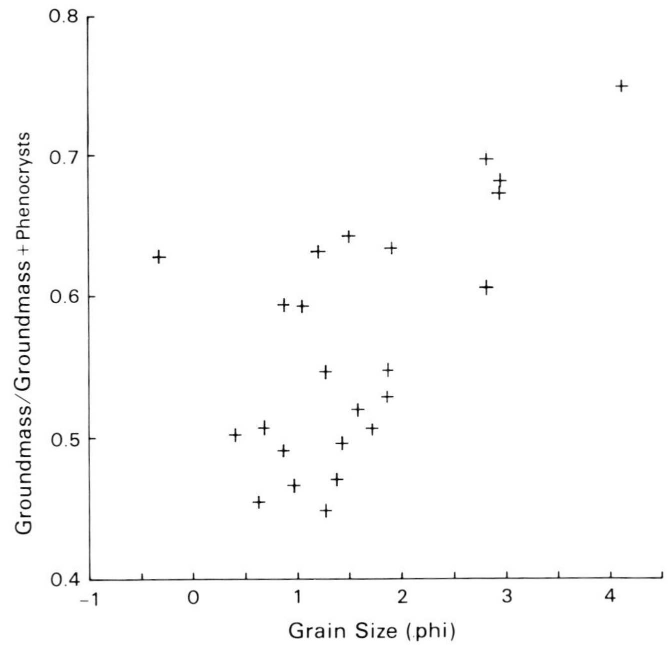


FIG. 12.—Graph showing increased proportion of groundmass relative to phenocrysts in VRFs with decreasing mean grain size. Correlation coefficient = +0.64, *n* = 24.

spar and devitrified VRFs containing shards. Because of this association and the lack of phenocrystic quartz in Datil andesites, detrital quartz in lower Datil sandstones is interpreted to be reworked from pyroclastic deposits, possibly derived from distant sources. Trace amounts of clinopyroxene and zircon were also noted in about 20 percent of the thin sections examined.

Nonvolcanic detritus is present in minor amounts throughout the lower Datil in most areas (Cather and Chapin, 1989). In sandstones, such detritus includes fragments of Paleozoic siltstone and limestone, and Precambrian quartzite, schist, and granite/gneiss rock fragments that were derived from Laramide uplifts.

*Matrix (0.0–50.6 percent, mean 19.7 ± 13.8 percent of total rock volume).—*

Matrix is a conspicuous component of most lower Datil sandstones; it is present in about 85 percent of the thin sections examined. Matrix occurs as fine-grained, low-birefringent, nearly amorphous masses between framework grains. The grain-supported nature of most matrix-bearing sandstones suggests that detrital components of matrix materials were infiltrated post-depositionally (e.g., Walker and others, 1978). Although largely detrital, matrix also contains authigenic constituents. Because Datil sandstones have been neither deeply buried nor tectonically deformed, the development of pseudomatrix and the widespread authigenesis and alteration of matrix such as that described by Galloway (1974) has not occurred.

Detrital components of matrix were analyzed using EDS in conjunction with the scanning electron microscope. Dominant constituents are silt-size alkali feldspar and silica minerals, with subordinate (<20 percent) plagioclase and mafic minerals. The composition of the detrital matrix indicates derivation largely from groundmass minerals in lower Datil andesites.

### *Mudstone Petrology*

Mudstone is a minor constituent of the lower Datil Group, comprising about 5 percent of the unit; claystone is rare. Significant amounts of mudstone occur only in lacustrine deposits of the lower Datil Group.

Detrital constituents of Datil mudstone consist of varying proportions of silt and clay. As indicated by optical and scanning electron microscopy, the silt component is composed largely (approximately 70 to 90 percent) of groundmass minerals (dominantly alkali feldspar and silica minerals), although lesser amounts of finely comminuted phenocrysts (plagioclase and mafic minerals) are also present. Smectite is the predominant detrital clay mineral in the lower Datil Group, as it is in most volcanic-derived deposits (Blatt and others, 1972). X-ray diffractometry also indicates the presence of chlorite in many Datil mudstones, although this may be diagenetic (Cather, 1986).

## MINERALOGIC FRACTIONATION AMONG LOWER DATIL VOLCANICLASTIC ROCKS

Because the great majority of lower Datil eruptive products are andesites that exhibit relatively uniform mineralogic and textural properties, a straightforward comparison of modal characteristics of lower Datil volcaniclastic rocks is possible. Furthermore, because of only small overlap between phenocryst and groundmass mineralogies in lower

Datil andesites, most detrital grains can be apportioned to either phenocryst or groundmass categories.

In this discussion, point-count data for sandstones were tabulated using the Gazzi-Dickinson method. By this point-counting method, grain-size effects on VRF abundance are diminished and the mineralogic composition of the sandstone is emphasized (Ingersoll and others, 1984). Point-count data for strongly altered sandstones and rare sandstones containing exotic(?) pyroclastic grains (alkali feldspar, quartz, and devitrified fragments containing shards) were excluded from the following analysis.

The range in relative abundance of dominant phenocryst minerals (plagioclase, amphibole, and biotite) is nearly uniform among lower Datil sandstones, conglomerate clasts, and correlative(?) lavas (Fig. 13). The average abundance of amphibole in sandstones, however, is slightly diminished relative to conglomerate clasts, especially in the finer grained sandstones (Fig. 13). These changes are consistent with the durabilities of phenocryst-derived grains noted earlier. The relative abundance of phenocryst minerals in lower Datil mudstones could not be determined due to inability to distinguish between phenocryst- and groundmass-derived plagioclase in silt-size grains.

Feldspar compositions in Datil andesite clasts and sandstones are fundamentally similar (Fig. 6). Statistical comparison of microprobe analyses for plagioclase indicates no significant differences between these rock types with regard to anorthite content of "cores" or rims; similarly, no statistically significant difference exists between orthoclase contents of "core" areas (Table 1). Orthoclase content of plagioclase rims in sandstones, however, is significantly less ($p = 0.02$) than those of andesite clasts. This difference may reflect the abrasional removal of potassic rims from

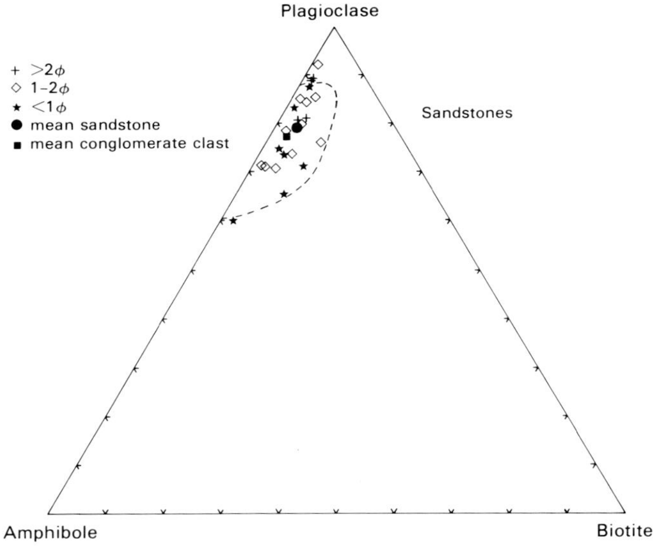


FIG. 13.—Amphibole-plagioclase-biotite compositions of 25 lower Datil sandstones. Symbols are keyed to mean grain size of sample: crosses > 2ϕ, diamonds 1–2ϕ, stars < 1ϕ. Dashed line encloses field for conglomerate clasts (see Fig. 3). Solid circle is average for lower Datil sandstones; solid square is average for andesites.

TABLE 1.—COMPARISON OF CORE AND RIM AREAS OF PLAGIOCLASE IN CONGLOMERATE CLASTS AND SANDSTONES

| | Clasts (n = 73) | Sandstone (n = 78) | Significance Level of Difference Between Means |
|---|---|---|---|
| $An_{(core)}$ | 32.2 ± 9.9 | 30.2 ± 6.7 | $p = 0.15$ |
| $An_{(rim)}$ | 28.9 ± 6.4 | 29.0 ± 4.9 | $p = 0.50$ |
| $Or_{(core)}$ | 4.2 ± 1.7 | 4.5 ± 1.8 | $p = 0.40$ |
| $Or_{(rim)}$ | 4.9 ± 1.5 | 4.4 ± 1.3 | $p = 0.02$ |

detrital-feldspar grains during transport. This hypothesis, however, is in need of further testing, due to the small number of samples utilized in the present study. No microprobe analyses were performed for feldspars in lower Datil mudstones or correlative(?) lavas.

Although the relative abundance of plagioclase, amphibole, and biotite is approximately constant among lower Datil sandstones, conglomerate clasts, and correlative(?) lavas, the ratio of phenocryst to groundmass minerals is substantially increased in the framework fraction of sandstones (Table 2, Fig. 14a), particularly in medium sandstones. This increase has occurred at the expense of groundmass and indicates relatively low durability of groundmass with respect to phenocrysts during transportation. Finely comminuted groundmass is the major detrital contributor to matrix in sandstones. When modal values for infiltrated matrix are counted as groundmass in lower Datil sandstones, a net deficiency of groundmass with respect to andesite remains, however (Fig. 14b). Based on these relations and the predominance of groundmass minerals in Datil mudstones, a hypothetical evolutionary sequence is depicted in Figure 14c.

## DISCUSSION

Several processes can produce crystal enrichment in volcaniclastic sandstones: 1) inheritance from magmas with high-crystal content; 2) fragmentation of juvenile pyroclasts (*sensu* Fisher and Schmincke, 1984, p. 90) by vesiculation and shattering during explosive eruptions, with subsequent elutriation of fine vitric ash; and 3) attrition of epiclasts and pyroclasts during transportation. Although all of these processes may have contributed to the observed mineralogic fractionation in the lower Datil Group, inheritance and eruptive fractionation of juvenile materials appear to have been subordinate to transportational effects.

Marsh (1981) modeled the influence of crystallinity on magma rheology and concluded that basaltic magmas with crystal contents greater than about 55 percent no longer behave as liquids and are incapable of erupting as lavas. For silicic magmas, the critical crystallinity necessary to cause immobility is significantly less than that for basalts, due to the higher viscosity of silicic melts. Average crystallinity of lower Datil andesite clasts (53 percent) and correlative(?) lavas (49 percent) approach the rheologic limit of Marsh (1981). Thus, it is difficult to envisage that the average crystal content of the framework fraction of lower Datil sandstones (73 percent) was inherited from precursor lavas or domes. Pyroclastic eruption of highly crystallized plutons (e.g., Steven and Lipman, 1976) probably did not contribute in a major way to the observed crystal concentrations in lower Datil sandstones, as indicated by the scarcity

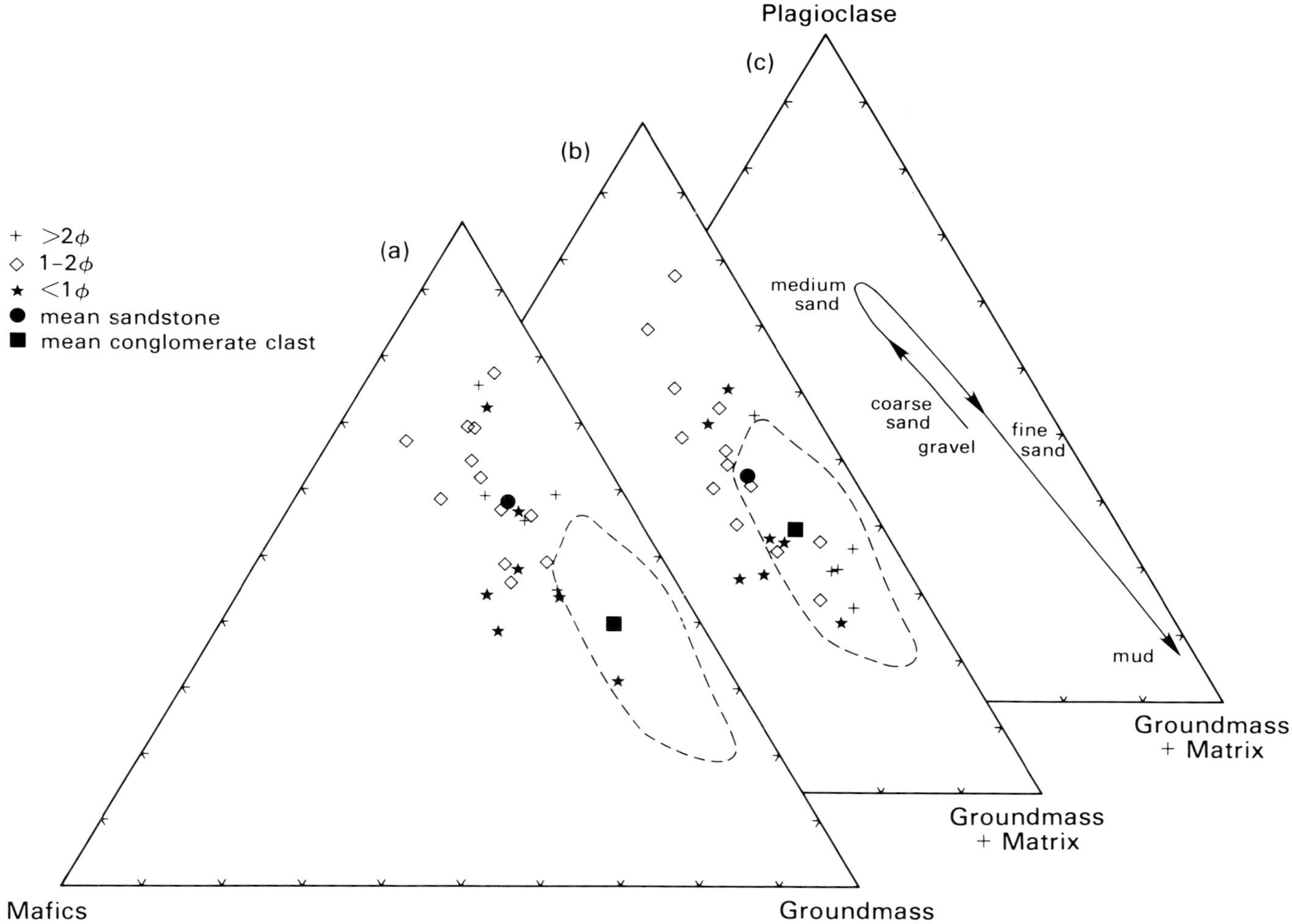


FIG. 14.—Petrographic data showing relative proportions of mafic minerals (amphibole, biotite, titanomagnetite, and pyroxene), plagioclase, and groundmass minerals. Symbols are keyed to mean grain size of sample. Dashed line encloses field for andesite clasts (Fig. 4). (a) Framework grains only; (b) framework grains plus matrix counted as groundmass; (c) hypothetical evolutionary sequence for lower Datil Group volcaniclastic sediments based on (b) and mineralogic data for mudstones (see text).

TABLE 2.—COMPARISON OF AVERAGE MODAL ANALYSES OF LOWER DATIL CONGLOMERATE CLASTS AND SANDSTONES

| | Clasts (%) (n = 28) | Sandstone (%*) (n = 25) | Significance Level of Difference Between Means |
|---|---|---|---|
| Plagioclase | 37.5 ± 7.2 | 56.1 ± 12.1 | $p < 0.001$ |
| Amphibole | 9.5 ± 3.6 | 11.6 ± 4.5 | $p = 0.06$ |
| Biotite | 1.7 ± 1.5 | 2.5 ± 1.7 | $p = 0.07$ |
| Opaques | 1.5 ± 1.0 | 2.3 ± 1.2 | $p = 0.01$ |
| Groundmass | 47.0 ± 8.1 | 26.7 ± 9.3** | $p < 0.001$ |

*Percent of framework grains.
**Does not include detrital matrix.

of vitric grains and the fact that ignibrites rarely exceed 60 weight-percent crystals (Cas and Wright, 1987, p. 338). Similarly, the dominance of lithic fragments and the low abundance of vitric detritus in the lower Datil Group argue against major derivation of crystals by eruptive comminution and aerodynamic segregation of juvenile pyroclastic materials, although the efficacy of this process is well documented for ignimbrites (e.g., Hay, 1959; Walker, 1972; Sparks and Walker, 1977).

Despite the scarcity of unambiguous juvenile pyroclasts (i.e., shards and pumice), the hydrous mineralogy of lower Datil andesites implies that a significant part of the volcanic detritus in the lower Datil Group was probably produced by eruptive fragmentation. The relative proportion of pyroclasts versus epiclasts (*sensu* Fisher, 1961) in the unit, however, is not known. The problem is exacerbated by the apparent lack of primary pyroclastic deposits in the study area. Thus, in the following discussion of the effects of transportation on the mineralogic evolution of lower Datil detritus, the relative importance of pyroclastic and epiclastic processes remains unknown.

The mechanical diminution of grain size during transport can be attributed mainly to two end-member processes: 1) wholesale breakage of particles caused by energetic collisions (impact shattering); and 2) abrasion due to low-energy collisions or impingement of smaller particles (sand blasting). Impact shattering and abrasion occur during both pyroclastic and aqueous transport, although the efficacy of grain diminution may be greater for pyroclastic transport due to lack of protective water films on particles (e.g., Mackie, 1897; Sorby, 1908).

Davies and others (1978) studied the downstream modification of volcaniclastic sediment in rivers draining the active volcano Fuego in Guatemala. Their method of petrographic analysis differed from ours, however, in that they tabulated only the various types of free crystals and rock fragments (i.e., traditional point-counting technique) and not the modal mineralogy of the sediments (Gazzi-Dickinson technique). Davies and others (1978) documented significant downstream increases in roundness and free crystals with concomitant decreases in mean grain size, sorting, and percentage of rock fragments. They emphasized the role of impact shattering of sand grains to explain the observed downstream destruction of rock fragments and the liberation of free crystals.

Impact shattering was likely an important process in the mineralogic evolution of lower Datil sand and silt, but probably not for coarser detritus. Examination of natural joints and fractures in lower Datil conglomerate clasts (as well as those induced by hammering) shows that breakage occurs quite equitably across both groundmass and phenocrysts. Thus, impact shattering of clasts probably did not contribute toward the observed segregation of phenocrysts from groundmass in the lower Datil Group. It is probable, however, that impact shattering may cause effective mineralogic disaggregation of smaller fragments (sand and silt), as the length of intercrystalline boundaries approaches the diameter of these smaller grains. In other words, the scale of crystal sizes in lower Datil andesites effectively imparts increasing mechanical anisotropy as grain size diminishes. Disaggregation along intercrystalline boundaries would cause the over-representation of phenocryst phases in sands and the dominance of groundmass minerals in silts and muds, and may have contributed significantly to the grain-size-related changes in the lower Datil Group.

The role of abrasion in the downstream rounding of coarse volcaniclastic detritus has been noted by several workers (Cameron and Blatt, 1971; Pearce, 1971; Davies and others, 1978). Cameron and Blatt (1971) showed that fluvial transportation does not produce rounding in felsic volcanic detritus that is finer than very coarse sand, a conclusion that is consistent with the protective effects of water films on the rounding of fine-grained sediments (less than about 0.5 mm) noted by early workers (Daubreé, 1879; Mackie, 1897; Sorby, 1908). Rounding of volcaniclastic pebbles and cobbles has been attributed to abrasion of rough edges during transport (Davies and others, 1978) and to *in situ* subaqueous sand blasting of clasts imbedded in the stream bed (Pearce, 1971).

Very coarse sand is the minimum-size fraction in the lower Datil Group in which significant amounts of rounded and well-rounded grains occur. Rounding generally increases with increasing clast size in Datil fluvial conglomerates, but is poorly developed or not present in medium sandstones and finer grained rocks. These observations suggest that abrasional rounding dominated grain-shape development in sediments coarser than about coarse sand and that production of angular fragments by impact shattering dominated in finer sediments.

The potential role of clast abrasion in the fractionation of groundmass from phenocrysts can be demonstrated experimentally. Ten clasts of lower Datil andesite were sawed into sections and polished, and random areas of each polished surface were protected with masking tape. Samples were then sand blasted for 30 seconds using a poorly sorted mixture of glass beads (0.5 mm and smaller in diameter) entrained in a high-velocity stream of air. With exception of one clast in which the groundmass was highly silicified, rapid and pronounced erosion of groundmass and amphibole phenocrysts occurred, leaving plagioclase and, to a lesser extent, biotite standing out in bold relief (Fig. 15). These results indicate that abrasional attrition of clast surfaces may cause crystal- and cleavage-controlled fragmentation of groundmass and amphibole, resulting in concentration of more durable phenocrysts (plagioclase, biotite) in the sand fraction. These experimental results probably best apply to abrasion during pyroclastic or eolian transport; the effectiveness of sand blasting of volcanic clasts in subaqueous environments is controversial (cf., Pearce, 1971, p. 1071–1072; Davies and others, 1978, p. 72, 74).

Chemical weathering in source regions appears to have had little or no effect on lower Datil detritus, as shown by the lack of discernible pre-depositional alteration in thin section (Cather, 1986). Even in humid Guatemala, pyroxene-andesite detritus derived from Fuego volcano shows no evidence of chemical weathering, because of the short residence time of sediments in source areas and in fluvial drainages (Davies and others, 1978). In contrast, grains derived solely by epiclastic processes in areas of inactive vol-

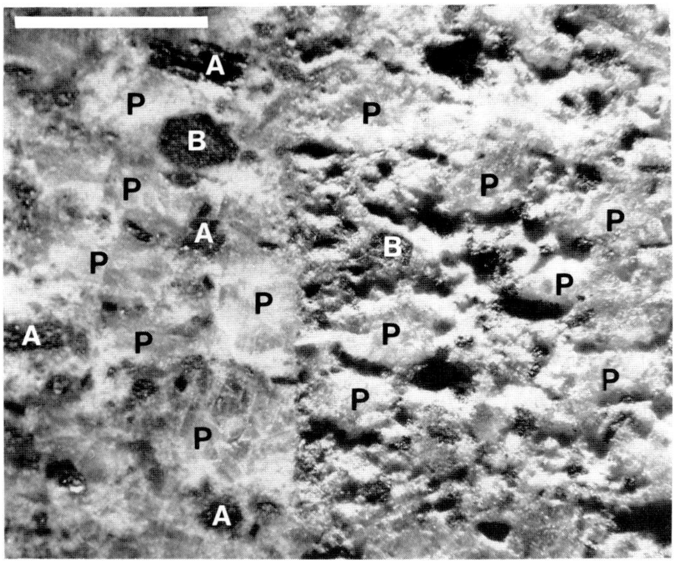


FIG. 15.—Photomicrograph of polished (left) and sand blasted (right) parts of andesite clast (plagioclase = *P*, amphibole = *A*, biotite = *B*), showing preferential removal of groundmass and amphibole phenocrysts, Scale bar is 2.0 mm; reflected-light source is at low angle from top. See text for discussion.

canism may undergo chemical weathering due to greater residence times in source-area soils (e.g., Cameron and Blatt, 1971). Evolution of detritus in the lower Datil Group occurred in a semi-arid climate, but is very similar to that described by Roobol (1976) for Upper Cretaceous hornblende andesites in central Jamaica. Thus, the sedimentary fractionation of andesitic detritus appears to be insensitive to climate.

## CONCLUSIONS

Differences in abundance of groundmass versus phenocrysts mark the most important change in detrital mineralogy that occurs among conglomerate, sandstone, and mudstone of the lower Datil Group. Sedimentary fractionation of phenocrysts from groundmass by physical processes (dominantly abrasion and impact shattering) results in grain-size-related control of mineralogic and, hence, bulk-chemical composition (Fig. 16). Conglomerate seems to adequately represent the composition of extrusive rocks in the source region, but does not record the unwelded pyroclastic detritus that was observed in some sandstones. Sandstone compositions are biased for the following reasons: 1) groundmass is more readily comminuted to silt size during transport, resulting in over-representation of phenocryst phases in sands; 2) opaque rims on biotite and amphibole are rapidly destroyed; 3) amphibole abundance is slightly diminished relative to plagioclase and biotite; and 4) K-rich rims of zoned plagioclase phenocrysts may be removed by abrasion. Volcaniclastic mudstone is a poor indicator of provenance because of over-abundant groundmass constituents. Thus, provenance determinations may be biased by grain-size effects in volcaniclastic deposits. For example, study of siltstones in distal environments might lead to unrealistically silicic interpretations of source-area volcanism.

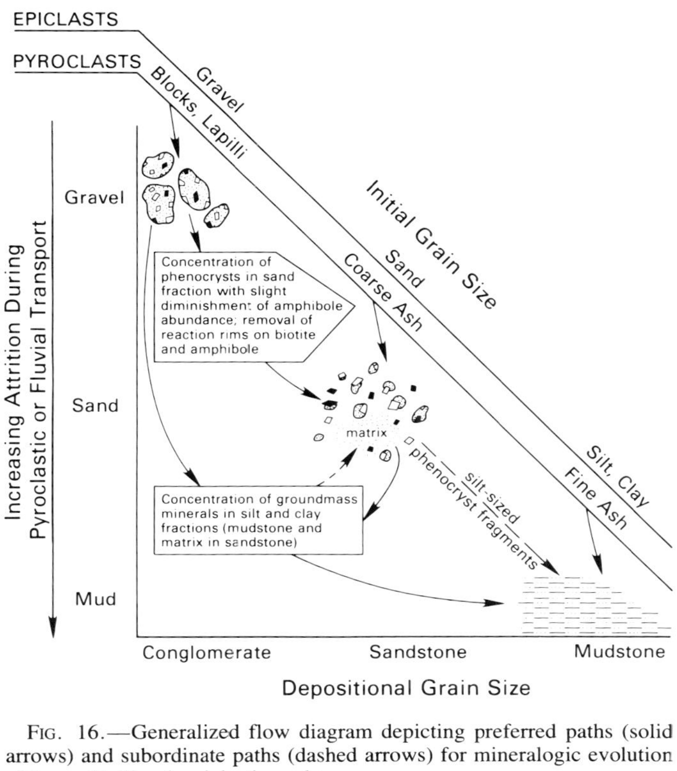


FIG. 16.—Generalized flow diagram depicting preferred paths (solid arrows) and subordinate paths (dashed arrows) for mineralogic evolution of lower Datil volcaniclastic rocks.

## ACKNOWLEDGMENTS

Financial support for this study was provided by the New Mexico Bureau of Mines and Mineral Resources (F. E. Kottlowski, Director) and the University of Texas Geology Foundation. The basis of the work was the Ph.D. dissertation of the senior author. We benefited from discussions with D. S. Barker, R. M. Chamberlin, C. E. Chapin, S. J. Johansen, R. L. Kugler, S. G. Lucas, G. R. Osburn, D. Smith, and G. A. Smith. The manuscript was improved by reviews from M. E. Cather, C. E. Chapin, R. V. Ingersoll, K. Marsaglia, W. C. McIntosh, and G. A. Smith. G. K. Hoops of the University of Texas at Austin provided wet-chemical analyses. Graphics and statistical calculations were greatly expedited by software written by L. E. Long. Figures were drafted by Rebecca Titus and the manuscript was typed by Lynne McNeil.

## REFERENCES

BLATT, H., MIDDLETON, G., AND MURRAY, R., 1972, Origin of sedimentary rocks: Prentice-Hall, Englewood Cliffs, 634 p.

BROULLIARD, L. A., 1984, Geology of the northeastern Gallinas Mountains, Socorro County, New Mexico: New Mexico Institute of Mining and Technology, Socorro, Unpubl. M.S. thesis, 161 p.

CAMERON, K. L., AND BLATT, H., 1971, Durabilities of sand size schist and "volcanic" rock fragments during fluvial transport, Elk Creek, Black Hills, South Dakota: Journal of Sedimentary Petrology, v. 41, p. 565–576.

CAS, R. A. F., AND WRIGHT, J. V., 1987, Volcanic Successions, Modern and Ancient: Allen and Unwin, London, 528 p.

CATHER, S. M., 1983, Lacustrine deposits of the Eocene Baca Formation, western Socorro County, New Mexico: New Mexico Geological Society, 34th Field Conference Guidebook, p. 179–185.

CATHER, S. M., 1986, Volcano-sedimentary evolution and tectonic implications of the Datil Group (latest Eocene-early Oligocene), west-central New Mexico: University of Texas at Austin, Austin, Unpubl. Ph.D. Dissertation, 484 p.

CATHER, S. M., AND CHAPIN, C. E., 1989, Field guide to upper Eocene and lower Oligocene volcaniclastic rocks of the northern Mogollon-Datil volcanic field, *in* Chapin, C. E., and Zidek, J., eds., Field Excursions to Volcanic Terranes in the Western United States: Southern Rocky Mountain Region: New Mexico Bureau of Mines and Mineral Resources Memoir 46, p. 60–68.

CATHER, S. M., AND JOHNSON, B. D., 1984, Eocene tectonics and depositional setting of west-central New Mexico and eastern Arizona: New Mexico Bureau of Mines and Mineral Resources Circular 192, 33.

CATHER, S. M., AND JOHNSON, B. D., 1986, Eocene depositional systems and tectonics in west-central New Mexico and eastern Arizona: American Association of Petroleum Geologists Memoir 41, p. 623–652.

DAUBRÉÉ, A., 1879. Études synthetiques de géologie expérimentále: Dunod, Paris, 828 p.

DAVIES, D. K., VESSELL, R. K., MILES, R. C., FOLEY, M. G., AND BONIS, S. B., 1978, Fluvial transport and downstream sediment modifications in an active volcanic region, *in* Miall, A. D., ed., Fluvial Sedimentology: Canadian Society of Petroleum Geologists Memoir 5, p. 61–84.

DICKINSON, W. R., 1970, Interpreting detrital modes of graywacke and arkose: Journal of Sedimentary Petrology, v. 40, p. 695–707.

DICKINSON, AND SUCZEK, C. A., 1979, Plate tectonics and sandstone compositions: American Association of Petroleum Geologists Bulletin, v. 63, p. 2164–2182.

EGGLER, D. H., AND BURNHAM, C. W., 1973, Crystallization and fractionation trends in the system andesite-$H_2O$—$CO_2$—$O_2$ at pressures to 10 Kb: Geological Society of America Bulletin, v. 84, p. 2517–2532.
EWART, A., 1979, A review of the mineralogy and chemistry of Tertiary-Recent dacitic, latitic, rhyolitic, and related salic volcanic rocks, *in* Barker, F., ed., Trondhjemites, Dacites, and Related Rocks: Elsevier, Amsterdam, p. 13–121.
EWART, A., 1982, The mineralogy and petrology of Tertiary-Recent orogenic volcanic rocks, with special reference to the andesitic-basaltic compositional range, *in* Thorpe, R. S., ed., Andesites-Orogenic Andesites and Related Rocks: John Wiley and Sons, New York, p. 25–95.
FARKAS, S. E., 1969, Geology of the southern San Mateo Mountains, Socorro and Sierra Counties, New Mexico: University of New Mexico, Albuquerque, Unpubl. Ph.D. Dissertation, 137 p.
FISHER, R. V., 1961, Proposed classification of volcaniclastic sediments and rocks: Geological Society of American Bulletin, v. 72, p. 1409–1414.
FISHER, R. V., AND SCHMINCKE, H.-U., 1984, Pyroclastic Rocks: Springer-Verlag, Berlin, 472 p.
FOLK, R. L., 1974, Petrology of sedimentary rocks: Hemphill Publishing Company, Austin, 182 p.
FORURIA, J., 1984, Geology of part of the southern San Mateo Mountains, Socorro and Sierra Counties, New Mexico: Colorado State University, Fort Collins, Unpubl. M.S. Thesis, 163 p.
GALLOWAY, W. E., 1974, Deposition and diagenetic alteration of sandstone in northeast Pacific arc-related basins: Implications for graywacke genesis: Geological Society of American Bulletin, v. 85, p. 379–390.
GAZZI, P., 1966, Le arenarie del flysch sopracretaceo dell' Appeninno modenese: Correlazioni con il flysch di Monghidoro: Mineralogica et Petrographica Acta, v. 12, p. 69–97.
GILL, J. B., 1981, Orogenic andesites and plate tectonics: Springer-Verlag, New York, 390 p.
GREEN, T. H., 1972, Crystallization of calcalkaline andesite under controlled high-pressure hydrous conditions: Contributions to Mineralogy and Petrology, v. 34, p. 150–166.
HARRELL, J. A., AND ERIKSSON, K. A., 1979, Empirical conversion equations for thin-section and sieve-derived size distribution parameters: Journal of Sedimentary Petrology, v. 49, p. 273–280.
HARRISON, R. W., 1980, Geology of the northeastern Datil Mountains, Socorro and Catron Counties, New Mexico: New Mexico Institute of Mining and Technology, Socorro, Unpubl. M.S. Thesis, 137 p.
HAY, R. L., 1959, Formation of crystal-rich glowing avalanche deposits of St. Vincent, B.W.I.: Journal of Geology, v. 67, p. 540–562.
INGERSOLL, R. V., BULLARD, T. F., FORD, R. L., GRIMM, J. P., PICKLE, J. D., AND SARES, S. W., 1984, The effect of grain size on detrital modes: A test of the Gazzi-Dickinson point-counting method: Journal of Sedimentary Petrology, v. 54, p. 103–116.
LUCAS, S. G., 1983, The Baca Formation and the Eocene-Oligocene boundary in New Mexico: New Mexico Geological Society, 34th Field Conference Guidebook, p. 187–192.
LUCAS, S. G., 1986, Oligocene mammals from the Black Range, southwestern New Mexico: New Mexico Geological Society, 37th Field Conference Guidebook, p. 261–263.
LUCAS, S. G., SCHOCH, R. M., MANNING, E., AND TSENTAS, C., 1981, The Eocene biostratigraphy of New Mexico: Geological Society of America Bulletin, v. 92, p. 951–967.
MACKIE, W., 1897, On the laws that govern the rounding of particles of sand: Transactions of the Edinburgh Geological Society, v. 7, p. 298–311.
MARSH, B. D., 1981, On the crystallinity, probability of occurrence, and rheology of lava and magma: Contributions to Mineralogy and Petrology, v. 78, p. 85–98.
NEW MEXICO GEOLOGICAL SOCIETY, 1982, New Mexico Highway Geologic Map, scale 1:1,000,000, 1 sheet.
OSBURN, G. R., AND CHAPIN, C. E., 1983, Nomenclature for Cenozoic rocks of northeast Mogollon-Datil volcanic field, New Mexico: New Mexico Bureau of Mines and Mineral Resources, Stratigraphic Chart 1, 2 sheets.
PEARCE, T. H., 1971, Short distance fluvial rounding of volcanic detritus: Journal of Sedimentary Petrology, v. 41, p. 1069–1072.
POWERS, M. C., 1953, A new roundness scale for sedimentary particles: Journal of Sedimentary Petrology, v. 28, p. 108–110.
ROOBOL, M. J., 1976, Post-eruptive mechanical sorting of pyroclastic material—An example from Jamaica: Geological Magazine, v. 113, p. 429–440.
SCHIEBOUT, J. A., AND SCHRODT, A. K., 1981, Vertebrate paleontology of the lower Tertiary Baca Formation of western New Mexico: Geological Society of America Bulletin, v. 92, p. 976–979.
SNYDER, D. O., 1971, Stratigraphic analysis of the Baca Formation, west-central New Mexico: University of New Mexico, Albuquerque, Unpubl. Ph.D. Dissertation, 158 p.
SORBY, H. C., 1908, On the application of quantitative methods to the study of the structure and history of rocks: Quarterly Journal of the Geological Society of London, v. 64, p. 171–232.
SPARKS, R. S. J., AND WALKER, G. P. L., 1977, The significance of vitric-enriched air-fall ashes associated with crystal-enriched ignibrites: Journal of Volcanology and Geothermal Research, v. 2, p. 329–341.
STEVEN, T. A., AND LIPMAN, P. W., 1976, Calderas of the San Juan volcanic field, southwestern Colorado: U.S. Geological Survey Professional Paper 958, 35 p.
WALKER, G. P. L., 1972, Crystal concentration in ignimbrites: Contributions to Mineralogy and Petrology, v. 36, p. 135–146.
WALKER, T. R., WAUGH, B., AND CRONE, A. J., 1978, Diagenesis of first-cycle alluvium of Cenozoic age, southwestern United States and northwestern Mexico: Geological Society of America Bulletin, v. 89, p. 19–32.

APPENDIX A.—POINT-COUNT DATA FOR LOWER DATIL GROUP CONGLOMERATE CLASTS

| Sample Number | Groundmass | Plagioclase | Alkali Feldspar | Quartz | Amphibole | Biotite | Clinopyroxene | Orthopyroxene | Opaque | Apatite | Zircon | 2° Calcite | 2° Silica | Vesicle | Total Points* |
|---|---|---|---|---|---|---|---|---|---|---|---|---|---|---|---|
| SNJ-431-CC | 43.5 | 41.2 | 0.0 | 0.0 | 8.6 | 4.3 | 0.0 | 0.0 | 4.3 | 0.0 | 0.0 | 0.0 | 0.0 | 0.5 | 209 |
| SNJ-1-CC | 43.7 | 39.3 | 0.0 | 0.0 | 10.7 | 2.9 | 0.0 | 0.0 | 0.5 | 0.0 | 0.5 | 1.9 | 1.9 | 2.8 | 206 |
| SNJ-394-CC | 44.2 | 40.9 | 0.0 | 0.0 | 9.8 | 0.0 | 0.0 | 0.0 | 0.0 | 0.0 | 0.0 | 0.5 | 0.5 | 2.7 | 215 |
| SNJ-539-CC | 42.9 | 41.5 | 0.0 | 0.0 | 9.3 | 0.0 | 1.0 | 1.0 | 3.4 | 0.0 | 0.0 | 4.8 | 0.0 | 0.5 | 205 |
| SNJ-B-CC | 35.0 | 44.7 | 0.0 | 0.0 | 9.2 | 1.8 | 0.0 | 0.0 | 1.8 | 0.0 | 0.0 | 2.8 | 5.1 | 4.8 | 217 |
| SNJ-C-CC | 47.6 | 36.1 | 0.0 | 0.0 | 9.1 | 1.0 | 0.0 | 0.0 | 1.0 | 0.5 | 0.0 | 0.0 | 4.8 | 2.8 | 208 |
| SNJ-1a-CC | 40.6 | 42.7 | 0.0 | 0.0 | 8.5 | 1.6 | 0.0 | 0.0 | 1.2 | 0.0 | 0.0 | 0.0 | 5.3 | 0.0 | 246 |
| SDS-615-CC | 51.5 | 33.2 | 0.0 | 0.0 | 10.9 | 3.5 | 0.0 | 0.0 | 0.4 | 0.4 | 0.0 | 0.0 | 0.0 | 0.0 | 229 |
| SNJ-701-CC | 54.0 | 33.8 | 0.0 | 0.0 | 10.3 | 0.9 | 0.0 | 0.0 | 0.9 | 0.0 | 0.0 | 0.0 | 0.0 | 5.7 | 213 |
| SNJ-194-CC | 44.0 | 46.2 | 0.0 | 0.4 | 3.8 | 3.4 | 0.0 | 0.0 | 0.4 | 0.0 | 0.0 | 0.4 | 1.3 | 0.4 | 234 |
| SNJ-195-CC | 36.2 | 46.7 | 1.4 | 0.0 | 5.7 | 2.9 | 0.0 | 0.0 | 2.4 | 1.0 | 0.0 | 0.5 | 3.3 | 4.1 | 210 |
| SJH-103-CC | 43.4 | 37.0 | 0.0 | 0.0 | 10.5 | 1.8 | 0.0 | 0.0 | 1.4 | 0.0 | 0.0 | 0.9 | 4.6 | 8.0 | 219 |
| SJH-236-CC | 44.4 | 43.4 | 0.0 | 0.0 | 10.2 | 0.0 | 0.0 | 0.0 | 0.5 | 0.5 | 0.0 | 0.0 | 0.5 | 4.8 | 196 |
| SNJ-10-CC | 42.5 | 44.8 | 0.0 | 0.0 | 3.5 | 2.2 | 2.2 | 0.0 | 2.5 | 0.0 | 0.0 | 0.0 | 0.0 | 0.5 | 402 |
| SDS-1016-CC | 46.1 | 32.0 | 0.0 | 0.0 | 19.9 | 0.0 | 0.0 | 0.0 | 2.5 | 0.0 | 0.0 | 0.0 | 3.7 | 0.8 | 241 |
| SDM-93-CC | 42.9 | 38.7 | 0.0 | 0.0 | 13.2 | 0.0 | 2.8 | 0.0 | 2.4 | 0.0 | 0.0 | 0.0 | 0.0 | 0.0 | 212 |
| BIMC-7 | 57.8 | 22.6 | 0.0 | 0.0 | 13.2 | 0.0 | 2.0 | 0.0 | 2.9 | 0.0 | 0.0 | 0.5 | 1.0 | 0.0 | 204 |
| SDS-30-CC | 50.2 | 32.4 | 0.0 | 0.0 | 13.0 | 1.0 | 0.0 | 0.0 | 0.5 | 0.5 | 0.0 | 1.0 | 1.4 | 2.4 | 207 |
| SBM-39-CC | 51.5 | 32.0 | 0.0 | 0.0 | 6.5 | 2.5 | 0.0 | 0.0 | 1.0 | 0.0 | 0.0 | 1.0 | 0.0 | 1.5 | 200 |
| SDS-5-CC | 38.6 | 45.8 | 0.0 | 0.0 | 12.0 | 0.8 | 0.0 | 0.0 | 0.8 | 0.0 | 0.0 | 2.0 | 0.0 | 3.1 | 249 |
| SBM-411-CC | 47.1 | 39.7 | 0.0 | 0.5 | 7.8 | 2.0 | 0.0 | 0.0 | 2.0 | 0.0 | 0.0 | 1.0 | 0.0 | 0.0 | 204 |
| SDM-220-CC | 50.0 | 39.6 | 0.0 | 0.0 | 4.9 | 2.5 | 0.0 | 0.0 | 2.0 | 0.0 | 0.0 | 0.0 | 0.0 | 1.0 | 202 |
| SDS-1128-CC | 70.0 | 20.5 | 0.0 | 0.0 | 5.0 | 0.5 | 1.0 | 0.0 | 1.0 | 0.0 | 0.0 | 2.0 | 0.0 | 0.0 | 200 |
| SJH-24-CC | 54.3 | 33.8 | 0.0 | 0.0 | 7.6 | 0.0 | 1.4 | 0.0 | 1.9 | 0.0 | 0.0 | 0.0 | 1.0 | 1.9 | 210 |
| SNJ-5-CC | 32.7 | 44.4 | 0.0 | 0.0 | 13.6 | 4.2 | 0.0 | 0.0 | 1.4 | 0.0 | 0.0 | 1.4 | 1.9 | 2.2 | 214 |
| SJH-1-CC | 44.1 | 41.8 | 0.0 | 0.0 | 8.3 | 4.5 | 0.0 | 0.0 | 0.8 | 0.0 | 0.0 | 15.6 | 0.0 | 1.1 | 263 |
| SNJ-A-CC | 61.0 | 23.9 | 0.0 | 0.0 | 13.2 | 0.0 | 0.0 | 0.0 | 1.5 | 0.0 | 0.0 | 0.0 | 0.5 | 0.0 | 205 |
| SDM-638-CC | 55.1 | 31.8 | 0.0 | 0.0 | 6.8 | 4.2 | 0.0 | 0.0 | 1.7 | 0.0 | 0.0 | 0.0 | 3.8 | 2.9 | 236 |
| | | | | | | | | | | | | | | | 6,256 |
| mean | 47.0 | 37.5 | tr | tr | 9.5 | 1.7 | tr | tr | 1.5 | tr | tr | 1.3 | 1.3 | 1.9 | |
| std. dev. | 8.1 | 7.2 | | | 3.6 | 1.5 | | | 1.0 | | | 3.0 | 1.8 | 2.1 | |
| high | 70.0 | 46.7 | | | 19.9 | 4.5 | | | 4.3 | | | 15.6 | 5.3 | 8.0 | |
| low | 32.7 | 20.5 | | | 3.5 | 0.0 | | | 0.0 | | | 0.0 | 0.0 | 0.0 | |

*Excludes vesicles.
2° = secondary.
All values are percent of total rock volume, exclusive of vesicles.

APPENDIX B.—POINT-COUNT DATA FOR LOWER DATIL GROUP SANDSTONES

| | | | | | | | | VRF | | | | | | | | | | | |
|---|---|---|---|---|---|---|---|---|---|---|---|---|---|---|---|---|---|---|---|
| Sample Number | Plagioclase | Alkali Feldspar | Quartz | Amphibole | Biotite | Clinopyroxene | Opaque | Plagioclase | Alkali Feldspar | Quartz | Amphibole | Biotite | Clinopyroxene | Opaque | Groundmass | Non-volcanic Grains | Matrix* | Mean Grain Size (ϕ) | Total Points |
| SBM-94-MS | 46.2 | 0.0 | 0.0 | 5.4 | 1.0 | 0.0 | 1.0 | 12.9 | 0.0 | 0.0 | 1.0 | 0.0 | 0.0 | 0.0 | 32.3 | 0.0 | 35.2 | 2.84 | 324 |
| SDS-700-SS | 70.4 | 0.0 | 0.5 | 7.0 | 1.4 | 0.5 | 0.5 | 3.8 | 0.0 | 0.0 | 0.5 | 0.0 | 0.0 | 0.5 | 14.1 | 1.0 | 24.1 | 4.14 | 290 |
| SNJ-64-MS | 51.3 | 0.0 | 0.0 | 10.4 | 2.1 | 0.0 | 2.1 | 7.8 | 0.0 | 0.0 | 2.6 | 0.0 | 0.0 | 0.5 | 23.3 | 0.0 | 43.8 | 2.97 | 345 |
| SDS-150-MF | 24.4 | 0.0 | 0.0 | 9.9 | 0.9 | 0.0 | 0.5 | 22.1 | 0.5 | 0.0 | 4.2 | 1.9 | 0.0 | 1.9 | 31.9 | 1.4 | 19.2 | 0.69 | 281 |
| SDM-105-SS | 13.6 | 0.0 | 0.0 | 8.7 | 1.0 | 0.0 | 1.0 | 24.3 | 0.0 | 0.0 | 15.1 | 0.5 | 0.0 | 1.5 | 34.5 | 0.0 | 16.4 | 0.63 | 250 |
| SJH-180-MF | 6.8 | 0.0 | 0.0 | 1.3 | 0.4 | 0.0 | 0.4 | 23.3 | 0.0 | 0.0 | 7.2 | 3.4 | 0.8 | 3.0 | 43.2 | 0.0 | 16.1 | −0.30 | 305 |
| SDM-91-SS | 17.4 | 0.0 | 0.0 | 5.5 | 0.5 | 0.5 | 1.6 | 25.0 | 0.0 | 0.0 | 7.2 | 1.2 | 0.0 | 2.1 | 39.0 | 0.0 | 15.4 | 0.41 | 253 |
| SDM-43-SS | 57.4 | 0.0 | 0.0 | 7.1 | 1.0 | 0.0 | 1.0 | 11.4 | 0.0 | 0.0 | 3.3 | 1.4 | 0.0 | 0.5 | 17.1 | 0.0 | 15.5 | 1.72 | 271 |
| SDS-10-SS | 37.6 | 0.0 | 0.0 | 13.2 | 1.5 | 0.0 | 1.5 | 10.7 | 0.0 | 0.0 | 4.9 | 0.0 | 0.0 | 1.5 | 30.7 | 0.0 | 24.1 | 1.51 | 290 |
| SNJ-196-SS | 30.9 | 0.0 | 0.0 | 7.3 | 0.0 | 0.0 | 2.4 | 23.7 | 0.0 | 0.0 | 2.9 | 1.0 | 0.0 | 1.0 | 27.5 | 0.5 | 2.6 | 0.87 | 229 |
| SDM-175-SS | 46.2 | 0.0 | 0.4 | 6.7 | 1.3 | 0.0 | 1.3 | 19.7 | 0.0 | 0.0 | 4.9 | 2.2 | 0.0 | 0.4 | 19.7 | 0.0 | 5.4 | 1.27 | 258 |
| SNJ-625-SS | 62.4 | 0.0 | 0.0 | 12.2 | 6.9 | 0.0 | 0.5 | 4.8 | 0.0 | 0.0 | 0.5 | 1.6 | 0.0 | 1.6 | 9.5 | 0.0 | 19.4 | 1.87 | 258 |
| SNJ-399-SS | 57.0 | 0.0 | 0.0 | 5.8 | 0.5 | 0.0 | 0.5 | 15.0 | 0.0 | 0.0 | 2.9 | 1.0 | 0.0 | 0.5 | 16.9 | 0.0 | 15.7 | 0.97 | 261 |
| SJH-78-SS | 35.5 | 0.0 | 0.0 | 7.1 | 3.0 | 0.0 | 2.0 | 7.6 | 0.0 | 0.0 | 10.2 | 2.5 | 0.0 | 0.5 | 30.5 | 0.0 | 24.0 | 0.89 | 279 |
| SJH-119-SS | 51.4 | 0.0 | 0.0 | 10.5 | 3.3 | 0.0 | 0.5 | 10.0 | 0.0 | 0.0 | 1.9 | 0.0 | 0.0 | 0.5 | 21.4 | 0.5 | 15.6 | 1.92 | 276 |
| SNJ-134-SS | 35.3 | 0.0 | 0.0 | 7.3 | 0.5 | 0.0 | 1.4 | 21.1 | 0.0 | 0.0 | 6.9 | 0.9 | 0.0 | 0.5 | 26.1 | 0.0 | 17.7 | 1.38 | 294 |
| SNJ-161-SS | 22.6 | 0.0 | 0.0 | 7.4 | 0.9 | 0.0 | 2.3 | 23.0 | 0.0 | 0.0 | 8.8 | 1.8 | 0.0 | 0.0 | 33.2 | 0.0 | 36.0 | 1.43 | 347 |
| SNJ-305-SS | 62.4 | 0.0 | 0.0 | 6.3 | 1.8 | 0.0 | 1.8 | 6.3 | 0.0 | 0.0 | 2.3 | 1.4 | 0.0 | 0.9 | 15.8 | 0.0 | 0.0 | 1.06 | 271 |
| SNJ-395-SS | 65.2 | 0.0 | 0.4 | 4.8 | 0.4 | 0.0 | 0.4 | 11.9 | 0.0 | 0.0 | 0.9 | 0.4 | 0.0 | 0.0 | 15.4 | 0.0 | 0.0 | 1.28 | 295 |
| SNJ-626-SS | 52.2 | 0.0 | 0.0 | 13.4 | 3.5 | 0.0 | 2.5 | 6.0 | 0.0 | 0.0 | 3.0 | 1.0 | 0.0 | 0.5 | 17.9 | 0.0 | 21.3 | 1.22 | 267 |
| SDS-29-SS | 34.0 | 0.0 | 0.0 | 6.7 | 1.0 | 0.5 | 2.9 | 21.1 | 0.0 | 0.0 | 1.9 | 0.0 | 0.0 | 1.9 | 30.1 | 0.0 | 32.2 | 1.88 | 311 |
| SJH-212-MF | 31.4 | 0.6 | 0.0 | 3.5 | 1.7 | 2.3 | 3.5 | 12.2 | 0.0 | 0.0 | 5.2 | 0.0 | 0.0 | 1.2 | 38.4 | 0.0 | 32.1 | 2.96 | 318 |
| SNJ-28-SS | 38.3 | 0.0 | 0.0 | 7.5 | 3.0 | 0.0 | 0.8 | 16.5 | 0.0 | 0.0 | 2.3 | 0.0 | 0.0 | 0.8 | 30.1 | 0.8 | 38.3 | 2.82 | 222 |
| SNJ-11-SS | 24.3 | 0.5 | 0.0 | 5.6 | 0.5 | 0.0 | 0.9 | 23.8 | 0.5 | 0.0 | 6.1 | 0.9 | 0.0 | 0.9 | 35.5 | 0.0 | 0.0 | 1.59 | 265 |
| SDS-195-SS** | 46.1 | 3.9 | 5.2 | 3.2 | 11.0 | 0.0 | 2.0 | 6.5 | 0.0 | 0.0 | 1.3 | 0.6 | 0.0 | 0.2 | 20.2 | 0.0 | 50.6 | 1.70 | 253 |
| SNJ-432-SS | 43.4 | 0.0 | 0.0 | 1.5 | 0.0 | 0.0 | 1.0 | 21.0 | 0.5 | 0.0 | 1.5 | 1.5 | 0.0 | 1.5 | 25.9 | 1.5 | 12.0 | 0.39 | 251 |
| SJH-10-SS** | 27.6 | 1.8 | 0.6 | 5.9 | 0.0 | 0.0 | 0.6 | 21.8 | 1.2 | 0.0 | 7.6 | 0.6 | 0.0 | 0.0 | 32.4 | 0.0 | 0.0 | 1.20 | 230 |
| | | | | | | | | | | | | | | | | | | | 7,494 |
| mean | 40.4 | tr | tr | 7.1 | 1.8 | tr | 1.4 | 15.3 | tr | 0.0 | 4.3 | 1.0 | tr | 0.9 | 26.4 | tr | 19.7 | | |
| std. dev. | 16.6 | | | 3.1 | 2.3 | | 0.9 | 7.2 | | | 3.4 | 0.9 | | 0.8 | 8.7 | | 13.8 | | |
| high | 70.4 | | | 13.4 | 11.0 | | 3.5 | 25.0 | | | 15.1 | 3.4 | | 3.0 | 43.2 | | 50.6 | | |
| low | 6.8 | | | 1.3 | 0.0 | | 0.4 | 3.8 | | | 0.5 | 0.0 | | 0.0 | 9.5 | | 0.0 | | |

*Percent of total rock volume; others are percent of framework grains.
**Samples containing exotic(?) pyroclastic grains.
VRF bracket indicates counting categories within volcanic rock fragments.
See Cather (1986) for modal data for diagenetic components.

# RECONSTRUCTION OF OLIGO-MIOCENE VOLCANICLASTIC DISPERSAL PATTERNS IN NORTH-CENTRAL NEW MEXICO USING SANDSTONE PETROFACIES

RAYMOND V. INGERSOLL AND WILLIAM CAVAZZA[1]

*Department of Earth and Space Sciences, University of California, Los Angeles, California 90024-1567*

ABSTRACT: Oligo-Miocene volcaniclastic sedimentation in northern New Mexico occurred before and during initiation of the Rio Grande rift. Reconstruction of dispersal patterns and paleogeography of extensive volcaniclastic aprons, using geochemical, paleocurrent, textural and petrofacies analyses, places important constraints on models for the tectonomagmatic initiation and evolution of the rift.

K-Ar ages and chemical analyses of volcanic clasts from conglomerate provide details concerning three primary volcanic centers: 1) San Juan Mountains (29-27 Ma, high-K andesite and rhyodacite); 2) Latir volcanic field (28-25 Ma, high-K andesite and rhyolite); and 3) the previously unrecognized Servilleta Plaza center (23-22 Ma, latite, high-K andesite and rhyodacite), which may have been a southern extension of the Latir field. These three volcanic centers provided detritus to the following units, respectively: 1) Esquibel Member of Los Pinos Formation, upper Abiquiu Formation, middle Picuris Formation, and Bishops Lodge Member of Tesuque Formation; 2) Cordito Member of Los Pinos Formation and uppermost Abiquiu Formation; and 3) Chama-El Rito Member of Tesuque Formation and upper Picuris Formation. Correlation of these units is based on both conglomerate and sandstone petrofacies.

Detailed point counting of volcanic lithic-grain varieties (vitric, granular, seriate, microlitic and lathwork) is especially useful in discriminating petrofacies. The Esquibel petrofacies has abundant plagioclase, seriate lithics and dense minerals, with low quartz. The Cordito petrofacies has abundant sanidine, quartz and vitric-to-granular lithics, with few dense minerals. The Plaza petrofacies is similar to the Esquibel, but with higher quartz and microlitic lithics, and lower total feldspar and dense minerals. Use of petrofacies simplifies the chaos of stratigraphic nomenclature and allows regional correlations of poorly exposed units.

Paleocurrent and textural data from these volcaniclastic petrosomes define volcaniclastic aprons derived from the above volcanic centers. The coordinated use of clast ages and petrology to define source characteristics, sandstone petrofacies to define petrosomes, and paleocurrent and textural analyses to define dispersal patterns provides a powerful method for reconstructing ancient volcaniclastic systems.

## INTRODUCTION

Cenozoic continental deposits of north-central New Mexico and southern Colorado record three major tectonomagmatic events: 1) amagmatic Laramide orogeny (latest Cretaceous through Eocene); 2) widespread Oligocene intermediate magmatism, concentrated in the San Juan and Ortiz-Cerrillos centers; and 3) latest Oligocene through Recent, fundamentally basaltic (and bimodal) magmatism and formation of the Rio Grange rift (see Ingersoll and others, 1990, for a summary). The transition from Oligocene intermediate magmatism to fundamentally basaltic (and bimodal) magmatism, as well as structural disruption, at 28 to 27 Ma, is probably the best date for rift initiation (Chapin, 1988; Ingersoll and others, 1990). Nonetheless, the detailed stratigraphic record of this latest Oligocene event is difficult to document due to later structural disruption, sedimentation and erosion. One of the best areas in which to document this tectonomagmatic transition is in north-central New Mexico and southern Colorado (Fig. 1).

Application of traditional chronostratigraphic and lithostratigraphic correlation methods is difficult within the study area due to the following characteristics of Oligo-Miocene volcaniclastic strata (see Ingersoll and others, 1990): 1) poorly fossiliferous, in general, although some vertebrate and palynological data exist; 2) marked facies changes from very coarse lahar deposits through alluvial conglomerate and sandstone to lacustrine and eolian facies; 3) broadly similar intermediate to silicic volcaniclastic compositions; and 4) greater contrasts in lithofacies within consanguineous depositional units than among units derived from separate magmatic centers.

In order to reconstruct paleoenvironments and dispersal patterns, a coordinated approach was developed, which utilized 1) major- and minor-element geochemistry and K-Ar dating to characterize coarse clasts, 2) sand/sandstone petrofacies to define consanguineous petrostratigraphic units (petrosomes), and 3) textural and paleocurrent data to define dispersal patterns. These data permit greater stratigraphic resolution, and ultimately, more refined paleoenvironmental and paleotectonic reconstruction (see Ingersoll and others, 1990).

## STRATIGRAPHY

Cenozoic stratigraphic nomenclature of the study area is extraordinarily complex (Fig. 2) due to interfingering of distantly and locally derived nonmarine units of widely differing provenance and lithology (Ingersoll and others, 1990). Rationalization of nomenclature is possible based on improved chronostratigraphy, detailed lithologic mapping, petrostratigraphic studies, and detailed sedimentology (e.g., May, 1979, 1984; Manley, 1981; Ekas and others, 1984; Ingersoll and others, 1987a, 1990).

### *Chronostratigraphy*

Chronostratigraphic control for Oligo-Miocene volcaniclastic units of the study area consists of three types of data: 1) paleontologic (primarily vertebrate and palynologic data), 2) stratigraphic (e.g., interfingering with units of known age and radiometric dates of underlying or overlying units), and 3) radiometric dates of clasts from the units. Ingersoll and others (1990) and Figure 2 summarize the paleonto-

[1]Present Address: Department of Mineralogical Sciences, University of Bologna, Piazza di Porta S. Donato 1, 40126 Bologna, Italy

Sedimentation in Volcanic Settings, SEPM Special Publication No. 45

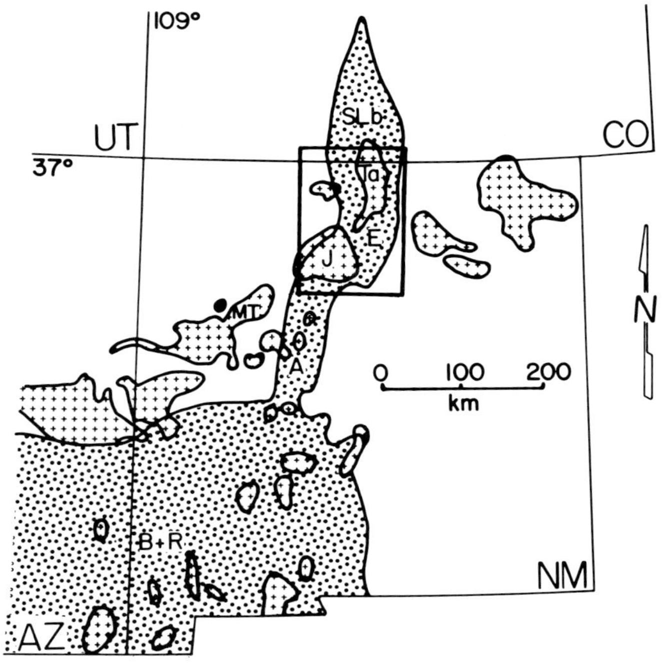


FIG. 1.—Location map of study area. Location of area shown in Figure 8 is indicated by the rectangle. Neogene and modern volcanic fields are indicated by cross patterns. Areas of basins of the Rio Grande rift, and the Basin and Range province are indicated by stipple patterns (uplifts within basins not shown). A = Albuquerque basin; B + R = Basin and Range province; E = Española basin; J = Jemez volcanic field; MT = Mount Taylor volcanic field; SLb = San Luis basin; Ta = Taos Plateau volcanic field. After Lucas and Ingersoll (1981).

logic and stratigraphic constraints; radiometric dates of clasts are discussed herein.

### *Lithostratigraphy*

Lucas and Ingersoll (1981) summarize the history of stratigraphic nomenclature in the study area. The volcaniclastic lithostratigraphic units studied in detail here are (listed from oldest to youngest and north to south; see Fig. 2): Esquibel Member of Los Pinos Formation, upper Abiquiu Formation, middle Picuris Formation and Bishops Lodge Member of Tesuque Formation; Cordito Member of Los Pinos Formation and uppermost Abiquiu Formation; and Chama-El Rito Member of Tesuque Formation, and upper Picuris Formation. Each of these units has been studied using all or some of the methods described below.

### *Petrostratigraphy*

The complex stratigraphic nomenclature summarized above is simplified through the use of sandstone petrofacies (e.g., Mansfield, 1971; Dickinson and Rich, 1972; Stanley, 1976; Ingersoll, 1983), which define three consanguineous petrostratigraphic units of common provenance (Ingersoll and others, 1987a). Each of these three volcaniclastic petrosomes (Esquibel, Cordito and Plaza; see Fig. 2) can be mapped in the field on the basis of conglomerate clast types, sandstone composition and overall lithology (color, grain size, etc.). Definitive petrostratigraphic correlation is accomplished through the use of sandstone petrofacies and statistical methods (discussed below). The resulting information provides many paleoenvironmental and paleotectonic insights, as discussed by Ingersoll and others (1990).

## METHODS

### *Paleocurrents and Textural Analysis*

Paleocurrent and grain-size measurements for the three petrosomes were obtained throughout the field area (Fig. 3). The three petrosomes exhibit general southward fining of grain size and southward paleocurrent indicators (primarily conglomerate imbrications, but also cross-bedding, bedding lineations and channel axes). In detail, however, the three petrosomes contrast, as follows: 1) Esquibel: the most widespread and volumetrically the greatest, with coarsest deposits near Antonito and systematic southward fining; coarse deposits east of Taos may be a separate petrosome (little is known about this isolated exposure). Although the likely source for the Esquibel is the San Juan field of Colorado (see below), paleocurrents (imbrications) near Antonito have a significant southwest component, probably reflecting the greater eastern extent of the San Juan field than at present. 2) Cordito: exposures are restricted to the central part of the study area, and grain size does not vary greatly; paleocurrents are strongly directed southwest, away from the Questa caldera, the likely source. 3) Plaza: the most proximal detritus in the study area occurs in the area of Servilleta Plaza; the mean diameter of the 10 coarsest clasts is 235 cm (Fig. 3). Paleocurrents are directed south and grain size decreases rapidly in the same direction. The previously unrecognized Plaza volcanic center (see Ingersoll and others, 1990) probably lies buried under the southern Taos Plateau northeast of Servilleta Plaza.

### *Conglomerate Ages and Chemistry*

Several volcanic clasts from conglomerate of each petrosome were selected for K-Ar dating, and major- and minor-element chemistry. These data are summarized in Table 1 (see Ingersoll and others, 1990, for details). Some of the samples were collected and analyzed before the three petrosomes were explicitly defined (e.g., Ekas and others, 1984), whereas other samples were collected to test the sandstone-defined petrosomes. The ages and chemistry of the clasts provide further support for the distinctive sources of the three petrosomes.

### *Sandstone Petrofacies*

*Sample collection and preparation.*—

Approximately 200 sandstone samples were collected throughout the study area. Highly weathered and highly cemented (usually calcite) samples were avoided. Some samples were moderately lithified sandstones, whereas others were loose sand. Medium-to-coarse sand/sandstone was collected in order to lessen the effect of grain size on composition (e.g., Ingersoll, 1978, 1983; Ingersoll and others,

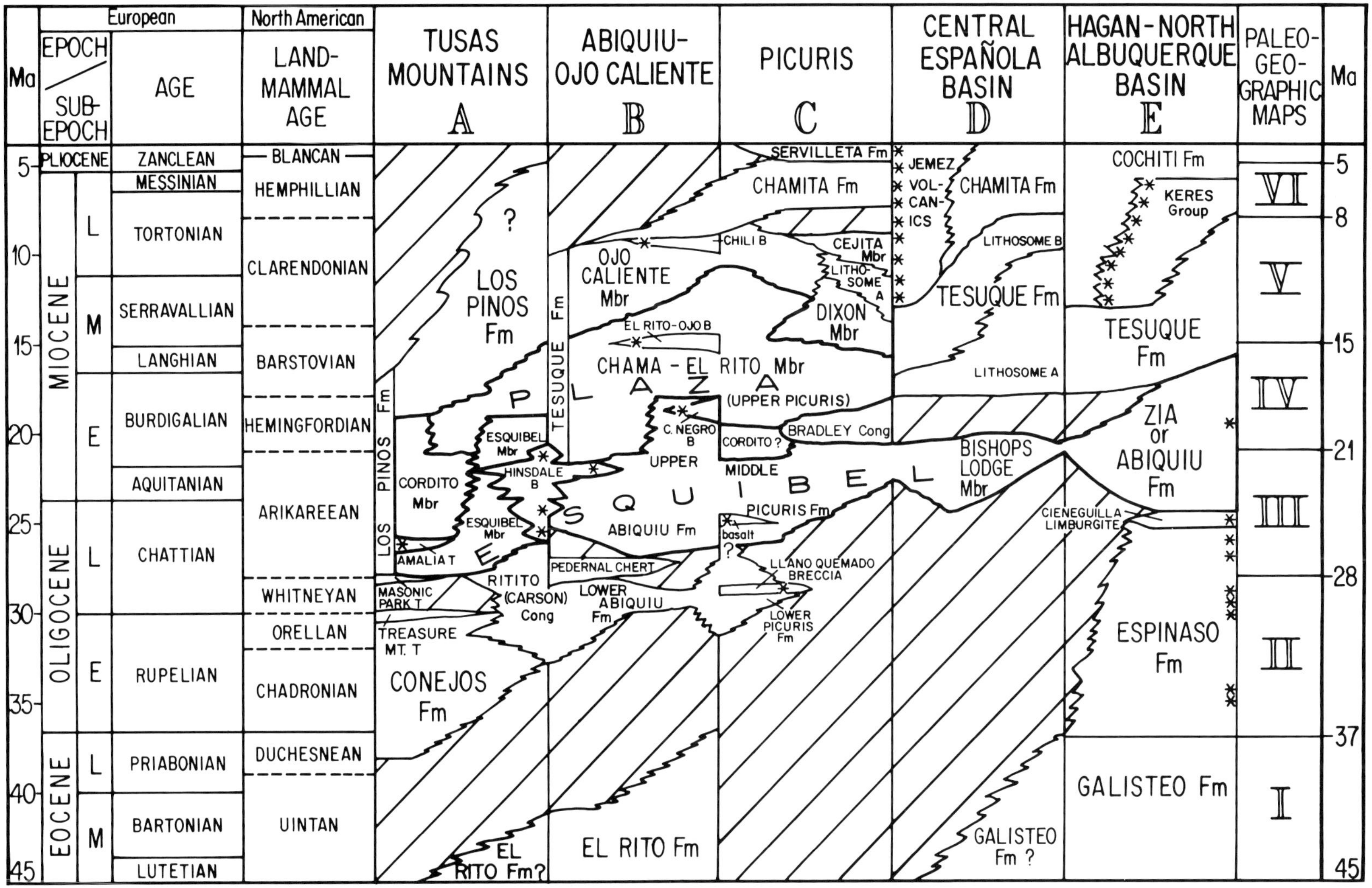


FIG. 2.—Chronostratigraphic chart of middle Eocene through lower Pliocene stratigraphic units within study area. Radiometric ages for Epoch, Subepoch and Age boundaries on left are from Palmer (1983). Correlations between these boundaries and North American Land-Mammal Ages are from Haq and Van Eysinga (1987). T = tuff. B = basalt. Asterisks indicate radiometric ages (conglomerate = maximum age; flow = cooling age). The Esquibel petrosome includes various members of the Los Pinos, Abiquiu, Picuris and Tesuque formations. The Plaza petrosome consists of an unnamed part of the Los Pinos Formation and members of the Tesuque and Picuris formations. The Cordito petrosome corresponds to the Cordito Member of the Los Pinos Formation. (From Ingersoll and others, 1990).

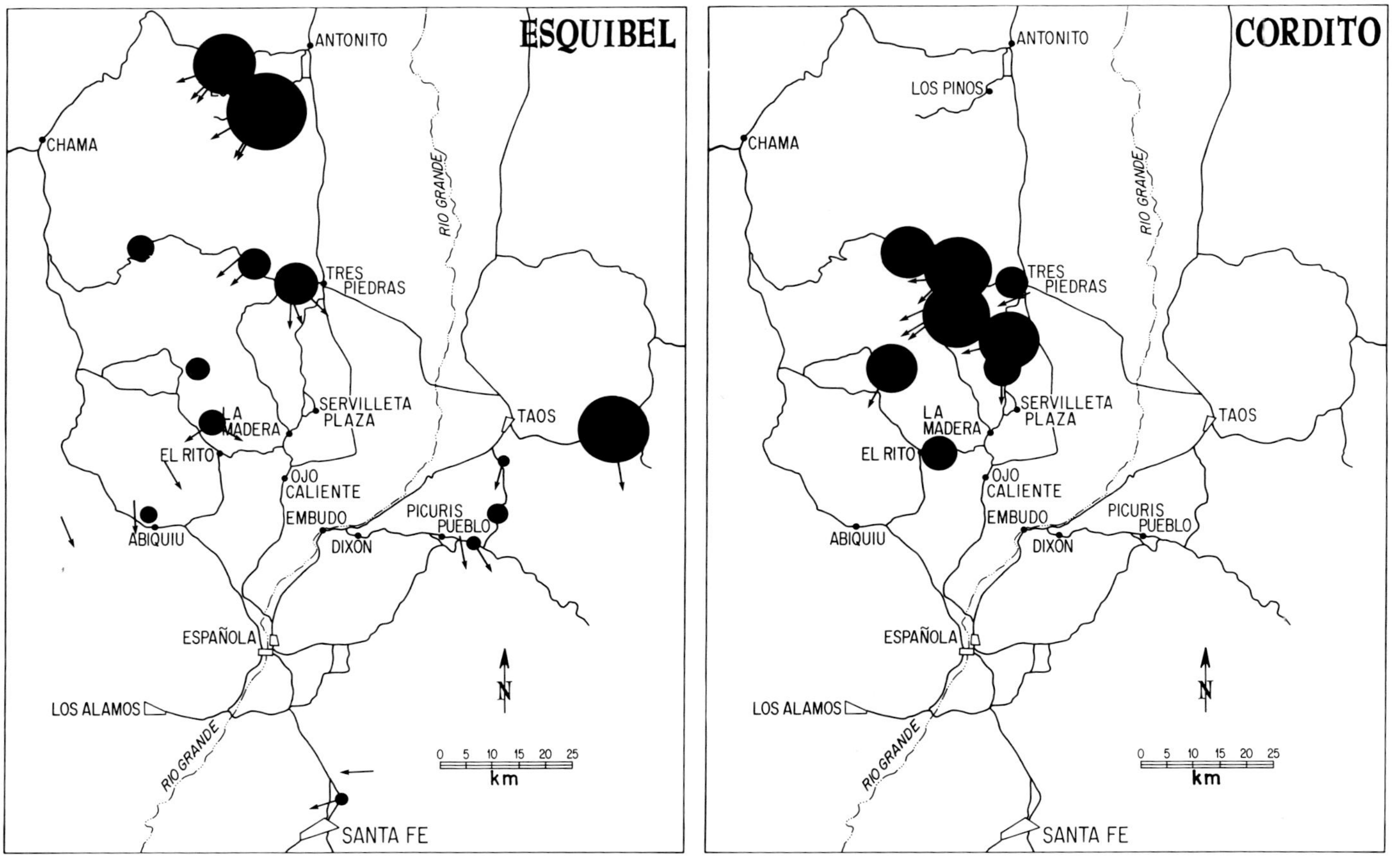


1984). Each sample was impregnated with epoxy, and standard thin sections were prepared. The thin sections were etched in concentrated hydrofluoric acid and stained for potassium feldspar.

We used a staining method modified slightly from previous studies. We found that staining for both potassium and plagioclase feldspar (e.g., Norman, 1974; Ingersoll, 1978, 1983) tended to obscure the identity of sanidine (especially prominent in the Cordito petrofacies). Therefore, we stained only for potassium feldspar; plagioclase feldspar was easily recognized due to the preservation of a residuum of material resulting from the short but intense etching. Volcanic quartz, albite and sanidine were readily distinguished (verified through several careful studies of optic figures of suitably oriented crystals).

*Point counting.*—

Each author counted 200 points on half of each thin section, using the categories listed in Table 2. The sample number and location were unknown at the time of counting. Results of the two authors were compared, discrepancies were discussed, and sections with significant discrepancies were recounted.

Identification of modal categories was based on the criteria of Dickinson (1970), Graham and others (1976), Ingersoll and Suczek (1979), and Ingersoll and others (1987b). In dividing Lv into objectively distinguishable subcategories, we found that many grains were difficult to categorize using Dickinson's four categories (vitric, felsitic, microlitic and lathwork). We noticed that Dickinson's (1970, p. 701) definition of "felsitic" includes two types, granular and seriate, and we found that counting these subtypes greatly facilitated point counting, categorization and discrimination of petrofacies. Therefore, our five categories of volcanic lithic fragments are defined as follows (partly after Dickinson, 1970; see Figs. 4 and 5):

(a) *Vitric* to vitrophyric grains and shards are glass or altered glass, which may be phyllosilicates, zeolites, feldspars, silica minerals, or combinations of these in microcrystalline aggregates; these textures are especially common in silicic volcanic rocks.

(b) *Granular* grains are anhedral microcrystalline mosaics, with uniform very fine grain size, composed mainly of quartz and feldspar, and represent mainly silicic-volcanic rocks, either lavas or tuffs; devitrified-vitric grains are gradational with granular grains.

(c) *Seriate* grains are anisometric mosaics, with wide ranges of grain sizes and shapes, composed mainly of quartz and feldspar, and represent mainly silicic-volcanic rocks, either lavas or tuffs, especially high-K varieties; this category includes all volcanic grains that do not fit any of the other categories.

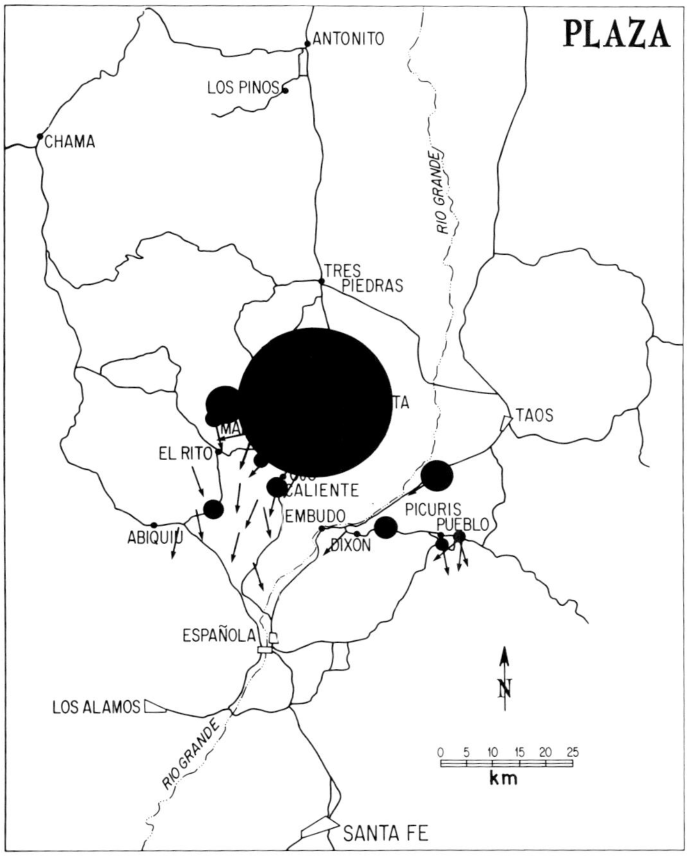


FIG. 3.—Mean paleocurrent directions (arrows) and mean maximum diameters of the 10 coarsest volcanic clasts (black circles) for the 3 volcaniclastic petrosomes. Coarsest mean diameter measured at Servilleta Plaza within the Plaza petrosome is 235 cm. Other black circles are scaled proportionately. (From Ingersoll and others, 1990).

(d) *Microlitic* grains contain subhedral to euhedral feldspar plates and prisms (less than 0.0625 mm in maximum dimension) in pilotaxitic, felted, trachytic, or hyalopilitic patterns of microlites, and represent mainly intermediate types of lavas.

(e) *Lathwork* grains contain plagioclase laths in intergranular and intersertal textures, representing mainly basaltic lavas; this category is gradational with microlitic.

TABLE 1.—PETROLOGY AND AGES OF CONGLOMERATE CLASTS FROM THREE VOLCANICLASTIC PETROSOMES (FROM INGERSOLL AND OTHERS, 1990)

| PETROSOME | PETROLOGY | K-Ar AGE (m.y.) |
|---|---|---|
| Esquibel | High-K Andesite | 29.2±0.7 |
| Esquibel | High-K Andesite | 28.2±0.7 |
| Esquibel | Rhyodacite | 27.8±0.7 |
| Esquibel | Rhyodacite | 27.5±0.7 |
| Esquibel | Rhyodacite | — |
| Esquibel | Rhyodacite | — |
| Esquibel | Rhyodacite | 28.0±0.7 |
| Esquibel | High-K Andesite | 27.2±0.6 |
| Esquibel | High-K Andesite | — |
| Cordito | High-K Andesite | 27.79±0.32 |
| Cordito | Rhyolite | 25.83±0.30 |
| Cordito | Rhyolite | 24.8±0.6 |
| Cordito | Rhyolite | 14.1±0.4* |
| Plaza | Rhyodacite | 22.6±0.5 |
| Plaza | Rhyodacite | 22.4±0.5 |
| Plaza | High-K Andesite | 21.8±0.5 |
| Plaza | Latite | 21.7±0.5 |

*Suspect age (feldspar concentrate; highly silicic [80% $SiO_2$])

The dominant textural type in each detrital grain as a whole determines the category. In the unusual situation where a grain has more than one textural type and none is dominant, the textural type under the cross-hair is counted.

## PETROGRAPHIC RESULTS

### *Detrital Modes*

Forty-eight sandstone/sand samples were point-counted using the categories of Table 2. Recalculated parameters for each petrofacies (as determined by discriminant analysis; see below) are summarized in Table 3 (raw counts are

TABLE 2.—RECALCULATED PETROGRAPHIC VARIABLES USED IN STEPWISE DISCRIMINANT ANALYSIS

| | | | |
|---|---|---|---|
| QFL%Q | where | Q | = total quartz (Qm + Qp) |
| QFL%F | | F | = total feldspar (K + P) |
| QFL%L | | L | = unstable aphanitic lithic grains |
| QmKP%Qm | where | Qm | = monocrystalline quartz |
| QmKP%K | | K | = potassium feldspar |
| QmKP%P | | P | = plagioclase feldspar |
| P/F | | | |
| Framework%M | where | M | = monocrystalline phyllosilicates |
| Framework%D | where | D | = monocrystalline dense minerals (nonphyllosilicate accessories) |
| Lv% Lvvg | where | Lv | = aphanitic volcanic-hypabyssal lithic grains |
| | | Lvvg | = vitric and granular Lv |
| Lv%Lvs | | Lvs | = seriate Lv |
| Lv%Lvml | | Lvml | = microlitic and lathwork Lv |

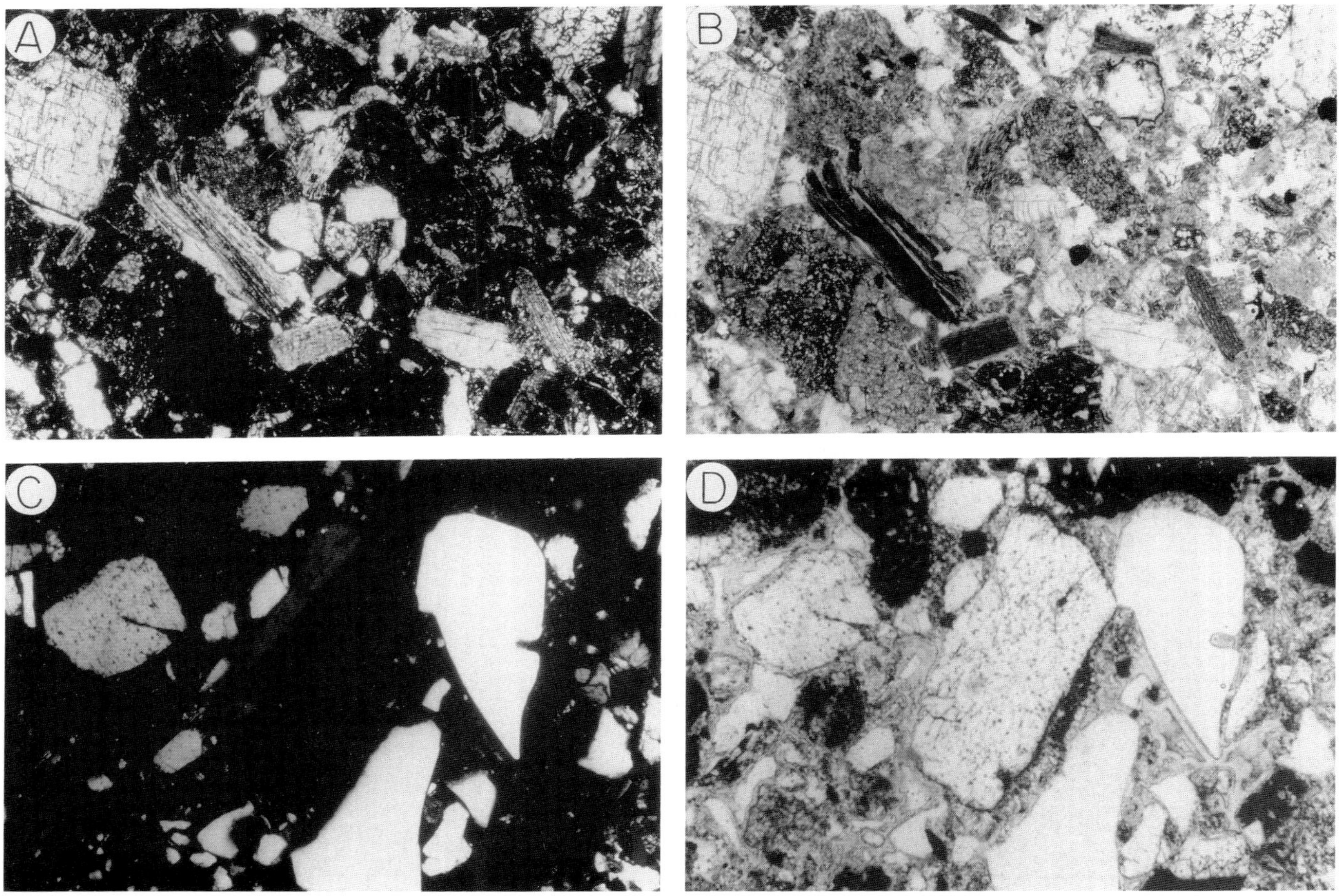


FIG. 4.—Photomicrographs showing typical Esquibel and Cordito petrofacies (Plaza generally resembles Esquibel). Horizontal dimension is 2.5 mm in all frames. (A) Esquibel petrofacies, with dominantly microlitic, seriate and granular volcanic-lithic fragments, plagioclase feldspar, dense minerals and phyllosilicates; crossed polars. (B) Same as (A); plane-polarized light. (C) Cordito petrofacies, with dominantly quartz, potassium feldspar, and vitric, granular and seriate volcanic-lithic fragments; crossed polars. (D) Same as (C); plane-polarized light.

available from the senior author). QFL modes are consistent with the overall transitional- and undissected-magmatic-arc provenance (e.g., Dickinson, 1985; Fig. 6). The Esquibel and Plaza petrofacies have similar monocrystalline components that plot within Dickinson's (1985) "circum-Pacific volcanoplutonic suites" on a QmKP triangle (Fig. 6); in contrast, the Cordito petrofacies is a potassium-feldspar-dominated suite that is unrepresented in other compilations of sandstone detrital modes. To our knowledge, the Cordito petrofacies has the highest QmKP% K value (51) and lowest P/F (0.20) of any unaltered sandstone in the world! More work on high-K, volcaniclastic sand and sandstone is needed in order to evaluate the uniqueness of the Cordito.

Lithic fragments are dominantly volcanic (LmLvLs% Lv is greater than 94 for all samples, with most values 100%); therefore, on plots on lithic types, the three petrofacies cluster in one corner. The subcategories of volcanic-lithic types (defined above) are important discriminators of the petrofacies. Again, the Cordito is unique in having a high proportion of vitric and granular volcanic lithics, in contrast to the more diverse lithic types of the Esquibel and Plaza (Fig. 6).

### *Discriminant Analysis*

Twelve recalculated petrographic variables (Table 3) were used in several runs of the BMDPC stepwise discriminant-analysis program (7M; Dixon, 1983). (Ingersoll, 1990, illustrates the application of discriminant analysis in petrofacies studies.) Each volcaniclastic lithofacies (formations and members of Fig. 2) was used to define the starting groups. Samples were moved from group to group and similar groups were combined until the discrimination stabilized in three groups, with 100 percent "correct" assignment of Cordito and Plaza petrofacies, and 96.2 percent "correct" assignment of Esquibel petrofacies (one Esquibel sample out of 26 was classified as Plaza, but it is clearly Esquibel based on geologic criteria). The four petrographic variables that best discriminate the three petrofacies are, in

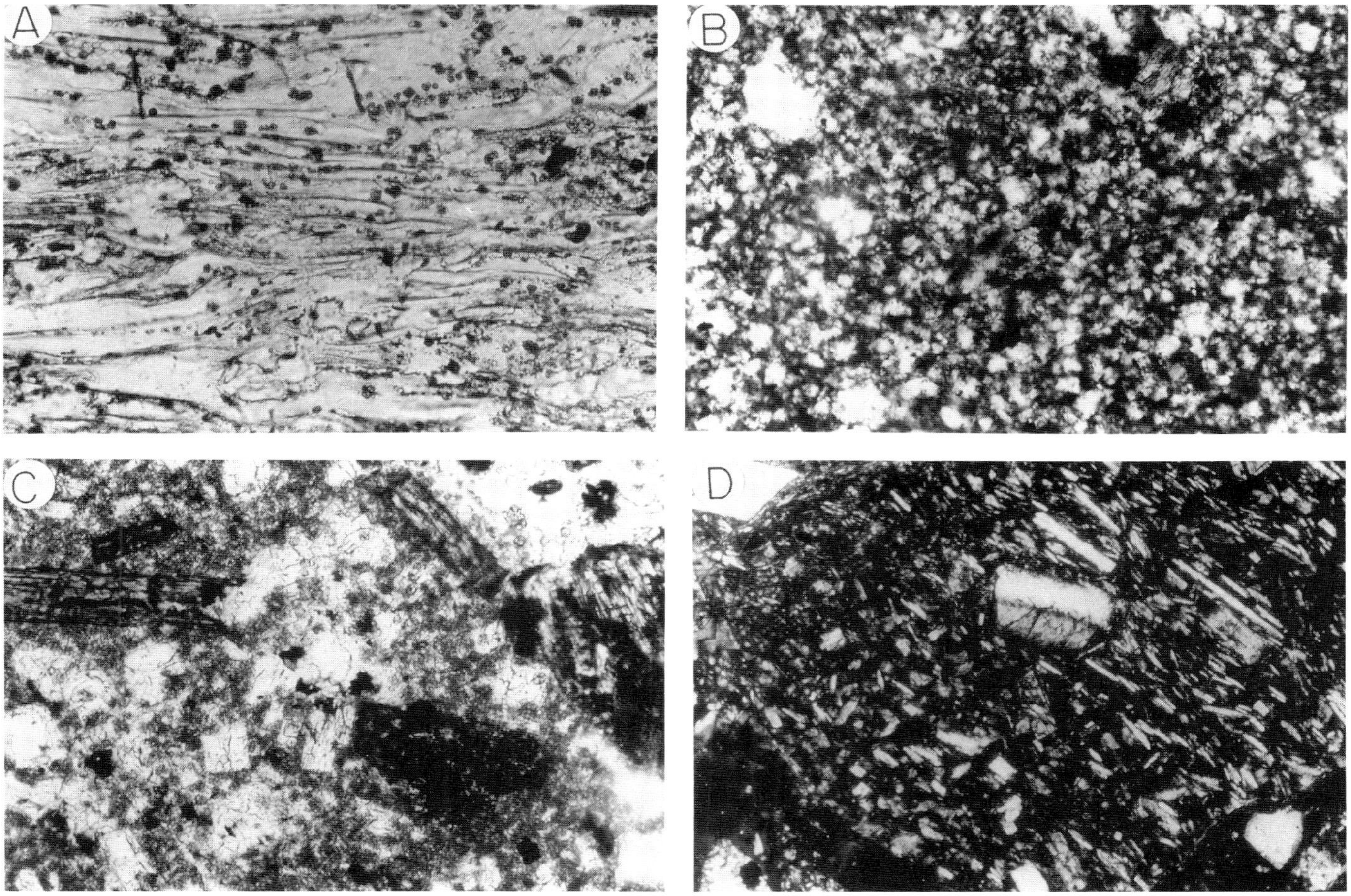


FIG. 5.—Photomicrographs showing textures of volcanic-lithic fragments from the study area. Horizontal dimension is 0.5 mm in all frames. (A) Vitric volcanic-lithic fragment, showing flow and compaction textures. Dark spots are potassium stain; plane-polarized light. (B) Granular volcanic-lithic fragment, made up of fine mosaic of feldspar and quartz; crossed polars. (C) Seriate volcanic-lithic fragment, showing wide range of crystal size and shape; plane-polarized light. (D) Microlitic volcanic-lithic fragment, illustrating the dominance of plagioclase microlites; crossed polars.

TABLE 3.—MEANS AND STANDARD DEVIATIONS FOR RECALCULATED VARIABLES FOR THREE PETROSOMES (FOUR DETERMINING VARIABLES ARE UNDERLINED)

| Petrofacies | Esquibel | Cordito | Plaza |
|---|---|---|---|
| No. of Samples | 26 | 17 | 7 |
| Variable | | | |
| QFL%Q | 6±8 | 18±11 | 11±13 |
| QFL%F | 41±16 | 28±13 | 27±10 |
| QFL%L | 53±15 | 54±22 | 62±7 |
| QmKP%Qm | 13±16 | 37±10 | 26±29 |
| QmKP%K | 7±7 | 51±12 | 9±13 |
| QmKP%P | 80±19 | 12±7 | 65±28 |
| P/F | 0.92±0.09 | 0.20±0.12 | 0.89±0.16 |
| Framework%M | 3±2 | 0±1 | 2±1 |
| Framework%D | 10±4 | 2±1 | 5±2 |
| Lv%Lvvg | 30±14 | 73±9 | 34±24 |
| Lv%Lvs | 45±12 | 24±10 | 25±14 |
| Lv%Lvml | 25±14 | 2±2 | 41±13 |

order of decreasing importance: P/F, Lv%Lvs, Framework%D, and Lv%Lvml. Table 4 and Figure 7 summarize the resulting two canonical variables.

The Esquibel petrofacies is characterized by abundant volcanic plagioclase, low quartz, moderate-to-abundant intermediate-to-silicic volcanic fragments (especially rhyodacitic with seriate textures), and abundant dense minerals (primarily pyroxenes and amphiboles).

The Cordito petrofacies is characterized by abundant sanidine, abundant volcanic quartz, dominantly vitric-silicic-volcanic fragments (primarily vitric and granular), and scarce dense minerals.

The Plaza petrofacies is broadly similar to the Esquibel, but it can be distinguished by lower total feldspar, slightly higher quartz, a higher proportion of intermediate (andesitic) volcanic-lithic fragments (primarily microlitic), and lower content of dense minerals.

## SOURCE AREAS AND DISPERSAL PATTERNS

The integration of chronostratigraphy, lithostratigraphy, petrostratigraphy, paleocurrents, textural analysis, radiometric ages, and clast geochemistry results in the paleogeographic reconstructions shown in Figure 8. The Esquibel petrosome best illustrates the transition from predominantly volcanic rocks near eruptive centers (San Juan Mountains), southward to proximal volcaniclastic facies

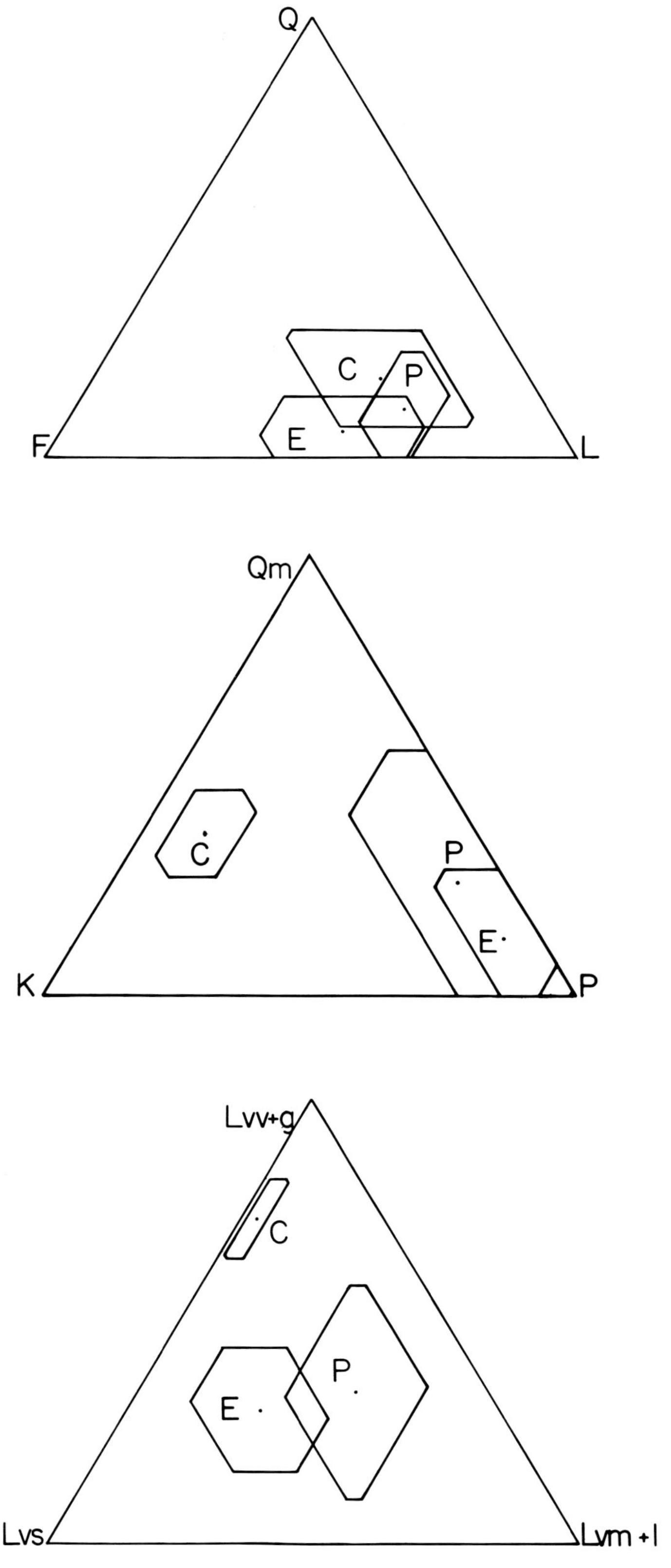


FIG. 6.—Triangular plots showing means and standard deviations for three petrosomes (E = Esquibel; C = Cordito; P = Plaza; see Tables 2 and 3).

TABLE 4.—SUMMARY CHARACTERISTICS OF CANONICAL VARIABLES

| | Canonical Variable 1 | Canonical Variable 2 |
|---|---|---|
| Cumulative proportion of total dispersion | 0.91915 | 1.00000 |
| Canonical correlations | 0.96462 | 0.73535 |
| | Coefficients for canonical variables | |
| Petrographic variable | | |
| P/F | 0.08366 | −0.01406 |
| Lv%Lvs | 0.04784 | 0.04384 |
| Framework%D | 0.01891 | 0.27493 |
| Lv%Lvml | 0.01823 | −0.07472 |

Note: Underlined numbers indicate dominant petrographic variable for each canonical variable.

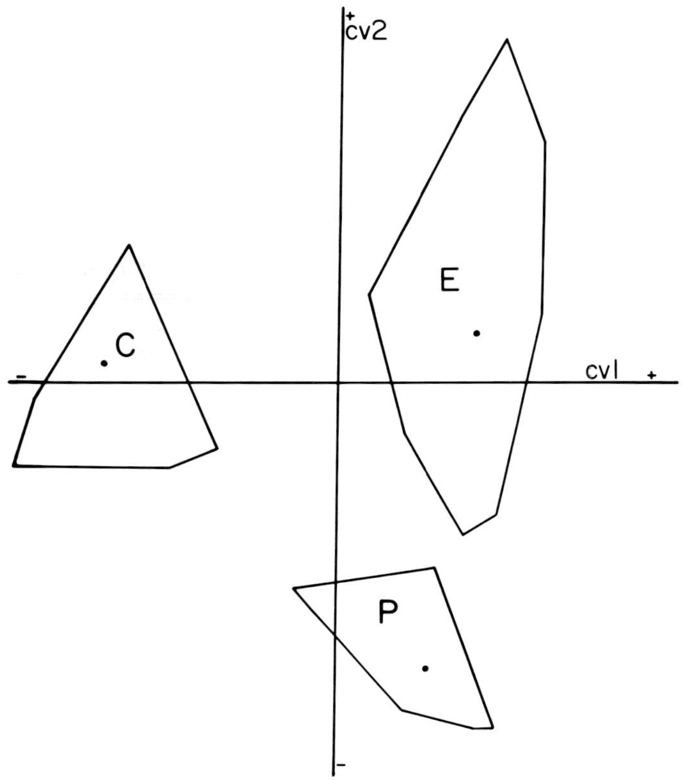


FIG. 7.—Compositions of petrosomes plotted in canonical-variable space (E = Esquibel; C = Cordito; P = Plaza; see Tables 3 and 4; compare to Fig. 6). Canonical variable 1 is horizontal (+ to the right); canonical variable 2 is vertical (+ is up). Each field includes all samples from that group. Black dots are mean values.

(lahar deposits, ignimbrites and ashfall deposits), and farther southward to medial and distal volcaniclastic facies (alluvial-channel and floodplain deposits; e.g., Vessell and Davies, 1981). The Plaza petrosome has similar proximal-to-distal volcaniclastic facies, but the eruptive center is presumed to be buried beneath the Taos Plateau. The Cordito petrosome shows less well-developed proximal-to-distal characteristics, partly due to exposure in a more limited area, and partly due to the highly explosive nature of eruption of the silicic Amalia Tuff from the Questa caldera (see

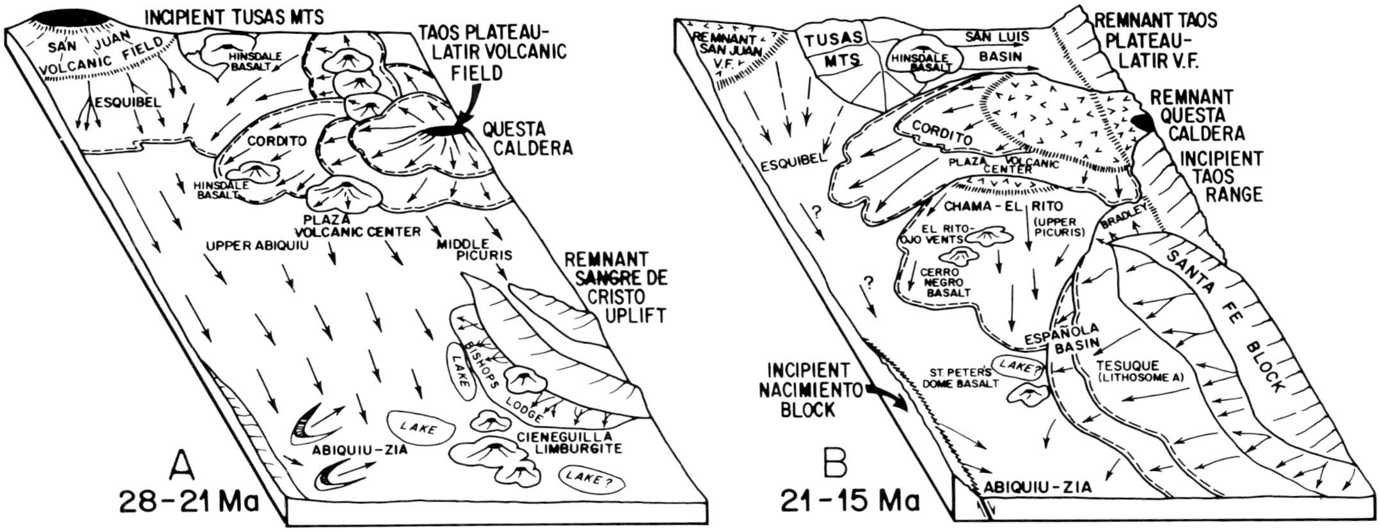


FIG. 8.—Paleogeographic reconstructions of study area (see Fig. 1). See Figure 2 for stratigraphic relations. (A) 28 to 21 Ma. (B) 21 to 15 Ma. (Modified from Ingersoll and others, 1990).

TABLE 5.—SOURCE AREAS FOR THREE PETROSOMES

| Petrosome | Magmatic Center | Associated Volcanic Unit(s) | Age | Reference |
|---|---|---|---|---|
| Esquibel | San Juan Mts. (Platoro, Summitville Calderas) | Treasure Mt. Tuff | 29.1–29.8 m.y. | Steven and Lipman (1976) |
| Esquibel | San Juan Mts. (Mount Hope Caldera) | Masonic Park Tuff | 28.2 m.y. | Steven and Lipman (1976) |
| Cordito | Latir field (Questa Caldera) | Amalia Tuff | 26.5 m.y. | Lipman and others (1986) |
| Plaza | Servilleta Plaza Center | None known | 21.7–22.6 m.y. | Ekas and others (1984) |

Ingersoll and others, 1990). With increasing distance south, the three petrosomes lose their distinctive characteristics due to mixing, both with each other and with locally derived nonvolcaniclastic detritus (transition from first-order to second-order petrofacies; see Ingersoll, 1990). Likely source areas for the three petrosomes are summarized in Table 5 (compare with Table 1).

One aspect of the QmKP modes of the Esquibel and Cordito petrosomes is worthy of special mention. In general, whole-rock $K_2O$ values are similar for rocks of the San Juan and Latir fields (e.g., Questa caldera), but rocks from the latter have higher proportions of alkali-feldspar phenocrysts than do rocks of the former (P.W. Lipman, pers. commun., 1988, 1990). "Silicic Latir rocks tend to be one-feldspar types (with sodic sanidine, or even anorthoclase), while the San Juans have two-feldspar rhyolites, with potassic feldspar accompanied by abundant plagioclase. This probably reflects hotter, drier magmas in the Latirs" (P.W. Lipman, pers. commun., 1990). In other words, the less abundant alkali feldspars from the San Juan field are more potassic than are the more abundant alkali feldspars from the Latir field. These feldspar characteristics, in combination with the more silicic nature of the Questa caldera eruption (i.e., the Amalia Tuff), explain the QmKP modes of the derivative sandstone petrofacies (Esquibel and Cordito, respectively). Mean Esquibel QmKP modes are 13-7-80, whereas mean Cordito QmKP modes are 37-51-12 (Table 3; Fig. 6). The relatively low-K content of Cordito alkali feldspars may explain the difficulty of staining for potassium. This example also illustrates the need for integrated study of whole-rock geochemistry and sandstone/volcanic petrography.

## CONCLUSIONS

The present study illustrates the usefulness of volcaniclastic sand/sandstone petrofacies in the reconstruction of ancient eruptive and dispersal systems. Volcaniclastic petrofacies define consanguineous, regionally correlatable units (petrosomes), thus simplifying the chaos of traditional lithostratigraphic nomenclature based on many local exposures of grossly similar units. The coordinated use of ages and petrology of volcanic clasts to define source characteristics, sandstone petrofacies to define petrosomes, and paleocurrent and textural analyses to define dispersal patterns provides a powerful method for reconstructing ancient volcaniclastic systems. This method should find wide application in other volcaniclastic settings.

## ACKNOWLEDGMENTS

We thank the following for support during the last 10 years: Donors of the Petroleum Research Fund, administered by the American Chemical Society; the Institute of Geophysics and Planetary Physics at Los Alamos National Laboratory; the Committee on Research of the Academic Senate of the Los Angeles Division of the University of California; the New Mexico Bureau of Mines and Mineral Resources; and the Department of Geology at the University of New Mexico.

Also, we thank W. S. Baldridge, G. H. Mack and G. A. Smith for critically reviewing the manuscript.

## REFERENCES

CHAPIN, C. E., 1988, Axial basins of the northern and central Rio Grande rifts, *in* Sloss, L. L., ed., Sedimentary Cover—North American Craton, U.S.: The Geology of North America, Volume D-2: Geological Society of America, Boulder, p. 165–170.

DICKINSON, W. R., 1970, Interpreting detrital modes of graywacke and arkose: Journal of Sedimentary Petrology, v. 40, p. 695–707.

DICKINSON, W. R., 1985, Interpreting provenance relations from detrital modes of sandstones, *in* Zuffa, G. G., ed., Provenance of Arenites: Boston, D. Reidel, p. 333–361.

DICKINSON, W. R., AND RICH, E. I., 1972, Petrologic intervals and petrofacies in the Great Valley sequence, Sacramento Valley, California: Geological Society of America Bulletin, v. 83, p. 3007–3024.

DIXON, W. J., ed., 1983, BMDP statistical software, 1983 printing with additions: Los Angeles, University of California Press, 733 p.

EKAS, L. M., INGERSOLL, R. V., BALDRIDGE, W. S., AND SHAFIQULLAH, M., 1984, The Chama—El Rito Member of the Tesuque Formation, Espanola basin, New Mexico: New Mexico Geological Society Guidebook 35, p. 137–143.

GRAHAM, S. A., INGERSOLL, R. V., AND DICKINSON, W. R., 1976, Common provenance for lithic grains in Carboniferous sandstones from Ouachita Mountains and Black Warrior basin: Journal of Sedimentary Petrology, v. 46, p. 620–632.

HAQ, B. U., AND VAN EYSINGA, F. W. B., 1987, Geological time table: Elsevier Science Publishers, Amsterdam, 1 sheet.

INGERSOLL, R. V., 1978, Petrofacies and petrologic evolution of the Late Cretaceous fore-arc basin, northern and central California: Journal of Geology, v. 86, p. 335–352.

INGERSOLL, R. V., 1983, Petrofacies and provenance of late Mesozoic forearc basin, northern and central California: American Association of Petroleum Geologists Bulletin, v. 67, p. 1125–1142.

INGERSOLL, R. V., 1990, Actualistic sandstone petrofacies: discriminating modern and ancient source rocks: Geology, v. 18, p. 733–736.

INGERSOLL, R. V., BULLARD, T. F., FORD, R. L., GRIMM, J. P., PICKLE, J. D., AND SARES, S. W., 1984, The effect of grain size on detrital modes: A test of the Gazzi-Dickinson point-counting method: Journal of Sedimentary Petrology, v. 54, p. 103–116.

INGERSOLL, R. V., CAVAZZA, W., BALDRIDGE, W. S., AND SHAFIQULLAH, M., 1987a, Oligocene-Miocene volcaniclastic petrofacies and basin evolution in northern New Mexico: Implications for initiation of Rio Grande rift: Geological Society of America Abstracts with Programs, v. 19, p. 712.

INGERSOLL, R. V., CAVAZZA, W., BALDRIDGE, W. S., AND SHAFIQULLAH, M., 1990, Cenozoic sedimentation and paleotectonics of north-central New Mexico: Implications for initiation and evolution of the Rio Grande rift: Geological Society of America Bulletin, v. 102, p. 1280–1296.

INGERSOLL, R. V., CAVAZZA, W., GRAHAM, S. A., AND INDIANA UNIVERSITY GRADUATE FIELD SEMINAR PARTICIPANTS, 1987b, Provenance of impure calclithites in the Laramide foreland of southwestern Montana: Journal of Sedimentary Petrology, v. 57, p. 995–1003.

INGERSOLL, R. V., AND SUCZEK, C. A., 1979, Petrology and provenance of Neogene sand from Nicobar and Bengal fans, DSDP sites 211 and 218: Journal of Sedimentary Petrology, v. 49, p. 1217–1228.

LIPMAN, P. W., MEHNERT, H. H., AND NAESER, C. W., 1986, Evolution of the Latir volcanic field, northern New Mexico, and its relation to the Rio Grande rift, as indicated by potassium-argon and fission track dating: Journal of Geophysical Research, v. 91, p. 6329–6345.

LUCAS, S. G., AND INGERSOLL, R. V., 1981, Cenozoic continental deposits of New Mexico: An overview: Geological Society of America Bulletin, v. 92, Part I, p. 917–932, Part II, p. 1807–1981.

MANLEY, K., 1981, Redefinition and description of the Los Pinos Formation of north-central New Mexico: Geological Society of America Bulletin, v. 92, Part I, p. 984–989.

MANSFIELD, C. F., 1971, Stratigraphic variation in sandstone petrology of the Great Valley Sequence in the southern Coast Ranges west of Coalinga, California: Geological Society of America Abstracts with Programs, v. 3, p. 157.

MAY, S. J., 1979, Neogene stratigraphy and structure of the Ojo Caliente-Rio Chama area, Espanola basin, New Mexico: New Mexico Geological Society Guidebook 30, p. 83–88.

MAY, S. J., 1984, Miocene stratigraphic relations and problems between the Abiquiu, Los Pinos, and Tesuque formations near Ojo Caliente, northern Espanola basin: New Mexico Geological Society Guidebook 35, p. 129–135.

NORMAN, M. B., 1974, Improved techniques for selective staining of feldspar and other minerals using amaranth: U.S. Geological Survey Journal of Research, v. 2, p. 73–79.

PALMER, A. R., 1983, The Decade of North American Geology 1983, geologic time scale: Geology, v. 11, p. 503–504.

STANLEY, K. O., 1976, Sandstone petrofacies in the Cenozoic High Plains sequence, eastern Wyoming and Nebraska: Geological Society of America Bulletin, v. 87, p. 297–309.

STEVEN, T. A., AND LIPMAN, P. W., 1976, Calderas of the San Juan volcanic field, southwestern Colorado: U.S. Geological Survey Professional Paper 958, 35 p.

VESSELL, R. K., AND DAVIES, D. K., 1981, Nonmarine sedimentation in an active fore arc basin, in Ethridge, F. G., and Flores, R. M., eds., Recent and Ancient Nonmarine Depositional Environments: Models for Exploration: Society of Economic Paleontologists and Mineralogists Special Publication 31, p. 31–45.

# PROVENANCE OF SANDS AND SANDSTONES FROM A RIFTED CONTINENTAL ARC, GULF OF CALIFORNIA, MEXICO

KATHLEEN M. MARSAGLIA

*Department of Geological Sciences, The University of Texas at El Paso, El Paso, Texas 79968*

ABSTRACT: Leg 64 of the Deep Sea Drilling Project recovered syn-rift and post-rift sediments deposited within the Gulf of California, along the southern tip of Baja California and within the Guaymas Basin. Although detrital modes of Leg 64 sand and sandstone are consistent with the dissected magmatic-arc tectonic setting of the Gulf of California, compositional differences between southern Baja California sites and Guaymas Basin sites reflect the contrasting plate-tectonic histories of the two regions.

Southern Baja California is a true rifted continental margin, in which Pliocene and Pleistocene sands deposited in offshore basins show a temporal change in the Lower Pliocene from volcaniclastic, rift-related sands to quartzo-feldspathic, drift-related sands. The microcrystalline volcanic component in the rift-related sands is primarily felsitic in texture, and probably derived from syn-rift volcanism.

In contrast, the Guaymas Basin is a transtensional basin floored by quasi-oceanic crust created by successive sill injections into sedimentary-basin fill. Quaternary sands deposited in the Guaymas Basin show no temporal changes in detrital composition, but do indicate multiple sources of sand. Volcaniclastic sands enter the basin from rivers draining mainland Mexico and Baja peninsula streams. The nature of microcrystalline textures in populations of volcanic-lithic fragments helps distinguish these two provenances. The andesitic-source terranes of Baja California provide predominantly microlitic-volcanic-lithic fragments, whereas the Sierra Madre Occidental volcanic province yields a hybrid mixture of volcanic-lithic fragments displaying felsitic and microlitic-microcrystalline textures similar to those observed in modern sands from the Rio Grande rift. Guaymas Basin sands have also been variably affected by hydrothermal alteration, which has significantly modified detrital modes of some sands by silicification of vitric-volcanic fragments and removal of potassium feldspar.

## INTRODUCTION

The Gulf of California is a young transtensional proto-oceanic gulf (Ingersoll, 1988) that has formed as the result of instepping of the East Pacific Rise across western Mexico (Fig. 1). Over the past 10 million years or so, Baja California has been transferred from the North American plate to the Pacific plate (Atwater, 1970; Mammerickx and Klitgord, 1982). In Cenozoic time, prior to 13 to 12 Ma, subduction occurred along all of western Mexico (Fig. 1A), and a continental magmatic arc spanned the length of the Baja California Peninsula. The migration of the Rivera triple junction south along the coast of Baja California sequentially transformed the subduction margin (trench) to a transform margin (Fig. 1B, C). At approximately 5 Ma, when the Rivera triple junction lay in the vicinity of the present Gulf, a major plate reorganization occurred; the plate boundary jumped eastward, transferring Baja California to the Pacific plate and initiating rifting and strike-slip motion in the present Gulf of California (see Stock and Molnar, 1988, and Stock and Hodges, 1989, for detailed discussions of plate configurations). The Gulf of California is unique among modern rift systems in that it has a large translational component; the East Pacific Rise enters the Gulf of California, where it splits into a series of long transforms and isolated, relatively short spreading centers (Fig. 1D). Most well-known rift systems began either as intracontinental rifts (e.g., Red Sea rift, East African rift, Rio Grande rift) or as backarc rifts along oceanic arc-trench systems (e.g., Mariana and Okinawa Troughs). In contrast, the rifting/translation event that shaped the Gulf of California has occurred within a continental-margin arc system.

Leg 64 of the Deep Sea Drilling Project was initiated to study the products of early seafloor spreading in this rift and the resulting young passive continental margins (Curray and others, 1982a). Early rift-related sediments are unusually accessible in the Gulf of California because this region has not yet undergone the extreme subsidence and burial typical of the Mesozoic passive margins that flank the Atlantic Ocean (Curray and Moore, 1982). Sites drilled during Leg 64 recovered syn-rift and post-rift sediments from the southernmost tip of Baja California, a pure rifted margin (Cabo San Lucas region, Fig. 1D), and from the Guaymas Basin, a pull-apart basin in the central portion of the Gulf of California (Fig. 1D).

The Gulf of California provides a unique neotectonic setting, and previous studies relating plate-tectonic setting to sand composition have not characterized detrital modes for sands and sandstones from rifted continental-arc systems (e.g., Dickinson and Suczek, 1979; Dickinson and others, 1983). The main purposes of this paper are to document detrital modes of sands recovered during Leg 64, to determine their probable provenance(s), and define the tectonic influence of rifting and subsequent seafloor spreading on sand composition.

## METHODS

Representative sand and sandstone samples from cores recovered on Leg 64 were collected at the DSDP core repository in La Jolla, California. At least one sample was taken from each 1.5-m section of each sand-bearing core. Unconsolidated samples were dry sieved in order to separate sand-size grains from finer material. Thin sections prepared from sandstone chips and epoxied grain mounts of the sieved sands were etched with concentrated hydrofluoric acid and stained with sodium cobaltinitrite and amaranth in order to differentiate potassium- from calcium-bearing feldspar grains.

The light mineralogy of 78 samples was determined petrographically using a Swift automated-stage point-counting system. The Gazzi-Dickinson point-counting method was used in order to lessen the dependence of composition on grain size (Dickinson, 1970; Ingersoll, 1978; Ingersoll and others, 1984). Five hundred points were counted for each thin section. Definitions of point-counting categories, raw point-count data and recalculated parameters for each sam-

Sedimentation in Volcanic Settings, SEPM Special Publication No. 45

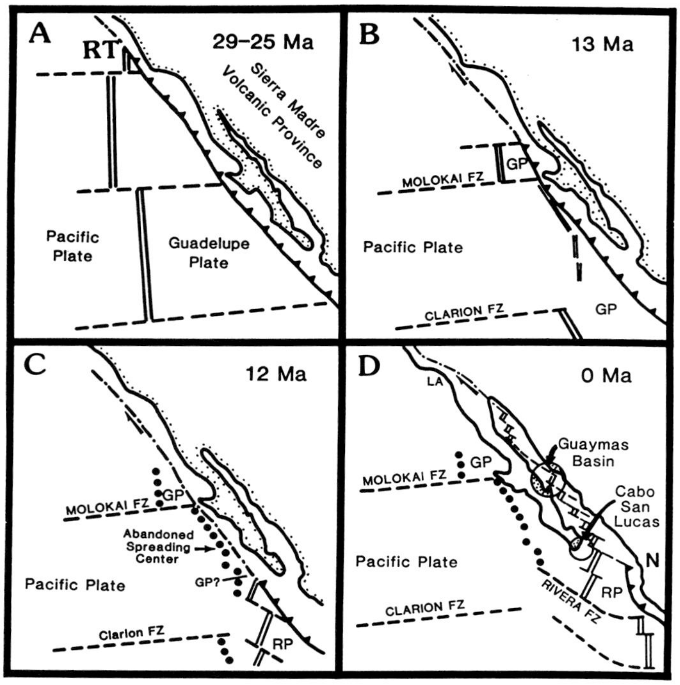


FIG. 1.—Plate-tectonic evolution of Baja California peninsula from 29 to 0 Ma, modified from Saunders and others (1987). Double lines indicate spreading axes, barbed lines represent subduction zones, and arrowed, dashed lines signify strike-slip fault systems. Outlines of Baja California and Gulf of California as they exist today are given as reference (LA = Los Angeles). A) During the late Oligocene (29–25 Ma), the Guadelupe Plate (GP) was subducted along the western Mexico margin arc-trench system, following the formation of the Rivera (trench-ridge-fault?) triple junction (RT). B) and C) The Rivera triple junction migrates southeast, progressively ending subduction and replacing it with strike-slip movement along this plate boundary (or pivoting subduction; Hausback, 1984). At approximately 5 Ma, the Pacific-Guadelupe spreading center was abandoned and the plate boundary jumped inland. D) Today, the East Pacific Rise (EPR) enters the Gulf of California and splits into a series of isolated spreading centers (N = Nayarit).

ple are given in Marsaglia (1989). Group means and standard deviations were tabulated using a Lotus 1-2-3 spreadsheet program (Table 1). Statistical analysis of the data base was performed using the UCLA BMDP stepwise discriminant-analysis program 7m.

## SOUTHERN BAJA CALIFORNIA SITES

*Stratigraphic framework and regional geology.—*

Sites 474, 475 and 476 were drilled on a transect across the boundary between continental and oceanic crust at the southern tip of Baja California (Figs. 2, 3). The following summary comes from site descriptions in the Initial Reports volume for Leg 64 (Curray and others, 1982a), to which the reader is referred for further information. Sites 475 and 476 are located on the lower continental slope, Site 475 in a slope basin and Site 476 on a slope terrace (Fig. 2). Sections penetrated here consist of lower Pliocene to Quaternary diatomaceous hemipelagic sediments, underlain by a thin glauconitic-phosphatic mudstone (not shown in Fig. 2), which caps a unit composed of predominantly metamorphic cobbles. The conglomeratic units appear to be nonmarine, and the glauconite-phosphate facies has been interpreted to record deposition on an isolated offshore bank; both may be of Late Miocene age. The conglomeratic unit was completely penetrated at Site 476, and deeply weathered granite was recovered below (Kelts and others, 1982). At Site 474, approximately 500 m of Upper Pliocene through Quaternary deep-water diatomaceous muds and distal turbidites rest atop oceanic-basaltic crust with a magnetic-anomaly age of 3.2 Ma (Fig. 2). An early Pleistocene thick slump deposit caps this section.

A general geologic map of the tip of Baja California is shown in Figure 3. A major structural feature of this region is the San Jose del Cabo Trough (Fig. 3), a graben that contains Miocene alluvial-fan deposits, Miocene to Pliocene shelf and basinal sediments, and Pleistocene alluvium (McCloy, 1984). Miocene pyroclastic rocks are exposed on the eastern side of the trough (Fig. 3); the surrounding horsts are composed of Cretaceous (potassium/argon dates of 109–70 Ma; Hausback, 1984) diorites, tonalites and granodiorites and Paleozoic metamorphic (metasedimentary and metavolcanic) rocks. Extensive submarine exposures of volcanic and pyroclastic rocks occur to the south of the drill sites (Normark and Curray, 1968), which imply "that considerable volcanism must have occurred through this now-subsided continental crust during and following rifting" (Curray and others, 1982b, p. 1090). Pebbles in the conglomerates at Sites 475 and 476 and isolated pebbles in the lower sections recovered at Site 474 include a suite of igneous, volcanic, sedimentary, metasedimentary and metavolcanic rocks consistent with local subaerially exposed lithologies, and those inferred from dredge-haul results (Curray and others, 1982b).

*Sand detrital modes.—*

Samples of Pliocene and Pleistocene sands from Sites 474, 475 and 476 were examined in this study. They are predominantly very fine to fine grained; most are quartzo-feldspathic, with low percentages of volcanic and metamorphic-lithic detritus, but some have high proportions of volcanic-lithic components (Table 1). Two distinct sand populations are indicated by the clustering of data on a standard QFL diagram (CABO1 and CABO2 in Fig. 4); however, these groups overlap on a QmKP diagram (Fig. 5). The lithic-rich CABO2 sands are predominantly composed of vitric- and felsitic-volcanic detritus and are restricted to the lowest Pliocene sections of Sites 475 and 476, except for an isolated Pleistocene sample at Site 474 (Fig. 2). The CABO1 sands are feldspathic arenites, with low, but somewhat variable, lithic proportions.

*Sand provenance and tectonic implications.—*

Curray and others (1982b) outlined the tectonic history of the southern Baja California region, based partly on the

TABLE 1.—GROUP MEANS AND STANDARD DEVIATIONS

| Group | | QFL% Q | F | L | LmLvLs% Lm | Lv | Ls | Fr% Bio | Fr% D | Fr% M | P/F |
|---|---|---|---|---|---|---|---|---|---|---|---|
| CABO1 | AVG | 29.5 | 64.7 | 5.8 | 48.1 | 51.5 | 0.3 | 13.5 | 2.2 | 6.0 | 0.69 |
| | STD | 4.5 | 4.8 | 2.0 | 18.8 | 18.8 | 1.6 | 14.3 | 1.3 | 7.8 | 0.04 |
| CABO2 | AVG | 19.4 | 36.2 | 44.4 | 14.0 | 86.0 | 0.0 | 3.2 | 1.0 | 1.0 | 0.73 |
| | STD | 3.8 | 5.7 | 9.2 | 3.6 | 3.6 | 0.0 | 4.1 | 0.6 | 0.6 | 0.05 |
| SONORA | AVG | 26.3 | 41.8 | 32.0 | 27.0 | 64.9 | 8.1 | 0.8 | 2.0 | 1.5 | 0.65 |
| | STD | 2.4 | 5.0 | 6.3 | 13.0 | 15.9 | 4.4 | 1.4 | 0.5 | 1.4 | 0.05 |
| BAJA | AVG | 5.3 | 52.1 | 42.6 | 3.8 | 94.9 | 1.4 | 8.3 | 6.8 | 0.1 | 0.96 |
| | STD | 1.9 | 3.7 | 4.2 | 1.5 | 2.1 | 1.2 | 6.6 | 3.1 | 0.3 | 0.01 |
| YAQUI | AVG | 15.4 | 37.5 | 47.1 | 18.7 | 78.2 | 3.1 | 2.4 | 3.4 | 0.7 | 0.80 |
| | STD | 2.3 | 3.5 | 4.7 | 5.2 | 5.7 | 1.6 | 2.9 | 1.7 | 0.6 | 0.04 |
| ALTERED | AVG | 27.6 | 43.6 | 28.8 | 30.2 | 58.2 | 11.6 | 0.2 | 4.6 | 0.0 | 0.91 |
| | STD | 3.7 | 5.3 | 5.2 | 5.6 | 12.0 | 13.2 | 0.4 | 3.9 | 0.0 | 0.09 |

| Group | | Qp/Q | QpLvmLsm% Qp | Lvm | Lsm | QmKP% Qm | K | P | LvfLvmiLvl% Lvf | Lvmi | Lvl | Lv% Lvv |
|---|---|---|---|---|---|---|---|---|---|---|---|---|
| CABO1 | AVG | 0.01 | 2.8 | 50.1 | 47.1 | 31.2 | 21.5 | 47.3 | 94.5 | 5.5 | 0.0 | 39.1 |
| | STD | 0.01 | 3.0 | 18.5 | 18.2 | 4.8 | 2.8 | 5.5 | 16.6 | 16.6 | 0.0 | 20.4 |
| CABO2 | AVG | 0.17 | 5.8 | 81.0 | 13.2 | 30.2 | 18.6 | 51.2 | 83.0 | 16.2 | 0.8 | 61.8 |
| | STD | 0.09 | 2.8 | 2.2 | 4.0 | 3.8 | 3.6 | 5.7 | 9.4 | 9.2 | 1.0 | 16.0 |
| SONORA | AVG | 0.12 | 9.0 | 60.4 | 30.6 | 35.4 | 23.0 | 41.6 | 67.8 | 27.9 | 4.4 | 55.7 |
| | STD | 0.04 | 3.0 | 16.2 | 13.9 | 3.5 | 3.5 | 2.6 | 24.6 | 20.1 | 5.3 | 13.5 |
| BAJA | AVG | 0.18 | 2.6 | 92.6 | 4.8 | 8.0 | 3.4 | 88.6 | 12.3 | 69.4 | 18.4 | 46.3 |
| | STD | 0.10 | 1.4 | 2.5 | 1.8 | 2.2 | 0.7 | 2.4 | 7.8 | 7.5 | 7.3 | 7.5 |
| YAQUI | AVG | 0.27 | 8.1 | 73.0 | 18.9 | 22.9 | 15.6 | 61.5 | 51.3 | 40.3 | 8.4 | 60.7 |
| | STD | 0.07 | 2.1 | 5.7 | 5.0 | 3.0 | 3.0 | 3.7 | 11.0 | 10.0 | 3.6 | 11.4 |
| ALTERED | AVG | 0.28 | 21.8 | 45.6 | 32.6 | 31.6 | 6.0 | 62.4 | 75.4 | 24.0 | 0.6 | 21.0 |
| | STD | 0.07 | 8.1 | 9.2 | 9.9 | 4.6 | 6.0 | 8.8 | 12.4 | 12.7 | 0.8 | 12.1 |

AVG = MEAN STD = STANDARD DEVIATION

Q = Total quartzose grains
F = Total feldspar grains
L = Unstable lithic grains
Lm = Metamorphic-lithic grains
Lv = Volcanic-hypabyssal (and volcaniclastic) lithic grains
Ls = Sedimentary-lithic grains
Qp = Polycrystalline-quartz grains
Lvm = Volcanic-hypabyssal and metavolcanic-lithic grains
Lsm = Sedimentary and metasedimentary-lithic grains
Lst = Sedimentary and polycrystalline-quartz grains
Qm = Monocrystalline quartz
P = Plagioclase feldspar
K = Potassium feldspar
Lvv = Vitric-volcanic lithic
Lvf = Volcanic lithic with felsitic texture
Lvmi = Volcanic lithic with microlitic texture
Lvl = Volcanic lithic with lathwork texture
M = Phyllosilicate minerals
D = Dense minerals
Bio = Bioclastic debris
Fr = Framework

ages and depositional environments of the sedimentary sequences recovered at Sites 474, 475 and 476. By approximately 8 Ma, the Rivera triple junction had reached the mouth of the modern Gulf of California, and fault-bounded crustal blocks were uplifted in the San Jose del Cabo Trough region. The sedimentary, metasedimentary, and volcanic rocks that covered the Cretaceous batholith were uplifted, eroded and reworked into subaerial conglomerates, which were deposited in the vicinity of Sites 475 and 476 during the change from a subduction to a transform tectonic setting. By 5.5 Ma, initial rifting at the tip of Baja California had begun, and pull-apart basins began to form along the Tertiary arc. Shallow-marine marls and glauconitic sediments accumulated in the proto-gulf atop attenuated continental crust.

The oldest sands studied from these DSDP sites were deposited approximately 3.5 Ma and these lower Pliocene sands represent early rift-related sedimentation. The compositional shift from CABO2 to CABO1 sands at Sites 474 and 475 may represent the transition from rift to passive-margin tectonics in this region. The isolated occurrence of CABO2 sand in the Pleistocene section of Site 474 could be explained by Pleistocene reworking of Pliocene volcanic-rich deposits exposed either onshore or within the walls of a submarine canyon.

Although they differ in their lithic compositions, CABO1 and CABO2 sands show similar proportions of quartz, plagioclase and K-feldspar grains, suggesting that they may have had a similar igneous source (Fig. 5). The quartzofeldspathic sands of the CABO1 group are probably derived from granitic basement rocks, but given the volcanic nature of CABO2 sands, two interpretations of Figure 5 are plausible. CABO2 sands may either be the products of dacitic eruptions or, more likely, they might represent mixing of two sand populations: one a monomineralic population derived from granitic rocks and the second, a felsic-vitroclastic population.

The highly volcanic nature of these sands (CABO2) implies either local reworking of syn-rift felsic-volcanic rocks within the now-submerged shelf regions of southern Baja California, or alternatively, erosion and reworking of pre-rift volcanic sequences during rifting. Because pre-rift, arc-related volcanic sequences would have probably produced a more varied suite of volcanic-lithic types dominated by

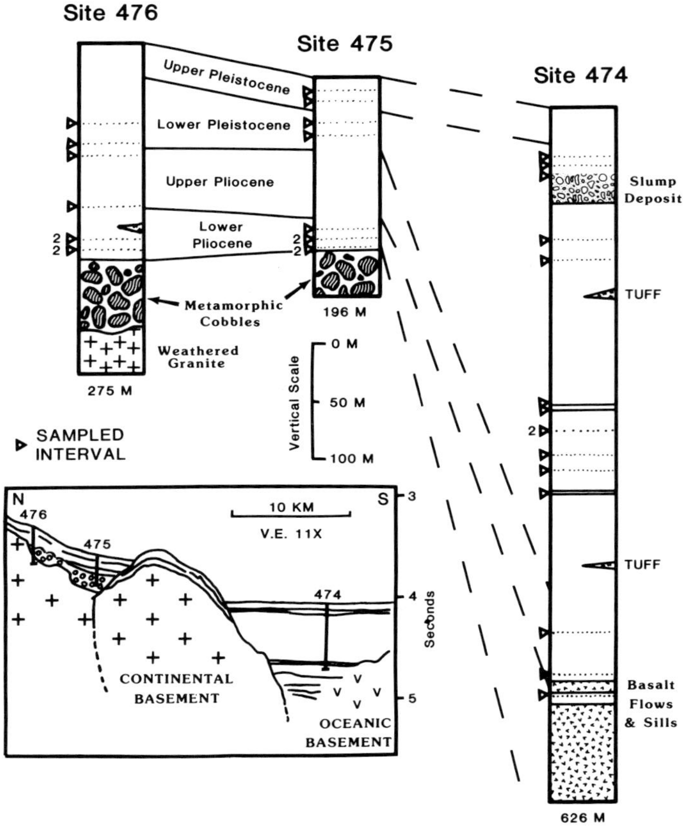


FIG. 2.—Schematic stratigraphic sections for Sites 474, 475, and 476, with sampled sandy intervals indicated on each column. Sampled intervals numbered "2" belong to the CABO2 group, and those without numbers form the CABO1 group (see text for compositional distinctions). Insert: sketch of north-south transect across southern Baja California continental margin constructed from seismic profiles. Sections, line drawing and biostratigraphy modified from Curray and others (1982b).

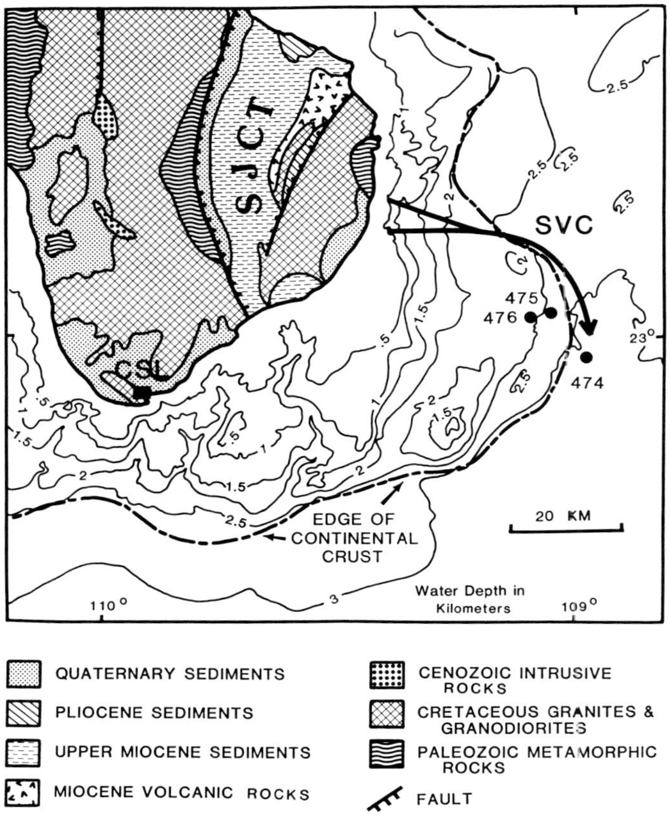


FIG. 3.—Geologic map of southern tip of Baja California Peninsula, modified from Curray and others (1982b): SJCT = San Jose del Cabo Trough; CSL = Cabo San Lucas; SVC = Salado-Vinorama Canyon system.

microlitic textures (discussed later), syn-rift felsic volcanism is the most likely source of the volcaniclastic (vitric and felsitic) components in CABO2 sands. Rhyolite-to-basaltic volcanism coeval with rifting in the southern Gulf of California has been documented in Nayarit, Mexico (Gastil and others, 1979; Curray and others, 1982b). Nayarit is presently southeast of Cabo San Lucas (Fig. 1D, N-Nayarit), but was proximal to DSDP Sites 474, 475 and 476 prior to rifting.

## GUAYMAS BASIN SITES

### *Stratigraphic framework and regional geology.—*

The actively spreading Guaymas Basin (Fig. 6) is presently the site of extremely high rates of sedimentation (up to 2,700 m/Ma), high-heat flow (20 HFU), and hydrothermal activity (van Andel, 1964; Kelts, 1982). The basin is floored by quasi-oceanic crust created by successive intrusions into the sedimentary basin fill (Einsele, 1985, 1986). The sedimentary cover on the basin plain varies from 2 to 600 m and thickens toward Mexico (Curray and others, 1982a).

Sites 479 and 480 are located in a slope basin; Sites 477 and 481 are located along spreading axes (on "zero-age" crust); Site 478 is on the flank of a spreading axis (on crust with a predicted age of 400 ka; Fig. 6). Sites 480 and 479 sampled upper Pliocene to upper Quaternary diatomaceous oozes, laminated mudstones, thin interbedded dolostones, and some sandy intervals; basinal sequences recovered at Sites 477, 478 and 481 include Quaternary diatom oozes and turbidites intruded by basalt, dolerite and gabbro sills (Fig. 7; Curray and others, 1982a). The intruded sediments show variable hydrothermal alteration up to lower-greenschist-facies metamorphism (Kelts, 1982).

The physiography of the Guaymas Basin, especially the distribution of submarine-canyon systems within the basin, suggests two possible continental sources of sand: Baja California and western Mexico. Submarine-channel systems along the western edge of the Guaymas Basin originate to the east of an unnamed peninsula near Concepcion Bay (Fig. 8). The geology of this peninsula (Fig. 8) is dominated by arc-related andesitic- and basaltic-volcanic rocks, the Lower to Middle Miocene Comondú Group (McFall, 1968; Hausback, 1984). Other units that crop out on the peninsula include Cretaceous granodiorite and quartz monzonite with

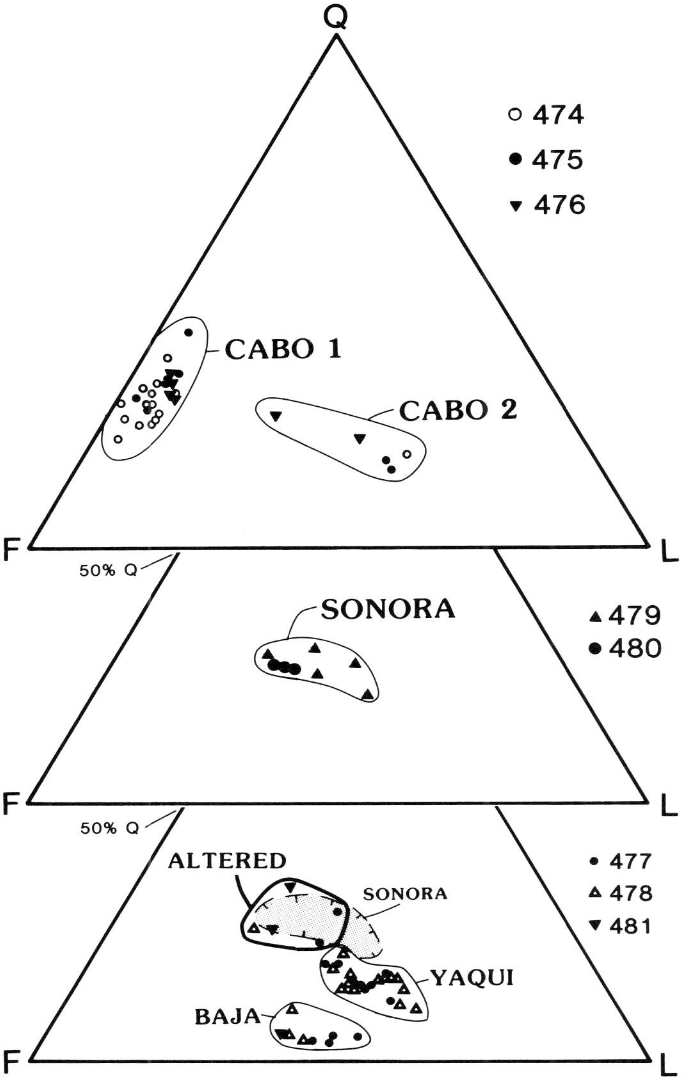


FIG. 4.—QFL plots of Leg 64 sands and sandstones. 1) Top: Sites 474, 475 and 476 off Cabo San Lucas. The two CABO provenance groups (CABO1 and CABO2) are distinct on this plot; 2) Middle: SONORA provenance group found in Sites 479 and 480; and 3) Lower: YAQUI and BAJA provenance group and sands from hydrothermally altered zones, found in basinal Sites 477, 478 and 481. Note overlap of ALTERED sand compositions with SONORA sands on this plot.

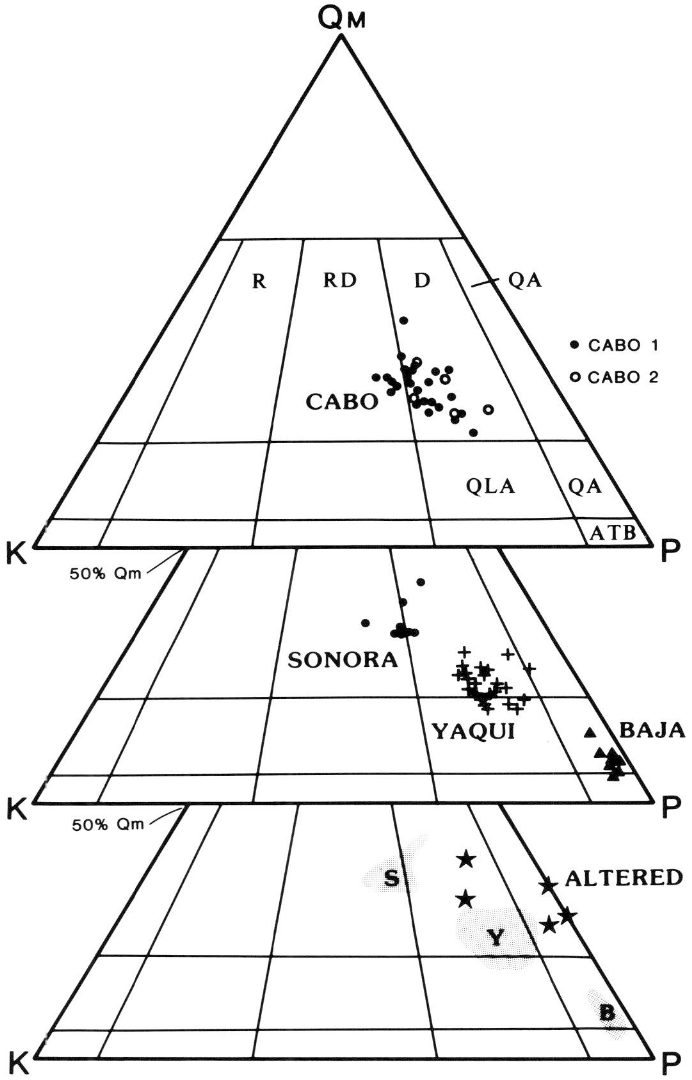


FIG. 5.—QmKP ternary diagrams for Leg 64 sands and sandstones. K and P axes have been transposed from conventional QmPK plots used by many workers in order to emphasize the likely igneous source-rock composition. Subdivisions on diagram are based on the QAPF ternary igneous classification scheme of Hyndman (1972) after Streckeisen (1967), where Q = silica minerals (quartz), A = alkali feldspar, P = plagioclase, F = feldspathoids (omitted due to absence as detritus). Fields are R (rhyolite [granite]), RD (rhyodacite [granite]), D (dacite [granodiorite]), QLA (quartz-latite-andesite [quartz-monzodiorite]), QA (quartz andesite [trondhjemite/tonalite/quartz diorite]), and ATB (andesite/tholeiitic basalt [diorite/norite/gabbro]). Source interpretations based on these fields assume that there has been no mixing of sources, and that transport and weathering effects on sand composition have been minimal; therefore, a tuffaceous submarine sandstone that resulted from a single eruption might yield the most accurate interpretation. Top: CABO1 and CABO2 sands show similar distributions. Middle: SONORA, YAQUI and BAJA groups from the Guaymas Basin plot discretely. Bottom: compositional fields from middle diagram shaded in order to compare with ALTERED sands.

small inclusions and pendants of older schists (McFall, 1968). Even though drainages are poorly developed on the arid peninsula, torrential rains associated with summer tropical cyclones feed flash floods that transport sediment seaward (McFall, 1968).

In contrast to streams draining eastern Baja California, the Yaqui and Sonora Rivers of western Mexico have extensive drainage basins (Fig. 9) and have constructed large delta systems (Fig. 6). In western Mexico rainfall is higher than on Baja California, especially in summer months, and shelf areas, where sediments may accumulate, are more extensive than those to the east on Baja California (Einsele and Kelts, 1982). Because the drainage areas for these rivers are large, the rivers traverse a wider variety of rock types than streams in the Concepcion Bay area, including Tertiary volcanic terranes, and Paleozoic and Mesozoic sedimentary, volcanic and intrusive rocks (Fig. 9). Source

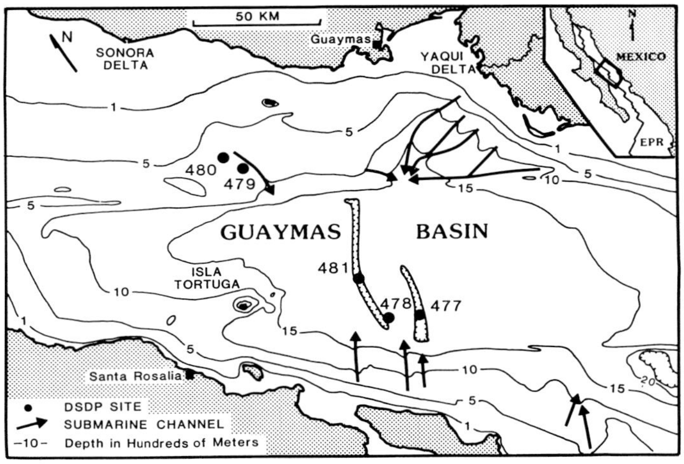


FIG. 6.—Bathymetry of Guaymas Basin, including major submarine channels and DSDP Leg 64 site locations. Two active spreading troughs (hachured = 2,000-m depth) are located in the middle of this basin; subparallel linear margins of basin created by transform fault boundaries to the northeast and southwest. Orientation of figure shown in inset at upper right; EPR = East Pacific Rise. Modified from Lonsdale (1985).

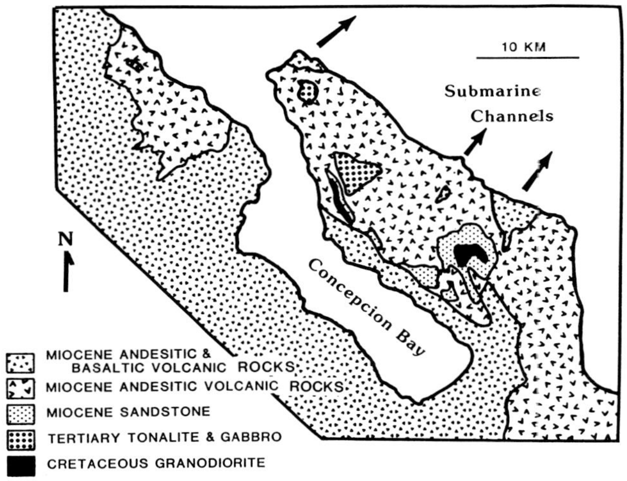


FIG. 8.—Geology of the Concepcion Bay area, Baja California. Submarine channels and unnamed peninsula east of Concepcion Bay are pictured in Figure 6 (bottom center of figure). Geological map modified from McFall (1968).

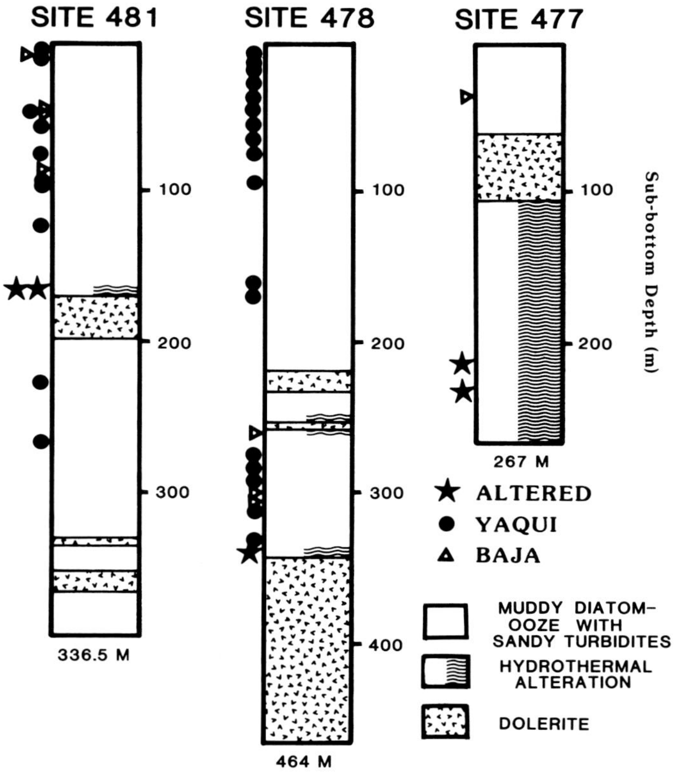


FIG. 7.—Simplified stratigraphic columns for Sites 477, 478 and 481. Subbottom maximum-penetration depths are listed below each column and sampled sand intervals have been coded according to petrographic groups defined in the text. For site locations, see Figure 6. Modified from Gieskes and others (1982).

regions for the Yaqui River extend into the Sierra Madre Occidental; outcrops in this drainage basin are mainly Tertiary volcanic rocks and assorted pre-Tertiary sequences. In contrast, the smaller Sonoran drainage basin is dominated by outcrops of Tertiary intrusive rocks (Fig. 9).

Other sources of volcanic sand include Isla Tortuga and submarine eruptions. Isla Tortuga is an active tholeiitic shield volcano formed on a fracture zone within the Guaymas Basin (Fig. 6). It is plausible that pyroclastic debris produced by phreatomagmatic eruptions of this volcano (Batiza, 1978) could have accumulated in the adjacent Guaymas Basin. Einsele and Kelts (1982) have proposed that subaqueous volcanic extrusions could have produced some of the volcanic sand found at Sites 477, 478, and 481.

Overall, however, a Yaqui Delta source for the majority of the coarse sediment recovered at Guaymas sites was favored by DSDP shipboard scientists (Curray and others, 1982a), even though the majority of the pro-delta sediments are presently trapped behind a structural high (Einsele and Kelts, 1982). This bathymetric high is interpreted by Lonsdale (1985) to mark the trace of a hinge fault between crustal blocks rather than a transform-related ridge. This high may have effectively dammed sediment from the Yaqui Delta throughout the Quaternary time interval recorded by DSDP cores in the Guaymas Basin. It is plausible, however, that eustatic sea-level changes overcame these tectonic effects, and that during Pleistocene lowstands in sea level the Yaqui River extended to the shelf margin and supplied sediments to a large submarine fan in the Guaymas Basin (Curray and others, 1982a).

*Sand detrital modes.*—

All the Guaymas Basin samples examined in this study are Quaternary in age, dominantly fine- to very fine-grained sand, and contain variable monomineralic and polymineralic components (Marsaglia, 1989). Four sand populations are indicated by clustering of compositional data on various

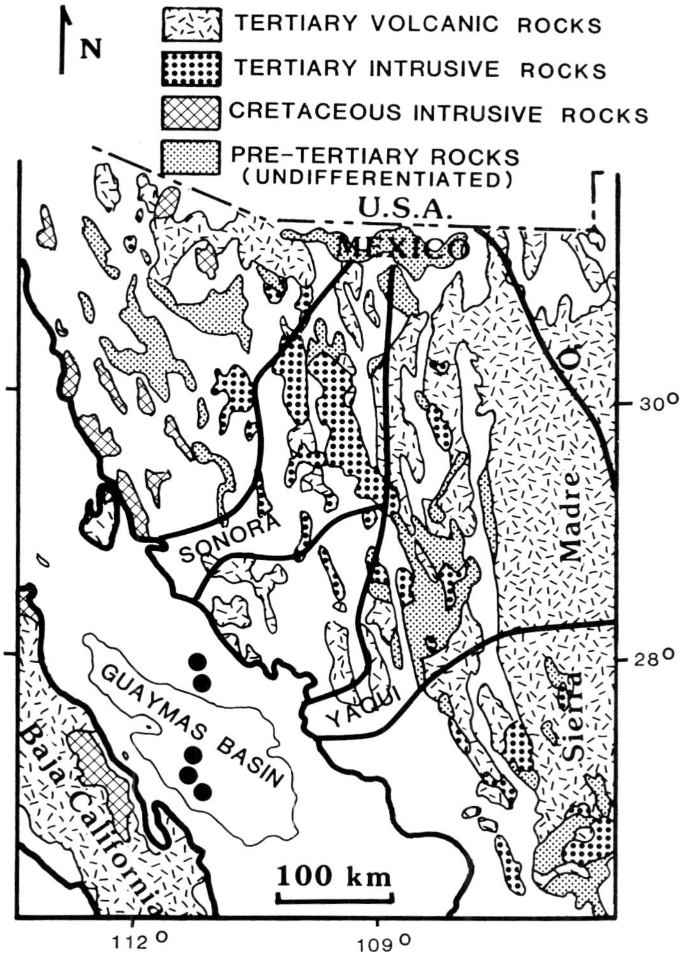


FIG. 9.—Generalized pre-Quaternary geological map of western Mexico in the region surrounding the Guaymas Basin. Geology greatly simplified from Geologic Map of North America (North American Geologic Map Committee, 1965). Pre-Tertiary rock category in legend excludes Cretaceous intrusive rocks, which are listed separately. Drainage basins for the Yaqui and Sonora River systems (heavy lines) approximated from river and stream systems shown on same source. Large dots approximate locations of DSDP sites.

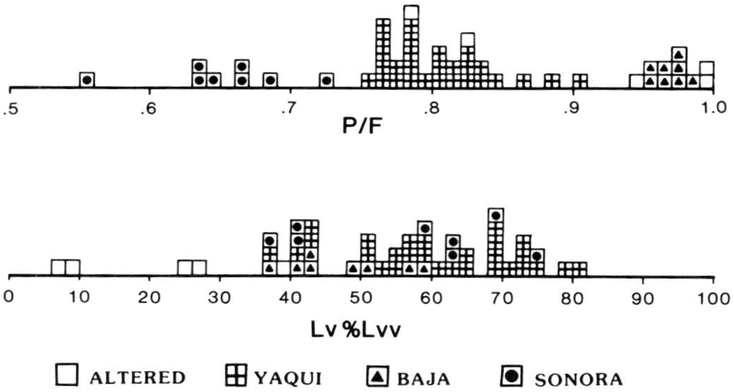


FIG. 10.—Histograms of P/F and Lv%Lvv for sands from Guaymas Basin sites. Top: Histogram of plagioclase-to-total-feldspar ratios plotted according to provenance group. SONORA, BAJA and YAQUI sands have discrete compositions, whereas hydrothermally altered sands (ALTERED) overlap YAQUI and BAJA ranges; two are devoid of potassium-feldspar grains. Bottom: Histogram of total volcanic-lithic-percent vitric-lithic fragments (Lv%Lvv), plotted by provenance group. Note considerable overlap among groups except for altered sands, which have distinctly lower vitric-volcanic percentages.

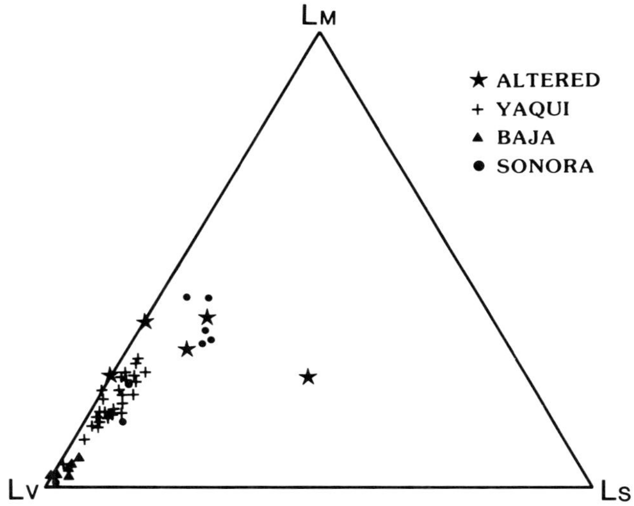


FIG. 11.—LmLvLs triangular plot of Guaymas Basin sands. Data plotted by provenance group. BAJA sands cluster at the volcanic pole, YAQUI sands contain a moderate metamorphic component, and SONORA sands are variable in composition, with the majority more metamorphic than the YAQUI group. Hydrothermally altered sands (ALTERED) are variable in composition and apparent Lm and Ls enrichments could be due to mistaken identifications of altered matrix in unconsolidated samples as lithic fragments.

ternary plots (Figs. 4, 5, 11, 12) and histograms (Fig. 10). Sands from Sites 479 and 480 constitute one provenance group, SONORA (13 samples), whereas sands from basinal sites (477, 478, 479) can be subdivided into three groups (Fig. 7): BAJA (8 samples), YAQUI (28 samples), and ALTERED (5 samples). SONORA sands are moderately quartzose (Fig. 4) and have variable lithic populations (Figs. 10–12), but can be compositionally distinguished from basinal sands by their high potassium-feldspar content (Fig. 5) and consequent low P/F ratio (Fig. 10). BAJA sands are compositionally distinct in that they are composed of roughly equal amounts of plagioclase-feldspar grains and volcanic-lithic fragments (Figs. 4, 5, 11, 12). YAQUI sands are intermediate in composition between the BAJA and SONORA groups in terms of QFL (Fig. 4) and lithic (Figs. 11, 12) parameters, and they are clearly distinguished by QmKP parameters (Fig. 5) and P/F ratios (Fig. 10). Although compositions of sands within the ALTERED and SONORA groups overlap on a QFL triangular diagram (Fig. 4, bottom), there are monomineralic and lithic compositional differences between these groups (Figs. 5, 10–12).

*Sand provenance and alteration.*—

If a single, large river system such as the Sonora or Yaqui was the *only* source of sand supplied to the Guaymas Basin, one might expect little variance in sand composi-

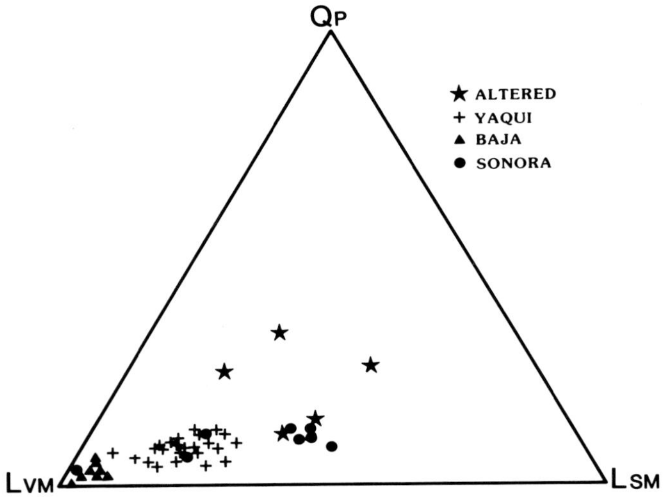


FIG. 12.—QpLvmLsm triangular plot of Guaymas Basin sands. Data plotted by provenance group. Hydrothermally altered sands (ALTERED) show enrichment of polycrystalline quartz (Qp) relative to other lithic types in some of the altered sands. BAJA sands cluster at the volcanic end of the diagram, YAQUI sands contain a moderate metamorphic and minor polycrystalline components, and SONORA sands are variable in composition, with the majority more metamorphic than the YAQUI group.

tions throughout the basin. Sand detrital modes presented here indicate multiple sand provenances (Table 1; Figs. 4, 5, 10–12).

The two sites (479, 480) drilled on the northeastern slope of the basin are proximal to a submarine channel downslope from the Sonora Delta (Fig. 6). The source of sands recovered at these sites (SONORA group) is likely to have been the Sonora River (Lonsdale, 1985). The high potassium-feldspar and mica contents of this group are consistent with the abundant plutonic outcrops in the Sonora River drainage basin; however, volcanic-rich end members of this group may be derived from volcanic terranes exposed south of the Sonora River drainage basin and north of the Yaqui River drainage basin (Fig. 9).

Sites on the basin floor (477, 478, and 481) may have received sand from western Mexico and/or from Baja California. The dominance of plagioclase feldspar over potassium feldspar and quartz (Fig. 5) and the microcrystalline volcanic-lithic proportions of sands in the BAJA group imply intermediate to mafic sources (discussed later). Although their volcanic component is consistent with a Baja California, an Isla Tortuga or a submarine-eruption provenance, Baja California is considered the most likely source of this compositional group because: 1) the volcanic detritus in these sands is altered and slightly rounded, implying subaerial exposure and stream transport; 2) these sands contain trace amounts of metamorphic and sedimentary lithics and moderate amounts of potassium feldspar and quartz (Marsaglia, 1989); and 3) a well-developed submarine-canyon system exists off Baja California in this region (Fig. 6).

Sands of the YAQUI group are found in cores from Sites 478 and 481 (Fig. 7). The intermediate composition of these sands is consistent with a Yaqui River source because the drainage basin of the Yaqui River includes fewer plutonic outcrops than that of the Sonora River (Fig. 9) and is lithologically more diverse than the smaller Baja California drainage systems (Figs. 8, 9).

Guaymas Basin sediments have experienced two modes of alteration (Gieskes and others, 1982; Kelts, 1982): some have undergone short-lived contact metamorphism around sills (e.g., Sites 478 and 481; Fig. 7), whereas others have been subjected to long-term hydrothermal alteration below thermally insulating sills, resulting in greenschist metamorphism (e.g., Site 477; Fig. 7). Site 477 exhibits the more advanced stages of metamorphism (chlorite-quartz-albite-sphene-epidote assemblage) accompanied by the removal of potassium from the solid phases (Gieskes and others, 1982; Kelts, 1982). Some unconsolidated sands which showed minor alteration were included in this study (ALTERED group).

Determining pre-alteration composition for the ALTERED sand group is not straightforward. An original BAJA-like composition is least probable due to the extreme differences between both their monomineralic and lithic proportions. Sands along sill-contact alteration zones of Sites 478 and 481 (Fig. 7) contain variable, somewhat depleted amounts of K-feldspar in comparison to SONORA samples, whereas samples from extensively altered zones in the lower section of Site 477 (Fig. 7) are devoid of potassium feldspar (Figs. 5, 10). ALTERED sands contain less vitric-volcanic-lithic detritus (Fig. 10) and high polycrystalline quartz with respect to other lithic types (Fig. 12). This implies that volcanic-vitric grains within ALTERED sands may have been replaced by polycrystalline quartz. Perhaps fortuitously, the only BAJA sample adjacent to a sill (Fig. 7, Site 475) also has a slightly higher QFL%Q content (Fig. 4, bottom) than the other BAJA samples. Unconsolidated hydrothermally altered sands also show an apparent increase in metamorphic- and sedimentary-lithic fragments (Fig. 11). Because these are disaggregated samples, petrographic criteria (e.g., localization of alteration, fabric transecting various lithic fragments) needed to define true metamorphic and sedimentary lithics are not preserved. These fragments may be "pseudolithic" fragments, therefore, in that they represent altered-matrix material.

A stepwise discriminant analysis (UCLA BMDP 7 m) of the SONORA and YAQUI groups was used to classify ALTERED samples (treated as unknowns) into either the SONORA or YAQUI groups. The main factors discriminating between the YAQUI and SONORA groups (as determined in the stepwise discriminant analysis) are, in decreasing order of importance: 1) QmPK%P, 2) Lv%Lvv, 3) QFL%Q, 4) Fr%M (framework-percent mica), and 5) Fr%D (framework-percent dense minerals). The discriminant program assigned four out of the five ALTERED sands to the YAQUI group, implying that a Yaqui River source is more probable for these altered sediments. Alternatively, if these sands belonged to the SONORA group, one might expect to find SONORA-like sands in the unaltered parts of the sections as well. Because SONORA-type sands are not found elsewhere at these sites, an original YAQUI sand composition is likely.

## DISCUSSION

*Detrital modes of Leg 64 sands related to the plate-tectonic setting of the Gulf of California.—*

Characteristic sandstone compositional parameters have been linked to certain plate-tectonic settings (Dickinson and Suczek, 1979; Dickinson and others, 1983). Traditionally, transform- and rift-related sediments have not been included in these analyses, in part because these tectonic settings result in sands of variable composition (R. V. Ingersoll, pers. commun.). Modern sands deposited in hybrid or transitional tectonic settings have also not been evaluated adequately (Mack, 1984).

During the last 10 million years or so, the active margin of western Mexico has been slivered and translated northward (Atwater, 1970; Mammerickx and Klitgord, 1982). Baja California is composed of Mesozoic and Tertiary arc-related rocks, and is similar to other suspect terranes presently coupled to the North American continent (Howell and others, 1985; Moore, 1985; Umhoefer, 1987). Ancient analogs of Guaymas Basin sediments may be present in deformed suture zones formed by amalgamation of ancient terranes. It is therefore of interest to see if the Guaymas Basin sands show compositional signatures unique to their hybrid arc/rift/transform origin.

On a QFL triangular diagram (Fig. 13), Leg 64 sands plot in the basement-uplift to dissected-to-transitional-magmatic-arc source fields, as defined by Dickinson and others (1983). Guaymas Basin sand compositions (BAJA, SONORA, YAQUI) fall in the dissected- and transitional-arc fields, and, indeed, the Guaymas Basin is a rifted and dissected arc. Interbedded ALTERED sands are compositionally displaced from their YAQUI source; the alteration has shifted their mean composition from the transitional-arc field to the dissected-arc field. Such a shift in composition can produce mistaken tectonic interpretations if the nature of the alteration is unrecognized (Fig. 13). Rift-related CABO2 sands from southern Baja California are similar to arc-derived sands in that they plot in arc fields on a QFL diagram (Fig. 13) and are volcaniclastic (Fig. 10), but unlike arc-derived sands (Marsaglia, 1989), the CABO2 sands are anomalously rich in felsitic-volcanic-lithic fragments (Fig. 14) and K-feldspar (Fig. 5). The CABO1 group from southern Baja California plots in the basement-uplift region (Fig. 13) and represents detritus shed from arc plutons in a completely dissected-arc terrain (see discussion of basement uplift provenance in Dickinson, 1985). The shift from CABO2 to CABO1 sand compositions upsection (arrow, Fig. 13) may be characteristic of rift-related marginal sequences.

*Comparison to the Sea of Japan.—*

One possible ancient analog to the Gulf of California is the Japan Sea, where a Cretaceous continental-arc system was split during an Oligo-Miocene rifting/translation event (Smith, 1982; Ishiwada and others, 1984; Lallemand and Jolivet, 1986). In contrast to the Gulf of California, the basins in the backarc region of Japan are predominantly extensional, with a smaller translational component (Lallemand and Jolivet, 1986). The Yamato Basin is an extensional basin that contains up to 100 m of sediment, half turbidite deposits and half pelagic sediments; the acoustic basement of the Yamato Basin does not have the characteristics of basalt and may be a volcaniclastic sequence (Ka-

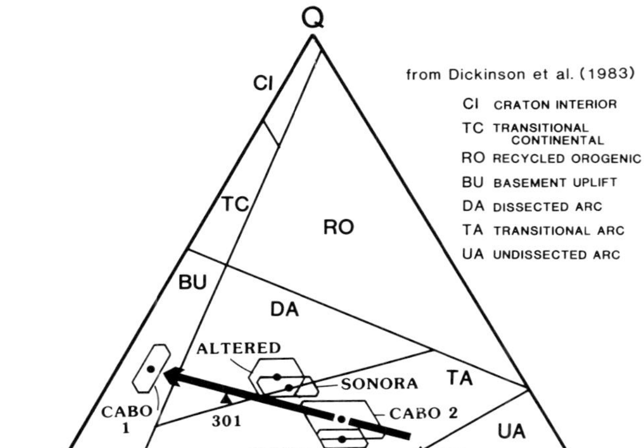


FIG. 13.—QFL triangular plot, polygons represent the fields of variation (standard deviations) about mean values (circles) for Leg 64 sand compositional groups. The provenance fields of Dickinson and others (1983) are outlined. These fields are based on mean sand compositions predominantly from ancient sandstone sequences (Dickinson and Suczek, 1979). In arc-related fields (UA, TA, DA), degree of arc dissection in source region for sands increases from right to left across the diagram. Arrow shows temporal change from syn-rift volcaniclastic sands (CABO2) to post-rift quartzo-feldspathic sands (CABO1). Triangles are site means for Sites 299 and 301 from the Japan Sea (Packer and Ingersoll, 1986). See text for discussion.

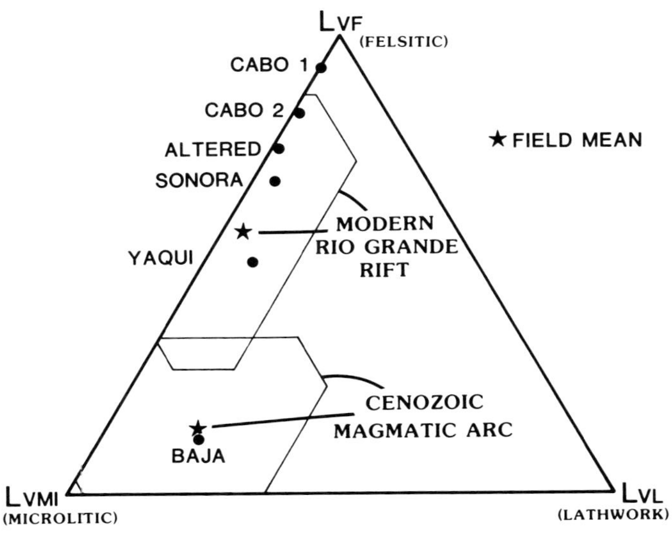


FIG. 14.—Triangular plot of microcrystalline volcanic-lithic types. Circles are Leg 64 group means; also shown are means (stars) and fields of variation (polygons) for Modern Rio Grande rift sands (R. V. Ingersoll, pers. commun.) and Cenozoic magmatic-arc sands (Marsaglia, 1989). See text for discussion.

rig and others, 1975). Other sediment-filled Miocene extensional-backarc basins along the western continental shelf of the Japan Islands have been deformed and uplifted by Pliocene to Recent east-west compression (Ishiwada and others, 1984); uplifted Miocene volcaniclastic sequences (Green Tuff) are altered to zeolites, chlorite and epidote (Sumi, 1977). These sedimentary sequences and their apparent hydrothermal alteration are somewhat analogous to the sediments and alteration seen in the Guaymas Basin.

Packer and Ingersoll (1986) presented detrital modes of sands deposited in the Japan Basin (Site 301) and in the Yamato Basin (Site 299; Fig. 13). These basins are separated by the Yamato Ridge and show contrasting sand compositions. The Japan Basin is near mainland Asia where remnants of Cretaceous arc basement are exposed, and thus sands deposited in this basin are more quartzo-feldspathic than those at Site 299. Site 301 sands are intermediate in composition between CABO1 (completely dissected arc) and SONORA (partially dissected arc) sands (Fig. 13). Mean QFL%Q's for Site 301 and the CABO1 and SONORA groups are roughly equivalent, only the lithic-to-feldspar ratios vary (Fig. 13). Both the Yamato basin (Site 299) and the Guaymas basin (BAJA group) receive sediments derived from a partially dissected magmatic-arc terrain. The central Japan arc has been uplifted and more deeply eroded than the Tertiary magmatic arc in central Baja California (Marsaglia, 1989), and thus the sands recovered at Site 299 are more quartzose than BAJA sands (Fig. 13). Although the Japan Sea and Gulf of California are comparable extensional backarc basins, they differ in that subduction sequentially ceased along Baja California (Fig. 1), whereas subduction and associated arc magmatism continued during and after the opening of the Japan Sea. Nonetheless, Gulf of California and Japan Sea sediments exhibit a similar range of sand detrital modes (Fig. 13).

*Characteristics of microcrystalline volcanic-lithic fragments.—*

Volcanic-lithic parameters for oceanic-arc, continental-arc, transtensional and rift settings should show distinct differences because these tectonic settings are characterized by different volcanic suites: ocean floors and oceanic-backarc basins—basalts; intra-oceanic arcs—basalts and basaltic andesites; continental-margin arcs—andesites and rhyolites; and rifts—rhyolites and basalts (bimodal; Martin and Piwinskii, 1972; Fisher and Schmincke, 1984; Cas and Wright, 1987). Although volcanic glass occurs within all compositional suites, lavas and pyroclastic deposits of varying composition yield sand-size microcrystalline volcanic detritus dominated by petrographically distinctive microscopic textures: basaltic—lathwork; andesitic—microlitic; and silicic—felsitic (Dickinson, 1970). Thus, the proportions of felsitic versus microlitic versus lathwork textures should be indicative of the volcanic suites exposed in source regions and, therefore, of the tectonic settings of basins. Microcrystalline volcanic-lithic fragments associated with intra-oceanic arcs are generally dominated by lathwork and microlitic textures, continental-margin arcs by primarily microlitic textures, and rift settings by felsitic and lathwork textures. Preferential destruction of one volcanic component by weathering and/or transport can also affect ultimate grain populations. Ingersoll (1984) has demonstrated that subaerial weathering of basalt in the Rio Grande rift resulted in very little sand-size sediment, suggesting that the sedimentological imprint of basalts in subaerial source regions (rift grabens) may be small and easily overwhelmed by silicic or intermediate components.

The Cenozoic volcanic history of western Mexico is complex. There have been at least three major periods of volcanism along the western coast of Mexico during the Tertiary (Gastil and others, 1979; Curray and others, 1982b). The first period of volcanism (≥35 Ma) produced a bimodal suite of basalt and high-silica ashflow tuffs from Nevada to Sinaloa, the second (22–12.5 Ma) produced typical continental-margin-arc volcanic rocks (hornblende andesite associated with smaller amounts of basalt and rhyolite) along the present axis of the Gulf of California, and the third (12.5–2.5 Ma) was associated with extension and produced basalt in central and southern Baja California (Comondú volcanics), and a rhyolite-to-basaltic suite in Nayarit (Mexico). The diversity of volcanic suites along Baja California suggests that a mixture of volcanic-lithic types would be delivered to basins adjacent to these various terranes.

The relative importance of volcanic-lithic sources, including syn-rift extensional and arc-volcanic rocks, can be evaluated by comparing the relative percentages of felsitic, microlitic and lathwork volcanic-lithic fragments in Leg 64 sands (Fig. 14). The population of volcanic-lithic fragments in sands from the southern tip of Baja California (CABO1 and CABO2) is dominated by felsitic textures (Fig. 14, Table 1); only the CABO2 sands have any measurable lathwork component, and in general, they have higher percentages of microlitic textures than the CABO1 sands. Given the unlikelihood of basaltic-sand production, as discussed earlier, this suite dominated by felsitic textures could be characteristic of a continental-rift setting.

Guaymas Basin sands (SONORA, YAQUI, BAJA and ALTERED) show a spectrum of microcrystalline volcanic compositions (Fig. 14). The microcrystalline volcanic lithics in BAJA sands exhibit predominantly microlitic and lathwork textures, and these sands are compositionally similar to Cenozoic oceanic and continental magmatic-arc sands (Fig. 14). The remaining sands derived from mainland Mexico (YAQUI, SONORA, and ALTERED groups) and off southern Baja California (CABO1 and CABO2) are dominantly felsitic, and more similar to the modern sands from the Rio Grande Rift than to Cenozoic arc-related sands (Fig. 14). Various volcanic suites are exposed in the Rio Grande rift region of north-central New Mexico, and these include both older arc-related volcanic sequences and younger rift-related ones (R. V. Ingersoll, pers. commun.). The mean volcanic compositions for YAQUI and SONORA sands fall on either side of the Rio Grande mean, whereas the CABO sands are dominated by felsitic detritus. I propose that the CABO sands represent a pure rift setting, the BAJA sands a pure continental-margin-arc setting, and the SONORA and YAQUI sands a hybrid mixture of arc and rift components. Although Tertiary bimodal volcanic rocks dominate the upper reaches of the Sonora and Yaqui drainage basins in the Sierra Madre Occidental, andesitic volcanics dominate the lower reaches, near the Gulf of California (Mina, 1957); thus, the volcanic-lithic composition of these sands reflects the hybrid tectonic nature of the source regions.

## CONCLUSIONS

Along the southern tip of Baja California, temporal changes from syn-rift volcaniclastic sands (CABO2) to post-rift quartzo-feldspathic sands (CABO1) are observed. The volcanic-rich CABO2 sands probably represent felsic syn-rift volcanic activity in the region rather than reworked pre-rift volcanic sections. In the Guaymas Basin, no temporal compositional changes were observed in the sedimentary sections sampled, but only upper Quaternary sections were analyzed. Instead, the Guaymas Basin sequence is characterized by a mixed-sand provenance and local sediment alteration. Sediment dispersal patterns in the Gulf of California are complex, and sands enter the Guaymas Basin from both the Pacific and North American plates along the transform margins of the pull-apart basin. This contrasts with southernmost Baja California, where the rifted margin of Baja California is (at present) the only source of sediment.

The spectrum of sand compositions in the Gulf of California (Fig. 13) indicates an arc system undergoing various degrees of dissection. Sand compositions in the Gulf of California span the basement-uplift to transitional-dissected-arc fields, similar to the range found in the Japan Sea, a possible ancient analog. These compositional differences can be related to the amount of arc-basement exposure in source regions, except for rift-related CABO2 sands. Microcrystalline volcanic-lithic populations within Leg 64 sands suggests that CABO2 sands are derived from a rift-related volcanic provenance, BAJA sands from an arc-related volcanic provenance, and SONORA and YAQUI sands from a hybrid arc/rift volcanic provenance. Hydrothermal alteration of sands within the Guaymas Basin has shifted their composition on a QFL triangular plot, yielding a false signal of increased arc dissection in the source region and increased sand maturity (more quartz, less feldspar and volcanic-lithic fragments).

## ACKNOWLEDGMENTS

This manuscript represents a portion of the author's dissertation, submitted to the University of California at Los Angeles. This research was partially supported by ARCO and the donors of the Petroleum Research Fund, administered by the American Chemical Society. Core samples were provided by the Ocean Drilling Program. Editorial comments from Richard Fisher, Raymond V. Ingersoll, Neil Lundberg, and James White greatly improved the text.

## REFERENCES

ATWATER, T., 1970, Implications of plate tectonics for the Cenozoic tectonic evolution of western North America: Geological Society of America Bulletin, v. 81, p. 3513–3536.

BATIZA, R., 1978, Geology, petrology, and geochemistry of Isla Tortuga, a recently formed tholeiitic island in the Gulf of California: Geological Society of America Bulletin, v. 89, p. 1309–1324.

CAS, R. A., AND WRIGHT, J. V., 1987, Volcanic Successions: Modern and Ancient, a Geological Approach to Processes, Products and Successions: Allen and Unwin, London, 528 p.

CURRAY, J. R., AND MOORE, D. G., 1982, Introduction to the Baja California passive-margin-transect symposium: Initial Reports of the Deep Sea Drilling Project, v. 64, pt. 2, p. 1067–1069.

CURRAY, J. R., AND SHIPBOARD SCIENTIFIC PARTY, 1982a, Initial Reports of the Deep Sea Drilling Project, v. 64, pt. 1, 507 p.

CURRAY, J. R., MOORE, D. G., KELTS, K. R., AND EINSELE, G., 1982b, Tectonics and geological history of the passive continental margin at the tip of Baja California: Initial Reports of the Deep Sea Drilling Project, v. 64, pt. 2, p. 1089–1116.

DICKINSON, W. R., 1970, Interpreting detrital modes of graywacke and arkose: Journal of Sedimentary Petrology, v. 40, p. 695–707.

DICKINSON, W. R., 1985, Interpreting provenance relations from detrital modes of sandstone, *in* ZUFFA, G. G., ed., Provenance of Arenites: NATO ASI Series, Series C, Mathematical and Physical Sciences, v. 148, p. 333–361.

DICKINSON, W. R., BEARD, L. S., BRAKENRIDGE, G. R., ERJAVEC, J. L., FERGUSON, R. C., INMAN, K. F., KNEPP, R. A., LINDBERG, F. A., AND RYBERG, P. T., 1983, Provenance of North America Phanerozoic sandstones in relation to tectonic setting: Geological Society of America Bulletin, v. 94, p. 222–235.

DICKINSON, W. R., AND SUCZEK, C. A., 1979, Plate tectonics and sandstone compositions: American Association of Petroleum Geologists Bulletin, v. 63, p. 2164–2182.

EINSELE, G., 1985, Basaltic sill-sediment complexes in young spreading centers: Genesis and significance: Geology, v. 13, p. 249–252.

EINSELE, G., 1986, Interaction between sediments and basalt injections in young Gulf of California-type spreading centers: Geologische Rundschau, v. 75, p. 197–208.

EINSELE, G., AND KELTS, K. R., 1982, Pliocene and Quaternary mud turbidites in the Gulf of California: Sedimentology, mass physical properties, and significance: Initial Reports of the Deep Sea Drilling Project, v. 64, pt. 2, p. 511–528.

FISHER, R. V., AND SCHMINCKE, H.-U., 1984, Pyroclastic Rocks: Springer-Verlag, New York, 472 p.

GASTIL, G., KRUMMENACHER, D., and MINCH, J., 1979, The record of Cenozoic volcanism around the Gulf of California: Geological Society of America Bulletin, v. 90, p. 839–857.

GIESKES, J. M., KASTNER, M., EINSELE, G., KELTS, K. R., AND NIEMITZ, J. W., 1982, Hydrothermal activity in the Guaymas Basin, Gulf of California: A synthesis: Initial Reports of the Deep Sea Drilling Project, v. 64, pt. 2, p. 1159–1167.

HAUSBACK, B. P., 1984, Cenozoic volcanic and tectonic evolution of Baja California Sur, Mexico, *in* Frizzell, JR., V. A., ed., Geology of the Baja California Peninsula: Society of Economic Paleontologists and Mineralogists, Pacific Section, v. 39, p. 219–236.

HOWELL, D. G., JONES, D. L., AND SCHERMER, E. R., 1985, Tectonostratigraphic terranes of the circum-Pacific region: *in* Howell, D. G., ed. Tectonostratigraphic Terranes of the Circum-Pacific Region: Circum-Pacific Council for Energy and Mineral Resources Earth Science Series, n. 1, p. 3–30.

HYNDMAN, D. W., 1972, Petrology of Igneous Rocks: McGraw-Hill, New York, 533 p.

INGERSOLL, R. V., 1978, Petrofacies and petrologic evolution of the Late Cretaceous fore-arc basin, northern and central California: Journal of Geology, v. 86, p. 335–352.

INGERSOLL, R. V., 1984, Felsic sand/sandstone within mafic volcanic provinces: Rift sand/sandstone shows no basaltic provenance: Geological Society of America Abstracts with Programs, v. 16, p. 548.

INGERSOLL, R. V., 1988, Tectonics of sedimentary basins: Geological Society of America Bulletin, v. 100, p. 1704–1719.

INGERSOLL, R. V., BULLARD, T. F., FORD, R. L., GRIMM, J. P., PICKLE, J. D., AND SARES, S. W., 1984, The effect of grain size on detrital modes: A test of the Gazzi-Dickinson point-counting method: Journal of Sedimentary Petrology, v. 54, p. 103–116.

ISHIWADA, Y., HONZA, E., AND TAMAKI, K., 1984, Sedimentary basins of the Japan Sea: Proceedings of the 27th International Geological Congress, v. 23, p. 43–65.

KARIG, D. E., AND SHIPBOARD SCIENTIFIC PARTY, 1975, Site 301: Initial Reports of the Deep Sea Drilling Project, v. 31, p. 409–437.

KELTS, K. R., 1982, Petrology of hydrothermally metamorphosed sediments at Deep Sea Drilling Site 477, southern Guaymas Basin rift, Gulf of California: Initial Reports of the Deep Sea Drilling Project, v. 64, pt. 2, p. 1123–1136.

KELTS, K. R., CURRAY, J. R., AND MOORE, D. G., 1982, Introduction and explanatory notes: Initial Reports of the Deep Sea Drilling Project, v. 64, pt. 1, p. 5–33.

LALLEMAND, S., AND JOLIVET, L., 1986, Japan Sea: A pull-apart basin?: Earth and Planetary Science Letters, v. 76, p. 375–389.

LONSDALE, P., 1985, A transform continental margin rich in hydrocarbons, Gulf of California: American Association of Petroleum Geologists Bulletin, v. 69, p. 1160–1180.

MACK, G. H., 1984, Exceptions to the relationship between plate tectonics and sandstone composition: Journal of Sedimentary Petrology, v. 54, p. 212–220.

MAMMERICKX, J., AND KLITGORD, K. D., 1982, Northern East Pacific Rise: Evolution from 25 Ma B.P. to the present: Journal of Geophysical Research, v. 87, p. 6751–6759.

MARSAGLIA, K. M., 1989, Petrography, provenance, and diagenesis of arc-related sands and sandstones recovered by the Deep Sea Drilling Project on circum-Pacific, Mediterranean and Caribbean Legs: University of California, Los Angeles, Unpubl. Ph.D. Dissertation, 394 p.

MARTIN, R. F., AND PIWINSKII, A. J., 1972, Magmatism and tectonic settings: Journal of Geophysical Research, v. 77, p. 4966–4975.

MCCLOY, C., 1984, Stratigraphy and depositional history of the San Jose del Cabo Trough, Baja California Sur, Mexico, *in* Frizzell, JR., V. A., ed., Geology of the Baja California Peninsula: Society of Economic Paleontologists and Mineralogists, Pacific Section, v. 39, p. 267–273.

MCFALL, C. C., 1968, Reconnaissance geology of the Concepcion Bay area, Baja California, Mexico: Stanford University Publications, Geological Sciences, v. 10, 25 p.

MINA, F., 1957, Bosquejo Geologico del Territorio sur de la Baja California: Boletín de la Asociación Mexicana de Geólogos Petroleros, v. 9, p. 139–269.

MOORE, T. E., 1985, Stratigraphy and tectonic significance of the Mesozoic tectonostratigraphic terranes of the Vizcaino peninsula, Baja California Sur, Mexico, *in* Howell, D. G., ed., Tectonostratigraphic Terranes of the Circum-Pacific Region: Circum-Pacific Council for Energy and Mineral Resources Earth Science Series, n. 1, p. 315–329.

NORMARK, W. R., AND CURRAY, J. R., 1968, Geology and structure of the tip of Baja California, Mexico: Geological Society of America Bulletin, v. 79, p. 1589–1600.

NORTH AMERICAN GEOLOGIC MAP COMMITTEE, 1965, Geologic map of North America: U.S. Geological Survey, Washington, 1:5,000,000.

PACKER, B. M., AND INGERSOLL, R. V., 1986, Provenance and petrology of Deep Sea Drilling Project sands and sandstones from the Japan and Mariana forearc and backarc regions: Sedimentary Geology, v. 51, p. 5–28.

SAUNDERS, A. D., ROGERS, G., MARRINER, G. F., TERRELL, D. J., AND VERMA, S. P., 1987, Geochemistry of Cenozoic volcanic rocks, Baja California, Mexico: Implications for the petrogenesis of post-subduction magmas: Journal of Volcanology and Geothermal Research, v. 32, p. 223–245.

SMITH, A. J., 1982, The Neogene to Recent geology of Japan and its surrounding seas: Proceedings of the Geologists' Association, v. 93, p. 161–178.

STOCK, J. M., AND HODGES, K. V., 1989, Transfer of Baja California to the Pacific plate: Tectonics, v. 8, p. 99–115.

STOCK, J. M., AND MOLNAR, P., 1988, Uncertainties and implications of the Late Cretaceous and Tertiary position of North America relative to the Farallon, Kula, and Pacific plates: Tectonics, v. 7, p. 1339–1384.

STRECKEISEN, A. L., 1967, Classification and nomenclature of igneous rocks: Neues Jahrbuch fur Mineralogie Abhandlung, v. 107, p. 144–214.

SUMI, K., 1977, Rock alteration of Neogene volcanic rocks in Northeast Japan, *in* Tanaka, K., and Nozawa, T., eds., Geology and mineral resources of Japan, 3rd edition, v. 1, Geology: Geological Survey of Japan, Tokyo, p. 402–408.

UMHOEFER, P. J., 1987, Northward translation of "Baja British Columbia" along the Late Cretaceous to Paleocene margin of western North America: Tectonics, v. 6, p. 377–394.

VAN ANDEL, TJ. H., 1964, Recent marine sediments of Gulf of California: American Association of Petroleum Geologists Memoir 3, p. 216–310.

# SUBJECT INDEX

## SUBJECT INDEX

Q



R



S